Periodic Table of the Elements with the Gmelin System Numbers

1 H 2																1 H 2	2 He 1
3 Li 20	4 Be 26											5 B 13	6 C 14	7 N 4	8 O 3	9 F 5	10 Ne 1
11 Na 21	12 Mg 27											13 Al 35	14 Si 15	15 P 16	16 S 9	17 Cl 6	18 Ar 1
19 * K 22	20 Ca 28	21 Sc 39	22 Ti 41	23 V 48	24 Cr 52	25 Mn 56	26 Fe 59	27 Co 58	28 Ni 57	29 Cu 60	30 Zn 32	31 Ga 36	32 Ge 45	33 As 17	34 Se 10	35 Br 7	36 Kr 1
37 Rb 24	38 Sr 29	39 Y 39	40 Zr 42	41 Nb 49	42 Mo 53	43 Tc 69	44 Ru 63	45 Rh 64	46 Pd 65	47 Ag 61	48 Cd 33	49 In 37	50 Sn 46	51 Sb 18	52 Te 11	53 I 8	54 Xe 1
55 Cs 25	56 Ba 30	57** La 39	72 Hf 43	73 Ta 50	74 W 54	75 Re 70	76 Os 66	77 Ir 67	78 Pt 68	79 Au 62	80 Hg 34	81 Tl 38	82 Pb 47	83 Bi 19	84 Po 12	85 At 8a	86 Rn 1
87 Fr 25a	88 Ra 31	89*** Ac 40	104 71	105 71													

* NH_4 23

**Lanthanides 39	58 Ce	59 Pr	60 Nd	61 Pm	62 Sm	63 Eu	64 Gd	65 Tb	66 Dy	67 Ho	68 Er	69 Tm	70 Yb	71 Lu

***Actinides	90 Th 44	91 Pa 51	92 U 55	93 Np 71	94 Pu 71	95 Am 71	96 Cm 71	97 Bk 71	98 Cf 71	99 Es 71	100 Fm 71	101 Md 71	102 No 71	103 Lr 71

A Key to the Gmelin System is given on the Inside Back Cover

Gmelin Handbook of Inorganic and Organometallic Chemistry

8th Edition

Gmelin Handbook of Inorganic and Organometallic Chemistry

8th Edition

Gmelin Handbuch der Anorganischen Chemie

Achte, völlig neu bearbeitete Auflage

PREPARED AND ISSUED BY
Gmelin-Institut für Anorganische Chemie der Max-Planck-Gesellschaft zur Förderung der Wissenschaften
Director: Ekkehard Fluck

FOUNDED BY
Leopold Gmelin

8TH EDITION
8th Edition begun under the auspices of the Deutsche Chemische Gesellschaft by R. J. Meyer

CONTINUED BY
E. H. E. Pietsch and A. Kotowski, and by Margot Becke-Goehring

Springer-Verlag Berlin Heidelberg GmbH 1994

GMELIN HANDBOOK

Dr. J. von Jouanne

Dr. L. Berg, Dr. H. Bergmann, Dr. J. Faust, J. Füssel, Dr. H. Katscher, Dr. R. Keim, Dipl.-Phys. D. Koschel, Dr. A. Kubny, Dr. P. Merlet, Dr. M. Mirbach, Prof. Dr. W. Petz, Dr. H. Schäfer, Dr. F. A. Schröder, Dr. A. Slawisch, Dr. W. Töpper

Dr. R. Albrecht, Dr. G. Bär, D. Barthel, Dr. N. Baumann, Dr. K. Behrends, Dr. W. Behrendt, D. Benzaid, Dr. R. Bohrer, K. D. Bonn, Dipl.-Chem. U. Boßlet, Dr. U. Busch, A.-K. Castro, Dipl.-Ing. V. A. Chavizon, A. Dittmar, Dipl.-Geol. R. Ditz, R. Dowideit, Dr. H.-J. Fachmann, B. Fischer, Dr. D. Fischer, Dr. K. Greiner, Dipl.-Bibl. W. Grieser, Dr. R. Haubold, Dipl.-Min. H. Hein, H.-P. Hente, H. W. Herold, U. Hettwer, G. Hoell, Dr. G. Hönes, Dr. W. Hoffmann, G. Horndasch, Dr. W. Huisl, Dr. M. Irmler, B. Jaeger, Dr. R. Jotter, Dipl.-Chem. P. Kämpf, Dr. B. Kalbskopf, H.-G. Karrenberg, Dipl.-Phys. H. Keller-Rudek, Dipl.-Chem. C. Koeppel, R. Kolb, Dr. M. Kotowski, E. Kranz, E. Krawczyk, Dipl.-Chem. I. Kreuzbichler, Dr. W. Kurtz, M. Langer, Dr. B. Ledüc, H. Mathis, E. Meinhard, M. Meßer, C. Metz, K. Meyer, E. Mlitzke, Dipl.-Chem. B. Mohsin, Dr. U. Neu-Becker, K. Nöring, Dipl.-Min. U. Nohl, Dr. U. Ohms-Bredemann, Dr. H. Pscheidl, Dipl.-Phys. H.-J. Richter-Ditten, E. Rudolph, Dr. B. Sarbas, Dr. R. Schemm, Dr. D. Schiöberg, V. Schlicht, Dipl.-Chem. D. Schneider, E. Schneider, A. Schwärzel, Dr. B. Schwager, R. Simeone, Dr. F. Stein, Dr. C. Strametz, Dr. G. Swoboda, Dr. D. Tille, A. Tuttas, Dipl.-Phys. J. Wagner, R. Wagner, M. Walter, Dr. E. Warkentin, Dr. C. Weber, Dr. A. Wietelmann, Dr. M. Winter, Dr. B. Wöbke, K. Wolff

GMELIN ONLINE

Dr. G. Olbrich

Dr. P. Kuhn

Dr. R. Baier, Dr. K. Beesk, Dr. B. Becker, Dipl.-Chem. E. Best, Dr. H.-U. Böhmer, Dipl.-Phys. R. Bost, Dr. A. Brandl, Dr. R. Braun, Dr. T. Buck, R. Hanz, Dipl.-Phys. C. Heinrich-Sterzel, Dr. S. Kalwellis-Mohn, Dr. A. Kirchhoff, Dr. M. Körfer, Dipl.-Chem. H. Köttelwesch, Dr. V. Kruppa, Dr. M. Kunz, Dipl.-Chem. R. Maass, Dr. A. Nebel, Dipl.-Chem. R. Nohl, Dr. M. Nohlen, H. Reinhardt, Dr. B. Rempfer, Dipl.-Ing. H. Vanecek, Dipl.-Phys. M. Weiss

Volumes published on "Sulfur" (Syst. No. 9)

Sulfur A 1 – History (in German) – 1942

Sulfur A 2 – Occurrence. Technology of Sulfur and Its Compounds. Colloidal Sulfur. Toxicity (in German) – 1953

Sulfur A 3 – The Element. Preparation in Pure State. Properties (in German) – 1953

Sulfur B 1 – Hydrides and Oxides of Sulfur (in German) – 1953

Sulfur B 2 – Sulfur-Oxygen Acids (in German) – 1960

Sulfur B 3 – Compounds (concluded) (in German) – 1963

Sulfur-Nitrogen Compounds 1 – Compounds with Sulfur of Oxidation Number VI (in German) – 1977

Sulfur Suppl. Vol. 1 – Thionyl Halides (in German) – 1978

Sulfur Suppl. Vol. 2 – Sulfur Halides (in German) – 1978

Sulfur Suppl. Vol. 3 – Sulfur Oxides (in German) – 1980

Sulfur Suppl. Vol. 4 a/b – Sulfanes – 1983

Sulfur-Nitrogen Compounds 2 – Compounds with Sulfur of Oxidation Number IV – 1985

Sulfur-Nitrogen Compounds 3 – Compounds with Sulfur of Oxidation Number IV – 1987

Sulfur-Nitrogen Compounds 4 – Compounds with Sulfur of Oxidation Number IV – 1987

Sulfur-Nitrogen Compounds 5 – Compounds with Sulfur of Oxidation Number IV – 1990

Sulfur-Nitrogen Compounds 6 – Compounds with Sulfur of Oxidation Number IV – 1990

Sulfur-Nitrogen Compounds 7 – Compounds with Sulfur of Oxidation Number IV – 1991

Sulfur-Nitrogen Compounds 8 – Compounds with Sulfur of Oxidation Number IV – 1991

Sulfur-Nitrogen Compounds 9 – Compounds with Sulfur of Oxidation Number II – 1993

Sulfur-Nitrogen Compounds 10 a – Compounds with Sulfur of Oxidation Number II – 1994 (present volume)

Gmelin Handbook of Inorganic and Organometallic Chemistry

8th Edition

S

Sulfur-Nitrogen Compounds

Part 10 a

Compounds with Sulfur of Oxidation Number II

With 14 illustrations

AUTHORS Norbert Baumann, Hans-Jürgen Fachmann, Reimund Jotter, Alfons Kubny

EDITORS Norbert Baumann, Hans-Jürgen Fachmann, Reimund Jotter, Alfons Kubny

NOMENCLATURE Ursula Hettwer

CHIEF EDITOR Alfons Kubny

System Number 9

Springer-Verlag Berlin Heidelberg GmbH 1994

LITERATURE CLOSING DATE: 1991
IN MANY CASES MORE RECENT DATA HAVE BEEN CONSIDERED

Library of Congress Catalog Card Number: Agr 25-1383

ISBN 978-3-662-06353-8 ISBN 978-3-662-06351-4 (eBook)
DOI 10.1007/978-3-662-06351-4

Originally published by Springer-Verlag Berlin · Heidelberg · New York · London · Paris · Tokyo · Hong Kong · Barcelona in 1994
Softcover reprint of the hardcover 8th edition 1994

Preface

The present volume describes acyclic sulfur-nitrogen compounds with sulfur of the oxidation number II.

The first chapter deals with sulfur imide (S=NH), N-organyl-sulfur imides (S=NR, R = organyl), and metal complexes of sulfur imides. N-Organyl-sulfur imides have not been isolated, but they can be trapped, for example, by cycloaddition and stabilized by coordination to transition metals.

The following chapter present a detailed survey of other S^{II}-N compounds with one-coordinate sulfur, e.g., dithionitrous acid (S=N–SH) and derivatives, the salts of the dithionitryl ion (1+) ($[S=N=S]^+$), and N-thionitroso-diorganyl-amines ($S=NNR_2$, R = organyl). Only the reactions of $[S=N=S]^+ AsF_6^-$, among the known $[S=N=S]^+$ salts, have been extensively studied. Cycloadditions with alkenes and triple bonded compounds produce 1,3,2-dithiazolium salts and their derivatives. $S=NN(CH_3)_2$ and $S=NN(C_6H_5)_2$ form transition metal complexes. The ligand is coordinated via the S atom in a monodentate manner.

The description of S^{II}-N compounds with two-coordinate sulfur fills a large part of this volume. Thiohydroxylamines ($HSNR_2$), thiooximes (HSN=R), and thionitrous acid (HSNO) are examples of this class of compounds. Next, sulfur amide halogenides ($XSNH_2$, X = F, Cl) and their derivatives (amino-halogeno-sulfanes), including numerous examples of N,N-diorganyl-substituted amino-halogeno-sulfanes ($XSNR_2$, X = F, Cl, Br, I; R = organyl) and salts of the cation $(XS)_2N^+$ (X = Cl, Br), are described. N,N-Diorganyl-substituted amino-chloro-sulfanes, which can be obtained directly from secondary amines and SCl_2, are important intermediates in syntheses. $ClSNR_2$ is added to olefines and alkynes to give 1,2-addition products. Reactions with secondary amines give diaminosulfanes. $(ClS)_2N^+$ is added to alkenes and alkynes to give sulfur-nitrogen compounds.

The last sections deal with sulfur amide hydroxide derivatives ($ROSNR'_2$) and S-phosphorus-substituted thiohydroxylamines ($R_2P(X)SNR'_2$, X = O, S). Compounds of the type $(RO)_2P(X)SNR'_2$ (X = O, S) are effective as pesticides in the case of X = O and accelerate the vulcanization of rubber in the case of X = S.

The nomenclature used for acyclic sulfur(II)-nitrogen compounds follows current IUPAC recommendations and the practice in Chemical Abstracts.

The literature up to 1991 has been fully evaluated, and in many cases more recent publications are included.

The series is continued with the description of sulfur diamide derivatives in the volume "Sulfur-Nitrogen Compounds", Part 10b, 1994.

Frankfurt am Main
November 1994

Alfons Kubny

Remarks on Abbreviations and Standards

Physical data are cited in short form using abbreviations:

D_m	measured density
D_x	density calculated from X-ray data
m.p.	melting point; dec.: melting with decomposition
b.p.	boiling point, given in °C/Torr
IR	infrared spectrum. The medium of measurement (the physical state or a solvent) is given in parentheses. For the absorption maxima (vibration frequencies ν in cm^{-1}), intensity and shape are abbreviated as usual: s (strong), vs (very strong), w (weak), vw (very weak), vvw (very very weak), m (medium), br (broad), sh (shoulder). The assigned bands are usually labeled with the symbols ν for stretching, δ for in-plane bending, γ for out-of-plane bending, ρ for rocking, and τ for torsion vibration; the symbols $_s$ and $_{as}$, e.g., in ν_s and ν_{as}, mean symmetric and antisymmetric.
Raman	Raman spectrum. For abbreviations, see IR.
UV	ultraviolet spectrum. The medium of measurement is set in parentheses. The wavelength of the absorption maximum λ_{max} is given in nm followed by the extinction coefficient ε or log ε, ε in $L \cdot cm^{-1} \cdot mol^{-1}$, in parentheses; sh means shoulder, br means broad.
PES	photoelectron spectroscopy
X(U)PS	X-ray (ultraviolet) photoelectron spectroscopy
NMR	nuclear magnetic resonance spectrum, noise decoupling is indicated by braces { }. Solvent and standard are given in parentheses. For δ (chemical shift in ppm) downfield shift is indicated by a positive sign as recommended by IUPAC. Standard substances (if not otherwise cited) are TMS (= $Si(CH_3)_4$) as internal standard for ^1H and ^{13}C NMR, neat CH_3NO_2 as external standard for ^{14}N NMR and ^{15}N NMR, $CFCl_3$ as internal standard for ^{19}F NMR, and 85% H_3PO_4 as external standard for ^{31}P NMR. The multiplicity of the signal is given in parentheses: s (singlet), d (doublet), t (triplet), q (quartet), quint (quintet), sept (septet), m (multiplet). The assignment is given behind the multiplicity; if necessary for clarity the assigned atoms are underlined. ^nJ (coupling constant in Hz) is fixed by the number of bonds involved n and the coupling nuclei (e.g., $^1J(^{13}C,H)$), or by listing the coupling groups (e.g., $^3J(CH_2,CH_3)$)
NQR	nuclear quadrupole resonance spectrum
ES(P)R	electron spin (paramagnetic) resonance
MS	mass spectrum. m/e = mass/charge; M^+ = molecular ion (1+); relative intensity in parentheses
EIMS	electron impact mass spectrum
FIMS	field ionization mass spectrum
CIMS	chemical ionization mass spectrum

Further abbreviations:

$i\text{-}C_3H_7$	isopropyl, $CH(CH_3)_2$
$i\text{-}C_4H_9$	isobutyl, $CH_2CH(CH_3)_2$
$s\text{-}C_4H_9$	*sec*-butyl, $CH(CH_3)C_2H_5$
$t\text{-}C_4H_9$	*tert*-butyl, $C(CH_3)_3$
$c\text{-}C_3H_5$	cyclopropyl
$c\text{-}C_6H_{11}$	cyclohexyl
$(CH_2)_2N$	1-aziridinyl
$(CH_2)_4N$	1-pyrrolidinyl
$(CH_2)_5N$	1-piperidinyl
$O(CH_2)_4N$	4-morpholinyl
C_5H_5N	pyridine
C_7D_8	toluene-d_8
tos	p-tolylsulfonyl, $SO_2C_6H_4CH_3\text{-}4$
DMF	dimethylformamide, $HC(O)N(CH_3)_2$
DMSO	dimethyl sulfoxide, $(CH_3)_2SO$
HMPA	hexamethylphosphoric triamide, $((CH_3)_2N)_3PO$
THF	tetrahydrofuran, C_4H_8O
TMS	tetramethylsilane, $Si(CH_3)_4$
HMDS	hexamethyldisiloxane, $((CH_3)_3Si)_2O$

Table of Contents

Page

Acyclic Sulfur(II)–Nitrogen Compounds and Derivatives

1 Sulfur Imide and Derivatives

1.1 Sulfur Imide, S=NH

S=NH, the parent compound of the sulfur imide compounds, has not yet been isolated in the free state. However, it was possible to stabilize the molecule by coordination as a six-electron ligand to two Fe atoms like in $Fe_2(CO)_6(\mu\text{-}S{=}NH)$ [1]; see p. 2. Several quantum-chemical calculations using semiempirical and ab initio approaches have been performed; see Table 1. According to these studies, S=NH has a closed-shell $1^1A'$ ground state which is 1.02 eV (23.4 kcal/mol) lower than the ground state of the isomer HS≡N. There is no low-lying barrier that can allow thermal isomerization [2]. The energy difference between the $1^1A'$ singlet and the $1^3A''$ triplet states of S=NH is very small (42 kJ/mol at the MP2/6-31G** [3] and 28 kJ/mol at the CEPA-SD level [2]) and classifies the molecule as a biradicaloid species. Optimization of the equilibrium geometry in the ground state showed S=NH to be a bent molecule having an HNS bond angle of 110° (SCF/6-31G** level) [3] or 107° (CEPA-SD level) [2]. The SN bond length is 1.59 Å at CEPA-SD level (N–H = 1.03) [2] and 1.54 Å at SCF/6-31G** level (N–H = 1.01) [3] (corresponding to a conventional SN double bond). The proton affinity was estimated to range from 153 (triplet) to 195 (singlet) kcal/mol at the MP4 level. Protonation at nitrogen is expected to be preferable over protonation at sulfur [4].

Table 1
Ab Initio and Semiempirical Quantum-Chemical Calculations[a)] of S=NH.

method	calculated molecular properties	Ref.
ab initio calculations		
SCF-MO + CEPA	geometry, μ, f, D_e, A_p, ν_i, ΔH_f°, pop	[2]
SCF-MO	geometry, E_t	[5]
SCF-MO + MP2	geometry, E_t, q, μ	[3]
SCF-MO	geometry, E_t	[6]
HF + MP4	geometry, E_r, A_p	[4]
SCF-MO + MP4	geometry, E_t, E_o	[7]
SCF-MO + MRD-CI	geometry, E_t, E_{trans}	[8]
SCF-MO	geometry	[9]
semiempirical calculations		
INDO	geometry	[6]
CNDO/S	E_t, q, IP, E_{trans}	[6]
MNDO	geometry, A_p, μ, IP	[10]
MNDO	geometry, q, μ, IP	[11]
MNDO	geometry, q	[9]

[a)] The following abbreviations and symbols are used: MO = molecular-orbital; HF = Hartree-Fock; MP = Møller-Plesset; SCF = self-consistent field; (C,M,I)NDO = (complete, modified, intermediate) neglect of differential overlap; CEPA = coupled electron-pair approximation; MRD-CI = multireference double replacement configuration interaction; E_t = total energy; IP = ionization potential; D_e = dissociation energy; ΔH_f° = heat of formation; E_r = relative energy; E_{trans} = transition energies; E_0 = zero point vibration energies; pop = population analysis; q = charge distribution; f = force constants; ν_i = fundamental vibrations; EA = electron affinity; A_p = proton affinity; μ = dipole moment.

References:

[1] Herberhold, M.; Bühlmeyer, W. (Angew. Chem. **96** [1984] 64; Angew. Chem. Int. Ed. Engl. **23** [1964] 80).
[2] Wasilewski, J.; Staemmler, V. (Inorg. Chem. **25** [1986] 4221/8).
[3] Fabian, J.; Mehlhorn, A. (Z. Chem. **27** [1987] 30/1).
[4] Redondo, P.; Largo, A. (J. Mol. Struct. **253** [1992] 261/73 [THEOCHEM **85**]).
[5] Collins, M. P. S.; Duke, B. J. (J. Chem. Soc. Dalton Trans. **1978** 277/9).
[6] Mehlhorn, A.; Sauer, J.; Fabian, J.; Mayer, R. (Phosphorus Sulfur **11** [1981] 325/34).
[7] Schmidt, M. W.; Truong, P. N.; Gordon, M. S. (J. Am. Chem. Soc. **109** [1987] 5217/27).
[8] Vetter, R.; Mehlhorn, A. (J. Mol. Struct. **206** [1990] 11/6 [THEOCHEM **65**]).
[9] Penkovsky, V. V.; Shermolovitch, Yu. G.; Solovyov, A. V.; Vovna, V. I.; Borisenko, A. V. (Phosphorus Sulfur Silicon Relat. Elem. **73** [1992] 1/4).
[10] Mehlhorn, A. (Collect. Czech. Chem. Commun. **53** [1988] 2116/27).

[11] Shermolovich, Yu. G.; Solov'ev, A. V.; Borodin, A. V.; Pen'kovskii, V. V.; Trachevskii, V. V.; Markovskii, L. N. (Zh. Org. Khim. **27** [1991] 1637/41; J. Org. Chem. USSR [Engl. Transl.] **27** [1991] 1433/6).

1.2 Iron Complex of Sulfur Imide

1.2.1 $Fe_2(CO)_6(\mu\text{-SNH})$

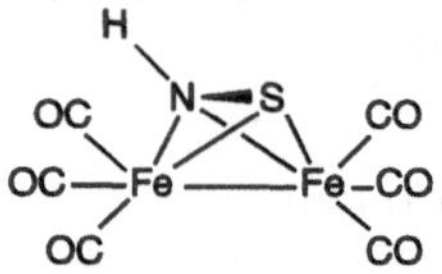

The complex was obtained in quantitative yield when $Fe_2(CO)_6(\mu\text{-SNSi(CH}_3)_3)$ (see p. 24) was chromatographed on silica gel with pentane as eluent under N_2 [1].

The orange, diamagnetic crystals melt at 46 to 48 °C. The crystals are volatile at room temperature and 10^{-2} mbar [1].

^1H NMR (5 to 10% in C_6D_6/TMS): δ = 2.55 ppm [2] (2.15 (broad s) [1]); ($CDCl_3$/TMS): no signal (H/D exchange) [1]. The 300-MHz ^1H NMR spectrum of the complex recorded by normal techniques and by using the perfect Hahn spin-echo pulse sequence is shown in a figure. ^{15}N satellites were observed giving $^1J(^{15}N,^1H)$ = 89.5 Hz [2].

^{13}C NMR (5 to 10% in C_6D_6/TMS): δ = 211.5 ppm (CO).

^{14}N NMR (5 to 10% in C_6D_6/external neat CH_3NO_2): δ = −375 ppm.

^{15}N NMR (5 to 10% in C_6D_6/external neat CH_3NO_2): δ = −374.7 ppm; $^1J(^{57}Fe,^{15}N)$ = 6.1 ± 0.08 Hz. The isotope-induced shift of the ^{15}N NMR signal $^1\Delta^{34/32}S$ was determined to be

0.036 ppm. The 30.4-MHz ^{15}N NMR spectrum measured by the refocused INEPT pulse sequence with ^{1}H decoupling in C_6D_6 is shown in a figure [2].

^{17}O NMR (5 to 10% C_6D_6/external H_2O): $\delta = 374.2$ ppm [2].

IR (CsI): ν (in cm^{-1}) = 3365 s, ν(NH); 984 m, δ(HNS); 760 w, ν(N–S); (pentane): ν(CO) = 2080 s, 2040 vs, 2000 vs, 1995 s, 1985 m, 1978 w [1].

MS (70 eV): m/e (rel. int. in %) = 327 (21) M^+, $Fe_2(CO)_n(HNS)^+$ (n = 5 to 1), 159 (100) $Fe_2(HNS)^+$, 158 (23) Fe_2NS^+, 144 (16) Fe_2S^+, 112 (10) Fe_2^+ [1].

The complex is readily soluble in nonpolar and polar organic solvents. It is stable in air for a short time. The crystals decompose slowly at room temperature.

Deuteration of the complex with D_2O in pentane gave $Fe_2(CO)_6(\mu\text{-SND})$.

The complex was methylated with CH_2N_2 (mole ratio 1 : 3) in diethyl ether at 0 °C to give $Fe_2(CO)_6(\mu\text{-SNCH}_3)$ [1].

$Fe_2(CO)_6(\mu\text{-SNH})$ reacted with an excess of Na dust in THF (mole ratio 3 : 1) at −78 °C to form **$Fe_2(CO)_6(\mu\text{-SNNa})$**. The conversion was complete at 50 °C. The excess of sodium was removed by filtration. $Fe_2(CO)_6(\mu\text{-SNNa})$ was not isolated (IR (THF): ν(CO) = 2040 s, 1987 vs, 1951 vs, 1934 cm^{-1} s) [2].

1.2.2 $Fe_2(CO)_6(\mu\text{-SND})$

The complex was obtained by deuterating $Fe_2(CO)_6(\mu\text{-SNH})$ with D_2O in pentane.

IR (CsI): ν (in cm^{-1}) = 2502 s, ν(ND); 806 m, δ(DNS); 729 w, ν(NS) [1].

References:

[1] Herberhold, M.; Bühlmeyer, W. (Angew. Chem. **96** [1984] 64; Angew. Chem. Int. Ed. Engl. **23** [1984] 80).

[2] Wrackmeyer, B.; Kupce, E.; Distler, B.; Dirnberger, K.; Herberhold, M. (Z. Naturforsch. **46b** [1991] 1679/83).

1.3 N-Organyl-sulfur Imides, S=N–R, R = organyl

N-Organyl-sulfur imides have not been isolated in substance. Undoubtedly, in a large number of sulfurization reactions of amines, sulfur imides act as intermediates. They can be trapped, for instance, by cycloaddition reactions. N-Organyl-sulfur imides can also be stabilized by coordination to transition metals.

The thionitroso group is polarized in the reverse direction from that of the nitroso group. Due to the positively charged sulfur atom, the thionitroso group tends to desulfurize forming the corresponding sulfur diimide. Only the sulfur atom is attacked by further oxidation to give the N-sulfinyl or N-thiosulfinyl amines.

Reference:

Mayer, R. (Phosphorus Sulfur Relat. Elem. **23** [1985] 277/96).

1.3.1 N-Alkyl-sulfur Imides, S=N–R, R = alkyl

N-Alkyl-sulfur imides are unstable. They were established as intermediates by trapping them with dienes in form of cycloadducts.

1.3.1.1 $S{=}NCH_3$

$S{=}NCH_3$ is formed as an intermediate in the reaction of *cis*- or *trans*-2-methyl-3-phenyl-oxaziridine with cyclohexene episulfide in $CHCl_3$ at room temperature leading to cyclohexene, benzaldehyde, N,N′-dimethyl-sulfur diimide, and a small amount of azomethane. The sulfur diimide and azomethane are assumed to be formed by reacting $S{=}NCH_3$ with oxaziridine or a second mole of $S{=}NCH_3$ [1].

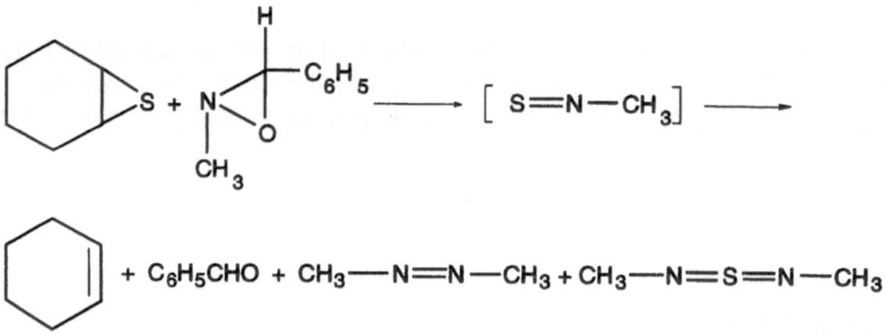

Evidence for the formation of $S{=}NCH_3$ was obtained by isolation of 2-methyl-3,6-dihydro-1,2-thiazine when the reaction was carried out in the presence of 1,3-butadiene [1].

The reaction of $CH_3NHSi(CH_3)_3$ with SCl_2 in the presence of $(C_2H_5)_3N$ (mole ratio 1 : 1 : 1) in diethyl ether at −10 to 20°C over a period of 1 h led to polymeric substances with the composition $(CH_3NS)_x$. The intermediate $ClSN(CH_3)Si(CH_3)_3$ evidently decomposed to $S{=}NCH_3$ which then formed polymeric reaction products [2].

The geometrical parameters of the molecule have been calculated by a modified and an extended version of the original semiempirical MNDO method. It was found that an average correction of 3 pm will be necessary to approach the SN bond lengths expected from experimental investigation. Bond lengths of S–N: 149.5 pm and C–N: 145.6 pm were estimated [3].

An ab initio calculation (MP2) showed the ground state of the molecule to be a singlet state (−472.61826 au). The lowest excited state is a triplet state (−472.5867 au). The vertical $S_1 \leftarrow S_0$ excitation energy is ca. 7000 cm^{-1}. The molecule is assumed to have a biradicaloid character [4].

The frontier orbital energies of $CH_3N{=}S$ were estimated by a semiempirical MNDO method to be E(HOMO) = −9.75 eV, E(LUMO) = −1.37 eV (the values are probably 0.5 to 1.5 eV lower) [3].

1.3.1.2 $S{=}NC_3H_7$-i. $S{=}NC_4H_9$-t

$S{=}NC_3H_7$-i and $S{=}NC_4H_9$-t were formed by thermal decomposition of $ClSN(C_3H_7\text{-i})Si(CH_3)_3$ and $ClSN(C_4H_9\text{-t})Si(CH_3)_3$, respectively. On heating the starting compounds in toluene at 100 °C for 1 h, the species were trapped by 2,3-dimethyl-1,3-butadiene forming 2-isopropyl-4,5-dimethyl-3,6-dihydro-1,2-thiazine and 2-tert-butyl-4,5-dimethyl-3,6-dihydro-1,2-thiazine, respectively. The yields were 53 and 55% [2].

References:

[1] Hata, Y.; Watanabe, M. (J. Org. Chem. **45** [1980] 1691/2).
[2] Markovskii, L. N.; Solov'ev, A. V.; Kaminskaya, E. I.; Borodin, A. V.; Shermolovich, Y. G. (Zh. Org. Khim. **26** [1990] 2083/6; J. Org. Chem. USSR [Engl. Transl.] **26** [1990] 1799/801).
[3] Mehlhorn, A. (Collect. Czech. Chem. Commun. **53** [1988] 2116/27).
[4] Fabian, J.; Mehlhorn, A. (Z. Chem. **27** [1987] 30/1).

1.3.2 N-(Hexachloro-3-cyclopenten-1-ylidene)-imidosulfur Ion (1+), $[S{=}N{=}C_5Cl_6]^+$ AsF_6^-

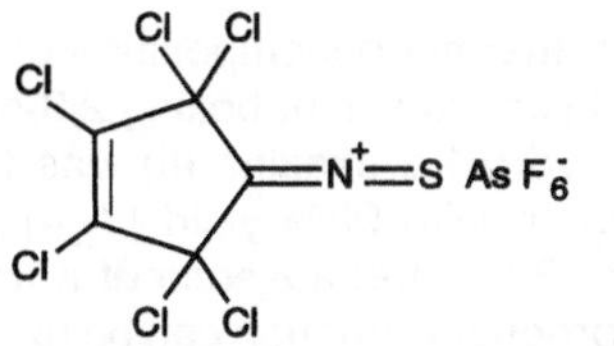

$[S{=}N{=}C_5Cl_6]^+$ AsF_6^- was synthesized by reacting $ClSN{=}C_5Cl_6$ with Ag^+ AsF_6^- (mole ratio 1 : 1) in SO_2 at −78 °C over a period of 16 h. After filtration of AgCl and removal of SO_2, the title compound was isolated in form of dark purple crystals in almost quantitative yield. Purification was achieved by recrystallization from an SO_2–$CFCl_3$ mixture.

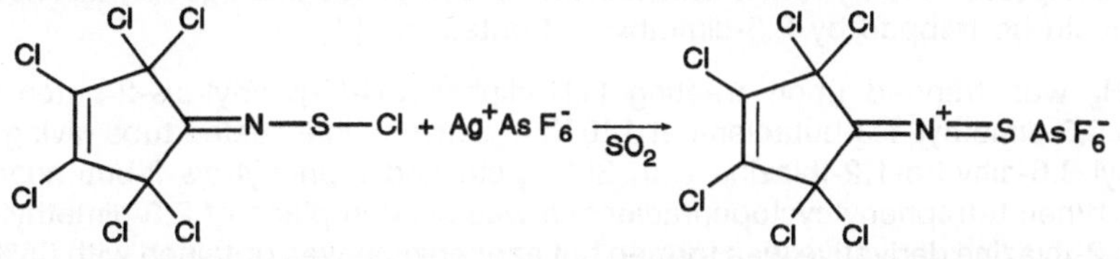

The salt decomposes rapidly in solvents other than SO_2. It is very moisture-sensitive.

UV–VIS (CH_2Cl_2): $\lambda_{max} = 562$ nm.

IR (ν in cm^{-1}) = 1750 m, 1581 m, 1561 m, 1259 s, 1164 s, 1063 m, 1041 sh, 860 m, 836 s, 811 vs, 702 vs ν_3 AsF_6^-, 658 m, 639 m, 439 m.

Reference:

Apblett, A.; Chivers, T.; Fait, J. F.; Vollmerhaus, R. (Can. J. Chem. **69** [1991] 1022/7).

1.3.3 N-Aryl-sulfur Imides, S=N–R, R = aryl

N-Aryl-sulfur imides are unstable. They were identified like the alkyl species as intermediates by trapping them with dienes in form of cycloadducts.

1.3.3.1 $S{=}N{-}C_6H_5$

The title compound was efficiently generated under very mild conditions from N-(phenylaminosulfanyl)phthalimide by stirring with an excess of triethylamine in acetone at room temperature (1,2-elimination). In order to trap $S{=}NC_6H_5$ simultaneously, 1,3-butadiene or 2,3-dimethyl-1,3-butadiene was added. The cycloadducts 2-phenyl-3,6-dihydro-1,2-thiazine (I) and 2-phenyl-4,5-dimethyl-3,6-dihydro-1,2-thiazine (II) were formed [1, 2]:

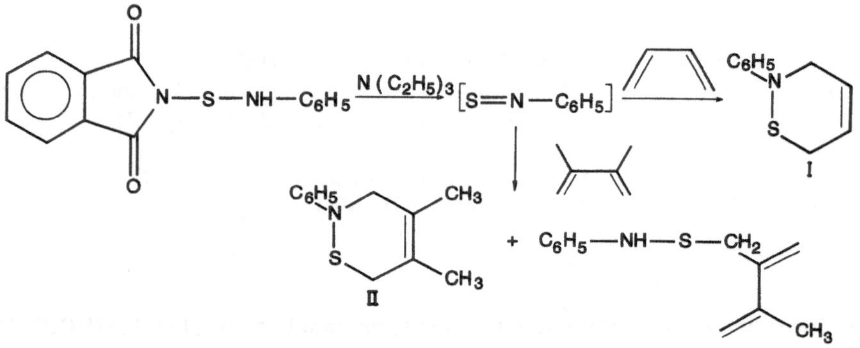

$S{=}NC_6H_5$ is supposed to be an intermediate during the thermal decomposition of C_6H_5–NH–S–NH–C_6H_5. When C_6H_5–NH–S–NH–C_6H_5 was heated in an excess of boiling 2,3-dimethyl-1,3-butadiene for 6 h, the 2-phenyl-4,5-dimethyl-3,6-dihydro-1,2-thiazine (II) was formed and then isolated by column chromatography and distillation with 92% yield [3, 4]. When C_6H_5–NH–S–NH–C_6H_5 was heated in benzene at 50 °C for 72 h in the absence of a trapping agent, azobenzene, aniline, and sulfur were obtained. The product formation can be explained by initial formation of $S{=}NC_6H_5$ followed by dimerization. The resulting species disproportionate to azobenzene and sulfur [5].

The thermal decomposition of C_6H_5–NH–S–$N(CH_2)_5$ involves the formation of C_6H_5–NH–S–NH–C_6H_5 (detected by NMR) which further decomposes analogously as shown above. $S{=}NC_6H_5$ could be trapped by 2,3-dimethyl-1,3-butadiene [4].

$S{=}NC_6H_5$ was trapped upon treating N,N′-diphenyl-N-(2-phenyl-*cis*-2-butenoyl)thiobisamine with 2,3-dimethyl-1,3-butadiene at 140 °C in benzene in a sealed tube giving 2-phenyl-4,5-dimethyl-3,6-dihydro-1,2-thiazine with 35% yield and 2-phenyl-*cis*-2-butenoanilide with 80% yield. When tetraphenylcyclopentadienone was used in place of 2,3-dimethyl-1,3-butadiene, no 1,2-thiazine derivative was formed but azobenzene was obtained with 65% yield [5].

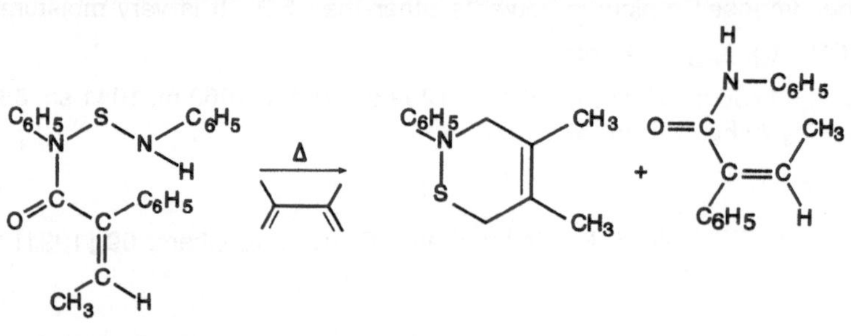

$S{=}NC_6H_5$ is probably an intermediate compound during the reaction of $C_6H_5N(Si(CH_3)_3)_2$ with SCl_2 in the presence of 2,3-dimethyl-1,3-butadiene. The trapping reaction leads to 2-phenyl-4,5-dimethyl-3,6-dihydro-1,2-thiazine [6].

The geometrical parameters of the molecule were calculated by a modified and extended version of the original semiempirical MNDO method. It was found that an average correction of 3 pm is necessary to approach the SN bond lengths, expected from experimental investigations. The following values were estimated: S=N = 150.5 pm, C–N = 140.4 pm [7].

The frontier orbital energies of $S{=}NC_6H_5$ were estimated to be E(HOMO) = −9.44 eV, E(LUMO) = −1.76 eV. The values probably are 0.5 to 1.5 eV lower [7].

UV-VIS Spectrum. $n-\pi^*$ Transition energies corresponding to 1234 and 491 nm were calculated by the semiempirical CNDO/S method. PPP calculations have been carried out for $\pi-\pi^*$ transitions. Values are given in a figure [8] (303 and 323 nm, from [9]).

1.3.3.2 S=N–R, R = $4\text{-}CH_3OC_6H_4$, $4\text{-}CH_3C_6H_4$, $4\text{-}BrC_6H_4$, $4\text{-}ClC_6H_4$, $4\text{-}NO_2C_6H_4$, $1\text{-}C_{10}H_7$, $3\text{-}C_5H_4N$

Formation

The compounds S=NR, where R = $4\text{-}ClC_6H_4$ and $4\text{-}BrC_6H_4$, supposedly are intermediates during the thermal decomposition of R–NH–S–NH–R. When R–NH–S–NH–R was heated in an excess of boiling 2,3-dimethyl-1,3-butadiene for 6 h, the 2-aryl-4,5-dimethyl-3,6-dihydro-1,2-thiazine was formed and subsequently isolated by column chromatography and distillation (R = $4\text{-}ClC_6H_4$, 32% yield; R = $4\text{-}BrC_6H_4$, 41% yield) [3]. The same reaction was studied by others for R–NH–S–NH–R where R = $4\text{-}CH_3OC_6H_4$, $4\text{-}BrC_6H_4$, $4\text{-}ClC_6H_4$, $3\text{-}NO_2C_6H_4$. The thermal decomposition in the presence of 2,3-dimethyl-1,3-butadiene was carried out in a sealed tube. The reaction mixture was kept at 50 °C for 72 h. S=NR were trapped by 2,3-dimethyl-1,3-butadiene, producing the corresponding 2-aryl-4,5-dimethyl-3,6-dihydro-1,2-thiazines. In the absence of a trapping agent, azobenzene, aryl amine, and sulfur were obtained. The S=NR species are also believed to be intermediates during the thermal decomposition of $R\text{–}NH\text{–}S\text{–}N(CH_2)_5$ which involves the corresponding R–NH–S–NH–R [4].

The title compounds S=NR (R = $4\text{-}CH_3OC_6H_4$, $4\text{-}CH_3C_6H_4$, $4\text{-}BrC_6H_4$, $4\text{-}ClC_6H_4$, $4\text{-}NO_2C_6H_4$, $1\text{-}C_{10}H_7$, $3\text{-}C_5H_4N$) are formed by 1,2-elimination from N-(arylaminosulfanyl)phthalimides. The educts were suspended in acetone (or, occasionally, in $CHCl_3$) at room temperature, and an excess of $N(C_2H_5)_3$ was added. To simultaneously trap the thionitroso compounds, a conjugated diene was added [1, 2]; see below.

S=NR (R = $4\text{-}CH_3OC_6H_4$, $4\text{-}ClC_6H_4$) probably are also formed as intermediates during the reaction of the very unstable $Cl_2S{=}NR$ (formed in situ) with 2,3-dimethyl-1,3-butadiene by 1,1-elimination. 2,3-Dimethyl-1,3-butadiene acted both as the dechlorinating agent and the trapping diene. The products isolated were the expected 1,2-thiazine derivatives and sulfenamides: see below. The instability of $Cl_2S{=}NR$ limits the synthetic utility of this reaction [2].

Reactions with Alkenes and 1,3-Dienes

The thionitrosoarenes were trapped with alkenes and conjugated dienes.

Alkenes. S=NR (R = $4\text{-}CH_3OC_6H_4$, $1\text{-}C_{10}H_7$, $3\text{-}C_5H_4N$) reacted with isobutene, $(CH_3)_2C{=}CH_2$, and α-methylstyrene, $C_6H_5C(CH_3){=}CH_2$, to yield the corresponding ene adducts $R\text{–}NH\text{–}S\text{–}CH_2\text{–}C(R'){=}CH_2$, R = $4\text{-}CH_3OC_6H_4$, R′ = CH_3, C_6H_5; R = $1\text{-}C_{10}H_7$, R′ = CH_3, C_6H_5; R = $3\text{-}C_5H_4N$, R′ = C_6H_5 [1, 2].

1,3-Dienes. The compounds S=NR (R = $4\text{-}CH_3OC_6H_4$, $4\text{-}CH_3C_6H_4$, $4\text{-}BrC_6H_4$, $4\text{-}ClC_6H_4$, $4\text{-}NO_2C_6H_4$, $1\text{-}C_{10}H_7$, $3\text{-}C_5H_4N$) reacted with **1,3-butadiene** to give the corresponding Diels-Alder adducts, 1,2-thiazine derivatives [1, 2].

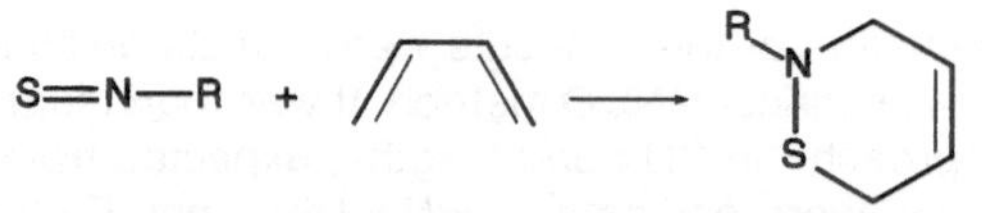

R = 4-$CH_3OC_6H_4$, 4-$CH_3C_6H_4$, 4-BrC_6H_4,
4-ClC_6H_4, 4-$NO_2C_6H_4$, 1-$C_{10}H_7$, 3-C_5H_4N

The reaction of S=NR (R = 4-$CH_3OC_6H_4$, 4-$CH_3C_6H_4$, 4-BrC_6H_4, 4-ClC_6H_4, 4-$NO_2C_6H_4$, 1-$C_{10}H_7$, 3-C_5H_4N) with **2,3-dimethyl-1,3-butadiene** gave two products, the corresponding Diels-Alder adduct and the ene adduct (see p. 6). Electron-donating substituents on the benzene ring favor the Diels-Alder reaction, while electron-withdrawing substituents favor the ene reaction [1, 2]. In contrast to these results other workers observed no ene products in the reactions of S=NR (R = 4-ClC_6H_4, 4-BrC_6H_4, 4-$CH_3OC_6H_4$, 3-$NO_2C_6H_4$, formation via RNH–S–NHR) with 2,3-dimethyl-1,3-butadiene [3, 4].

S=NR (R = 4-$CH_3OC_6H_4$) generated in the presence of **1,4-diphenyl-1,3-butadiene** gave the cis Diels-Alder adduct (I) with 60% yield. S=NR (R = 3-C_5H_4N) did not react cleanly. The cis product (I) was obtained with low yield [2].

S=NR (R = 4-$CH_3OC_6H_4$, 4-BrC_6H_4) reacted with **(E,E)-2,4-hexadiene** to give the cis diastereoisomers (II). S=NR (R = $CH_3OC_6H_4$, 4-$CH_3C_6H_4$, 4-BrC_6H_4, and 4-ClC_6H_4) reacted with **(E,Z)-2,4-hexadiene** to yield the corresponding diastereomers with trans geometry (III). The trans products were accompanied by cis isomers (due to the presence of (E,E)-diene as impurity in the commercial (E,Z)-diene sample). (E,E)- and (E,Z)-2,4-hexadienes react while retaining their diene stereochemistry. The reaction of S=NR (R = 4-$OCH_3C_6H_4$, 4-BrC_6H_4) with a 1:1 mixture of (E,E)-2,4-hexadiene and (E,Z)-2,4-hexadiene was studied. The methoxy derivative gave only the cis isomer (II) with a yield of >96%; the bromo derivative yielded a mixture of cis (II) (80%) and trans (III) isomers [2, 10].

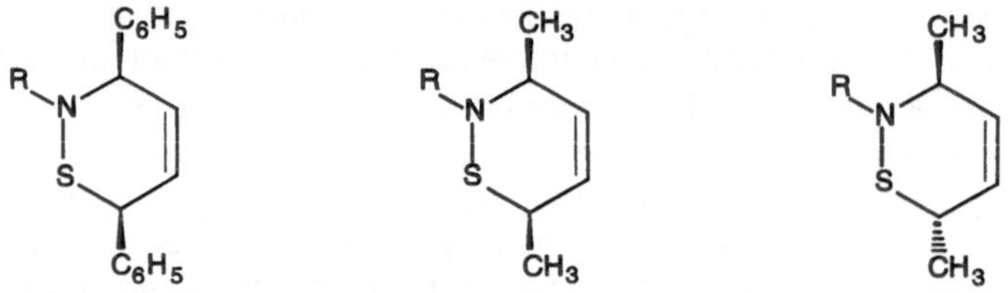

I II III

S=$NC_6H_4OCH_3$-4 reacted with the bicyclic 1,3-dienes 1,1′-bicyclopentenyl (IV) and 1,1′-bicyclohexenyl (V) to yield the single stereoisomeric products (VI) and (VII) with cis structure [2, 10].

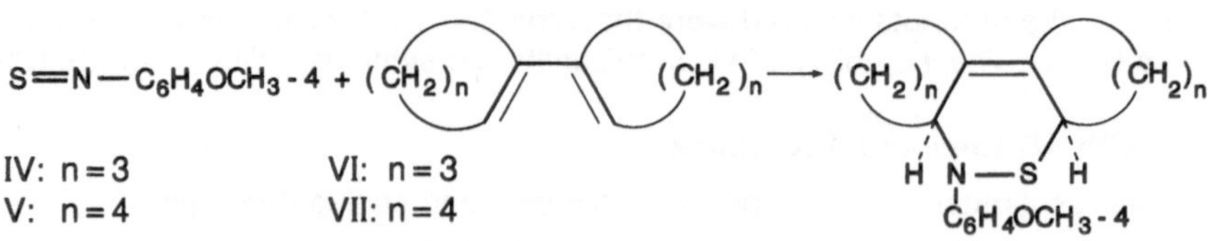

IV: n = 3 VI: n = 3
V: n = 4 VII: n = 4

1.3.3.3 S=N–C_6H_4NO-2

S=NC_6H_4NO-2 was formed reversibly as a short-lived intermediate (k = 100 to 200 s^{-1}) by flash photolysis of 2,1,3-benzothiadiazole 2-oxide in ethanol and cyclohexane. The flash photolysis was performed in aerated and degassed 10^{-5} M solutions using light of wavelengths above 230 nm. The quantum yield for the formation was found to be unaffected by the presence of oxygen. Independent of this reaction, 1,3-dihydro-2,1,3-benzothiadiazole 2,2-dioxide was also formed. The results in cyclohexane are similar as those in ethanol. Due to the solvent, 2,1,3-benzothiadiazole 2-oxide was observed instead of 1,3-dihydro-2,1,3-benzothiadiazole 2,2-dioxide. The photolysis was also performed at low temperatures in a variety of media (diethyl ether, isopentane, ethanol, 5 : 5 : 2 (EPA), poly(vinyl chloride) (PVC), argon). At room temperature and at low temperatures, identical photochemical processes occur. The relative amounts of the two photoproducts 2,1,3-benzothiadiazole 2-oxide and $S{=}NC_6H_4NO$-2 are approximately 3 : 1 at room temperature and 1 : 2 at low temperatures, while the degree of conversion is small. The quantum yield for the formation of the transient is 0.014 (measured by ferrioxalate actinometry). Photolysis at 400 nm at low temperature (85 K in EPA or PVC) causes also the photochemical reversion to starting material [11]. The activation parameters for the thermal reversion of the title compound at room temperature are listed in Table 2.

Table 2
Thermodynamic Data on the Thermal Reversion of $S{=}NC_6H_4NO$-2.
Temperatures vary between 15 and 51 °C in ethanol and between 10 and 35 °C in cyclohexane. The 10^{-5} M solutions were not degassed [11].

	96% ethanol	cyclohexane
E_a in kcal/mol	12.8 (0.8)	16.1 (0.6)
A in s^{-1}	2.77×10^{11}	1.06×10^{14}
$\Delta S^{\neq}$ in cal·K^{-1}· mol^{-1} at 25 °C	−8.1 (2.5)	3.7 (1.9)
$\Delta H^{\neq}$ in kcal/mol at 25 °C	12.2 (0.8)	15.5 (0.6)
$\Delta G^{\neq}$ in kcal/mol at 25 °C	14.7 (0.2)	14.4 (0.2)

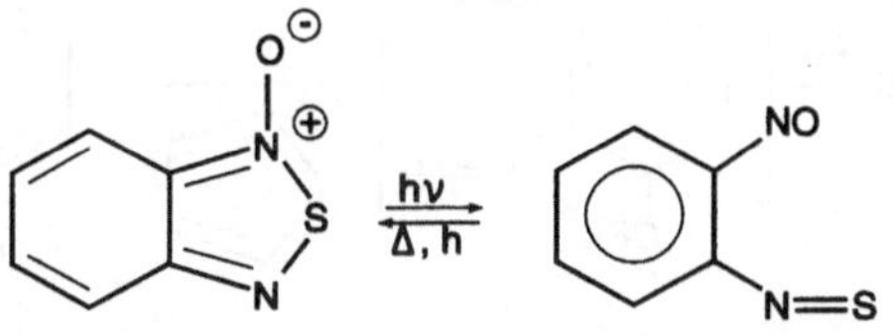

The UV absorptions of the transient λ_{max} (in nm) in different media and at different temperatures: ethanol (298 K): 467, cyclohexane (298 K): 493, PVC (85 K): 485[a], EPA (85 K): 485[a], argon (20 K): 486[a]. ([a] λ_{max} of overlapping absorption bands of benzo[c]-1,2,5-thiadiazole 1-oxide and 2-thionitrosonitrosobenzene.)

The flash spectrum in ethanol is considerably blue-shifted (467 nm) relative to that of cyclohexane (493 nm), probably indicating that different rotamers are present in polar and nonpolar solvents at room temperature [11].

The wavelengths for n–π* transitions were calculated by the semiempirical CNDO/S method: λ_{max} = 1375, 787, 500 nm. PPP Calculations have been carried out for π–π* transitions (shown in a figure): λ_{max} range from 350 to 370 nm [8]. The theoretical transition energies [8] do not agree with those obtained from the flash spectrum [11].

1.3.3.4 $S{=}N{-}C_6H_3Br$-5-CN-2

The species was formed as an intermediate during thermal decomposition of 3-azido-6-bromo-2,1-benzisothiazole in neat 2,3-dimethyl-1,3-butadiene at 69 °C by rapid dissolution leading to the formation of the cycloadduct [12].

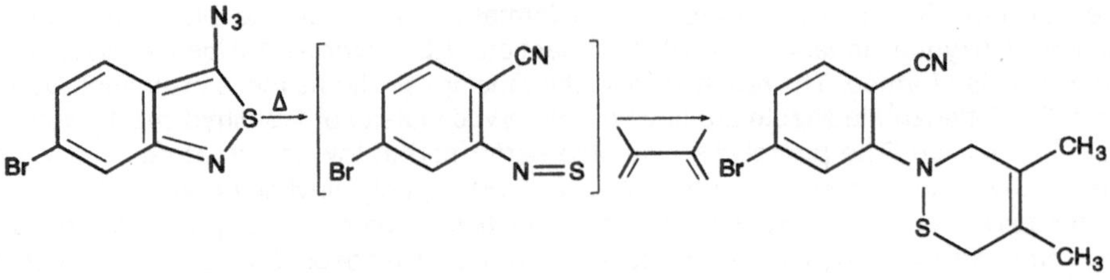

1.3.3.5 S=N–C_6H_3CN-2-CF_3-6

The title compound was formed by photolyzing 3-azido-7-trifluoromethyl-2,1-benzisothiazole in ether at room temperature with a 450 W high-pressure Hg lamp within less than 2 h. The species was detected by IR (ν(CN) = 2225 cm^{-1}) and mass spectroscopy (m/z = 216). It could not be isolated in a pure form and after some hours transformed into 2-H_2N-3-CF_3–C_6H_3CN-1. When the photolysis was repeated in the presence of an excess of cyclopentadiene, the major product was the cycloadduct (60% yield) along with a minor amount of the same aniline. A similar cycloadduct was obtained by reaction with 1-methoxy-1,3-cyclohexadiene. The dimerization product of the title compound (with loss of S), R–N=S=N–R, R = 2-CN-6-$CF_3C_6H_3$, was obtained by the decomposition of 3-azido-7-trifluoromethyl-2,1-benzisothiazole. The decomposition was slow in low-boiling solvents and rapid at 180 °C in o-dichlorobenzene [12].

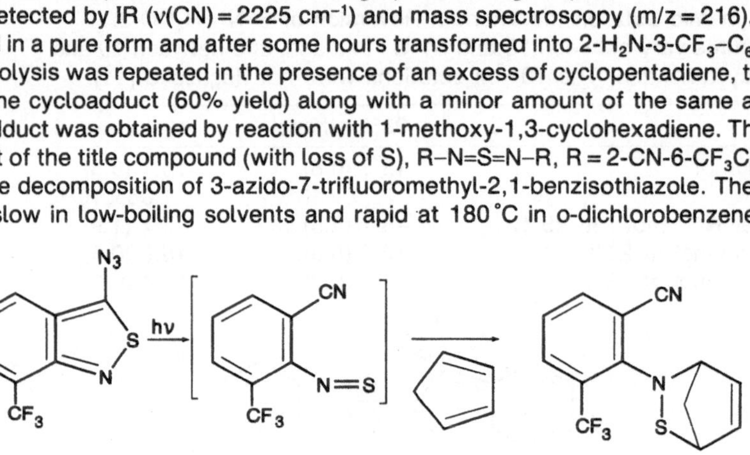

1.3.3.6 S=N–$C_6H_2(C_4H_9\text{-t})_2$-2,4-$CH_3$-6

S=N–$C_6H_2(C_4H_9\text{-t})_2$-2,4-$CH_3$-6 was observed as an intermediate in the photolysis of S=S=N–$C_6H_2(C_4H_9\text{-t})_2$-2,4-$CH_3$-6 with light from a medium-pressure mercury lamp in an EPA matrix (diethyl ether, isopentane, ethanol) at 77 K. H_2N–$C_6H_2(C_4H_9\text{-t})_2$-2,4-$CH_3$-6, 2,4-(t-$C_4H_9)_2$-6-$CH_3C_6H_2$–N=S=N–$C_6H_2(C_4H_9\text{-t})_2$-2,4-$CH_3$-6, and S_8 were the products isolated. The change in the electronic spectrum was monitored intermittently. A very intense band at 473 nm appeared which could be assigned to the title compound. It was also assumed to be an intermediate during the thermolysis of S=S=N–$C_6H_2(C_4H_9\text{-t})_2$-2,4-$CH_3$-6 [13].

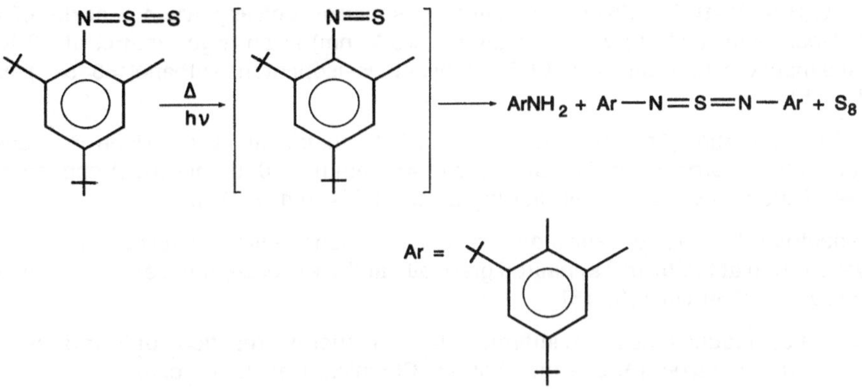

1.3.3.7 S=N-$C_6H_2(C_4H_9\text{-t})_3$-2,4,6

S=N-$C_6H_2(C_4H_9\text{-t})_3$-2,4,6 presumbly is an intermediate in the reaction of 4,6,7a-2,4,6-tri-tert-butyl-7aH-1,2,3-benzodithiazole with $P(C_6H_5)_3$ in ethereal solution between −78 °C and room temperature. The reaction gave 2,4,6-tri-tert-butyl-aniline, N,N′-bis(2,4,6-tri-tert-butyl-phenyl)-sulfur diimide, and N-sulfinyl-2,4,6-tri-tert-butyl-benzenamine. Evidence for the formation of S=N-$C_6H_2(C_4H_9\text{-t})_3$-2,4,6 was obtained when the reaction was carried out under a stream of oxygen. The sulfur diimide was not produced, but the yield of the N-sulfinyl-2,4,6-tri-tert-butyl-benzenamine increased. It was concluded that S=N-$C_6H_2(C_4H_9\text{-t})_3$-2,4,6 was trapped by oxygen to give O=S=N-$C_6H_2(C_4H_9\text{-t})_3$-2,4,6 [14].

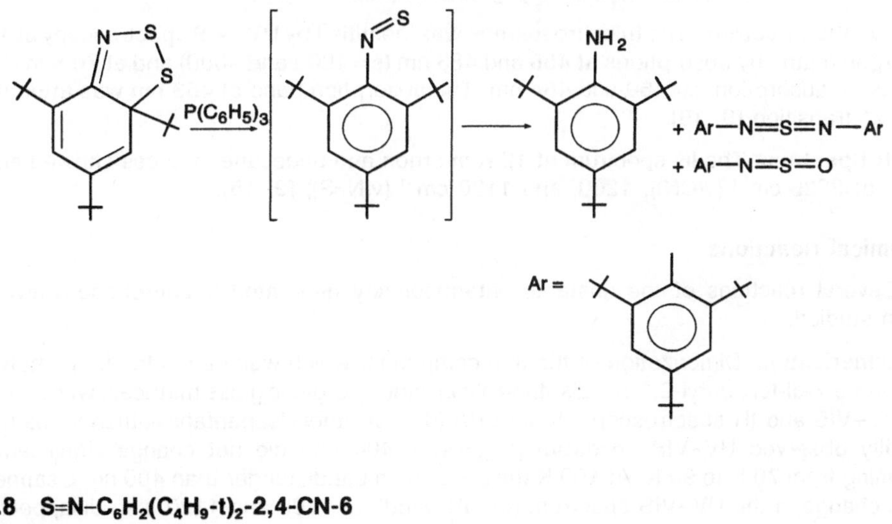

1.3.3.8 S=N-$C_6H_2(C_4H_9\text{-t})_2$-2,4-CN-6

Formation

S=N-$C_6H_2(C_4H_9\text{-t})_2$-2,4-CN-6 was generated by the photochemical and thermal reactions of 3-azido-5,7-di-tert-butyl-2,1-benzisothiazole (I). The species was detected in matrices at cryogenic temperatures and monitored by UV-VIS and IR spectra [9, 15]. The title compound could be trapped in several reactions; see the chapter "Chemical Reactions" below.

$S{=}N{-}C_6H_2(C_4H_9\text{-}t)_2$-2,4-CN-6 was detected spectroscopically when 3-azido-5,7-di-tert-butyl-2,1-benzisothiazole (I) was irradiated ($\lambda \geq 320$ nm) in an argon matrix at 12 K, in an undecane matrix at 12 K, and in an EPA (diethyl ether–isopentane–ethanol = 5 : 5 : 2) matrix at 70 K [9, 15].

UV–VIS spectrum (Ar matrix at 12 K and EPA matrix at 70 °C): Upon irradiating I, a spectrum with absorption maxima at 460 and 484 nm ($\varepsilon = 4000$ and 4600) appeared at the expense of the absorption of the starting azide at 354 nm [9, 15].

IR spectrum (Ar and undecane matrices at 12 K): Upon irradiating I, an absorption band due to a cyano group at 2225 cm^{-1} appeared gradually at the expense of the absorption due to the azide group at 2130 cm^{-1} [9, 15].

The title compound acts as an intermediate in the thermal reactions of 3-azido-5,7-di-tert-butyl-2,1-benzisothiazole (I); see the chapter "Chemical Reactions" below.

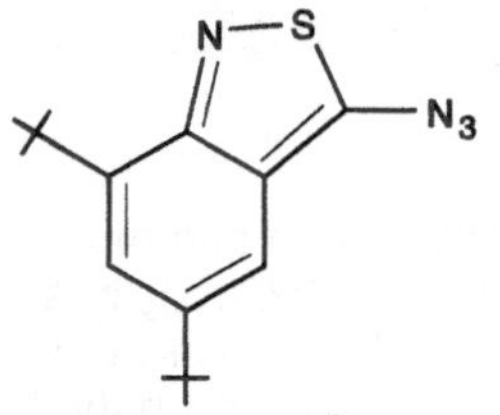

I

Spectra

The title compound has a singlet electronic ground state, because a triplet signal could not be observed in the ESR spectrum in various organic glasses, such as EPA, 2-methyltetrahydrofuran, or methylcyclohexane at cryogenic temperature [9].

UV–VIS Spectrum. The thionitrosoarene was identified by UV–VIS spectroscopy at 12 K in an argon matrix by absorptions at 456 and 483 nm ($\varepsilon = 4000$ and 4600) and at 70 K in an EPA matrix by absorptions at 460 and 484 nm. The absorption band at 483 nm was attributed to a π–π^* transition [9, 15].

IR Spectrum. The IR spectrum at 12 K in argon and undecane matrices showed absorptions at 2225 cm^{-1} (ν(CN)), 1200, and 1120 cm^{-1} (ν(N=S)) [9, 15].

Chemical Reactions

Several reactions of the unstable, intermediately generated thionitrosoderivative have been studied.

Dimerization. Dimerization of the title compound, which was generated by photolysis of 3-azido-5,7-di-tert-butyl-2,1-benzisothiazole in various organic glass matrices, was monitored by UV–VIS and IR spectroscopy. In an EPA (diethyl ether–isopentane–ethanol) matrix, the initially observed UV–VIS spectrum ($\lambda_{max} = 460$, 484 nm) did not change significantly on warming from 70 K to 90 K. At 100 K the absorption bands longer than 400 nm disappeared. This change in the UV–VIS spectrum (the IR bands at 1200 and 1120 also disappeared) is assumed to be due to dimerization of the title compound. The UV–VIS spectrum of the title compound disappeared at ca. 100 K in rigid matrices such as EPA, undecane, or 3-methylpentane, while it disappeared at 72 K in soft matrices like isopentane. These results indicate that the softer the organic glass, the lower the disappearance temperature of the thionitroso intermediate which undergoes dimerization. A possible structure of the dimer was assigned to be sulfur diimide N-sulfide ArNSN(S)Ar [9].

$$2\,ArN{=}S \longrightarrow Ar{-}N\underset{S}{\overset{S}{\diagdown\diagup}}N{-}Ar \ \text{ or } \ Ar{-}N{=}S{=}N{-}Ar \xrightarrow{-S} Ar{-}N{=}S{=}N{-}Ar$$

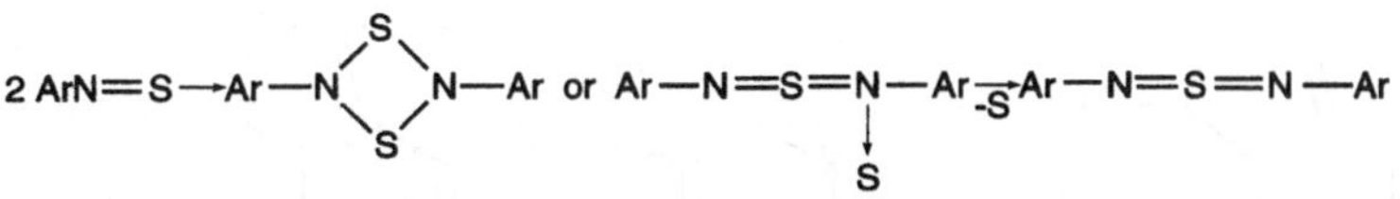

$$Ar = C_6H_2(C_4H_9\text{-}t)_2\text{-}2,4\text{-}CN\text{-}6$$

The photolysis of 3-azido-5,7-di-tert-butyl-2,1-benzisothiazole in hexane (with a medium-pressure mercury lamp) for 45 min at 0 °C under nitrogen yielded dark pink crystals of N-(2,4-di-tert-butyl-6-cyanophenyl)-N'-(5,7-di-tert-butyl-2,1-benzisothiazole-3-yl)-sulfur diimide (2%), N,N'-bis(2,4-di-tert-butyl-6-cyanophenyl)-sulfur diimide (78%), and 2,4-di-tert-butyl-6-cyano-aniline (16%) via the thionitroso intermediate [9, 15].

The thermal reaction of 3-azido-5,7-di-tert-butyl-2,1-benzisothiazole in hexane or ethanol gave the same products as those in the photoreaction indicating that the reaction proceeds via the thionitroso intermediate. Heating at 60 °C for 1 h in hexane (ethanol) produced 72% (18%) of sulfur diimide, 14% (68%) of the amine, and 11% (10%) of unsymmetrical sulfur diimide. The thermal reaction also proceeded slowly at room temperature in hexane. The unsymmetrical sulfur diimide was produced by thermally or photochemically reacting 3-azido-5,7-di-tert-butyl-2,1-benzisothiazole. The most plausible mechanism for the formation of the unsymmetrical sulfur diimide is the 1,3-dipolar cycloaddition reaction of S=N–C₆H₂(C₄H₉-t)₂-2,4-CN-6 with the azide followed by loss of nitrogen from an unstable thiatetrazoline intermediate [9, 16]; see the following scheme.

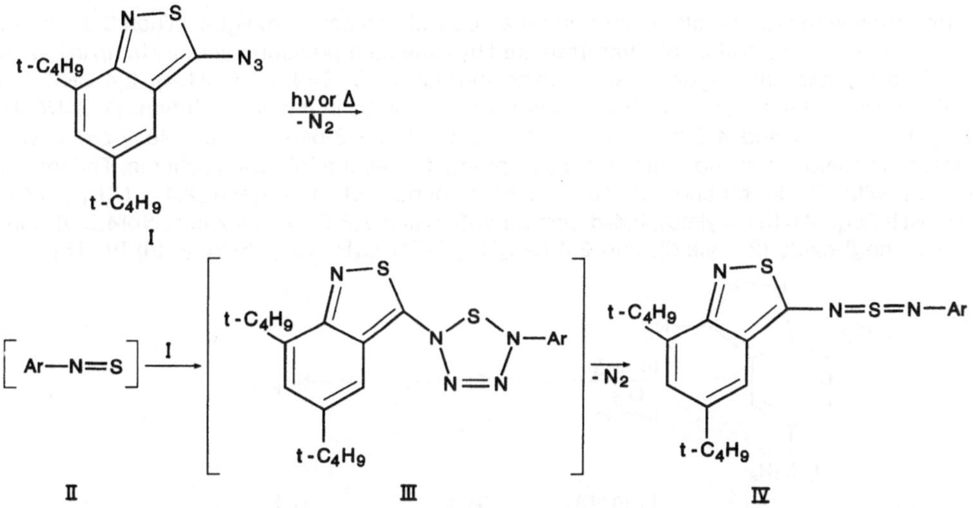

[2+4] Cycloaddition Reactions. The thermal reaction of S=N–C₆H₂(C₄H₉-t)₂-2,4-CN-6 (formed via 3-azido-5,7-di-tert-butyl-2,1-benzisothiazole) with 2,3-dimethyl-1,3-butadiene in hexane for 3 d at room temperature led to the [2+4] cycloadduct and the products from thermolysis or photolysis without reactant. The thermal reaction with cyclic dienes resulted in the formation of thermolysis products only [9].

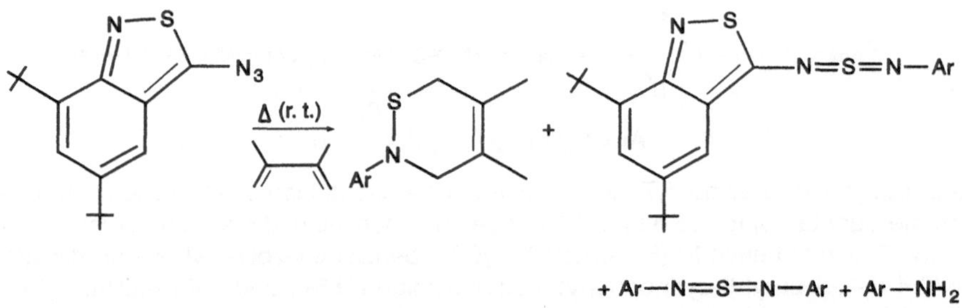

[2+3] Cycloaddition Reactions. Diazo compounds having higher 1,3-dipolar reactivity than aryl azides reacted with the title compound. When 3-azido-5,7-di-tert-butyl-2,1-benzisothiazole (I) was allowed to react with an excess of diphenyldiazomethane (IIIa) and 9-diazofluorene (IIIb) in hexane at room temperature, the corresponding imines (VIa and VIb) were formed; see the following scheme. When the reaction was carried out with an excess amount of diazo compounds having bulky alkyl groups like 2,2,6,6-tetramethyldiazocyclohexane (IIIc) and 2,2,5,5-tetramethyldiazocyclopentane (IIId), the corresponding thiocarbonylimines (Vc and Vd) could be isolated. The reaction with 2-diazo-1,1,3,3-tetramethylindan (IIIe) gave the corresponding 1,2,3,4-thiatriazoline (IVe). In all these cases the products can be explained by the initial [2+3] cycloaddition reaction of the title compound (II) with the diazo compounds (III) to give the corresponding 1,2,3,4-thiatriazoline (IV) [9, 16]; see the scheme on p. 15.

Reaction with O_2. The title compound reacted with molecular oxygen. When 3-azido-5,7-di-tert-butyl-2,1-benzisothiazole was irradiated by a medium-pressure mercury lamp in hexane at 0 °C for 30 min and oxygen was bubbled through it, $O{=}S{=}NC_6H_2(C_4H_9\text{-t})_2$-2,4-CN-6 was obtained. Upon thermolysis at 40 °C in the presence of O_2 in hexane solution, $O{=}S{=}NC_6H_2$-$(C_4H_9\text{-t})_2$-2,4-CN-6 and 4,6-di-tert-butyl-7a-cyano-7aH-1,2,3-benzoxathiazole 2-oxide were isolated in addition to the normal thermolysis products; see the following scheme. The thermal reaction with O_2 in ethanol at 40 °C over a period of 3 h gave 2,4-$(\text{t-}C_4H_9)_2$-6-CN-C_6H_2-NH-$SO_3C_2H_5$ (34% yield), 2,4-di-tert-butyl-6-cyano-7,8,9-oxathiazabicyclo[4.3.0]-nona-2,4,9-triene 8-oxide (2% yield), and 2,4-$(\text{t-}C_4H_9)_2$-6-CN-C_6H_2-NH_2 (54% yield) [9, 16].

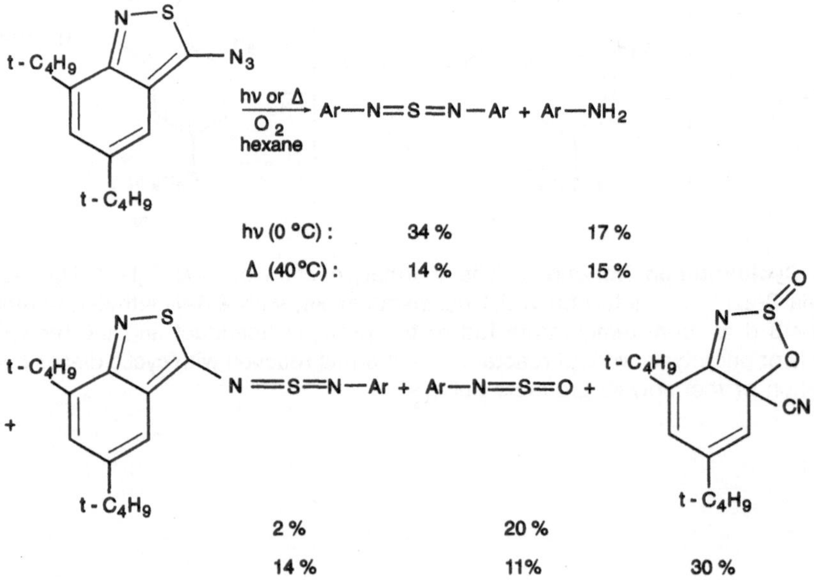

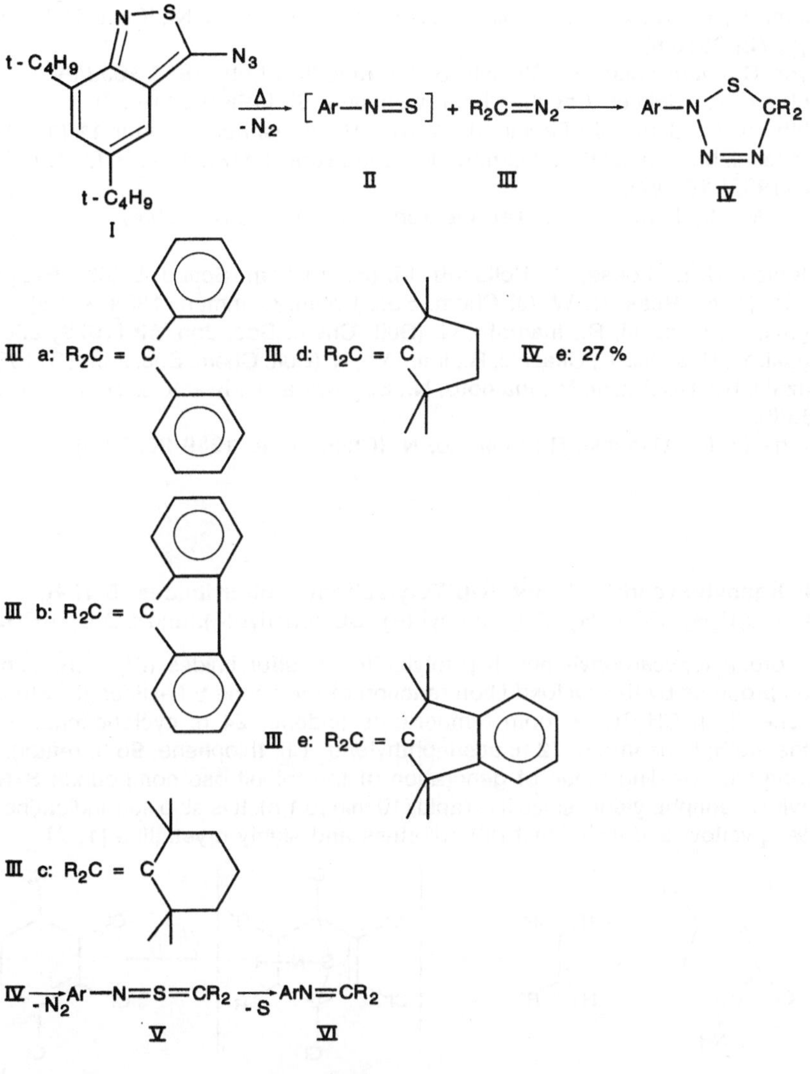

V c: 98 % VI a: 78 %

V d: 50 % VI b: 80 %

Reaction with Phenylthiirane. The thermal reaction of the title compound with phenylthiirane in hexane at 30 °C over a period of 2 days yielded 2,4-(t-C_4H_9)$_2$-6-CN–C_6H_2–N=S=O (31%), 2,4-(t-C_4H_9)$_2$-6-CN–C_6H_2–NH_2 (22%), and N-(2,4-di-tert-butyl-6-cyanophenyl)-N′-(5,7-di-tert-butyl-2,1-benzisothiazole-3-yl)-sulfur diimide (5%) [9, 16].

References:

[1] Bryce, M. R.; Taylor, P. C. (J. Chem. Soc. Chem. Commun. **1988** 950/1).
[2] Bryce, M. R.; Taylor, P. C. (J. Chem. Soc. Perkin Trans. I **1990** 3225/35).
[3] Tavs, P. (Angew. Chem. Int. Ed. Engl. **5** [1966] 1048/9).
[4] Davis, F. A.; Skibo, E. B. (J. Org. Chem. **41** [1976] 1333/6).

[5] Minami, T.; Yamataka, K.; Ohshiro, Y.; Agawa, T.; Yasuoka, N.; Kasai, N. (J. Org. Chem. **37** [1972] 3810/8).
[6] Mayer, R.; Domschke, G.; Bleisch, S. (Tetrahedron Lett. **1978** 4003/6).
[7] Mehlhorn, A. (Collect. Czech. Chem. Commun. **53** [1988] 2116/27).
[8] Mehlhorn, A.; Sauer, J.; Fabian, J.; Mayer, R. (Phosphorus Sulfur **11** [1981] 325/34).
[9] Takahashi, M.; Okazaki, R.; Inamoto, N.; Sugawara, T.; Iwamura, H. (J. Am. Chem. Soc. **114** [1992] 1830/7).
[10] Bryce, M. R.; Taylor, P. C. (Tetrahedron Lett. **30** [1989] 3835/6).

[11] Pedersen, C. L.; Lohse, C.; Poliakoff, M. (Acta Chem. Scand. B **32** [1978] 625/31).
[12] Joucla, M. F.; Rees, C. W. (J. Chem. Soc. Chem. Commun. **1984** 374/5).
[13] Inagaki, Y.; Okazaki, R.; Inamoto, N. (Bull. Chem. Soc. Jpn. **52** [1979] 2002/7).
[14] Inagaki, Y.; Hosogai, T.; Okazaki, R.; Inamoto, N. (Bull. Chem. Soc. Jpn. **53** [1980] 205/9).
[15] Okazaki, R.; Takahashi, M.; Inamoto, N.; Sugawara, T.; Iwamura, H. (Chem. Lett. **1989** 2083/6).
[16] Takahashi, M.; Okazaki, R.; Inamoto, N. (Chem. Lett. **1989** 2087/90).

1.3.4 N-Organyloxycarbonyl- and N-p-Tolylsulfonyl-sulfur Imides, S=N–R, $R = CO_2C_2H_5$, $CO_2C_6H_5$, CO_2menthyl-(–), CO_2fenchyl-(+), and $SO_2C_6H_4CH_3$-4

The N-organyloxycarbonyl- and N-p-tolylsulfonyl-sulfur imides (IV) were generated as elimination products by the cycloaddition reaction of thiophene S,N-ylides (I) with an nucleophilic alkene (II) in CH_2Cl_2 at room temperature (indene, 24 h; cyclohexene, 3 h, reflux; norbornene, 48 h; benzofuran, 72 h; acenaphthylene, 1 h; thiophene, 96 h, reflux). The most efficient and high-yielding mode of generation of the thionitroso compounds S=N–R is the reaction with acenaphthylene, since it is rapid (10 min to 1 h). It is also self-indicating, since the dienophile is yellow and its adduct (III) colorless and highly crystalline [1, 2].

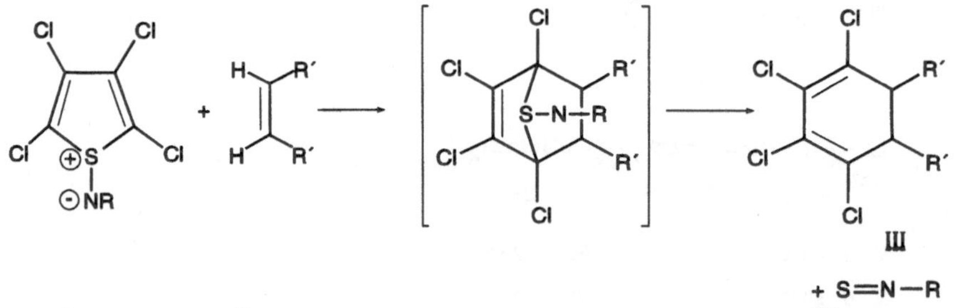

I II IV

In the absence of a suitable trap, $S{=}NCO_2C_2H_5$ in CH_2Cl_2 gave a complex mixture of unidentified products. In benzene, toluene, or cumene solution $S{=}NCO_2C_2H_5$ decomposed to give $(C_2H_5O_2CNH)_2S$ with moderate yield (18%) [1, 2].

The highly reactive molecules can be trapped by dienes (to give thiazines by [4+2] cycloaddition) or by enes (to give N-organyloxycarbonylthiohydroxylamines by ene reaction). The ene reaction of thionitroso compounds was found to be totally regiospecific. If the dienophile behaves as an ene (e.g., cyclohexene or indene), the thionitroso compound is efficiently trapped. With other dienophiles (e.g., norbornene and acenaphthylene) an ene or diene trap may be used separately [1, 2].

Reactions with Dienes. When the thiophene S,N-ylide (I, R = $CO_2C_2H_5$), acenaphthylene, and buta-1,3-diene were stirred together in CH_2Cl_2, the expected acenaphthylene adduct (III) was formed quantitatively and the S=$NCO_2C_2H_5$ adduct (V, R = $CO_2C_2H_5$) was obtained with 82% yield. Using cyclohexa-1,3-diene the analogous bicyclic thiazine (VI, R = $CO_2C_2H_5$) was obtained with 40% yield [1, 2].

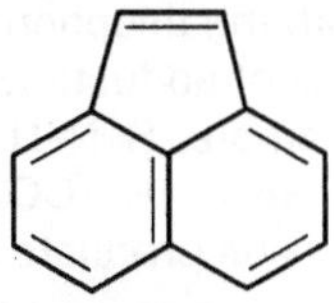
acenaphthylene

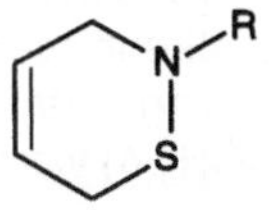
V

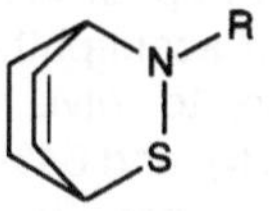
VI

S=$NCO_2C_2H_5$ reacted with cyclopentadiene between −78 °C and room temperature in CH_2Cl_2 to give the adduct VII with 20% yield and the 1,3,2,4-dithiadiazine (VIII) with 14% yield [1, 2].

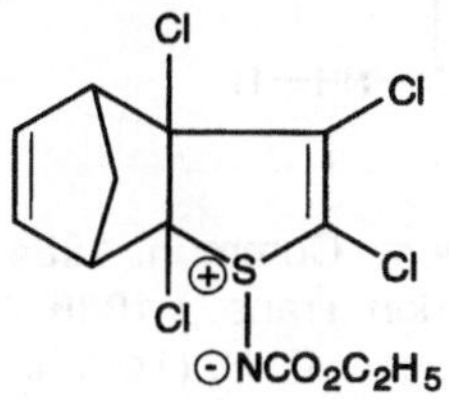
VII

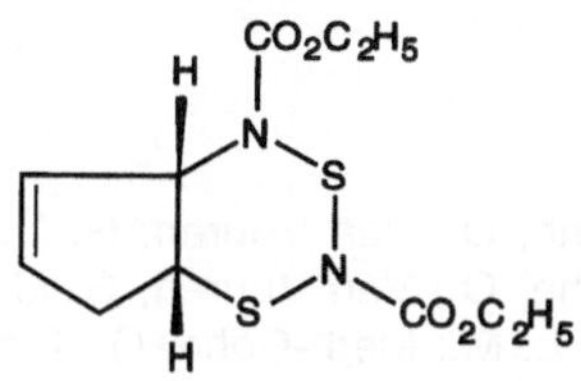
VIII

The reaction of S=$NCO_2C_6H_5$ with 2,3-dimethyl-1,3-buta-diene gave the Diels-Alder adduct IX and the ene adduct X both with 43% yield. With isoprene the two Diels-Alder adducts XI and XII with 25% yield each and the ene adduct XIII with 50% yield were obtained [1, 2].

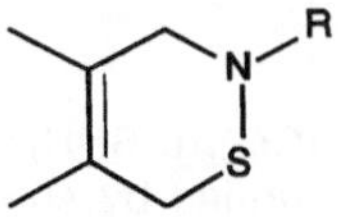
IX

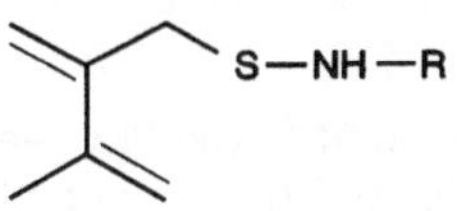
X

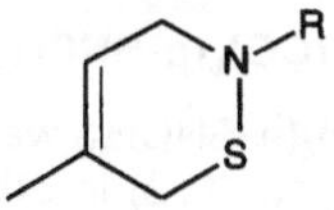
XI

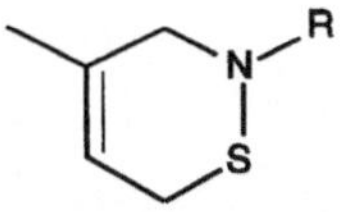
XII

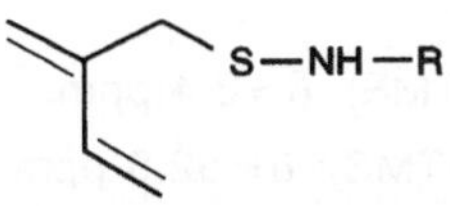
XIII

S=NCO_2menthyl-(−) and S=NCO_2fenchyl-(+) prepared from the thiophene S,N-ylide (I, R = CO_2menthyl-(−), CO_2fenchyl-(+)) and acenaphthylene were trapped with 1,4-diphenyl-1,3-buta-diene in CH_2Cl_2 at room temperature (stirring for ca. 12 h). The thionitroso adducts, thiazines XIV, were obtained with no diastereo differentiation [3].

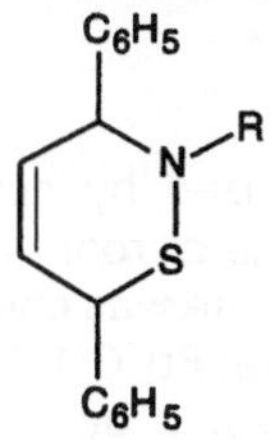
XIV

The thionitroso compounds S=N–R (R = $CO_2C_2H_5$, $CO_2C_6H_5$, $SO_2C_6H_4CH_3$-4) gave no defined adducts with thebaine, 9,10-dimethylanthracene, or β-pinene. When, e.g., an equimolar amount of S=$NCO_2C_6H_5$ was generated in the presence of thebaine, phenyl carbamate was produced and half of the thebaine was recovered [2].

Reactions with Alkenes. The thionitroso compounds reacted with alkenes to form ene adducts. The alkene could be used as a dienophile for reacting with the thiophene S,N-ylide (I) or as an ene trap for the liberated thionitroso compound. The yields of adducts were generally higher when acenaphthylene and the alkene were used. For example, S=N–R reacted with cyclohexene to give the corresponding ene adducts with 98% (R = $CO_2C_2H_5$), 72% (R = $CO_2C_6H_5$), and 64% (R = $SO_2C_6H_4CH_3$-4) yields. Several other ene products are listed in a table in the paper [2].

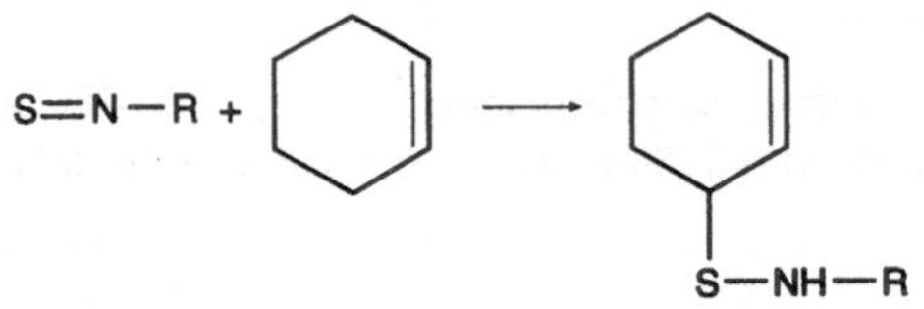

References:

[1] Meth-Cohn, O.; Van Vuuren, G. (J. Chem. Soc. Chem. Commun. **1984** 1144/6).
[2] Meth-Cohn, O.; Van Vuuren, G. (J. Chem. Soc. Perkin Trans. I **1986** 245/50).
[3] Dillen, J. L. M.; Meth-Cohn, O.; Moore, C.; Van Rooyen, P. H. (Tetrahedron **44** [1988] 3127/38).

1.4 Iron and Ruthenium Complexes with N-Alkyl- and N-Aryl-sulfur Imides

1.4.1 $Fe_2(CO)_6(\mu\text{-}SNCH_3)$

$Fe_2(CO)_6(\mu\text{-}SNCH_3)$ was obtained with 92% yield by reacting $Fe_2(CO)_6(\mu\text{-}SNH)$ with CH_2N_2 (mole ratio ca. 1 : 3) in ether at 0 °C for 0.5 h. The complex was purified by sublimation.

The orange, diamagnetic crystals melt at 30 to 32 °C. They are volatile at room temperature at 10^{-2} mbar.

^1H NMR ($CDCl_3$/TMS): $\delta = 3.4$ ppm.

^{13}C NMR ($CDCl_3$/TMS): $\delta = 52.5$ ppm (CH_3).

IR (pentane): ν(CO) (in cm^{-1}) = 2074 s, 2034 vs, 1994 vs, 1988 s, 1980 sh, 1972 sh.

MS: m/e = 341 M^+.

The complex is stable in air. It is readily soluble in nonpolar and polar organic solvents [1].

1.4.2 $Fe_2(CO)_6(\mu\text{-}SNC_4H_9\text{-}t)$

$Fe_2(CO)_6(\mu\text{-}SNC_4H_9\text{-}t)$ was synthesized by stirring a mixture of $Fe(CO)_5$ and t-C_4H_9-N=S=NC_4H_9-t (mole ratio 1 : 1) in hexane at room temperature under irradiation with diffuse sunlight for 20 h. After filtration, the filtrate was chromatographed on alumina with hexane as eluent. The orange solid was sublimed at 60 °C/1 Torr to obtain the orange complex with 3% yield [3]. The complex was also obtained by stirring a mixture of $Fe_2(CO)_9$ and t-C_4H_9-N=S=NC_4H_9-t (mole ratio 1 : 1) in hexane (or benzene) at room temperature in the dark for 3 d.

The complex was isolated by column chromatography on silica gel with n-hexane as eluent. It was further purified by sublimation in vacuum at 45 °C. The yield was 7%. Other crystalline products were $Fe_3(CO)_9(\mu\text{-}SNC_4H_9\text{-}t)S$ and a mixture of $Fe_2(CO)_6S_2$ and $Fe_3(CO)_9S_2$ [4, 5].

The complex melts at 48 to 50 °C [3].

The complex is thermally stable and fairly volatile. It is well soluble in aromatic and paraffinic hydrocarbons [3].

A molecular structure was proposed in which the SNC_4H_9-t fragment is rotated by 90° with respect to the Fe–Fe bond so that the sulfur and nitrogen are bridged to the two iron atoms. In this structure the SNC_4H_9-t ligand acts as a six-electron donor and thus has a single SN bond [5]. The structural features of the complex are similar to those of $Ru_2(CO)_6(\mu\text{-}SNC_4H_9\text{-}t)$; see p. 22.

^1H NMR (C_7D_8/TMS): δ (in ppm) = 0.68 (s) [5]; (C_6H_6/TMS): 0.62 (s) [3]; (5 to 10% in C_6D_6/TMS): 0.66; ($CDCl_3$/TMS): 1.13; $^3J(^{15}N,^1H) = 4.5 \pm 0.3$ Hz (from basic INEPT) [2].

^{13}C NMR (5 to 10% in C_6D_6/TMS): δ (in ppm) = 32.0, 61.3 (t-C_4H_9); 210.1 (CO).

^{14}N NMR (5 to 10% in C_6D_6/external neat CH_3NO_2): δ = −272 ppm.

^{15}N NMR (5 to 10% in C_6D_6/external neat CH_3NO_2): δ = −274.4 ppm; $^1J(^{57}Fe,^{15}N) = 6.0 \pm 0.5$ Hz.

^{17}O NMR (5 to 10% in C_6D_6/external H_2O): δ = 373.7 ppm [2].

IR (Nujol): ν (in cm^{-1}) = 2069 (s), 2030 (vs), 1999 (vs), 1996.5 (s), 1988.5 (m), 1980 (w), ν(CO); 1175 ν(CH); 705 m, ν(S–N); unassignable absorptions at 641 m, 619 w, 593 m, 570 s, 492 w, 477 w, 450 w, 440 w, and 362 w [3]; (hexane): ν(CO) = 2077 s, 2032 vs, 2000 sh, 1966 vs, 1986 sh, 1979 sh, 1954 w [5].

Raman (benzene): ν (in cm^{-1}) = 360 s, 314 m, 260 w, ν(FeS)/ν(FeN); 208 s, ν(FeFe) [5].

The mass spectrum showed the molecular ion (m/e = 383, t-$C_4H_9NSFe_2(CO)_6^+$), the fragment ions t-$C_4H_9NSFe_2(CO)_n^+$ (n = 0 to 5) formed by successive loss of 6 carbonyl ligands, and the ligand t-$C_4H_9NS^+$. Other prominent peaks are $NSFe_2^+$ and SFe_2^+ [3].

1.4.3 $Fe_2(CO)_6(\mu\text{-}SNC_6H_5)$

$Fe_2(CO)_6(\mu\text{-}SNC_6H_5)$ was prepared by heating a mixture of $Fe_2(CO)_9$ and $O{=}S{=}NC_6H_5$ (mole ratio ca. 1 : 2) in benzene at 40 to 45 °C for 3 h. Alumina chromatography using a hexane–benzene mixture as an eluent followed by recrystallization from hexane gave an orange crystalline complex with 1.6% yield. The diamagnetic complex is less soluble in paraffinic than in aromatic solvents [3].

The complex melts at 65 to 67 °C [3].

IR: ν in cm^{-1} = 2071 s, 2034 vs, 1999 vs, 1998.5 vs, 1995 m, 1985 w, ν(CO); 721 ν(SN); 755, 682, δ(CH out of plane); unassigned absorptions at 642 m, 617 s, 583 s, 561 s, ca. 475 w, br, 445 w, and 397 w, br [3].

The mass spectrum showed the molecular ion $C_6H_5NSFe_2(CO)_6^+$, fragment ions $C_6H_5NSFe_2(CO)_n^+$ (n = 0 to 5) formed by successive loss of 6 carbonyl ligands, and the ligand ion $C_6H_5NS^+$. Other prominent peaks are $CNSFe_2^+$, SFe_2^+, and $C_6H_5NSFe_2^{2+}$ [3].

1.4.4 $Fe_2(CO)_6(\mu\text{-}SNC_6H_4CH_3\text{-}4)$

The title complex was formed by reacting $Fe_2(CO)_9$ with 4-$CH_3C_6H_4N{=}S{=}NC_6H_4CH_3$-4. For reaction conditions, see $Fe_2(CO)_6(\mu\text{-}SNC_4H_9\text{-t})$, p. 18. It was not isolated and studied in solution only [5].

IR (hexane): ν(CO) = 2078 s, 2042 vs, 2003 cm^{-1} vs [5].

1.4.5 $Fe_2(CO)_7(\mu\text{-}SNC_4H_9\text{-t})$

The complex was obtained by stirring a mixture of $Fe_2(CO)_9$ and t-$C_4H_9N{=}S{=}NC_4H_9$-t (mole ratio 1:1) in hexane (or benzene) at room temperature for 3 d while excluding light. The resulting red solution was filtered and chromatographed on a silica gel column with n-hexane as eluent. The various fractions were concentrated to small volumes in vacuum and set aside at −30°C to obtain crystals; see $Fe_2(CO)_6(\mu\text{-}SNC_4H_9\text{-t})$, p. 18. When crystallization did not occur, the fractions were concentrated to dryness. Subsequent sublimation of the resulting residues yielded the title complex with 0.1% yield and an oily mixture with $Fe_2(CO)_6$-(t-C_4H_9NC(O)S). By monitoring the reaction with ^1H NMR spectroscopy, it was found that the title complex is the main product and $Fe_2(CO)_6(SNC_4H_9\text{-t})$ a by-product formed in a 7:1 molar ratio. The title complex was isolated in amounts sufficient to give an IR spectrum [5].

^1H NMR (C_7D_8/TMS): δ = 0.96 (s) ppm [5].

IR (hexane): ν(CO) = 2084 s, 2046 vs, 2012 vs, 2006 vs, 1703 m (1688 s in KBr) [5].

A structure containing a bridging carbonyl group was proposed for the title complex. The $S{=}NC_4H_9$-t ligand is positioned parallel to the Fe–Fe axis acting as a four-electron donor. An NS double bond is required [5].

The complex is unstable in $CHCl_3$ [5].

1.4.6 $Fe_2(CO)_7(\mu\text{-}SNC_6H_4CH_3\text{-}4)$

The title complex was formed in the reaction of $Fe_2(CO)_9$ with 4-$CH_3C_6H_4N{=}S{=}NC_6H_4CH_3$-4 but not isolated. It was studied in solution only. For experimental details of the preparation, see the analogous preparation of $Fe_2(CO)_7(\mu\text{-}SNC_4H_9\text{-t})$ [5].

IR (hexane): ν(CO) = 2082 s, 2046 vs, 2010 vs, 2007 vs, 1970 w, 1718 s [5].

1.4.7 $Fe_3(CO)_9(\mu_3\text{-}S)(\mu_3\text{-}SNC_4H_9\text{-t})$

The complex was obtained by stirring a mixture of $Fe_2(CO)_9$ and t-$C_4H_9N{=}S{=}NC_4H_9$-t (mole ratio 1:1) in hexane (or benzene) at room temperature for 3 d while excluding light. The complex was isolated with 1% yield by column chromatography on silica gel with n-hexane as eluent. The other products were $Fe_2(CO)_6(\mu\text{-}SNC_4H_9\text{-t})$ and a mixture of $Fe_2(CO)_6S_2$ and $Fe_3(CO)_9S_2$ [4, 5].

The complex decomposed in vacuum. It is unstable in $CHCl_3$ [5].

The crystals are monoclinic having the space group $P2_1/n - C^5_{2h}$ (No. 14) with a = 11.039(1), b = 11.525(1), c = 15.996(1) Å, β = 96.241(6)°; Z = 4; R = 0.067 for 1637 observed independent reflections. The molecular structure is shown in **Fig 1**. Selected bond lengths, bond angles, and dihedral angles are listed in Table 3 and Table 4, pp. 21/2. The molecular structure consists of three $Fe(CO)_3$ groups with a single Fe–Fe bond and a t-C_4H_9NS fragment with a single N–S bond and an S atom. The t-C_4H_9NS fragment and the sulfur atom each bridge the three Fe atoms in such a way that the skeleton, formed by the three Fe, two S, and one N atoms, has an approximately prismatic configuration [4, 5].

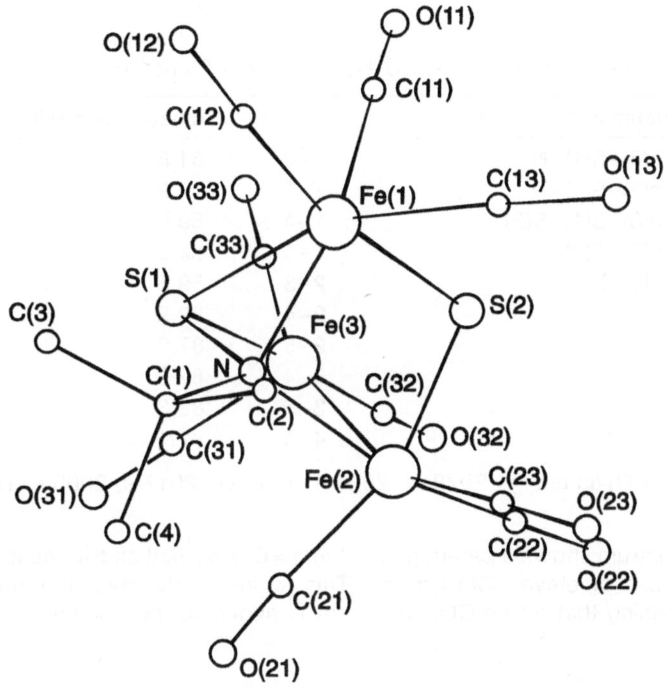

Fig. 1. Molecular structure of $Fe_3(CO)_9(\mu_3\text{-}S)(\mu_3\text{-}SNC_4H_9\text{-}t)$ [5].

Table 3
Selected Bond Lengths and Bond Angles of $Fe_3(CO)_9(\mu_3\text{-}S)(\mu_3\text{-}SNC_4H_9\text{-}t)$ [5].

bond length	in Å	bond angle	in °
N–S(1)	1.70(2)	S(1)–Fe(1)–N	46.5(4)
Fe(2)–Fe(3)	2.614(4)	S(1)–Fe(1)–S(2)	82.4(2)
Fe(1)–S(1)	2.227(5)	N–Fe(1)–S(2)	81.6(4)
Fe(1)–S(2)	2.314(6)	N–Fe(2)–Fe(3)	80.3(4)
Fe(2)–S(2)	2.237(6)	Fe(3)–Fe(2)–S(2)	54.7(2)
Fe(3)–S(1)	2.259(5)	N–Fe(2)–S(2)	83.1(4)
Fe(3)–S(2)	2.253(6)	S(1)–Fe(3)–Fe(2)	75.5(2)
Fe(1)–N	2.08(1)	S(1)–Fe(3)–S(2)	83.1(2)
Fe(2)–N	2.09(1)	S(2)–Fe(3)–Fe(2)	54.1(2)
N–C(1)	1.51(2)	Fe(1)–S(1)–N	62.2(5)
		Fe(1)–S(1)–Fe(3)	97.1(2)
		Fe(3)–S(1)–N	100.2(5)

Table 3 (continued)

bond angle	in °	bond angle	in °
Fe(1)–S(2)–Fe(2)	90.7(2)	S(1)–N–Fe(2)	103.7(7)
Fe(1)–S(2)–Fe(3)	94.8(2)	S(1)–N–Fe(1)	71.4(5)
Fe(2)–S(2)–Fe(3)	71.2(2)	Fe(1)–N–Fe(2)	101.8(6)

Table 4
Dihedral Angles of $Fe_3(CO)_9(\mu_3\text{-}S)(\mu_3\text{-}SNC_4H_9\text{-}t)$ between Planes [5].

plane	atoms defining the plane	planes	dihedral angle in °
1	Fe(2)–Fe(3)–S(1)–N	1–2	61.8
2	Fe(1)–Fe(2)–S(2)–N	1–3	59.2
3	Fe(1)–Fe(3)–S(1)–S(2)	1–4	89.6
4	Fe(2)–Fe(3)–S(2)	1–5	84.9
5	Fe(1)–S(1)–N	2–3	59.6
		2–4	89.2
		2–5	87.9
		3–4	82.6
		3–5	89.3
		4–5	7.1

IR (hexane): ν(CO) (in cm^{-1}) = 2078 m, 2061 s, 2033 vs, 2017 s, 2005 s, 1997 s, 1974 w, 1866 w [5].

The mass spectrum shows a parent peak at m/e = 667 as well as fragments resulting from the successive loss of eleven CO groups. This suggests the overall formula $Fe_4(CO)_{11}$-$(SNC_4H_9\text{-}t)S$, meaning that an $Fe(CO)_2$ fragment is added to the original compound [5].

1.4.8 $Ru_2(CO)_6(\mu\text{-}SNC_4H_9\text{-}t)$

$Ru_2(CO)_6(\mu\text{-}SNC_4H_9\text{-}t)$ was synthesized by heating a solution of $Ru_3(CO)_{12}$ and t-C_4H_9-N=S=NC_4H_9-t (mole ratio 1 : 1.5) in hexane at 69 °C for 8 h. The crude product was purified by thin-layer chromatography on silica gel with cyclohexane–THF (9 : 1) as eluent and recrystallization from pentane at –30 °C. The crystals were isolated with 70% yield [6].

The bright yellow crystals melt at 49 °C [6].

The molecular structure of $Ru_2(CO)_6(\mu\text{-}SNC_4H_9\text{-}t)$ was determined by X-ray structure analysis. The crystals belong to the triclinic space group $P\bar{1}-C_i^1$ (No. 2) with a = 9.610(2), b = 10.237(3), c = 16.832(3) Å, α = 75.35(2)°, β = 84.01(1)°, γ = 88.66(2)°; V = 1593.2 Å³; Z = 4; D_x = 1.971, D_o = 1.96 g/cm; R = 0.034 for 5753 independent reflections. The crystal structure contains two crystallographically independent molecules. The Ru_2NS group forms a tetrahedrane structure. The six-electron SNC_4H_9-t ligand is side-on-coordinated. The N–S bond is arranged perpendicular to the Ru–Ru axis. The $Ru(CO)_3$ unit forms a trigonal pyramid with Ru at the top. The SN distance is consistent with an S–N single bond. The molecular structure is shown in **Fig 2**. The selected bond lengths and bond angles given are mean values [6].

^1H NMR ($CDCl_3$/TMS at –50 °C): δ = 1.03 ppm (s, CH_3).

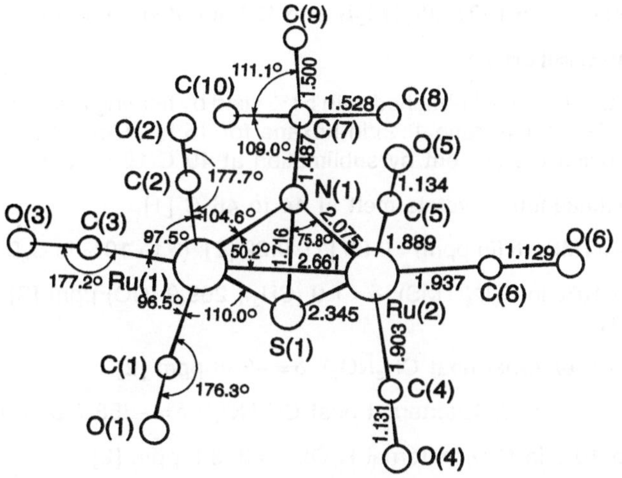

Fig. 2. Molecular structure of $Ru_2(CO)_6(\mu\text{-}SNC_4H_9\text{-}t)$ [6].

^{13}C NMR ($CDCl_3$/TMS at −50 °C): δ (in ppm) = 31.6 ($C(\underline{C}H_3)_3$), 57.5 ($\underline{C}(CH_3)_3$), 187.5, 198.5, 198.6 (CO). A coalescence temperature of ca. 30 °C was determined for the CO signals. The chemical shift δ = 195.2 ppm for the CO groups was measured at 90 °C. In solution the molecule has a fluctuating structure. All six CO ligands are equivalent in the time average.

IR (hexane): ν (in cm^{-1}) = 2082 m, 2055 s, 2000 vs, 1998 sh, 1987 m, 1977 w, ν(CO); (CsI): 795 w, ν(S–N).

The mass spectrum shows the successive cleavage of six CO ligands from the molecular ion (m/e = 475 with respect to ^{102}Ru).

The complex is stable in air. It can be sublimed at 10^{-2} Torr and is soluble in alkanes. The Ru complex (like the corresponding Fe complex) is diamagnetic [6].

References:

[1] Herberhold, M.; Bühlmeyer, W. (Angew. Chem. **96** [1984] 64; Angew. Chem. Int. Ed. Engl. **23** [1984] 80).

[2] Wrackmeyer, B.; Kupce, E.; Distler, B.; Dirnberger, K.; Herberhold, M. (Z. Naturforsch. **46b** [1991] 1679/83).

[3] Otsuka, S.; Yoshida, T.; Nakamura, A. (Inorg. Chem. **7** [1968] 1833/5).

[4] Meij, R.; Van der Helm, J.; Stufkens, D. J.; Vrieze, K. (J. Chem. Soc. Chem. Commun. **1978** 506/7).

[5] Meij, R.; Stufkens, D. J.; Vrieze, K.; Brouwers, A. M. F.; Schagen, J. D.; Zwinselman, J. J.; Overbeek, A. R.; Stam, C. H. (J. Organomet. Chem. **170** [1979] 337/54).

[6] Herberhold, M.; Bühlmeyer, W.; Gieren, A.; Hübner, T.; Wu, J. (Z. Naturforsch. **42b** [1987] 65/70).

1.5 Iron Complexes with N-Trialkylsilyl- and N-Trimethylstannanyl-sulfur Imides

1.5.1 $Fe_2(CO)_6(\mu\text{-}SNSi(CH_3)_3)$

$Fe_2(CO)_6(\mu\text{-}SNSi(CH_3)_3)$ was obtained with 51% yield by heating a solution of $Fe_3(CO)_{12}$ and $(CH_3)_3SiN{=}S{=}NSi(CH_3)_3$ (mole ratio 1 : 2) in hexane for 15 h under reflux. The complex was isolated after removing the solvent by sublimation at 40 °C/10^{-2} mbar [1].

The orange, diamagnetic crystals melt at 44 to 46 °C [1].

^1H NMR ($CDCl_3$/TMS): δ (in ppm) = 0.25 [1], 0.22 [2]; (5 to 10% in C_6D_6/TMS): −0.16 [2].

^{13}C NMR (5 to 10% in C_6D_6/TMS): δ = 1.0 (CH_3), 209.9 (CO) ppm [2]; ($CDCl_3$/TMS): δ = 1.03 ppm (CH_3) [1].

^{14}N NMR (CD_2Cl_2/external neat CH_3NO_2): δ = −358 ppm [2].

^{15}N NMR (5 to 10% in C_6D_6/external neat CH_3NO_2): δ = −358.7 ppm [2].

^{17}O NMR (5 to 10% in C_6D_6/external H_2O): δ = 373.8 ppm [2].

^{29}Si NMR (5 to 10% in C_6D_6/external TMS): δ = 35.5 ppm [2].

There were no significant changes in the ^1H and ^{29}Si NMR spectra at low temperatures (down to −50 °C) [2].

IR (pentane): ν(CO) (in cm^{-1}) = 2072 s, 2032 vs, 1992 vs, 1986 s, 1976 sh, 1966 sh [1].

MS (70 eV): m/e = 399 M^+ [1].

The crystals are stable in air. They are readily soluble in nonpolar and polar organic solvents.

The complex is sensitive to hydrolysis. It converts quantitatively to $Fe_2(CO)_6(\mu\text{-}SNH)$ when chromatographed on silica gel with pentane as eluent [1].

1.5.2 $Fe_2(CO)_6(\mu\text{-}SNSi(CH_3)_2C_4H_9\text{-}t)$

$Fe_2(CO)_6(\mu\text{-}SNSi(CH_3)_2C_4H_9\text{-}t)$ was synthesized by addition of $t\text{-}C_4H_9(CH_3)_2SiCl$ to a THF solution of $Fe_2(CO)_6(\mu\text{-}SNNa)$. After removing the solvent, the residue was extracted with hexane. From the hexane solution an oily, dark brown liquid was obtained with 85% yield. Further purification by sublimation again gave the complex as an oily liquid [2].

^1H NMR (5 to 10% in C_6D_6/TMS): δ (in ppm) = -0.19, 0.71; ($CDCl_3$/TMS): 0.15, 0.90 [2].

^{13}C NMR (5 to 10% in C_6D_6/TMS): δ (in ppm) = −2.3 ($SiCH_3$), 18.7, 26.1 ($t\text{-}C_4H_9$), 209.9 (CO) [2].

^{14}N NMR ($CDCl_3$/external neat CH_3NO_2): δ = −368 ppm [2].

^{17}O NMR (5 to 10% in C_6D_6/external H_2O): δ = 372.8 ppm [2].

^{29}Si NMR (5 to 10% in C_6D_6/external TMS): δ = 35.6 ppm [2].

IR (hexane): ν(CO) (in cm^{-1}) = 2075 s, 2032 vs, 1999 vs, 1988 vs, 1977 m, 1947 w [2].

1.5.3 $Fe_2(CO)_6(\mu\text{-}SNSn(CH_3)_3)$

$Fe_2(CO)_6(\mu\text{-}SNSn(CH_3)_3)$ was prepared by refluxing a mixture of $Fe_3(CO)_{12}$ and $(CH_3)_3SnN{=}S{=}NSn(CH)_3$ (mole ratio 1 : 2) in hexane for 15 h. After removing the solvent in a vacuum, the complex was obtained with 27% yield from the dark brown residue as brown crystals by sublimation at 55 °C/10^{-3} Torr [2].

^{1}H NMR ($CDCl_3$/TMS): $\delta = -0.02$ ppm, $^{2}J(^{119}Sn,^{1}H) = 56.5$ Hz [2].

^{13}C NMR (5 to 10% in C_6D_6/TMS): δ (in ppm) = −3.4 (CH_3), $^{1}J(^{119}Sn,^{13}C) = 375.4$ Hz; 210.7 (CO) [2].

^{14}N NMR ($CDCl_3$/external neat CH_3NO_2): $\delta = -353$ ppm [2].

^{17}O NMR (5 to 10% in C_6D_6/external H_2O): $\delta = 372.9$ ppm [2].

^{119}Sn NMR (5 to 10% in C_6D_6/external TMS): $\delta = 207.0$ ppm. The ^{119}Sn NMR signal was rather broad at room temperature. The broadening of the ^{119}Sn resonance has to be attributed to exchange processes (a Hahn spin echo extended INEPT experiment showed no ^{15}N satellites) [2].

IR (hexane): ν(CO) (in cm^{-1}) = 2069 w, 2025 vs, 1982 s, 1978 s, 1963 w [2].

The crystals turn black and sinter irreversibly at 82 °C. The binuclear complex $[Fe(CO)_4\text{-}Sn(CH_3)_2]_2$ has been identified as one of the decomposition products in solution by NMR spectroscopy and by recording the EI mass spectra [2].

References:

[1] Herberhold, M.; Bühlmeyer, W. (Angew. Chem. **96** [1984] 64; Angew. Chem. Int. Ed. Engl. **23** [1984] 80).

[2] Wrackmeyer, B.; Kupce, E.; Distler, B.; Dirnberger, K.; Herberhold, M. (Z. Naturforsch. **46b** [1991] 1679/83).

2 Disulfur Nitride, Nitrogen Disulfide, SNS

The SNS species, a radical, was detected IR spectroscopically only in an Ar matrix. No other unambiguous spectroscopic identification has yet been reported.

2.1 Formation

S_2N was produced besides other sulfur-nitrogen species in a microwave-discharged argon–nitrogen–sulfur vapor stream and subsequently trapped in solid argon at 12 K [1].

S_2N was also formed when $(SN)_x$ vapor diluted in Ar was passed through microwave discharge at 40 to 80 W [1].

S_2N was produced by near-ultraviolet (400 to 254 nm) photolysis of the less stable isomer NSS in a solid Ar matrix [1].

The S_2N molecule is believed to be formed in a discharge of a mixture of N_2 and CS_2 [2]; however, this species was later identified to be the NS radical in very highly excited vibrational states up to v = 20 [3].

An N_2–SCl_2 atomic flame is also believed to contain S_2N or SN [4].

The cathodic reduction of $(C_4H_9)_4N^+$ $S_4N_5^-$ in 0.15 M $(n\text{-}C_4H_9)_4N^+$ BF_4^-–CH_2Cl_2 on Pt at −40 °C and room temperature gave ESR spectra which were assigned to $S_4N_5^{2-}$, SN_2^-, and a species containing one N atom, possibly S_2N. There are three lines in the ESR spectrum of equal intensities with the coupling constant a_N = 1.13 mT. S_2N probably forms by thermal decomposition of $S_4N_5^{2-}$ at room temperature along with SN_2^- via $S_4N_5^- + e^-$ (−40 °C) → $S_4N_5^{2-} \xrightarrow{r.t.} 2\ SN_2^- + S_2N$ [5].

The formation of S_2N was discussed in conjunction with the polymerization of tetrahydrofuran with the initiators S_2N^+ AsF_6^- and $(ClS)_2N^+$ AsF_6^- [6].

2.2 Molecular Properties

2.2.1 Molecular Structure

The valence angle of C_{2v} SNS was calculated from isotopic frequency ratios in the IR spectrum of pairs of isotopic molecules. Central isotopic substitution gave a lower limit and terminal substitution an upper limit for the bond angle owing to the effect of normal cubic anharmonicity. The average value for the valence angle of S_2N considering anharmonicity cancellation is 153 ± 5° [1].

Using SCF and CISD methods the bent SNS isomer ($^2A_1$) was found to have the lowest energy structure among the species SNS bent ($^2A_1$), SNS linear ($^2\Pi_u$), NS_2 ring ($^2B_1$), NSS bent ($^2A'$), and SNS bent ($^4A_2$). Depending on the basis set and the level of theory employed, the calculated valence angle of SNS ranges from 150° to 155°. Single- and double-configuration interaction calculations predict a valence angle of 151 ± 1° [7]. The following values of the SN bond length r_e in Å and the SNS bond angle ∢ in ° of the SNS bent isomer were obtained by various methods: r_e = 1.543, ∢ = 155.3 (SCF-DZ+P); r_e = 1.530, ∢ = 154.5 (SCF-TZ+2P); r_e = 1.562, ∢ = 150.8 (CISD-DZ+P); r_e = 1.546, ∢ = 151.3 (CISD-TZ+2P); r_e = 1.601, ∢ = 145.8 (CAS-SCF-DZ+P); r_e = 1.602, ∢ = 146.0 (CAS-SCF-DZ+P) [7]; r_e = 1.529 (SCF-6-31G*) [8], r_e = 1.582, ∢ = 148.0 (CCSD) [9], and r_e = 1.55, ∢ = 151.5 (FCI) [10].

For S_2N a quasi-linear structure with a barrier height to linearity of 2.5 to 3.0 kcal/mol (CISD) [7] or about 2.0 kcal/mol (MRD-CI) has been calculated [10].

2.2.2 IR Spectrum

IR spectra were recorded of the products from a discharged sulfur vapor–nitrogen–argon stream deposited on a CsI window at 12 ± 1 K. S_2N absorbed at 1225.2 cm^{-1} which was assigned to the antisymmetric stretching vibration (showing a triplet and a doublet pattern with mixed sulfur and mixed nitrogen isotopes, respectively). The absorption bands at 1884.9 and 2449 cm^{-1} were assigned to the $\nu_1 + \nu_3$ combination and 2 ν_3 overtone bands, respectively. The symmetric stretching mode, which is expected to be a weak band near 660 cm^{-1}, could not be observed due to spectral interference from sulfur species in this region [1].

The fundamental vibrations were calculated by different theoretical methods (see Table 5), e.g., SCF(DZ+P): $\nu_1 = 742$, $\nu_2 = 272$, and $\nu_3 = 1228$ cm^{-1} [7].

2.2.3 Electronic Spectrum

The electronically low-lying doublet states of the SNS molecule were studied with the SCF, CISD, and CAS-SCF methods. At the SCF level the stability of the wave functions and the energetics of the doublet states were discussed in terms of the MO Hessian. The SCF wave functions for the three lowest lying doublet states (bent $^2A_1$, linear $^2\Pi_u$, ring $^2B_1$) were found to be stable, while the remaining six upper doublet states (bent $^2B_2$ and $^2A_2$, ring $^2A_2$ and $^2A_1$, linear two $^2\Pi_g$) are unstable [3].

Nine quartet states, six bent ($^4A_2$, $^4B_2$, two $^4A_1$, and two $^4B_1$) and three linear ($^4\Pi_g$ and two $^4\Pi_u$) ones, have been investigated by the SCF, CISD, and CAS-SCF methods. At the TZ+2P CISD and CISD+Q levels of theory, the energetic stability of the nine states was found to decrease in the order $^4A_2 > {}^4B_2 > {}^4A_1 > {}^4\Pi_g > {}^4B_1 > {}^4A_1 > {}^4B_1 > {}^4\Pi_u > {}^4\Pi_u$. Four of these quartet states lie within 1.8 eV of the $^2A_1$ ground state of S_2N. Thus, one or more of these quartet states should be observable. At the SCF level, two wave functions, $^4A_2$ and $^4\Pi_g$, were found to be stable, while the functions for the remaining seven displayed instability [11].

Among the low-lying, symmetry-allowed transitions ($X^2A_1 \rightarrow {}^2A_1$, $^2B_1$, and $^2B_2$), those with excitation energies of 1.87 eV ($X^2A_1 \rightarrow {}^2B_2$) and 2.87 eV ($X^2A_1 \rightarrow {}^2B_2$) are fairly intense (f = 0.005 and 0.002) (calculated using ab initio MRD-CI) and likely to be observed in the electronic spectrum of S_2N [10].

2.3 Quantum-Chemical Calculations

Several molecular properties of S_2N and its isomers have been calculated by different methods. A survey of the methods and the calculated properties is given in Table 5, p. 28.

Table 5
Quantum-Chemical Calculations for S_2N[a].

method (basis set)	calculated molecular properties	Ref.
ab initio SCF (6-31G*, STO-3G*)	geometry, E_t	[8]
CCSD	equilibrium and two excited doublet states ($^2A_1$, $^2B_2$, and $^2A_2$): geometry, E_t, ν_i, IP	[9]
ab initio MRD-CI	geometry, E_t, IP, EA, μ electronic spectrum	[10]
ab initio SCF (DZ+P, TZ+2P) CISD (DZ+P, TZ+2P) CAS-SCF (DZ+P)	geometry, E_t, IP, ν_i, μ of 5 S_2N isomers (SNS bent ($^2A_1$), SNS linear ($^2\Pi_u$), NS_2 ring ($^2B_1$), NSS bent ($^2A'$), and SNS bent ($^4A_2$))	[7]
ab initio SCF (DZ+P, TZ+2P) CISD (DZ+P, TZ+2P) CAS-SCF (DZ+P)	geometry, E_t, IP, ν_i, μ of 9 low-lying doublet states (three bent ($^2A_1$, $^2B_2$, $^2A_2$), three ring ($^2B_1$, $^2A_2$, and $^2A_1$), and three linear ($^2\Pi_u$ and two $^2\Pi_g$))	[3]
ab initio SCF (DZ+P, TZ+2P) CISD (DZ+P, TZ+2P) CAS-SCF (DZ+P)	geometry, E_t, ν_i, μ of 9 low-lying quartet states (six bent ($^4A_2$, $^4B_2$, two $^4A_1$ and two $^4B_1$) and three linear ($^4\Pi_g$, two $^4\Pi_u$))	[11]

[a] The following abbreviations and symbols are used: SCF = self-consistent field; CCSD = coupled cluster approximation with single- and double replacements; CISD = single and double replacement configuration interaction; CAS = complete active space; MRD-CI = multireference double replacement configuration interaction; DZ = double zeta basis set; TZ = triple zeta basis set; P = polarization; E_t = total energy; IP = ionization potential; EA = electron affinity; ν_i = fundamental vibration; μ = dipole moment.

References:

[1] Hassanzadeh, P.; Andrews, L. (J. Am. Chem. Soc. **114** [1992] 83/91).
[2] Amano, T.; Amano, T. (44th Symp. Mol. Spectros., Columbus, Ohio, 1989, Abstr. RF1).
[3] Yamaguchi, Y.; Xie, Y.; Alberts, I. L.; Grev, R. S.; Schaefer, H. F., III (J. Chem. Phys. **93** [1990] 5053/61).
[4] Pannetier, G.; Dessaux, O.; Arditi, I.; Goudmand, P. (C. R. Hebd. Seances Acad. Sci. **259** [1964] 2198/9).
[5] Fritz, H. P.; Bruchhaus, R. (Electrochim. Acta **29** [1984] 947/50).
[6] Fairhurst, S. A.; Hulme-Lowe, A.; Johnson, K. M.; Suttcliffe, L. H.; Passmore, J.; Schriver, M. (Magn. Reson. Chem. **23** [1985] 828/31).
[7] Yamaguchi, Y.; Xie, Y.; Grev, R. S.; Schaefer, H. F., III (J. Chem. Phys. **92** [1990] 3683/7).
[8] Gimarc, B. M.; Warren, D. S. (Inorg. Chem. **30** [1991] 3276/80).
[9] Kaldor, U. (Chem. Phys. Lett. **185** [1991] 131/5).
[10] Sannigrahi, A. B.; Peyerimhoff, S. D. (Int. J. Quantum Chem. **41** [1992] 413/9).
[11] Yamaguchi, Y.; Alberts, I. L.; Xie, Y.; Schaefer, H. F., III (J. Chem. Phys. **94** [1991] 1277/87).

3 Dithionitrous Acid and Derivatives

3.1 Dithionitrous Acid, S=N–SH

The title molecule is an unknown species which has not yet been isolated. Therefore, the only information available on this system comes from theoretical studies. A survey of the literature on quantum-chemical calculations is given in Table 6.

Table 6
Quantum-Chemical Calculations for S=N–SH.

method (basis set)[a)]	calculated molecular properties[b)]	Ref.
ab initio SCF (HF/4-31G*)	geometry	[1]
ab initio MP3 (6-31G**)	E_{rel}, pop, q_i, E/Z energy difference	[1]
ab initio (M3*)	geometry, torsional potentials	[2, 3]
MNDO	geometry, ν_i, E_{rel}, μ, EA	[4]
MNDO	geometry, q_i, μ	[5]

[a)] SCF = self-consistent field; HF = Hartree-Fock; MP3 = 3th order Møller-Plesset perturbation method; MNDO = modified neglect of differential overlap. – [b)] E_{rel} = relative energy; pop = orbital populations; q_i = charge densities; ν_i = fundamental vibrations; μ = dipole moment; EA = electron affinity.

The isomers S=N–S–H and S=S=NH (see "Sulfur-Nitrogen Compounds" Part 6, 1990, pp. 298/9) are relatively close in energy; however, S=N–S–H is less stable [1]. The Z and E isomers are predicted to be stable, but the gauche form represents an unstable transition state [2, 3]. According to MP3/6-31G** the E isomer should be slightly more stable than the Z isomer [1]. Geometrical data obtained by ab initio SCF method are given in **Fig. 3**.

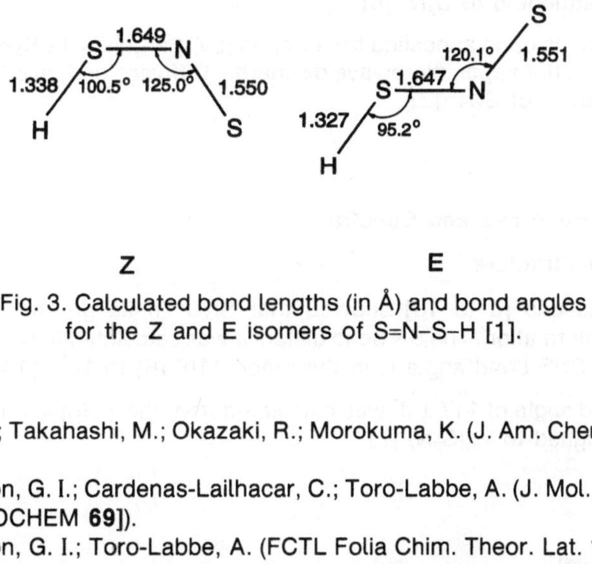

Fig. 3. Calculated bond lengths (in Å) and bond angles for the Z and E isomers of S=N–S–H [1].

References:

[1] Nakamura, S.; Takahashi, M.; Okazaki, R.; Morokuma, K. (J. Am. Chem. Soc. **109** [1987] 4142/8).

[2] Cardenas-Jiron, G. I.; Cardenas-Lailhacar, C.; Toro-Labbe, A. (J. Mol. Struct. **210** [1990] 279/89 [THEOCHEM **69**]).

[3] Cardenas-Jiron, G. I.; Toro-Labbe, A. (FCTL Folia Chim. Theor. Lat. **17** [1989] 177/90).

[4] Mehlhorn, A. (Collect. Czech. Chem. Commun. **53** [1988] 2116/27).

[5] Shermolovich, Yu. G.; Solov'ev, A. V.; Borodin, A. V.; Pen'kovskii, V. V.; Trachevskii, V. V.; Markovskii, L. N. (Zh. Org. Khim. **27** [1991] 1637/41; J. Org. Chem. USSR [Engl. Transl.] **27** [1991] 1433/6).

3.2 Dithionitrite Ion (1−), SNS^-

SNS^- has not yet been isolated in a pure form. Based on UV and IR spectroscopic analyses, SNS^- is assumed to form by several routes.

3.2.1 Formation

The SNS^- species is probably formed by electrochemical reduction of S_3N^-. Controlled-potential electrolysis of $((C_6H_5)_3P)_2N^+$ S_3N^- on a mercury pool electrode in CH_3CN–0.1 M $(C_2H_5)_4N^+$ ClO_4^- at −2.2 V (vs. Ag|0.1 M $AgClO_4$–CH_3CN) gave n values of 1.4 to 1.8, and the UV–visible spectrum of the electrolyzed solution showed a single band at 375 nm. The 375-nm species is attributed to SNS^- formed by the reaction $S_3N^- + n\ e^- \rightarrow SNS^- + S_x^{2-}$. The electrochemical results suggest that it should be possible to prepare S_2N^- by treating S_3N^- (or S_4N^-) with a suitable reducing agent [1].

It was suggested that photo-induced reactions of S_3N^- might produce S_2N^- by cleavage of a sulfur–sulfur bond. Illumination of sulfur–NH_3 solutions [2] and S_7NH–NH_3 solutions with added amide [3] with white light caused gradual formation of S_4N^-, S_3N^-, and S_2N^-. An absorption band in the UV–visible spectrum at 390 nm was assigned to S_2N^- [3].

The S_2N^- ion was also assumed to be an intermediate during the formation of S_4N^- by thermolysis of a solution of $((C_6H_5)_3P)_2N^+$ $S_3N_3^-$ in CH_3CN under reflux [1, 4, 5] via $S_3N_3^- \rightarrow S_2N^- +$ "NSN" (or $S^o + N_2$); $S_2N^- + 2\ S^o \rightarrow S_4N^-$.

The S_2N^- ion was formed by electrolysis of $S_4N_4O_2$ on a mercury pool in CH_3CN–0.1 M $(C_2H_5)_4N^+$ ClO_4^- at −1.7 V (vs. Ag|0.1 M $AgClO_4$–CH_3CN) with an n value of 2.8. The final solution showed two absorption bands in the UV–visible spectrum at 465 nm attributed to S_3N^- and 375 nm attributed to S_2N^- [6].

S_2N^- probably forms upon annealing the trapped (solid argon at 12 K) reaction products of an argon–nitrogen–sulfur vapor microwave discharge. Diffusion and reaction of SN and S^- are the most likely sources of S_2N^- [7].

3.2.2 Molecular Properties and Spectra

3.2.2.1 Molecular Structure

Ab initio calculations [8 to 10] showed that SNS^- in its ground state (X^1A_1) has a symmetrical C_{2v} conformation. The SN bond distance was calculated to lie in the range 1.57 [8] to 1.66 Å [9]. The SNS bond angle is in the range 110° [8] to 124° [10].

An average bond angle of 117±3° was calculated from the isotopic ratios of the observed IR frequencies assigned to $\nu_{as}(S_2N)$ [7].

3.2.2.2 IR Spectrum

A weak absorption band at 893.9 cm^{-1} (showing doublet and triplet patterns with mixed isotopic nitrogen and sulfur, respectively) was observed in an Ar matrix at 12 K. The band was assigned to the antisymmetric stretching vibration of S_2N^-, $\nu_{as}(S_2N)$ [7].

3.2.2.3 Raman Spectrum

Sulfur–NH_3 solutions at room temperature showed Raman lines at 858 cm^{-1} [2] and after adding amide (4 M; K^+ $NH_2^-/S = 0.6$) at 858 and 839 cm^{-1} [3], attributed tentatively to S_2N^- [2, 3].

3.2.2.4 UV Spectrum

The absorption in CH_3CN observed at $\lambda_{max} = 375$ nm [1] (or in NH_3 at $\lambda_{max} = 390$ nm [2]) was assigned to the $2a_2 \rightarrow 3b_1$ ($X^1A_1 \rightarrow {}^1B_2$) transition [9].

The electronic spectrum calculated by the transition state method shows maxima at 672 ($X^1A_1 \rightarrow {}^1B_1$; $4a_1 \rightarrow 2b_1$) and 421 nm ($XA_1^1 \rightarrow {}^1B_2$; $1a_2 \rightarrow 2b_1$) [8]. The most intense bands obtained by FCI (estimated full CI) are at 645 ($X^1A_1 \rightarrow {}^1B_1$), 470 ($X^1A_1 \rightarrow {}^1B_2$), and at 364 ($X^1A_1 \rightarrow {}^1B_2$) nm [9].

3.2.3 Quantum-Chemical Calculations

The electronic properties and geometric parameters given in Table 7 were the subject of quantum-chemical calculations.

Table 7
Quantum-Chemical Calculations for SNS^-.

method	calculated molecular properties[a)]	Ref.
ab initio SCF-HFS	geometry, pop, q, electronic spectrum	[8]
ab initio SCF	geometry, q, pop, bond orders, valencies	[10]
ab initio MRD-CI	geometry, E_t	[11]
ab initio MRD-CI	geometry, E_t, electronic spectrum	[9]

[a)] E_t = total energy; pop = population analysis; q = charge distribution.

3.2.4 Reaction

S_2N^- probably reacts with sulfur in boiling CH_3CN to yield S_4N^- [1, 4].

References:

[1] Chivers, T.; Hojo, M. (Inorg. Chem. **23** [1984] 2738/42).
[2] Dubois, P.; Lelieur, J. P.; Lepoutre, G. (Inorg. Chem. **28** [1989] 2489/91).
[3] Dubois, P.; Lelieur, J. P.; Lepoutre, G. (Inorg. Chem. **27** [1988] 3032/8).
[4] Chivers, T.; Schmidt, K. J.; McIntyre, D. D.; Vogel, H. J. (Can. J. Chem. **67** [1989] 1788/94).
[5] Chivers, T.; Laidlaw, W. G.; Oakley, R. T.; Trsic, M. (J. Am. Chem. Soc. **102** [1980] 5773).
[6] Chivers, T.; Hojo, M. (Inorg. Chem. **23** [1984] 4088/93).

[7] Hassanzadeh, P.; Andrews, L. (J. Am. Chem. Soc. **114** [1992] 83/91).
[8] Trsic, M.; Laidlaw, W. G. (J. Mol. Struct. **123** [1985] 259/65).
[9] Sannigrahi, A. B.; Peyerimhoff, S. D. (Chem. Phys. Lett. **175** [1990] 279/81).
[10] Behera, L.; Kar, T.; Sannigrahi, A. B. (J. Mol. Struct. **209** [1990] 111/24 [THEOCHEM **68**]).

[11] Sannigrahi, A. B.; Peyerimhoff, S. D. (Int. J. Quantum Chem. **41** [1992] 413/9).

3.3 S-Organyl Dithionitrites

3.3.1 S=N–S–$C_{10}H_{15}$-1, $C_{10}H_{15}$-1 = 1-adamantyl

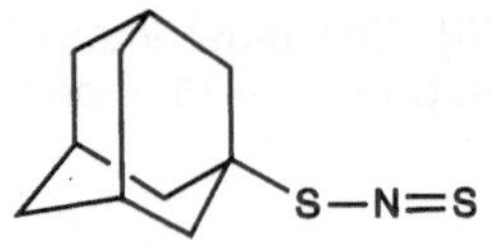

1-Adamantyl dithionitrite was prepared by adding an equimolar amount of $((CH_3)_3Si)_2N$–S–$C_{10}H_{15}$-1 in ether to a stirred solution of freshly distilled SCl_2 in ether at −70 °C within 15 min. After warming the reaction mixture to 20 °C and removing the solvent, 31% of the title compound could be isolated.

The brownish red substance decomposed at 96 °C.

^1H NMR (C_6D_6/TMS): δ (in ppm) = 1.42 (m) and 1.83 (m).

The substance is stable at 20 °C in an atmosphere of argon for several hours and decomposes when heated to 100 °C.

3.3.2 S=N–S–$N(CH_2)_4O$

4-Thionitrososulfanyl-morpholine was prepared by adding dropwise a solution of an equimolar amount of $((CH_3)_3Si)_2N$–S–$N(CH_2)_4O$ in ether to a stirred solution of freshly distilled SCl_2 in ether at −70 °C. After warming the reaction mixture to 20 °C and removing the volatile components, 31% of the crystalline compound was isolated.

The yellow needles melt at 110 to 115 °C.

The title compound is stable at 20 °C for several hours. S=N–S–$N(CH_2)_4O$ decomposed to give S_4N_4 in quantitative yield, when a hexane–benzene (2 : 1) solution was boiled for 30 min and was filtered off after cooling to 20 °C. In the filtrate $O(CH_2)_4N$–S–S–$N(CH_2)_4O$ was isolated with 74% yield. The decomposition of the title compound in ether at 20 °C was also monitored by ^{15}N NMR spectroscopy. When the solution was left at 20 °C for 10 h, the decomposition of the title compound quantitatively yielded various compounds. S_4N_4 was identified among them.

Heating the title compound at 110 °C decomposed it, thereby forming a mixture of compounds, the main components of which were $O(CH_2)_4N$–S–S–$N(CH_2)_4O$ and S_4N_4.

Reference:

Shermolovich, Yu. G.; Solov'ev, A. V.; Borodin, A. V.; Pen'kovskii, V. V.; Trachevskii, V. V.; Markovskii, L. N. (Zh. Org. Khim. **27** [1991] 1637/41; J. Org. Chem. USSR [Engl. Transl.] **27** [1991] 1433/6).

4 Salts of the Dithionitryl Ion (1+), $[S{=}N{=}S]^+$

4.1 Preparation and Formation of S_2N^+ Salts

4.1.1 $S_2N^+ AsF_6^-$

Preparation from S_4N_4, S_8, and AsF_5

The title compound was prepared by reacting S_4N_4 with sulfur and AsF_5 in SO_2 in the presence of a trace amount of bromine via $1/2\ S_8 + S_4N_4 + 6\ AsF_5 \rightarrow 4\ S_2N^+ AsF_6^- + 2\ AsF_3$. This is the only route which gives high yields of a crystalline, soluble S_2N^+ salt. The synthesis was carried out in flame-dried, three-bulb Pyrex glass vessels incorporating a medium sintered glass frit and closed with a Teflon-stemmed glass valve. It is important to use a well-dried apparatus and reagents, and to add the required amount of AsF_5 and traces of bromine before the reaction mixture warms up to room temperature. SO_2, AsF_5 (2% [1] or 10% [2] excess), and a trace amount of Br_2 were successively condensed onto a mixture of S_8 (10% deficit) and S_4N_4. The mixture was stirred for 12 h (20 h [3]) at room temperature [1 to 3]. After removing the volatiles, the brownish yellow solid was treated with S_8 in SO_2 for another 12 h to convert $SN^+ AsF_6^-$ (a minor product of the reaction) to $S_2N^+ AsF_6^-$ [1, 2]. In comparison to earlier papers [2, 3] the purification was improved. After filtration the soluble product was recrystallized four times by slowly removing the solvent. CCl_3F and SO_2 were added and the solution cooled to −78 °C. After filtration the solvent was recondensed onto the solid (this process was repeated five times). Removal of the solvent gave highly crystalline $S_2N^+ AsF_6^-$ (90% yield based on S_4N_4). The remaining traces of sulfur were removed by sublimation at ca. 60 °C/10^{-3} Torr after 14 h [1]. In another paper a different purification procedure was published: After stirring the reaction mixture for 12 h the solvent was removed, and the crude olive green product was stirred in CH_2Cl_2 for 24 h. Washing with CH_2Cl_2 increased the yield by 20%. The impurities were filtered off, and the product was washed briefly with more CH_2Cl_2 by back condensation. The yield was 95% [4].

The enthalpy of the reaction $1/2\ S_8 + S_4N_4 + 6\ AsF_5 \rightarrow 4\ S_2N^+ AsF_6^- + 2\ AsF_3$ was estimated to be $-(104 \pm 2)$ kJ/mol based on the value of the enthalpy of formation of $S_2N^+ AsF_6^-$, $\Delta H^\circ_{f,298} = -(1413.8 \pm 1.9)$ kJ/mol [2]. The enthalpy change of the reaction $S_8 + 2\ S_4N_4 + 12\ AsF_5 \rightarrow 8\ S_2N^+ AsF_6^- + 4\ AsF_3$ was derived to be -836.8 ± 16 kJ/mol (i.e. thermodynamically allowed) from $\Delta H^\circ_{f,298}$ $(S_2N^+ AsF_6^-) = -(1413.8 \pm 1.9)$, ΔH_f $(AsF_3) = -858.14$, ΔH_f $(AsF_5) = 1237 \pm 0.8$, and ΔH_f $(S_4N_4) = 469 \pm 1$ kJ/mol [1].

Preparation from S_4N_4 and $S_4^{2+} (AsF_6^-)_2$

The title compound was prepared by condensing SO_2 onto a frozen mixture of S_4N_4 and S_4^{2+} $(AsF_6^-)_2$ (mole ratio ca. 1 : 2) at −78 °C. The reaction proceeds according to the equation $S_4N_4 + 2\ S_4^{2+} (AsF_6^-)_2 \rightarrow 4\ S_2N^+ AsF_6^- + 1/2\ S_8$. After warming to room temperature, the mixture was stirred for 24 h. After standing 48 h without stirring at room temperature, the reaction mixture was filtered. The product was purified by repeated recrystallizations. A yellow crystalline solid was obtained with ca. 60% yield (based on S_4N_4) [1].

Preparation from $SN^+ AsF_6^-$ and S_8

The title compound was prepared by the reaction of $SN^+ AsF_6^-$ with stoichiometric [1] or excess of S_8 [1, 4, 5] in SO_2 via $8\ SN^+ AsF_6^- + S_8 \rightarrow 8\ S_2N^+ AsF_6^-$. SO_2 was condensed at −78 °C onto a frozen mixture of $SN^+ AsF_6^-$ and S_8 (mole ratio 8 : 1) in an NMR tube. The reaction mixture was warmed to room temperature. ^{14}N NMR spectroscopy showed the reaction to be complete within 5 min. $S_2N^+ AsF_6^-$ was quantitatively separated after removing the solvent.

The reaction with excess S_8 also proceeded quantitatively (50% yield based on $SN^+ AsF_6^-$ [5]). The authors recommend using a stoichiometric amount of S_8 in order to prevent contamination of $S_2N^+ AsF_6^-$ [1].

The enthalpy change of the reaction $SN^+ AsF_6^- + 1/8\ S_8 \rightarrow S_2N^+ AsF_6^-$ was estimated to be $\Delta H = -66 \pm 28$ kJ/mol and the entropy change $\Delta S = -1.6\ J \cdot mol^{-1} \cdot K^{-1}$ ($\Delta G = -65.5$ kJ/mol at room temperature) [1].

Preparation from $S_8^{2+} (AsF_6^-)_2$ and NaN_3

$S_2N^+ AsF_6^-$ was prepared with low yield by reacting $S_8^{2+} (AsF_6^-)_2$ with NaN_3 in SO_2 (mole ratio ca. 1:2). After warming to room temperature and stirring for 20 h, the solution was filtered. Several days later the solution was filtered again and the solvent removed. Yellow crystals were obtained with 20% yield based on $S_8^{2+} (AsF_6^-)_2$ and the equation $S_8^{2+} (AsF_6^-)_2 + NaN_3 \rightarrow S_2N^+ AsF_6^- + NaAsF_6 + N_2 + 3/4\ S_8$ [3].

Preparation by Reduction of $(ClS)_2N^+ AsF_6^-$ with $SnCl_2$

$S_2N^+ AsF_6^-$ was synthesized with 94% yield by stirring molar equivalents of $(ClS)_2N^+ AsF_6^-$ and $SnCl_2$ in liquid SO_2 for 18 h. After removing the solvent bilayer (SO_2 and $SnCl_4$) and pumping the system to dryness, the crude product was purified by washing with ice-cold CH_2Cl_2 [7].

Preparation from $[S_3N_2]_2^{2+} (AsF_6^-)_2$ and $S_4^{2+} (AsF_6^-)_2$

The title compound was prepared by condensing SO_2 onto a frozen mixture of $[S_3N_2]_2^{2+} (AsF_6^-)_2$ and $S_4^{2+} (AsF_6^-)_2$ (mole ratio 1:1). After stirring at room temperature for 15 min the solution was filtered and the red-brown soluble product purified by repeated recrystallizations. The yellow crystalline solid was obtained with 70% yield [1].

Formation from $(K^+)_2\ [N{=}S{=}N]^{2-}$

$S_2N^+ AsF_6^-$ was formed by the reaction of $(K^+)_2\ [N{=}S{=}N]^{2-}$ and fluorosulfonic acid in SO_2 in the presence of AsF_5 at 20 °C via $(K^+)_2\ [N{=}S{=}N]^{2-} + FSO_3H + AsF_5 \rightarrow NH_4^+$, N_2, S_2N^+, $[S(NH_2)_2]^{2+}$, AsF_6^-. The cation S_2N^+ was detected by ^{14}N NMR spectroscopy [8].

Formation by Thermal Decomposition of $(BrS)_2N^+ AsF_6^-$

The thermal decomposition of $(BrS)_2N^+ AsF_6^-$ in a dynamic vacuum at room temperature for 115 h gave $S_2N^+ AsF_6^-$ and Br_2 together with unchanged $(BrS)_2N^+ AsF_6^-$ [9].

Formation from $S_3N_2^{2+} (AsF_6^-)_2$

$S_2N^+ AsF_6^-$ and $S_3N_2F^+ AsF_6^-$ were identified as decomposition products of $S_3N_2^{2+} (AsF_6^-)_2$ in SO_2 at ca. −30 °C [6]. Later it was shown by ^{14}N NMR spectroscopy that the $S_3N_2^{2+}$ cation in SO_2 solution completely dissociates to SN^+ and S_2N^+ in a 1:1 ratio at room temperature [10, 11]. This behavior is consistent with lattice energy and ab initio calculations which show that the enthalpy change of the reaction $S_3N_2^{2+}$ (g) $\rightarrow$ SN^+ (g) + S_2N^+ (g) is $\Delta H = -400$ kJ/mol. The entropy change of the dissociation of $S_3N_2^{2+} (AsF_6^-)_2$ was estimated by the method of Latimer to be approximately zero [11]. Dissociation is favored due to orbital symmetry and the nature of the reaction products (correlation of MOs, energy levels by extended Hückel calculation). The dissociation is symmetry-allowed. The ab initio total energies favor fragmentation compared to ring formation. The decomposition is apparently driven by nuclear-nuclear repulsions. The energy change is −120 (6-31G*/MP2), −95 (6-31G*), and −22 kcal/mol (STO-3G*) [12].

Formation from $[S_3N_2]_2^{2+}$ $(AsF_6^-)_2$ or $(S_3N_2)_2N^+$ AsF_6^- and AsF_5

Equimolar amounts of S_2N^+ and NS^+ were obtained by oxidation of $[S_3N_2]_2^{2+}$ $(AsF_6^-)_2$ or $(S_3N_2)_2N^+$ AsF_6^- with AsF_5 in SO_2 in the presence of traces of Br_2 (the ^{14}N NMR spectrum showed that the total integration of SN^+ and S_2N^+ was eight times that obtained from the spectrum recorded prior to adding Br_2) [1].

Formation from $S_3N_2F^+$ AsF_6^- and AsF_5

Fluoride ion abstraction from $S_3N_2F^+$ AsF_6^- by AsF_5 in SO_2 at −30 °C also generated S_2N^+ and SN^+ (ratio 1 : 1) [6, 11].

Formation from S_4N_4 and AsF_5, $S_4N_4^{2+}$ $(AsF_6^-)_2$, or S_4^{2+} $(AsF_6^-)_2$

A small amount of S_2N^+ AsF_6^- was formed when a large excess of AsF_5 and a trace of Br_2 were successively condensed at −196 °C onto S_4N_4 in SO_2 and the reaction mixture kept at room temperature for 14 d. The yield was 3% (main product SN^+ AsF_6^-) [1].

A small amount of S_2N^+ AsF_6^- was obtained by the reaction of S_4N_4 with $S_4N_4^{2+}$ $(AsF_6^-)_2$ (mole ratio 1.5 : 1) at −70 °C followed by oxidation with AsF_5 in the presence of a trace of Br_2 in SO_2 [1].

Small amounts of S_2N^+ AsF_6^- were also detected in the bulk product ($[S_3N_2]_2^{2+}$ $(AsF_6^-)_2$) of the reaction of S_4N_4 with S_4^{2+} $(AsF_6^-)_2$ (mole ratio 1 : 1) [1].

Formation from $S_4N_4^{2+}$ $(AsF_6^-)_2$ and S_8^{2+} $(AsF_6^-)_2$, S_8, or S_4^{2+} $(AsF_6^-)_2$

Reactions of $S_4N_4^{2+}$ $(AsF_6^-)_2$ with S_8^{2+} $(AsF_6^-)_2$, S_8, and S_4^{2+} $(AsF_6^-)_2$, respectively, in SO_2 after 3 weeks reaction time gave a small amount of S_2N^+ AsF_6^- besides $S_4N_4^{2+}$ $(AsF_6^-)_2 \cdot n\ SO_2$ ($n \leq 1$) [1].

Formation from $(S_5N_5AsF_6)_x$ and AsF_5

A trace amount of S_2N^+ was detected in the ^{14}N NMR spectrum of the reaction mixture of $(S_5N_5AsF_6)_x$ with AsF_5 in the presence of Br_2 in SO_2 [1].

4.1.2 S_2N^+ SbF_6^-

S_2N^+ SbF_6^- was synthesized with 75% yield by stirring molar equivalents of $(ClS)_2N^+$ SbF_6^- and $SnCl_2$ in liquid SO_2 for 18 h. After removing the solvent bilayer (SO_2 and $SnCl_4$) and pumping the system to dryness, the crude product was purified by washing with ice-cold CH_2Cl_2 [7].

4.1.3 S_2N^+ $SbCl_6^-$

S_2N^+ $SbCl_6^-$ was prepared by the reactions of S_7NH, $S_7N{-}BCl_2$, and $S_6(NH)_2$-5,8 with $SbCl_5$ in liquid SO_2 (S_7NH: mole ratio 1 : 2; $S_6(NH)_2$-5,8: mole ratio 1 : 4; $S_7N{-}BCl_2$: mole ratio 1 : 2). After shaking for 48 h at room temperature, the reaction mixture was filtered and the solvent slowly removed. ($S_7N{-}BCl_2$ was synthesized in situ from S_7NH and BCl_3 at −48 °C after ca. 1 h.) S_2N^+ $SbCl_6^-$ was separated as large orange crystals [13].

S_2N^+ $SbCl_6^-$ was also obtained by the reaction of $S_3N_3Cl_3$, $SbCl_5$, and S_8 (mole ratio 1 : 3 : 3/8) in $SOCl_2$ at room temperature. The reaction mixture was stirred for ca. 24 h and then filtered. The insoluble solid was extracted with SO_2 yielding the title compound as a pale yellow, insoluble solid. The overall yield was 28%. The synthesis was also carried out in CH_2Cl_2. The reaction mixture was refluxed for 3 h and then cooled to 0 °C. The solid was purified by extraction with SO_2. The yield was 35%. An attempt to prepare the compound in CH_3NO_2 yielded a red tar from which S_2N^+ $SbCl_6^-$ could not be readily separated. Besides S_2N^+ $SbCl_5$ also $S_3N_2Cl^+$ $SbCl_6^-$ and $(ClS)_2N^+$ $SbCl_6^-$ were isolated. In CH_2Cl_2 the adduct $S_4N_4 \cdot SbCl_5$ was also formed [14].

S_2N^+ $SbCl_6^-$ was prepared by stirring a mixture of $(ClS)_2N^+$ $SbCl_6^-$ and $SnCl_2$ (mole ratio 1 : 1) in SO_2 at room temperature for 36 h. The overall yield of the highly insoluble yellow solid was 70% [15].

4.1.4 S_2N^+ $CF_3SO_3^-$

S_2N^+ $CF_3SO_3^-$ was synthesized with 70% yield by stirring molar equivalents of $(ClS)_2N^+$ $CF_3SO_3^-$ and $SnCl_2$ in liquid SO_2 for 18 h. After removing the solvent bilayer (SO_2 and $SnCl_4$) and pumping the system to dryness, the crude product was purified by washing with ice-cold CH_2Cl_2 [7].

The compound was also obtained by reacting $S_3N_3Cl_3$ and Ag^+ $CF_3SO_3^-$ in the mole ratio 1 : 3 (acting as a source of SN^+ $CF_3SO_3^-$) with sulfur in liquid SO_2 at −70 °C. After warming to room temperature the olive green solution was stirred for 4 h until it turned bright yellow. After filtration and reducing the solvent, small thin yellow platelets of S_2N^+ $CF_3SO_3^-$ were obtained. The yield was 37%. The compound is extremely air-sensitive [7].

4.1.5 S_2N^+ $AlCl_4^-$

S_2N^+ $AlCl_4^-$ was prepared by stirring a mixture of $(ClS)_2N^+$ $AlCl_4^-$ and $SnCl_2$ (mole ratio 1 : 1) in CH_2Cl_2 at room temperature for 24 h. The overall yield of the yellow solid was 86% [15].

S_2N^+ $AlCl_4^-$ was formed by the reaction of S_4N_4 with $AlCl_3$ in CH_2Cl_2. Yellow, isometricly shaped crystals of the compound grew at the vessel walls when a mixture of S_4N_4 and freshly sublimed $AlCl_3$, covered with a layer of CH_2Cl_2, was allowed to stand at room temperature for several days. The crystals are very sensitive to moisture. The main product of this reaction is the adduct $S_4N_4 \cdot AlCl_3$ [16].

S_2N^+ $AlCl_4^-$ was generated quantitatively upon adding sulfur to a solution of SN^+ $AlCl_4^-$ in CH_2Cl_2 as shown by ^{14}N NMR spectroscopy [17].

A small amount of S_2N^+ $AlCl_4^-$ was formed by the reaction of SN^+ $AlCl_4^-$ with Se (mole ratio 1 : 1) in CH_2Cl_2. $S_2N_2SeCl^+$ $AlCl_4^-$ was the main product. S_2N^+ was detected by ^{14}N NMR spectroscopy [17].

4.1.6 $S_2N^+ FeCl_4^-$

$S_2N^+ FeCl_4^-$ was synthesized with 72% yield by stirring molar equivalents of $(ClS)_2N^+ FeCl_4^-$ and $SnCl_2$ in liquid SO_2 for 18 h. After removing the solvent bilayer (SO_2 and $SnCl_4$) and pumping the system to dryness, the crude product was purified by washing with ice-cold CH_2Cl_2 [7].

References:

[1] Awere, E. G.; Passmore, J. (J. Chem. Soc. Dalton Trans. **1992** 1343/50).
[2] O'Hare, P. A. G.; Awere, E. G.; Parsons, S.; Passmore, J. (J. Chem. Thermodyn. **21** [1989] 153/8).
[3] Banister, A. J.; Hey, R. G.; MacLean, G. K.; Passmore, J. (Inorg. Chem. **21** [1982] 1679/80).
[4] Banister, A. J.; Lavender, I.; Rawson, J. M.; Clegg, W. (J. Chem. Soc. Dalton Trans. **1992** 859/65).
[5] Apblett, A.; Banister, A. J.; Biron, D.; Kendrick, A. G.; Passmore, J.; Schriver, M.; Stojanac, M. (Inorg. Chem. **25** [1986] 4451/2).
[6] Padma, D. K.; Mews, R. (Z. Naturforsch. **42b** [1987] 699/702).
[7] Ayres, B.; Banister, A. J.; Coates, P. D.; Hansford, M. I.; Rawson, J. M.; Rickard, C. E. F.; Hursthouse, M. B.; Abdul Malik, K. M.; Motevalli, M. (J. Chem. Soc. Dalton Trans. **1992** 3097/103).
[8] Haas, A.; Mischo, T. (Z. Anorg. Allg. Chem. **606** [1991] 191/9).
[9] Brooks, W. V. F.; MacLean, G. K.; Passmore, J.; White, P. S.; Wong, C. M. (J. Chem. Soc. Dalton Trans. **1983** 1961/8).
[10] Passmore, J.; Schriver, M. J. (Inorg. Chem. **27** [1988] 2749/51).

[11] Brooks, W. V. F.; Cameron, T. S.; Grein, F.; Parsons, S.; Passmore, J.; Schriver, M. J. (J. Chem. Soc. Chem. Commun. **1991** 1079/81).
[12] Gimarc, B. M.; Warren, D. S. (Inorg. Chem. **30** [1991] 3276/80).
[13] Faggiani, R.; Gillespie, R. J.; Lock, C. J. L.; Tyrer, J. D. (Inorg. Chem. **17** [1978] 2975/8).
[14] Banister, A. J.; Kendrick, A. G. (J. Chem. Soc. Dalton Trans. **1987** 1565/7).
[15] Banister, A. J.; Rawson, J. M. (J. Chem. Soc. Dalton Trans. **1990** 1517).
[16] Thewalt, U.; Berhalter, K.; Mueller, P. (Acta Crystallogr. B **38** [1982] 1280/2).
[17] Apblett, A.; Chivers, T.; Fait, J. F. (Inorg. Chem. **29** [1990] 1643/8).

4.2 Molecular Structure

4.2.1 Theoretical Structure

The equilibrium geometrical parameters of S_2N^+ were calculated by the ab initio SCF-MO method [1 to 3] and MRD-CI [4]. An S–N bond distance of 1.474 Å (6-31G* level) [1, 3, 9] and 1.478 Å (3-21G* and 3-21+G* level) [1] was obtained for the linear molecule (SNS bond angle of 180°) in agreement with the experimental values.

4.2.2 X-Ray Diffraction Structures

4.2.2.1 $S_2N^+ AsF_6^-$

The structure of crystalline $S_2N^+ AsF_6^-$ was determined by X-ray diffraction analysis. The crystals belong to the monoclinic space group $C2/m-C_{2h}^3$ (No. 12) with the parameters a = 9.5542(13), b = 6.5968(14), c = 5.3017(11) Å, β = 91.15(2)°; V = 334.1(2) Å³; Z = 2. The calculated density is D_x = 2.66 g/cm³. The analysis was performed at 293 K; R = 0.041 for 238 unique reflections. The structure consists of discrete SNS^+ cations and approximately octahedral AsF_6^- anions. The SNS^+ cation is linear. The SN bond distance is 1.480(3) Å, longer than or equal to the corresponding bond distances in $S_2N^+ SbCl_6^-$ or $S_2N^+ AlCl_4^-$; see below. The cation has six weak F–S contacts at approximately 90% of the sum of the van der Waals radii. The packing within the crystal is illustrated in the paper. A correction for thermal effects using the "riding model" [8] gave an SN distance of 1.510 Å (1.490 Å in $S_2N^+ SbF_6^-$, 1.494 and 1.517 Å for the two S–N distances in the $S_2N^+ AlCl_4^-$ salt) [5].

4.2.2.2 $S_2N^+ SbCl_6^-$

The structure of the molecule was determined by X-ray diffraction. The orthorhombic crystal has the space group $I222-D_2^8=V^8$ (No. 23) with a = 9.299(3), b = 7.976(3), c = 7.070(2) Å; Z = 2; V = 530 Å³; D_x = 2.61 g/cm³; 356 independent reflections gave R_1 = 0.0377 and R_2 = 0.0375. The S_2N^+ cation is linear (local symmetry $D_{\infty h}$) with a short SN bond length of 1.463(4) Å. The $SbCl_6^-$ anion is approximately octahedral; the bond distances Sb–Cl(1) (2.352(2) Å) and Sb–Cl(2) (2.363(2) Å) are significantly different and of the cis Cl–Sb–Cl angles only Cl(1)–Sb–Cl(2) (88.6(4)°, 91.4(2)°) differs significantly from 90°. The packing within the crystal is illustrated in the paper. A simple valence-bond structure such as $S{=}N^+{=}S$ does not adequately represent the molecule, as the bonds have some triple-bond character [6].

4.2.2.3 $S_2N^+ AlCl_4^-$

The molecular structure of the compound was determined by X-ray analysis at 292 K. The crystals are orthorhombic, space group $Pnma-D_{2h}^{16}=V_h^{16}$ (No. 62), with a = 10.908(2), b = 7.291(2), c = 11.069(2) Å; Z = 4; the calculated density is D_x = 1.863 g/cm³; R = 0.073 with 1036 unique reflections. The S_2N^+ cation is linear with an SNS bond angle of 179.5(3)°. The SN bond distances are 1.464(5) and 1.472(5) Å, respectively. The SN bonds possibly have triple-bond character. The crystal structure is illustrated in the paper. Coordinates and temperature factors are listed in the paper [7].

References:

[1] Behera, L.; Kar, T.; Sannigrahi, A. B. (J. Mol. Struct. **209** [1990] 111/24 [THEOCHEM **68**]).

[2] Gimarc, B. M.; Warren, D. S. (Inorg. Chem. **30** [1991] 3276/80).

[3] Brooks, W. V. F.; Cameron, T. S.; Grein, F.; Parsons, S.; Passmore, J.; Schriver, M. J. (J. Chem. Soc. Chem. Commun. **1991** 1079/81).

[4] Sannigrahi, A. B.; Peyerimhoff, S. D. (Int. J. Quantum Chem. **41** [1992] 413/9).

[5] Johnson, J. P.; Passmore, J.; White, P. S.; Banister, A. J.; Kendrick, A. G. (Acta Crystallogr. C **43** [1987] 1651/3).

[6] Faggiani, R.; Gillespie, R. J.; Lock, C. J. L.; Tyrer, J. D. (Inorg. Chem. **17** [1978] 2975/8).

[7] Thewalt, U.; Berhalter, K.; Müller, P. (Acta Crystallogr. B **38** [1982] 1280/2).
[8] Busing, W. R.; Levy, H. A. (Acta Crystallogr. **17** [1963] 142/6).
[9] Pyykkö, P. (Chem. Phys. Lett. **162** [1989] 349/54).

4.3 Electronic Structure

The extended Hückel MO theory shows that SNS^+ has 16 valence electrons in 8 MOs. Both bonding- and nonbonding-degenerate π MOs are fully occupied, and one member of each π set has a″ symmetry. The six remaining occupied valence MOs are symmetric (a′) [1].

An ab initio SCF study showed that the LMO (localized molecular-orbital) structure has the same three basis sets 4-31G, 3-21G*, and 3-21G*. The structure is described by six τ bonds (three between each NS pair of atoms) and two LPOs (lone-pair orbitals) on S. The presence of an NS triple bond in S_2N^+ is consistent with its bond length (1.48 Å) which is very close to that in FSN (1.45 Å). The LMO valency for the NS bond is predicted to lie between 0.6 and 0.7, indicating its highly polar character. The NS bond order is predicted to be about 1.5 (Mulliken population analysis) and 2.0 (Löwdin population analysis), indicating that the bond ionicity in S_2N^+ is highly exaggerated in the MPA scheme [2].

The electronic ground structure of S_2N^+ was calculated by the ab initio SCF [2] and ab initio MRD-CI [3] methods. The unit positive charge in S_2N^+ was found to be localized on S: 3-21+G*: S = +0.654, N = −0.308. Based on 4-31G calculations, both bonded and nonbonded overlap populations in S_2N^+ are negative. In this case the higher basis sets seem to yield more reliable results: 3-21+G* $OP_{NS} = 0.266$. The bond order increases when the basis set in the LPA (Löwdin population analysis) scheme is extended. The numerical values indicate that it may be a highly polar triple bond [2].

The vertical singlet-to-triplet excitation energy (in eV) was calculated by ab initio HF (6-31G*) to 0.56 and by MP2 to 3.87 [6].

The total energy of S_2N^+ was calculated by the ab initio SCF-MO theory at different levels: STO-3G* = −839.91515 [1], 6-31G* = −849.09024 [1, 4, 7], 6-31G*/MP2 = −849.56030 [1], −849.1575 hartrees (FCI: −849.64369) [3]. The total energy of S_2N^+ was also calculated by ab initio MO with a minimal Slater basis set to be −847.30044 hartrees (−23 055.045 eV) [5].

References:

[1] Gimarc, B. M.; Warren, D. S. (Inorg. Chem. **30** [1991] 3276/80).
[2] Behera, L.; Kar, T.; Sannigrahi, A. B. (J. Mol. Struct. **209** [1990] 111/24 [THEOCHEM **68**]).
[3] Sannigrahi, A. B.; Peyerimhoff, S. D. (Int. J. Quantum Chem. **41** [1992] 413/9).
[4] Brooks, W. V. F.; Cameron, T. S.; Grein, F.; Parsons, S.; Passmore, J.; Schriver, M. J. (J. Chem. Soc. Chem. Commun. **1991** 1079/81).
[5] Clark, D. T.; Adams, D. B. (J. Chem. Soc. Faraday Trans. II **68** [1972] 1819/24).
[6] Pyykkö, P. (Chem. Phys. Lett. **162** [1989] 349/54).
[7] Grein, F. (Can. J. Chem. **71** [1993] 335/40).

4.4 Spectra

^{14}N NMR of $\mathbf{S_2N^+ AsF_6^-}$ (SO_2/CH_3NO_2 external reference, 22°C, 14.45 MHz): δ = −91 ppm ($\Delta\nu^{1/2}$ = 8 Hz) [1 to 3]. ^{14}N NMR of $\mathbf{S_2N^+ AlCl_4^-}$ (CH_2Cl_2/CH_3NO_2 external reference, 14.45 MHz): δ = −87 ppm ($\Delta\nu^{1/2}$ = 30 Hz) [4].

The fundamental vibrational frequencies (in cm^{-1}) were calculated using ab initio HF (6-31G*) to $\nu_1 = 813$ δ(g), $\nu_2 = 460$ (π), $\nu_3 = 1592$ δ(u) [10].

The IR and Raman data of $S_2N^+ AsF_6^-$ and $S_2N^+ SbCl_6^-$ were interpreted in terms of an ionic structure. The data with assignments are listed in Table 8.

Table 8
IR and Raman[a] Spectra (ν in cm^{-1}) of $S_2N^+ AsF_6^-$ and $S_2N^+ SbCl_6^-$.

$S_2N^+ AsF_6^-$ [5]		$S_2N^+ SbCl_6^-$ [6]		
IR (Nujol mull)	Raman (solid)	IR (Nujol mull)	Raman (solid)	assignment [6]
1494 (m)		1498 (m)		$\nu_3(S_2N^+)$
1088 (w)				$\nu_1 + \nu_2(S_2N^+)$
818 (w)				AsF_6^-
	798 (18, br)		766 (17)	$2\nu_2(S_2N^+)$
	779 (3, ?)		747 (2)	$2\nu_2(^{32}S^{34}SN^+)$
	696 (100)		688 (44)	$\nu_1(S_2N^+)$
			680 (8)	$\nu_1(^{32}S^{34}SN^+)$
		374 (m)		$\nu_2(S_2N^+)$
	682 (30)		333 (100)	$\nu_1((A_{1g})AsF_6^-, SbCl_6^-)$
697 (vs)		320 (s)		$\nu_3((T_{1u})AsF_6^-, SbCl_6^-)$
391 (s)				$\nu_4((T_{1u})AsF_6^-)$
	573 (8)		293 (12)	$\nu_2((E_g)AsF_6^-, SbCl_6^-)$
			283 (8)	$\nu_2((E_g)AsF_6^-, SbCl_6^-)$
			180 (26)	
	368 (10)		175 (48)	$\nu_5((T_{2g})AsF_6^-, SbCl_6^-)$
			167 (4)	
			78 (100)	lattice mode
			62 (76)	lattice mode

[a] 5145-Å exciting line; intensities in parentheses.

The following IR and Raman spectral data were given without assignments.

$S_2N^+ SbF_6^-$. IR (Nujol): ν_{max} (in cm^{-1}) = 1500 (s), 805 (w), 650 (vs, br), 392 (vs), 285 (vs) [7].

$S_2N^+ SbCl_6^-$. IR (Nujol): ν_{max} (in cm^{-1}) = 1485 (vs), 372 (vs), and 340 (vs) [8]; 1480 (s), 376 (vs), 345 (vs) [9]. Raman (solid): ν_{max} (in cm^{-1}) = 770 (vw), 708 (vw, sh), 692 (w), 333 (s), 298 (w), 190 (sh), 180 (m), 80 (s), 68 (sh), 58 (vs), 28 (s) [8].

$S_2N^+ AsF_6^-$. IR (Nujol): ν (in cm^{-1}) = 1494 (s), 470 (s, br), 380 (s) [9]; 1495 (vs), 820 (w), 700 (vs, br), 395 (vs) [7].

$S_2N^+ CF_3SO_3^-$. IR (Nujol): ν_{max} (in cm^{-1}) = 1250 (s, br), 1220 (s, br), 1180 (s, br), 1030 (s, br), 760 (s), 635 (vs), 580 (m), 520 (m), 440 (s), 380 (s) [7].

$S_2N^+ AlCl_4^-$. IR (Nujol): ν_{max} (in cm^{-1}) = 494 (s), 470 (s, br), 380 (s) [9].

$S_2N^+ FeCl_4^-$. IR (Nujol): ν_{max} (in cm^{-1}) = 1480 (s), 802 (w), 380 (vs), 345 (m) [7].

References:

[1] Passmore, J.; Schriver, M. J. (Inorg. Chem. **27** [1988] 2749/51).

[2] Brooks, W. V. F.; Cameron, T. S.; Grein, F.; Parsons, S.; Passmore, J.; Schriver, M. J. (J. Chem. Soc. Chem. Commun. **1991** 1079/81).

[3] Haas, A.; Mischo, T. (Z. Anorg. Allg. Chem. **606** [1991] 191/9).

[4] Apblett, A.; Chivers, T.; Fait, J. F. (Inorg. Chem. **29** [1990] 1643/8).
[5] Banister, A. J.; Hey, R. G.; MacLean, G. K.; Passmore, J. (Inorg. Chem. **21** [1982] 1679/80).
[6] Faggiani, R.; Gillespie, R. J.; Lock, C. J. L.; Tyrer, J. D. (Inorg. Chem. **17** [1978] 2975/8).
[7] Ayres, B.; Banister, A. J.; Coates, P. D.; Hansford, M. I.; Rawson, J. M.; Rickard, C. E. F.; Hursthouse, M. B.; Abdul Malik, K. M.; Motevalli, M. (J. Chem. Soc. Dalton Trans. **1992** 3097/103).
[8] Banister, A. J.; Kendrick, A. G. (J. Chem. Soc. Dalton Trans. **1987** 1565/7).
[9] Banister, A. J.; Rawson, J. M. (J. Chem. Soc. Dalton Trans. **1990** 1517).
[10] Pyykkö, P. (Chem. Phys. Lett. **162** [1989] 349/54).

4.5 Enthalpy of Formation and Thermochemical Properties of $S_2N^+ AsF_6^-$

The standard molar enthalpy of formation of crystalline $S_2N^+ AsF_6^-$ from the elements in the standard states, $\Delta H^\circ_{f,298} = -(1413.8 \pm 1.9)$ kJ/mol at $p^0 = 101.325$ kPa, was determined by combining the standard molar enthalpy change $\Delta H^\circ = -(2265.6 \pm 1.5)$ kJ/mol for the reaction $S_2N^+ AsF_6^-$ (cr) + (11/2) F_2 (g) = 2 SF_6 (g) + (1/2) N_2 (g) + AsF_5 (g) obtained from calorimetric measurements and the values $\Delta H^\circ_f(SF_6) = -(1221.2 \pm 0.4)$ kJ/mol and $\Delta H^\circ_f(AsF_5) = -(1237.0 \pm 0.8)$ kJ/mol [1].

The SN bond dissociation energy and the SN bond enthalpy were estimated from a Born-Haber cycle based on the formation of $S_2N^+ AsF_6^-$ from its elements via the gas-phase reaction $SN^+ + S \rightarrow S_2N^+$. This Haber-Born cycle is formulated $\Delta H_f(S_2N^+ AsF_6) = \Delta H_f(S(g)) + \Delta H_f(SN^+(g)) + \Delta H_f(AsF_6^-(g)) - BDE$ (SN in S_2N^+) $- U(S_2N^+ AsF_6^-)$, where BDE is the bond dissociation energy and U the lattice energy. The **SN bond dissociation energy** (BDE) in S_2N^+ (i.e., ΔH (S_2N^+ (g) $\rightarrow$ SN^+ (g) + S (g)) was determined to be 441 ± 11 kJ/mol from this equation in which $U(S_2N^+ AsF_6^-)$ is 509.4 kJ/mol. The **SN bond enthalpy term** (BET) in S_2N^+ (enthalpy assuming that a given bond contributes to the total atomization enthalpy) was determined from the average of BDE (SN in S_2N^+) = 441 ± 11 kJ/mol and BDE (SN in SN^+) = 511 kJ/mol to be 476 kJ/mol [2].

A linear relationship between the sulfur–nitrogen bond enthalpy term and bond length was derived from experimental data of several S–N compounds. By using the equation BET(SN) = 2134.3 − 1126.4 d(SN), where d(SN) = 1.510 Å, a value of 433 kJ/mol was estimated [2].

The **molar entropy** S° of $S_2N^+ AsF_6^-$ was estimated using Bartletts relationship for $A^+ B^-$ salts, $S^\circ = 1.85\ V_m$ where $V_m = 167$ Å^3, giving 308.8 $J \cdot K^{-1} \cdot mol^{-1}$ [3].

The **lattice energy** U of $S_2N^+ AsF_6^-$ was calculated to be 509.4 kJ/mol by an extended calculation using X-ray structural data [2, 4] and estimated from the molar volume of $S_2N^+ AsF_6^-$ ($V_m = 167$ Å^3) to be 576 kJ/mol [5].

References:

[1] O'Hare, P. A. G.; Awere, E. G.; Parsons, S.; Passmore, J. (J. Chem. Thermodyn. **21** [1989] 153/8).
[2] Parsons, S.; Passmore, J. (Inorg. Chem. **31** [1992] 526/8).
[3] Awere, E. G.; Passmore, J. (J. Chem. Soc. Dalton Trans. **1992** 1343/50).
[4] Brooks, W. V. F.; Cameron, T. S.; Grein, F.; Parsons, S.; Passmore, J.; Schriver, M. J. (J. Chem. Soc. Chem. Commun. **1991** 1079/81).
[5] Parsons, S.; Passmore, J.; Schriver, M. J.; Sun, X. (Inorg. Chem. **30** [1991] 3342/8).

4.6 Mass Spectrum of $S_2N^+ CF_3SO_3^-$

The chemical ionization mass spectrum shows the following peaks: m/e (rel. int. in %) = 227 (8) $[S_2N \cdot CF_3SO_3]^+$, 149 (8) $CF_3SO_3^+$, and 78 (53) S_2N^+.

Reference:

Ayres, B.; Banister, A. J.; Coates, P. D.; Hansford, M. I.; Rawson, J. M.; Rickard, C. E. F.; Hursthouse, M. B.; Abdul Malik, K. M.; Motevalli, M. (J. Chem. Soc. Dalton Trans. **1992** 3097/103).

4.7 Reactions

Of the reactions of known S_2N^+ salts, only that of $S_2N^+ AsF_6^-$ has been extensively studied in the literature. Other S_2N^+ salts may not have the same simple chemistry as that of $S_2N^+ AsF_6^-$ due to the involvement of the anion. The relative reactivities of different S_2N^+ salts with C_6H_5CN decrease in the order $AsF_6^- > SbCl_6^- > CF_3SO_3^- > SbF_6^- > AlCl_4^-$. $S_2N^+ AsF_6^-$ undergoes the cleanest cycloaddition reactions with nitriles under mild conditions giving high yields. $S_2N^+ CF_3SO_3^-$ also reacts under mild conditions with high yield, but the extreme air sensitivity of this reagent is likely to hinder its applications. A high yield was achieved with $S_2N^+ SbCl_6^-$ under more forceful conditions, and in some cases this salt may prove to be an alternative reagent to $S_2N^+ AsF_6^-$ [1]. The reaction of poorly soluble $S_2N^+ SbCl_6^-$ with p-$C_6H_4(CN)_2$ in refluxing CH_2Cl_2 yielded only a variety of intractable products, but in contrast the room-temperature reaction of $S_2N^+ AsF_6^-$ with o-, m-, or p-$C_6H_4(CN)_2$ in liquid SO_2 gave o-, m-, and p-$[C_6H_4(\overline{CNSNS})_2]^{2+} (AsF_6^-)_2$ [2].

4.7.1 Reactions of $S_2N^+ AsF_6^-$

Hydrolysis

When a ground sample of $S_2N^+ AsF_6^-$ was exposed to moist air for 1 h, decomposition was observed (ν(N–H) and ν(S–O) bands in the IR spectra) [3, 4].

Reactions with Halogens

F_2. $S_2N^+ AsF_6^-$ (supported on nickel crucibles) reacted spontaneously with F_2 at approximately 0.2 MPa to give only gaseous products like SF_6 and AsF_5. No evidence was found for the formation of NF_3, AsF_3, or lower sulfur fluorides via $S_2N^+ AsF_6^-(cr) + (11/2)\,F_2(g) \rightarrow 2\,SF_6(g) + (1/2)\,N_2(g) + AsF_5(g)$. The standard molar enthalpy change for the reaction was measured to be $-(2265.6 \pm 1.5)$ kJ/mol [3].

The reaction of $S_2N^+ AsF_6^-$ with F_2 (mole ratio ca. 1 : 1) in SO_2F_2, performed by stirring for 4 h at −78 °C and then for 20 h at room temperature, gave SF_6, $SF_3^+ AsF_6^-$, $(F_2S)N^+ AsF_6^-$, and unreacted $S_2N^+ AsF_6^-$ [5].

Cl_2. When SO_2 and Cl_2 were consecutively condensed onto frozen $S_2N^+ AsF_6^-$ (mole ratio 63.3 : 3.96 : 2.16) and the mixture allowed to warm to room temperature while shaking, $(ClS)_2N^+ AsF_6^-$ separated after 1 h with 99% yield [5].

Br_2. $S_2N^+ AsF_6^-$ (4.88 mmol) reacted with Br_2 (5.34 mmol) in a similar manner as Cl_2 in SO_2 (150 mmol), yielding $(BrS)_2N^+ AsF_6^-$ with 99% yield [5].

I_2. $S_2N^+ AsF_6^-$ did not react with I_2 in SO_2 under similar reaction conditions as those in the reactions with Cl_2 or Br_2 [5].

Reaction with XeF_2

When SO_2F_2 (150.7 mmol) and XeF_2 (2.89 mmol) were consecutively condensed onto frozen $S_2N^+ AsF_6^-$ (2.73 mmol) in a well-dried and fluorinated Monel vessel and the mixture was held at 0 °C with agitation for 1 h and then for 2 h at room temperature, a mixture of $S_2N^+ AsF_6^-$ and $(F_2S)_2N^+ AsF_6^-$ was formed with release of the volatiles SO_2F_2 and SiF_4 [5].

Condensing XeF_2 (5.67 mmol) onto a frozen mixture of $S_2N^+ AsF_6^-$ (2.32 mmol) and SO_2F_2 (92.2 mmol) in a Monel vessel, stirring and shaking the mixture for 0.5 h at 0 °C, and further stirring at room temperature for 12 h yielded $(F_2S)_2N^+ AsF_6^-$ essentially quantitatively. The volatiles contained SO_2F_2 and a trace of SF_6 [5].

CAUTION: A violent reaction occurred on mixing solid XeF_2 and $S_2N^+ AsF_6^-$ at room temperature without a moderating solvent [5].

Reaction with S_4N_4

A 1 : 1 molar ratio of $S_2N^+ AsF_6^-$ and S_4N_4 reacted in liquid SO_2 under stirring at 0 °C over a period of 4 d producing crystalline $(S_3N_2)_2N^+ AsF_6^-$ with ca. 60% yield [6].

A reaction mixture of $S_2N^+ AsF_6^-$ and S_4N_4 in SO_2, which was allowed to stand at room temperature for 1 month, yielded $S_4N_3^+ AsF_6^-$ and S_4N_4 essentially quantitatively. The following reaction mechanism was discussed [6]:

$$S_2N^+ AsF_6^- + S_4N_4 \rightarrow (S_3N_2)_2N^+ AsF_6^-, \; 2\,(S_3N_2)_2N^+ AsF_6^- \rightarrow 2\,S_4N_3^+ AsF_6^- + S_4N_4$$

Reaction with CsN_3

$S_2N^+ AsF_6^-$ reacted with CsN_3 (mole ratio ca. 1 : 1) in SO_2 (15 min at −20 °C, then warming to room temperature) to produce $(SN)_x$ via $S_2N^+ AsF_6^- + CsN_3 \rightarrow 2/x\,(SN)_x + N_2 + CsAsF_6$. The yield of $(SN)_x$ is dependent upon the reaction conditions; lower temperatures and dilute solutions favor $(SN)_x$ rather than S_4N_4. $(SN)_x$ was produced in an impure form with 86% yield. The reaction probably proceeds via an open-chain form of S_2N_2 which may polymerize to $(SN)_x$ [7].

Reaction with $SN^+ AsF_6^-$

Evaporating a clear yellow solution of $S_2N^+ AsF_6^-$ and $SN^+ AsF_6^-$ in SO_2 over a period of 3 weeks in a cold room at 0 to 5 °C yielded $S_3N_2^{2+} (AsF_6^-)_2$. The symmetry-allowed reaction of S_2N^+ and SN^+ is a normal-electron-demand cycloaddition. Due to strong Coulomb repulsions, the reaction is highly endothermic in the gas phase with an energy of 95 to 99 kcal/mol (ab initio SCF calculation). In solution $S_3N_2^{2+}$ decomposes back to S_2N^+ and SN^+. The activation barrier to dissociation is about 45 kJ/mol (10.9 kcal/mol) (ab initio 3-21G* calculation). In the solid state the high lattice energy of $S_3N_2^{2+} (AsF_6^-)_2$ forces the reaction into the $S_3N_2^{2+}$ minimum [8, 23] as follows:

$$S_2N^+ AsF_6^- + SN^+ AsF_6^- \underset{\text{solution}}{\overset{\text{solid state}}{\rightleftharpoons}} S_3N_2^{2+} (AsF_6^-)_2$$

This reaction was also observed by [9].

Cycloaddition Reactions with Alkenes

$S_2N^+ AsF_6^-$ reacted quantitatively with alkenes via a concerted symmetry-allowed cycloaddition to yield 1 : 1 stoichiometric cations (I) which react with excess olefin to give 2 : 1 cations (II) via a second 4+2 cycloaddition [10].

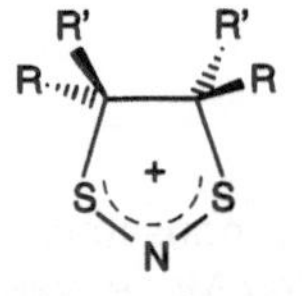

I

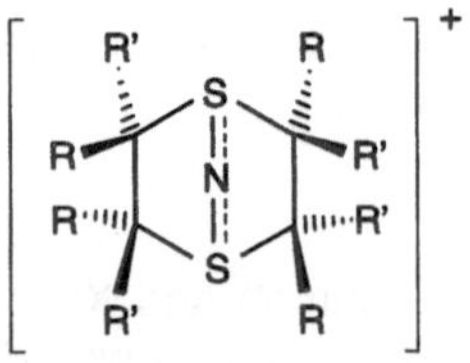

II

a: R = R′ = H,
b: R = H, R′ = CH_3

Ethene. The reaction of $S_2N^+AsF_6^-$ with ethene in liquid SO_2 at room temperature gave mixtures of the 1,3,2-dithiazolium (Ia) and 1,4-dithia-7-azanorbornylium (IIa) cations as shown by ^{1}H NMR spectroscopy. Using an excess of ethene (mole ratio 1 : 4.7) quantitatively yielded the 1,4-dithia-7-azanorbornylium cation (IIa, R = R′ = H) [10].

***trans*-2-Butene.** The title compound reacted with *trans*-2-butene (mole ratio 1 : 1) in SO_2 at room temperature to give the 1,3,2-dithiazolium cation (Ib, R = H, R′ = CH_3) with essentially quantitative yield (94%) as a poorly crystalline, red solid [10].

Norbornene. S_2N^+ AsF_6^- reacted with norbornene (mole ratio 1 : 1) in liquid SO_2 at room temperature yielding the corresponding 1,3,2-dithiazolium cation with quantitative yield. The reaction of S_2N^+ AsF_6^- with an excess of norbornene gave the respective 1,4-dithia-7-azanorbornylium cation (II) (shown by ^{1}H NMR) [10].

Cycloaddition Reactions with Triple Bond Compounds

Thiazyl Halides, XS≡N, X = F, Cl

The S_2N^+ cation and XSN (F, Cl; formed in situ by the reaction of SN^+ AsF_6^- and CsF and by dissolving $S_3N_3Cl_3$ in SO_2) produce 1-halo-1,2,4,3,5-trithiadiazolium salts (I). The cycloaddition is assumed to be of "type I" (normal electron demand) in the classification of Sustmann [11]. The principal orbital interaction is between $\pi^*(SNS^+)$ and π(XSN). The reaction enthalpy was calculated to be −340 kJ/mol [4].

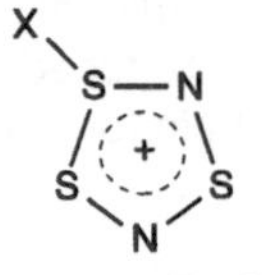

I X = F, Cl

FS≡N. Condensing FS≡N and SO_2 on $S_2N^+AsF_6^-$ (mole ratio 1.02 : 1) at −196 °C and warming the mixture to −30 °C yielded $S_3N_2F^+$ AsF_6^- as a colorless solid with essentially quantitative yield [12]. Independently, the reaction of S_2N^+ AsF_6^- and FS≡N (generated in situ from SN^+ AsF_6^- and CsF) in SO_2 was found to give $S_3N_2F^+$ AsF_6^- as a red-black crystalline solid with quantitative yield [4, 13].

ClS≡N. S_2N^+ AsF_6^- and $S_3N_3Cl_3$ (ratio 3 : 1, impure $S_3N_3Cl_3$, dissolved in SO_2 gave >80% ClS≡N monomer) reacted in SO_2 over a period of ca. 24 h to give $S_3N_2Cl^+$ AsF_6^- as a soluble, red-brown solid with essentially quantitative yield [4, 13].

Alkynes, RC≡CR. Nitriles, RC≡N

S_2N^+ AsF_6^- reacts with alkynes like HC≡CH, HC≡CCH_3, HC≡CCF_3, CH_3C≡CCH_3, CF_3C≡CCF_3, $(CH_3)_3SiC$≡$CSi(CH_3)_3$, C_6H_5C≡CC_6H_5, and CH_3CO_2C≡CCO_2CH_3 giving 1,3,2-dithiazolium salts (I) (see Table 9, p. 46) and with nitriles like HC≡N, IC≡N, $(CH_3)_3CC$≡N, CF_3C≡N, $(CH_3)_2NC$≡N, C_6H_5C≡N giving 1,3,2,4-dithiadiazolium salts (II) (see Table 10, p. 47).

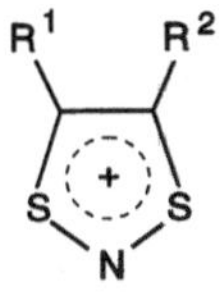

I

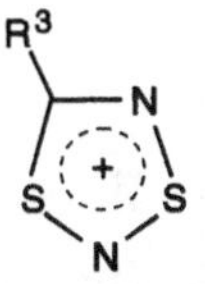

II

R^1, R^2, R^3 see Tables 9 and 10

The cycloaddition is thermally allowed. The π bonding in SNS^+ is related to that in 1,3-dipoles. Quantum-mechanical calculations showed a reverse electron-demand interaction between LUMO(SNS^+) and HOMO(triple bond). The cycloaddition was classified to be of "type III" following the classification of Sustman [11]. This assumption was confirmed by the rate of reaction which increased as the LUMO(SNS^+)–HOMO(triple bond) energy gap decreased [4, 14].

The reactions of $CF_3C{\equiv}CCF_3$ and RC≡N (R = CH_3, C_6H_5, 4-$O_2NC_6H_4$, and 3,5-$(O_2N)_2C_6H_3$) with S_2N^+ AsF_6^- were monitored by ^{19}F or ^1H NMR as a function of time. The linear variation of $lg\{[X]/[S_2N^+]\}$ (X = triply bonded species) with time implies that all reactions are overall second-order (first-order in X and S_2N^+) [4, 14]. Relative rate constants (k_{rel}) were also obtained for a variety of pairs of nitriles and/or alkynes (see Tables 9 and 10). The rates of cycloaddition reactions of S_2N^+ AsF_6^- with alkynes and nitriles increased as the ionization potential of the triple bond decreased (the logarithm of the second-order rate constants is approximately proportional to $-E_i$ (ionization potential) of the triple bond). Steric effects are more significant in the cycloadditions of alkynes than in those of nitriles which show little or no steric effect [4, 14].

The cycloaddition products were generally prepared by condensing an excess amount of the alkyne onto frozen solutions of S_2N^+ AsF_6^- in SO_2 (an equimolar amount of nonvolatile compounds dissolved in SO_2 was added at room temperature to the S_2N^+ AsF_6^- solution). After the volatile materials had been removed, the solid product was purified by recrystallization from SO_2–SO_2ClF (1 : 3 w/w). Experimental details are given in Tables 9 and 10, pp. 46 and 47.

The reaction of S_2N^+ AsF_6^- with [4-NCC_6H_4CNSSN]$^+$ Cl^- gave [4-NCC_6H_4CNSSN]$^+$ AsF_6^- together with a red-green insoluble product (shown to be $[S_3N_2]^+$ Cl^- by IR spectroscopy) and a red, soluble, volatile component (not the anticipated [4-SNSNC–C_6H_4–CNSSN]$^{2+}$ Cl^- AsF_6^-). The reaction with (4-NCC_6H_4CNSSN)$_2$ produced [4-NCC_6H_4CNSSN]$^+$ AsF_6^- and a red oil (presumably S_4N_2) which on standing gave S_4N_4 and sulfur [25].

Multifunctional Compounds, HC≡C–C≡N, $(CN)_2$, o-, m-, p-$C_6H_4(C{\equiv}N)_2$, 4-N≡CC_6H_4–C_6H_4C≡N-4′, 1,3,5-$(N{\equiv}C)_3C_6H_3$, K[C$(CN)_3$]

HC≡C–C≡N. S_2N^+ reacts with HC≡C–C≡N (mole ratio 1 : 1.9) in SO_2 at room temperature to give [HCSNSCC≡N]$^+$ AsF_6^- (I) with 96% yield after 20 h as colorless crystals. Neither the alternative 1 : 1 cycloadduct [HC≡CCNSNS]$^+$ nor the dication [HCSNSCCNSNS]$^{2+}$ were formed. The facile cycloaddition of SNS^+ to the CC triple bond in HC≡CC≡N is in accord with type-III behavior (the ionization of the CC and CN triple bonds occur at 11.60 and 14.03 eV, respectively). A second cycloaddition of SNS^+ to the cyano group was complete after ten weeks of heating in SO_2 at 50°C to give [SNSNC–CSNSCH]$^{2+}$ $(AsF_6^-)_2$ (II) [4, 14].

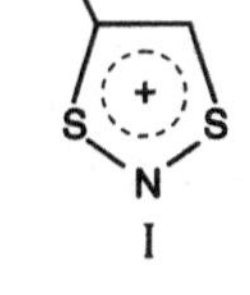

I

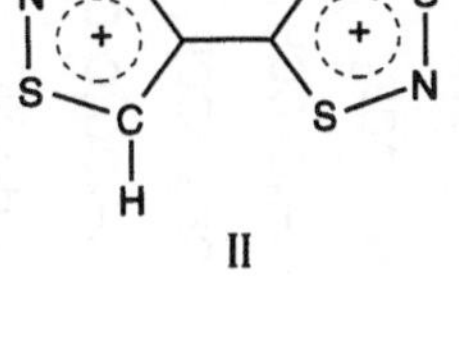

II

Table 9
Reactions of S_2N^+ AsF^-_6 with Alkynes, $R^1C{\equiv}CR^2$, in SO_2.

reactant	mole ratio S_2N^+ AsF_6^-: reactant	k_{rel} [a)]	reaction time	product (cation) (yield in %)	Ref.
$HC{\equiv}CH$	1:0.97	100	1 h	$[HCSNSCH]^+$	[4, 14 to 16]
$HC{\equiv}CCH_3$	1:1	100	6 h	$[HCSNSCCH_3]^+$	[14 to 16]
$HC{\equiv}CCF_3$	1:0.92	5	<5 min	$[HCSNSCCF_3]^+$ (100)	[4, 14]
$CH_3C{\equiv}CCH_3$	—	30	—	$[CH_3CSNSCCH_3]^+$	[14]
$CF_3C{\equiv}CCF_3$	1:2.39	0.1	4 h or 7 d	$[CF_3CSNSCCF_3]^+$ (95)	[4, 14, 17]
$CH_3CO_2C{\equiv}CCO_2CH_3$	1:0.96	30	<5 min	$[CH_3CO_2CSNSCCO_2CH_3]^+$ (95)	[4, 14]
$(CH_3)_3SiC{\equiv}CSi(CH_3)_3$	1:1	—	20 h	$[(CH_3)_3SiCSNSCSi(CH_3)_3]^+$ (100)	[4]
$C_6H_5C{\equiv}CC_6H_5$	1:1	2000	3 min	$[C_6H_5CSNSCC_6H_5]^+$	[4, 24]

a) k_{rel} = rate of cycloaddition to SNS^+ for alkynes (relative to CH_3CN) from NMR integrations.

Table 10
Reactions of $S_2N^+ AsF_6^-$ with Nitriles, RC≡N, in SO_2.

reactant RC≡N R	mole ratio $S_2N^+ AsF_6^-$: reactant	k_{rel} [a)]	reaction time	product (cation) (yield in %)	Ref.
H	1:4	—	<1 h	$[HCSNSN]^+$ (93)	[4]
I	1:1	—	<1 h	$[ICSNSN]^+$ (89)	[4, 18]
CH_3	1:4.5	1	1 h	$[CH_3CSNSN]^+$ (94)	[4, 14 to 16, 18]
CF_3	1:2.75	0.01	~14 d	$[CF_3CSNSN]^+$ (97)	[4, 14, 18]
$(CH_3)_3C$	1:1.03	2	<1 h	$[(CH_3)_3CCSNSN]^+$ (89)	[4, 14]
$(CH_3)_2N$	—	100	—	$[(CH_3)_2NCSNSN]^+$	[13, 14]
C_6H_5	1:1	2	b)	$[C_6H_5CSNSN]^+$	[1, 14, 24]
C_6F_5	—	—	3 h	$[C_6F_5CSNSN]^+$ (94)	[1]
4-$NO_2C_6H_4$	1:1	0.1	20 h	$[4\text{-}NO_2C_6H_4CSNSN]^+$ (88)	[4, 14, 24]
2,5-$(CH_3)_2C_6H_3$	1:1	5	15 min	$[2,5\text{-}(CH_3)_2C_6H_3CSNSN]^+$ (76)	[4, 24]
3,5-$(NO_2)_2C_6H_3$	1.1:1	0.03	3 d	$[3,5\text{-}(NO_2)_2C_6H_3CSNSN]^+$ (71)	[4, 14, 24]
$[CSNSCH]^+ AsF_6^-$	1:1	—	10 w at 45 °C	$[SNSNC–CSNSCH]^{2+} [AsF_6]^{2-}$ (76)	[26]
$[4\text{-}C_6H_4–CNSSN]^+ AsF_6^-$	1:1	—	18 h	$[4\text{-}SNSNC–C_6H_4–CNSSN]^{2+} (AsF_6^-)_2$ (92)	[19, 25]
$[3\text{-}C_6H_4–CNSSN]^+ AsF_6^-$	1:1	—	18 h	$[3\text{-}SNSNC–C_6H_4–CNSSN]^{2+} (AsF_6^-)_2$ (94)	[25]

a) k_{rel} = rate of cycloaddition to SNS^+ for nitriles (relative to CH_3CN) from NMR integrations. – b) Without solvent, 5 min at 20 °C, 90% yield; in liquid SO_2 for 1 h at 20 °C, 95% yield [1] (2 h, room temperature, 86% [24]).

$(CN)_2$. The reaction of SNS^+ with $(CN)_2$ (mole ratio 1.3:1) at 50°C for 5 days gave $[SNSNC–CNSNS]^{2+}$ $(AsF_6^-)_2$ (III) with 97% yield as a highly crystalline solid. Attempts to prepare the intermediate 1:1 cycloadduct $[N{\equiv}CCNSNS]^+$ AsF_6^- only produced $[SNSNC–CNSNS]^{2+}$ $(AsF_6^-)_2$ with possible traces of $[N{\equiv}CCNSNS]^+$ AsF_6^- [14, 26]. The reaction of SNS^+ with a large excess of $(CN)_2$ in SO_2 for 1 week at 80°C gave also mainly $[SNSNC–CNSNS]^{2+}$ with only minor amounts of the 1:1 cycloadduct $[NCCNSNS]^+$. $[SNSNC–CNSNS]^{2+}$ is the thermodynamically and kinetically favored product [26].

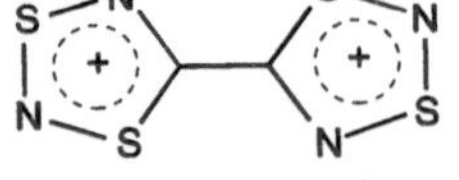

III

$Hg(CN)_2$. Stirring a mixture of S_2N^+ AsF_6^- and $Hg(CN)_2$ (mole ratio 2:1) in liquid SO_2 for 24 h gave $[Hg(CNSNS)_2]^+$ AsF_6^- in 96% yield [27].

o-, m-, p-$C_6H_4(CN)_2$. S_2N^+ AsF_6^- was reacted with o-$C_6H_4(CN)_2$, m-$C_6H_4(CN)_2$, and p-$C_6H_4(CN)_2$ (mole ratio 2:1) in liquid SO_2 at ambient temperature for 18 h to give o-, m-, and p-$[C_6H_4(CNSNS)_2]^{2+}$ $(AsF_6^-)_2$ (IV, V, VI), respectively. The yields were 75, 85, and 95%, respectively [2].

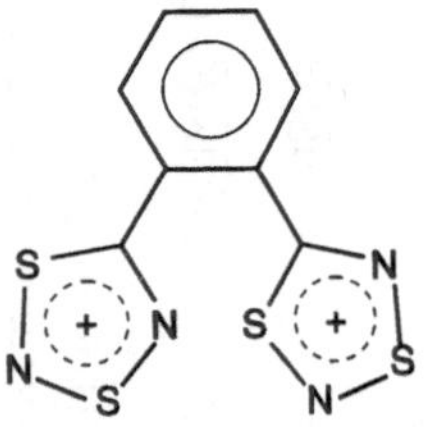

IV

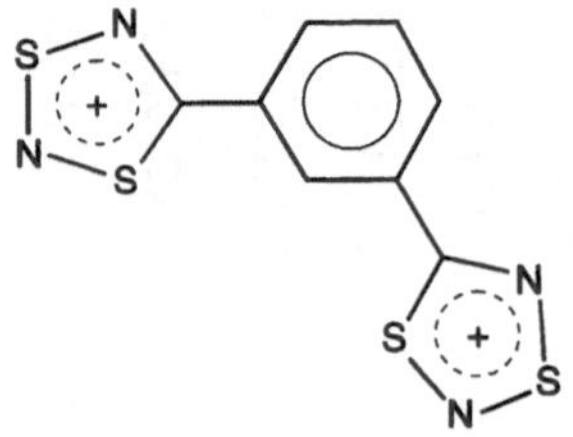

V

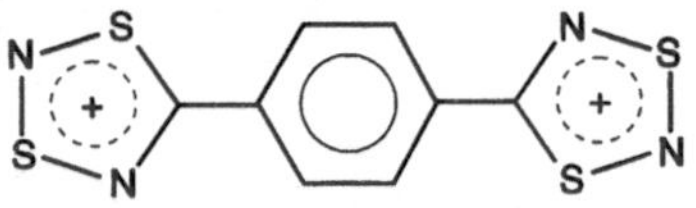

VI

4-$N{\equiv}CC_6H_4$–$C_6H_4C{\equiv}N$-4′. S_2N^+ AsF_6^- was reacted with 4-$N{\equiv}CC_6H_4$–$C_6H_4C{\equiv}N$-4′ in the mole ratio 1:1 in liquid SO_2 over a period of 36 h to give $[4\text{-}(SNSNC)C_6H_4–C_6H_4(CNSNS)]^{2+}$ $(AsF_6^-)_2$ with 79% yield. The crude orange precipitate was washed with CH_2Cl_2 [20].

1,3,5-$(N{\equiv}C)_3C_6H_3$. Stirring a mixture of S_2N^+ AsF_6^- and 1,3,5-$(N{\equiv}C)_3C_6H_3$ (mole ratio 3:1) in liquid SO_2 at room temperature for 5 d gave $[sym\text{-}C_6H_3(CNSNS)_3]^{3+}$ $(AsF_6^-)_3$ with 70% yield. The colorless, crystalline precipitate was washed with small amounts of SO_2 and subsequently with CH_2Cl_2 [20].

K^+ $[C(CN)_3]^-$. The reactions of 1, 2, and 3 equivalents of S_2N^+ AsF_6^- with K^+ $[C(CN)_3]^-$ in liquid SO_2 for 1.5, 16, and 16 h, respectively, generated the three cycloaddition products $(N{\equiv}C)_2C{=}CNSNS$ (VII), $[(N{\equiv}C)C(CNSNS)_2]^+$ AsF_6^- (VIII), and $[C(CNSNS)_3]^{2+}$ $(AsF_6^-)_2 \cdot SO_2$ (IX) with 72, 59, and 84% yield, respectively. The reaction is supposed to be induced by $[SNS]^+/[C(CN)_3]^-$ attraction, followed by electron donation from lower-lying, filled MOs of $[C(CN)_3]^-$ with the correct symmetry to the $[SNS]^+$ LUMOs. The addition of a second SNS^+ molecule proceeds via an intermediate complex in which the SNS^+ cation is bonded to the ring nitrogen which undergoes intramolecular cyclization to give the 2:1 product. The formation of the 3:1 product proceeds analogously. The addition of the SNS^+ molecule is stereospecific. It occurs only on one side of the cyano functionality and hence leads to only one product [21].

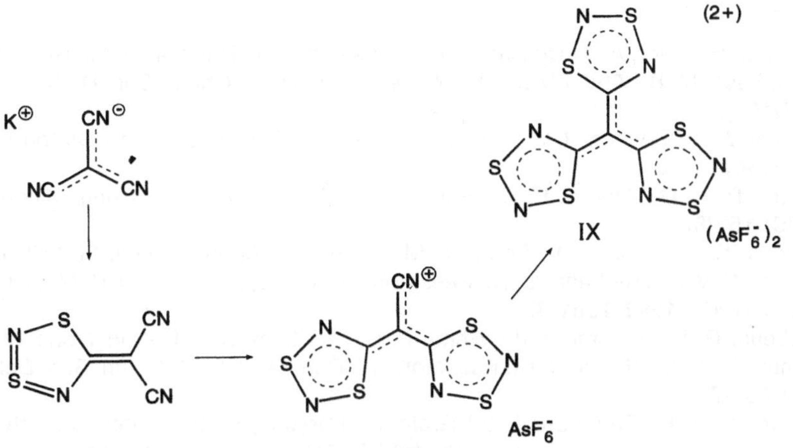

VII VIII

Polymerization of THF

S_2N^+ AsF_6^- was found to be an efficient initiator for the polymerization of tetrahydrofuran. The reaction involves the production of $S_3N_2^+$ which has been identified by ESR spectroscopy. A mechanism for the polymerization reaction is presented in the paper [22].

4.7.2 Reactions of S_2N^+ $SbCl_6^-$, S_2N^+ SbF_6^-, S_2N^+ $CF_3SO_3^-$, and S_2N^+ $AlCl_4^-$ with RC≡N, R = C_6H_5, C_6F_5

The salt S_2N^+ $SbCl_6^-$ did not decompose or react with $C_6H_5C{\equiv}N$, $CH_3C{\equiv}N$, or $C_6F_5C{\equiv}N$ at room temperature, but did react under refluxing conditions (with an excess of nitrile) to provide the desired 1,3,2,4-dithiadiazolium cation $[R\overline{CNSNS}]^+$ with high yield (>70%). It reacted with RC≡N (R = C_6H_5, C_6F_5) (without solvent) at ca. 100 °C over a period of 4 h. The olive green mixture was further heated until no more solid material was observed (ca. 30 min). After cooling a lime green solid precipitated. Purification by washing with CH_2Cl_2 gave $[R\overline{CNSNS}]^+$ $SbCl_6^-$ with 75% yield (R = C_6H_5) and 76% yield (R = C_6F_5) [1].

S_2N^+ SbF_6^- reacted with $C_6H_5C{\equiv}N$ in SO_2 at room temperature over a period of 5 h (or neat $C_6H_5C{\equiv}N$ at 20 °C) to give $[C_6H_5\overline{CNSNS}]^+$ SbF_6^- with 50% yield (35% yield). The material was washed with CH_3CN–CH_2Cl_2 (2 : 1) at 0 °C giving a yellow solid [1].

Stirring a mixture of S_2N^+ $CF_3SO_3^-$ and $C_6H_5C{\equiv}N$ (mole ratio 1 : 1) in SO_2 at room temperature for 2 h gave, after removing the solvent, yellow-brown solid $[C_6H_5\overline{CNSNS}]^+$ $CF_3SO_3^-$ with 74% yield. Other conditions gave the following yields: neat, 1 h, 20 °C: 70%; CH_2Cl_2, 2 h, 40 °C: 30% [1].

The salt S_2N^+ $AlCl_4^-$ reacted with excess of $C_6H_5C{\equiv}N$ in CH_2Cl_2 at ambient temperature over a period of 2 h to give a mixture of solid $[S_3N_2]_2^{2+}$ $(AlCl_4^-)_2$ and $(S_3N_2Cl)_2$ and liquid C_6H_5CN [1].

References:

[1] Ayres, B.; Banister, A. J.; Coates, P. D.; Hansford, M. I.; Rawson, J. M.; Rickard, C. E. F.; Hursthouse, M. B.; Abdul Malik, K. M.; Motevalli, M. (J. Chem. Soc. Dalton Trans. **1992** 3097/103).
[2] Banister, A. J.; Rawson, J. M.; Clegg, W.; Birkby, S. L. (J. Chem. Soc. Dalton Trans. **1991** 1099/104).
[3] O'Hare, P. A. G.; Awere, E. G.; Parsons, S.; Passmore, J. (J. Chem. Thermodyn. **21** [1989] 153/8).
[4] Parsons, S.; Passmore, J.; Schriver, M. J.; Sun, X. (Inorg. Chem. **30** [1991] 3342/8).
[5] Brooks, W. V. F.; MacLean, G. K.; Passmore, J.; White, P. S.; Wong, C. M. (J. Chem. Soc. Dalton Trans. **1983** 1961/8).
[6] MacLean, G. K.; Passmore, J.; White, P. S. (J. Chem. Soc. Dalton Trans.**1984** 211/7).
[7] Kennett, F. A.; MacLean, G. K.; Passmore, J.; Rao, M. N. S. (J. Chem. Soc. Dalton Trans. **1982** 851/7).
[8] Brooks, W. V. F.; Cameron, T. S.; Grein, F.; Parsons, S.; Passmore, J.; Schriver, M. J. (J. Chem. Soc. Chem. Commun. **1991** 1079/81).
[9] Awere, E. G.; Passmore, J. (J. Chem. Soc. Dalton Trans. **1992** 1343/50).
[10] Burford, N.; Johnson, J. P.; Passmore, J.; Schriver, M. J.; White, P. S. (J. Chem. Soc. Chem. Commun. **1986** 966/8).

[11] Padwa, A. (1,3-Dipolar Cycloaddition Chemistry, John Wiley, New York 1984).
[12] Padma, D. K.; Mews, R. (Z. Naturforsch. **42b** [1987] 699/702).
[13] Passmore, J.; Schriver, M. J. (Inorg. Chem. **27** [1988] 2749/51).
[14] Parsons, S.; Passmore, J.; Schriver, M. J.; White, P. S. (J. Chem. Soc. Chem. Commun. **1991** 369/71).
[15] MacLean, G. K.; Passmore, J.; Schriver, M. J.; White, P. S.; Bethell, D.; Pilkington, R. S.; Sutcliffe, L. H. (J. Chem. Soc. Chem. Commun. **1983** 807/8).
[16] MacLean, G. K.; Passmore, J.; Rao, M. N. S.; Schriver, M. J.; White, P. S.; Bethell, D.; Pilkington, R. S.; Sutcliffe, L. H. (J. Chem. Soc. Dalton Trans. **1985** 1405/16).
[17] Awere, E. G.; Burford, N.; Mailer, C.; Passmore, J.; Schriver, M. J.; White, P. S.; Banister, A. J.; Oberhammer, H.; Sutcliffe, L. H. (J. Chem. Soc. Chem. Commun. **1987** 66/9).
[18] Burford, N.; Passmore, J.; Schriver, M. J. (J. Chem. Soc. Chem. Commun. **1986** 140/2).
[19] Banister, A. J.; Lavender, I.; Rawson, J. M.; Whitehead, R. J. (J. Chem. Soc. Dalton Trans. **1992** 1449/50).
[20] Aherne, C.; Banister, A. J.; Luke, A. W.; Rawson, J. M.; Whitehead, R. J. (J. Chem. Soc. Dalton Trans. **1992** 1277/82).

[21] Banister, A. J.; Lavender, I.; Rawson, J. M.; Clegg, W. (J. Chem. Soc. Dalton Trans. **1992** 859/65).
[22] Fairhurst, S. A.; Hulme-Lowe, A.; Johnson, K. M.; Sutcliffe, L. H.; Passmore, J.; Schriver, M. J. (Magn. Reson. Chem. **23** [1985] 828/31).
[23] Grein, F. (Can. J. Chem. **71** [1993] 335/40).
[24] Passmore, J.; Sun, X.; Parsons, S. (Can. J. Chem. **70** [1992] 2972/9).
[25] Banister, A. J.; Lavender, I.; Rawson, J. M.; Clegg, W.; Tanner, B. K.; Whitehead, R. J. (J. Chem. Soc. Dalton Trans. **1993** 1421/9).
[26] Parsons, S.; Passmore, J.; White, P. S. (J. Chem. Soc. Dalton Trans. **1993** 1499/507).
[27] Banister, A. J.; Lavender, I.; Lawrence, S. E.; Rawson, J. M.; Clegg, W. (J. Chem. Soc. Chem. Commun. **1994** 29/30).

5 Dinitrogen Sulfide, SNN

5.1 Formation

Formation by Thermolysis of 5-Phenyl-1,2,3,4-thiatriazole

Flash vacuum pyrolysis of 5-phenyl-1,2,3,4-thiatriazole produces SNN and C_6H_5CN.

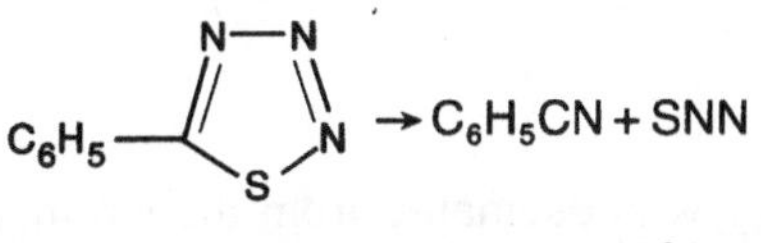

It was carried out at temperatures between 300 and 400 °C/10^{-3} to 10^{-5} Torr in an apparatus enabling contact times of ca. 1 ms. SNN can be isolated as a solid (together with C_6H_5CN) at the cold end of a liquid nitrogen cryostat at 77 K or in an Ar matrix at 10 to 20 K. The yield of SNN, measured by its IR absorbance relative to that of benzonitrile, critically depends on the contact time in the oven and hence on the oven design [1, 2].

Photoelectro-spectroscopic studies showed that flash vacuum pyrolysis of the flowing 5-phenyl-1,2,3,4-thiatriazole vapor at 310 °C leads to nearly complete (>95%) decomposition of the educt [3].

SNN was generated in a flow through gas infrared cell by thermolysis of 5-phenyl-1,2,3,4-thiatriazole. The educt was vaporized at 100 °C and pumped through a quartz tube and IR absorption cell (16 m path length). The transient SNN was only observed at the maximum pumping capacity of the system (200 L/min) [9]. It should be noted that in the thermal decomposition of 5-phenyl-1,2,3,4-thiatriazole the SNN species was not detected in earlier studies [4].

SNN was also formed by flash vacuum pyrolysis of some other 5-substituted 1,2,3,4-thiatriazoles (R = C_6D_5, 4-$CH_3C_6H_4$, and C_6H_5O). In the case of R = OC_6H_5, SNN and C_6H_5CNO were observed at pyrolysis temperatures as low as 200 °C [2].

Formation by Photolysis of 5-Phenyl-1,2,3,4-thiatriazole

SNN was also formed together with C_6H_5CN and traces of C_6H_5NCS by Ar matrix photolysis of 5-phenyl-1,2,3,4-thiatriazole at 310 ± 11 nm. The photolysis was faster with 254-nm light, although prolonged irradiation at this wavelength destroyed SNN [1].

Formation by Microwave Discharge in an Ar–N_2–Sulfur Vapor

The species SNN was formed by microwave discharge in an Ar–N_2–sulfur vapor. A 2% mixture of N_2 in Ar was passed through a 1-mm-orifice discharge tube seeded with sulfur vapor in equilibrium with the solid. The reaction products were trapped in solid argon at 12 K. SNN was detected by IR spectroscopy in experiments where the sulfur reservoir temperature was 50 °C. The yield became higher as the sulfur reservoir temperature increased to 70 °C [5].

Formation by the Reaction of N_3^- with CS_2

SNN was postulated to form in the gas-phase ion-molecule reaction of N_3^- with CS_2 in a flowing afterglow apparatus via $N_3^- + CS_2 \rightarrow NCS^- + SNN$. NCS^- was the exclusive product analyzed by a quadrupole mass filter and detected with an electron multiplier. N_3^- was generated in two different ways: from NH_2^- and N_2O and from $(CH_3)_3SiCH_2N_3$ and F^-. The rate constant for the reaction of N_3^- with CS_2 was found to be $(3.8 \pm 0.4) \times 10^{-11}$ $cm^3 \cdot molecule^{-1} \cdot s^{-1}$ giving a reaction coefficient $k_{expt}/k_{ADO} = 0.029$, where $k_{ADO} = 1.33 \times 10^{-9}$ $cm^3 \cdot molecule^{-1} \cdot s^{-1}$ [6].

SNN presumably is the carrier in a CS_2–N_2 laser. Its properties are different from those reported in later publications. The molecule was described as being stable and unreactive at room temperature, showing no reaction with charcoal soda lime or steel wool, and passing freely through pump oil. It was found to have a distinct, yet not unpleasant odor [7].

5.2 Heat of Formation

The heat of formation $\Delta H^{\circ}_{f,298}$ was estimated from the ion molecule reaction $N_3^- + CS_2 \rightarrow NCS^- + N_2S$ using known thermodynamic data to be ≤87.3 kcal/mol [6]. This is in fair agreement with a value of 47.6 kcal/mol calculated by the MNDO (RHF approximation) method [8].

5.3 Molecular Properties and Spectra

5.3.1 Molecular Structure and Bonding

Ab initio MO calculations (see Table 13, p. 54) have shown that SNN is a linear molecule with $C_{\infty v}$ symmetry. The bond lengths were determined from the ground state rotational constants of the two sulfur isotopomers ^{32}SNN and ^{34}SNN. An r_o structure with r_{NN} = 113.876(19) and r_{NS} = 157.751(14) pm was derived by a least-square fit of the rotational constants. A partial r_s structure with r_{NN} = 113.876(19), r_{NS} = 157.75(14) pm was calculated using Kraitchman's equations [9].

The experimental bond lengths are in good agreement with MP3/6-31G* calculations. HF/6-31G*: r_{NN} = 108.6, r_{NS} = 165.2 pm; MP3/6-31G*: r_{NN} = 112.529, r_{NS} = 159.816 pm [9]. For other calculations of bond lengths, see Table 13.

Ab initio SCF calculations indicated that the NN and NS bonds are triple and single polar bonds, respectively [10]. In contrast to the theoretical charge distribution [10, 11], the bonding, especially the relative energy of the S–N and N–N π bonds, strongly favors protonation at the sulfur atom, implying that a resonance structure of the type N≡N–S is dominant [11].

The S–N bond was found to be relatively weak. A value for the S–N bond energy of 68 kcal/mol was estimated from the total energy at the MP3 level [3].

5.3.2 Rotational and Centrifugal Distortion Constants

Rotational and centrifugal distortion constants were obtained by analyzing the fine-structure of the ν(N–N) band (ν_3) due to the ^{34}SNN and ^{32}SNN isotopomers [9]. Center frequencies, rotational and centrifugal distortion constants of the cold bands due to ^{32}SNN and ^{34}SNN, and four hot bands of ^{32}SNN are listed in Table 11.

The equilibrium B value was calculated from experimental data to be 0.215809(17) cm^{-1} [9] (0.21456 cm^{-1} [9] and 0.210 [3] based on MP3/6-31G* optimized geometry).

Table 11
Center Frequencies, Rotational and Centrifugal Distortion Constants (in cm^{-1}) of Vibrational Bands of SNN in cm^{-1} [9].

	$^{32}S\ 3_0^1$	$^{34}S\ 3_0^1$	$^{32}S\ 3_0^1\ 1_1^1$	$^{32}S\ 3_0^1\ 1_2^2$
ν_0 a)	2047.591873(98)	2047.40897(20)	2049.79785(24)	2046.64066(38)
B″	0.2164302(15)	0.2111677(40)	0.2161738(49)	0.2158447(84)
D″	$6.543(20)\times10^{-8}$	$6.275(104)\times10^{-8}$	$7.004(155)\times10^{-8}$	$7.771(392)\times10^{-8}$
B′	0.2153589(15)	0.2101264(39)	0.2151253(48)	0.2147920(82)
D′	$6.654(20)\times10^{-8}$	$6.386(110)\times10^{-8}$	$7.035(167)\times10^{-8}$	$8.015(426)\times10^{-8}$

	$^{32}S\ 3_0^1\ 2_1^{1e}$	$^{32}S\ 3_0^1\ 2_1^{1f}$	$^{32}S\ 3_0^1\ 2_2^{2f}$	vibration-rotation constants b)	
ν_0	2040.20330(22)	2040.20237(23)	2032.89883(30)	α_1	$0.256(6)\times10^{-3}$
B″	0.2175763(29)	0.2178537(32)	0.2189661(88)	α_2	$-1.285(6)\times10^{-3}$
D″	$6.844(54)\times10^{-8}$	$6.644(62)\times10^{-8}$	$7.190(427)\times10^{-8}$	α_3	$1.071(3)\times10^{-3}$
B′	0.2165226(28)	0.2168016(32)	0.2179342(86)		
D′	$6.950(56)\times10^{-8}$	$6.752(64)\times10^{-8}$	$7.344(460)\times10^{-8}$		

a) All center frequencies were adjusted by +0.00197 cm^{-1} on account of calibration data. – b) Derived from the B values of the ground state, 3_1, 2_1, and 1_1.

5.3.3 IR Spectrum

Three normal vibrations, ν_1 S–N stretch, ν_2 S–N bend, and ν_3 N–N stretch, are expected to be observed in the IR spectrum for a linear SNN molecule with symmetry $C_{\infty v}$.

The ν(N–N) absorption (ν_3) was observed in the high-resolution gas-phase IR spectrum at 2047.59 cm^{-1}. The ν_3 band shows a rotational structure typical for a linear molecule [9].

In a solid Ar matrix at 12 K, the ν_3 band was observed at 2040.2 cm^{-1} (showing a quartet at 2040.2, 2009.3, 2009.3, and 1972.3 cm^{-1} with mixed isotopic nitrogen) [5].

The ν(NN) band, observed at 2030 cm^{-1} in the IR spectrum of the product of 400 °C-pyrolyzed and Ar-matrix-isolated (at 20 K) 5-phenyl-1,2,3,4-thiatriazole [1], was shifted due to interaction with other species, including the precursor [5].

The SN stretching and SN bending frequencies (ν_1 and ν_2) could not be observed in the IR spectrum, because their intensities were too low [2, 5, 9] (a very weak band at 752 cm^{-1} was tentatively attributed to 2ν(SN) [2]). The SN bending frequency, $\nu_2 \approx 343$ cm^{-1}, was calculated from the equilibrium B value obtained from the vibration-rotation parameters [9]. The experimental value for the SN bending frequency is probably too low, being 100 cm^{-1} below the two most reliable theoretical estimates obtained by the CAS-SCF/DZP and CCSD/DZP methods [11].

The vibrational frequencies of SNN were calculated with the ab initio SCF-HF/31G* method as follows: 470.8, 582.6, and 2597.4 cm^{-1} [9] or 469, 584, and 2601 cm^{-1} [12] (see also [3]). Better results were achieved with the CCSD/DZP or CAS-SCF/DZP method (461, 712, and 2167 cm^{-1} and 447, 661, and 2081 cm^{-1}) [11].

5.3.4 Photoelectron Spectrum

The He I photoelectron spectrum of SNN was obtained by measuring the He I PES of the pyrolysis products of 5-phenyl-1,2,3,4-thiatriazole. Subtraction of the C_6H_5CN and N_2 spectra from the 310 °C-pyrolysis spectrum gave an SNN spectrum characterized by two sharp ionization bands assigned to the ground $^2\Pi(p_S)$ and first excited $^2\Sigma(\sigma_{NN})$ states of the SNN^+ ion. Another broad band was tentatively assigned to the second excited $^2\Pi(\pi_{NN})$ ion state. The experimental and calculated ionization energies of SNN are listed in Table 12. The assignment of the spectrum was made by comparing it with the spectrum of the electronic analog OCS [3].

Table 12
The Experimental Adiabatic and Vertical Ionization Energies E_i(ad) and E_i(vert) and the Calculated Ionization Energies E_i in eV for SNN.

ion state	MO	E_i(ad)	E_i(vert)	E_i (MP3)	E_i (GFAOV[a)])	E_i (GFAOV)	E_i (MNDO)
$^2\Pi$	3π	10.55	10.60	9.91	10.02	10.26	10.4
$^2\Sigma$	9σ	15.36	15.36	15.58	16.31	15.87	—
$^2\Pi$	2π	~15.7	~16.2	—	17.49	16.25	—
Ref.	[13]	[3]	[3]	[3]	[13]	[13]	[8]

a) GFAOV = Green's function approximation for outer-valence MO.

The first and second bands show evidence of fine-structure (S–N stretching vibrations of ~500 and 820 ± 50 cm^{-1} for the ground and first excited states of SNN^+) [3].

Many-particle effects are important for describing the electronic structure and geometry of SNN. With the exception of the ionization from the highest MO 3π, the ionic states can be described correctly only by using many-configuration wave functions [13].

5.4 Quantum-Chemical Calculations

The electronic properties and geometric parameters have been the subject of several quantum-chemical treatments. Table 13 gives a survey of publications.

Table 13
Quantum-Chemical Calculations for SNN.

method[a)]	calculated molecular properties	Ref.
ab initio SCF + CISD (CCSD, CAS-SCF)	E_t, geometry, ν_i	[11]
ab initio HF + MP2	E_t, geometry, ν_i	[12]
ab initio SCF	geometry, q, pop,	[10]
ab initio SCF	E_t, geometry, q, pop, μ	[14]
ab initio SCF	E_t, geometry	[15]
ab initio SCF + MP3	E_t, geometry, ν_i, q	[3]
ab initio HF + MP3	geometry, ν_i	[9]
ab initio HFS	E_t, pop, q, geometry	[16]
ab initio SCF	E_t, geometry, q, μ	[17]
GFAOV + ADC(3)	E_i	[13]
MP2 + MP4	geometry	[13]
MNDO	ΔH_f°, μ	[8]

a) MPn = nth order Møller-Plesset perturbation theory; SCF = self-consistent field; HFS = Hartree-Fock-Slater method; MNDO = modified neglect of differential overlap; CISD = single and double replacement configuration interaction; CCSD = coupled cluster approximation with single and double replacements; CAS = complete active space; GFAOV = Green's function approximation for outer-valence MO; ADC = algebraic diagrammatic construction; E_t = total molecular energy; IP = ionization energy; q = charge distribution; pop = orbital population; ΔH_f° = enthalpy of formation; μ = dipole moment; ν_i = vibrational frequencies.

Ab initio calculations showed that the theoretical isomer NSN has an energy that is 138 [14] or 44 kcal/mol [16] higher than that of SNN.

The Laplacian of the charge density of SNN was used to predict the structures of hydrogen-bonded gas-phase complexes of the type N_2S–HF. For this complex a three-membered ring structure is predicted. A Laplacian analysis predicts the linear SNN–HF structure to be 11.3 kJ/mol less stable than the ring structure [18].

5.5 Mass Spectrum

The molecular ion SNN^+ (m/e = 60) was observed in the mass spectrum of the pyrolysis products of 5-phenyl-1,2,3,4-thiatriazole at 100 to 300 °C. The pyrolyzed products were directly inserted into the ion source of a modified triple-sector mass spectrometer. The molecular ion signal of 5-phenyl-1,2,3,4-thiatriazole decreased and the signals at m/e = 103 (C_6H_5CN), 28 (N_2), 32 (S), and 60 (N_2S) increased with increasing temperature. A high-resolution mass measurement around m/e = 60 gave an exact mass of 59.9779 in agreement with a calculated value of 59.9782. Reducing the ionization potential from 70 to 10 to 14 eV simplified the spectrum, leaving the major components C_6H_5CN (m/e = 103, 77, 76), SNN (m/e = 60), S_2 (m/e = 64; exact mass 63.9420), H_2S (m/e = 34), and a molecule with m/e = 135 (presumably C_6H_5NCS) [1].

5.6 Thermolysis

SNN is stable at 77 K in an Ar matrix. Upon slow warming to 160 K it decomposed (disappearence of the IR absorbance). After further warming to room temperature, benzonitrile and sulfur were formed. The sample turned opaque, and sulfur was visibly formed [1, 2]. Results from low-voltage mass spectra [1] and photoelectron spectra [3] of the pyrolysis products of 5-phenyl-1,2,3,4-thiatriazole showed that the bimolecular process 2 SNN → 2 N_2 + S_2 occurs. S_2 was a secondary product in the mass spectra [1] as well as in the photoelectron spectra [3], but monoatomic sulfur was absent.

References:

[1] Wentrup, C.; Fischer, S.; Maquestiau, A.; Flammang, R. (J. Org. Chem. **51** [1986] 1908/10).
[2] Wentrup, C.; Kambouris, P. (Chem. Rev. **91** [1991] 363/73).
[3] Bender, H.; Carnovale, F.; Peel, J. B.; Wentrup, C. (J. Am. Chem. Soc. **110** [1988] 3458/61).
[4] Holm, A.; Carlsen, L.; Larsen, E. (J. Org. Chem. **43** [1978] 4816/22).
[5] Hassanzadeh, P.; Andrews, L. (J. Am. Chem. Soc. **114** [1992] 83/91).
[6] Kass, S. R.; DePuy, C. H. (J. Org. Chem. **50** [1985] 2874/7).

[7] Powell, F. X. (Chem. Phys. Lett. **33** [1975] 393/5).
[8] Sensarma, S.; Turner, A. G. (Inorg. Chim. Acta **64** [1982] L161/L162).
[9] Brown, R. D.; Elmes, P. S.; McNaughton, D. (J. Mol. Spectrosc. **140** [1990] 390/400).
[10] Behera, L.; Kar, T.; Sannigrahi, A. B. (J. Mol. Struct. **209** [1990] 111/24 [THEOCHEM **68**]).

[11] Davy, R. D.; Schaefer, H. F. (J. Am. Chem. Soc. **113** [1991] 1917/22).
[12] Pyykkö, P. (Chem. Phys. Lett. **162** [1989] 349/54).
[13] Zakzhevskii, V. G. (Zh. Neorg. Khim. **36** [1991] 2320/4; Russ. J. Inorg. Chem. [Engl. Transl.] **36** [1991] 1307/10).
[14] Chaban, G. M.; Klimenko, N. M.; Charkin, O. P. (Izv. Akad. Nauk SSSR Ser. Khim. **1990** 794/802; Bull. Acad. Sci. USSR Div. Chem. Sci. [Engl. Transl.] **1990** 704/11).
[15] Gimarc, B. M.; Warren, D. S. (Inorg. Chem. **30** [1991] 3276/80).
[16] Laidlaw, W. G.; Trsic, M. (Inorg. Chem. **20** [1981] 1792/4).
[17] Collins, M. P. S.; Duke, B. J. (J. Chem. Soc. Dalton Trans. **1978** 277/9).
[18] Carroll, M. T.; Chang, C.; Bader, R. F. W. (Mol. Phys. **63** [1988] 387/405).

6 N-Thionitroso-diorganylamines and Derivatives

6.1 N-Thionitroso-diorganylamines

6.1.1 N-Thionitroso-dimethylamine, $S{=}NN(CH_3)_2$

6.1.1.1 Preparation

The title compound was prepared by two different methods, one involving a sulfuration (oxidation) and the other a reduction.

By Reaction of Sulfur with $H_2NN(CH_3)_2$. When a mixture of 1,1-dimethylhydrazine (1 mole) and powdered rhombic sulfur (1 g-atoms) in ether was stirred at room temperature for 6 days, the title compound was formed with a low yield of 15%. The reaction mixture was filtered, and the undissolved sulfur was recovered. The purple filtrate was evaporated to dryness under reduced pressure. Purification of the residue was achieved by recrystallization from ether at −78 °C [1].

By Reduction of $O{=}S{=}NN(CH_3)_2$. To an ethereal solution of $O{=}S{=}NN(CH_3)_2$ (0.25 mole) cooled to −75 °C a 1 M solution of $LiAlH_4$ (0.063 mole) in ether was added dropwise over a period of 1 h. After warming to room temperature and stirring for 1 h, the reaction mixture was filtered and the filtrate concentrated by distillation at room temperature and reduced pressure (0.01 Torr) until no further distillate came over. The residue was recrystallized from ether at −78 °C. The crystalline product was obtained with a low yield of 24% [1].

6.1.1.2 Molecular Properties. Spectra

6.1.1.2.1 Geometrical Structure

The molecular geometry has been calculated by the semiempirical MNDO method giving d(N=S) = 1.527 Å, d(N−N) = 1.296 Å, valence angle S−N−N = 121.9° [2]. The N=S bond distance was estimated by INDO method to be 1.582 Å [3].

6.1.1.2.2 ^1H NMR Spectrum

^1H NMR (10% in CCl_4): δ (in ppm) = 3.6 (s), 4.1 (s). The distance between the two singlets varies with the solvent and temperature. In a saturated D_2O solution at 25 °C, the peaks were only 23 Hz apart at a measuring frequency of 60 MHz. In a 10% CCl_4 solution the signals were separated by 29 Hz at 25 °C and 25 Hz at 70 °C. On further heating, the sample decomposed before coalescence could be observed. The ^1H NMR spectrum indicated restricted rotation of the N−N bond [1].

6.1.1.2.3 Ionization Potentials E_i. Photoelectron Spectrum

The He I photoelectron spectrum (PES) was assigned with the aid of MO energies (ε_i) calculated by the semiempirical MNDO method [2]:

E_i (PES) in eV	8.0	8.9	11.2	11.8
ε_i (MNDO) in eV	−9.48	−9.92	−12.63	−13.58
assignment	n_s	π	σ	π

The energies of the frontier orbitals (calculated by CNDO/2S) are presented in a figure in [3].

6.1.1.2.4 UV–VIS Spectrum

UV-VIS (cyclohexane): λ_{max} (in nm) (ε in $L \cdot mol^{-1} \cdot cm^{-1}$) = 705 (1.5), 587 (27.3) R, 306 (11900) K; (CCl_4): 685 (1.8), 576 (38) R, 309 (12300) K; (ethanol): 680 (1.0), 533 (17.5) R, 306 (10800) K [1].

The R band (probably, $n \rightarrow \pi^*$) shows a hypsochromic shift of 54 nm in going from cyclohexane to ethanol (strong evidence for extensive hydrogen bonding in ethanol) [1]. The weak absorption at 705 nm found in [1] was assigned to an S–T transition [3]. The $n \rightarrow \pi^*$ transition energies were calculated by CNDO/S to be 1.523 μm^{-1} (657 nm), 1.706 μm^{-1} (586 nm), and 2.452 μm^{-1} (408 nm) [3].

Solutions of the title compound in nonpolar solvents are purple to blue. Solutions in polar solvents are orange to red [1].

6.1.1.2.5 IR Spectrum

The IR spectrum shows strong bands at 1475, 1342, and 1105 cm^{-1} and weaker bands at 2941, 1447, 1258, 1064, and 916 cm^{-1} [1].

6.1.1.3 Melting Point

The deep purple crystals melt at 20 to 21 °C [1].

6.1.1.4 Solutions

The bluish purple solution in cyclohexane or in other nonpolar solvents is stable for weeks, but the orange-red solution in H_2O is destroyed within a few minutes. $S{=}NN(CH_3)_2$ can be extracted from aqueous solution with ether to give a purple ethereal solution [1].

Cryoscopic molecular weight measurements indicated that the compound is monomeric in benzene solution [1].

6.1.1.5 Chemical Reactions

6.1.1.5.1 Thermolysis

$S{=}NN(CH_3)_2$ is stable at low temperatures (ca. −30 °C). Neat liquid samples decompose within a few hours at room temperature. The decomposition products were identified to be elemental sulfur and $(CH_3)_2S$. The nitrogen-containing products were not identified.

Sometimes the decomposition was accompanied by a strong detonation (probably destruction of azides) [1].

6.1.1.5.2 Reactions with Acids and Bases

Treating the title compound with acidic materials such as BF_3, CH_3COOH, or mineral acids led to quite rapid decomposition with deposition of sulfur. Pyridine, PPh_3, and even aqueous NaOH do not significantly accelerate or alter the course of the decomposition [1].

6.1.1.5.3 Reduction

The title compound reacts with $LiAlH_4$ in ether at 0 °C to yield $H_2NN(CH_3)_2$ which is converted to $4\text{-}NO_2C_6H_4NHC(O)NHN(CH_3)_2$ by reacting with $4\text{-}NO_2C_6H_4N{=}C{=}O$ [1].

6.1.1.5.4 Reaction with 3,6-Dicarbomethoxy-1,2,4,5-tetrazine

$S{=}NN(CH_3)_2$ reacted with 3,6-dicarbomethoxy-1,2,4,5-tetrazine (I) (mole ratio 1 : 1) in anhydrous CH_2Cl_2 at room temperature to yield the triazole (II) with 73% yield [4].

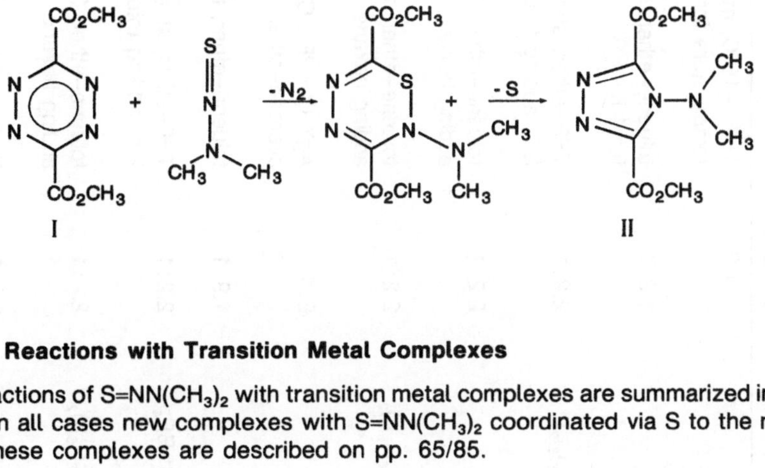

I II

6.1.1.5.5 Reactions with Transition Metal Complexes

The reactions of $S{=}NN(CH_3)_2$ with transition metal complexes are summarized in Table 14, pp. 60/3. In all cases new complexes with $S{=}NN(CH_3)_2$ coordinated via S to the metal were formed. These complexes are described on pp. 65/85.

6.1.1.6 Applications

$S{=}NN(CH_3)_2$ was tested in rats by sc. injections for carcinogenic activity. No tumors arose by injections with $S{=}NN(CH_3)_2$. In contrast, $O{=}NN(CH_3)_2$ was active in the same strain of rats and in the same mode of application in which $S{=}NN(CH_3)_2$ was given [13].

Table 14
Reactions of $S{=}NN(CH_3)_2$ with Transition Metal Complexes (1,5-C_8H_{12} = 1,5-cyclooctadiene, C_7H_8 = norbornadiene).

complex	mole ratio $S{=}NN(CH_3)_2$: complex	reaction conditions	main product	Ref.
$Cr(CO)_5(THF)$	1:1.2	THF, −15 °C, at room temperature for 1 h	$Cr(CO)_5(SNN(CH_3)_2)$	[5, 6]
$RuHCl(CO)(P(C_6H_5)_3)_3$	1.9:1	toluene–ether, stirring for 4 h	$RuHCl(CO)(P(C_6H_5)_3)_2(SNN(CH_3)_2)$	[7]
$[RuH(CO)(CH_3CN)_2(P(C_6H_5)_3)_2]^+$ SbF_6^-	2.2:1	THF–ether, stirring for 2 h, adding toluene	*cis,trans*-$[RuH(CO)(P(C_6H_5)_3)_2(SNN(CH_3)_2)_2]^+$ SbF_6^-	[7]
$RuCl(CO)(C_6H_5)(P(C_6H_5)_3)_2$	2.2:1	toluene–ether, 20 min, adding hexane	$RuCl(CO)(C_6H_5)(P(C_6H_5)_3)_2(SNN(CH_3)_2)$	[7]
$RuCl(CO)(4\text{-}CH_3C_6H_4)(P(C_6H_5)_3)_2$	2.2:1	toluene–ether, 20 min, adding hexane	$RuCl(CO)(4\text{-}CH_3C_6H_4)(P(C_6H_5)_3)_2(SNN(CH_3)_2)$	[7]
$[(\eta^6\text{-}C_6H_6)RuCl_2]_2$	8:1	AgX (X = PF_6, ClO_4), acetone–ether	$[(\eta^6\text{-}C_6H_6)Ru(SNN(CH_3)_2)_3]^{2+}$ $(X^-)_2$	[8]
$OsHCl(CO)(P(C_6H_5)_3)_3$	1.9:1	toluene–ether, for 4 h	$OsHCl(CO)(P(C_6H_5)_3)_2(SNN(CH_3)_2)$	[7]
$[OsH(H_2O)(CO)(P(C_6H_5)_3)_3]^+$ BF_4^-	2.2:1	THF–ether, stirring for 2 h, adding toluene	*cis,trans*-$[OsH(CO)(P(C_6H_5)_3)_2(SNN(CH_3)_2)_2]^+$ BF_4^-	[7]
$OsCl(CO)(4\text{-}CH_3C_6H_4)(P(C_6H_5)_3)_2$	2.2:1	toluene–ether, 20 min, adding hexane	$OsCl(CO)(4\text{-}CH_3C_6H_4)(P(C_6H_5)_3)_2(SNN(CH_3)_2)$	[7]
$OsCl(CS)(4\text{-}CH_3C_6H_4)(P(C_6H_5)_3)_2$	2.2:1	toluene–ether, 20 min, adding hexane	$OsCl(CS)(4\text{-}CH_3C_6H_4)(P(C_6H_5)_3)_2(SNN(CH_3)_2)$	[7]
$OsCl(CO)(4\text{-}CH_3C_6H_4)(P(C_6H_5)_3)_2L$ L = SO_2, 2,1,3-benzothiadiazole, 2,1,3-benzoselenadiazole			$OsCl(CO)(4\text{-}CH_3C_6H_4)(P(C_6H_5)_3)_2(SNN(CH_3)_2)$	[9]

$[RhCl(CO)_2]_2$	8 : 1	AgX (X = PF_6, ClO_4), acetone–ether	*cis*-$[Rh(CO)_2(SNN(CH_3)_2)_2]^+ X^-$	[8]
$[RhCl(CO)_2]_2$		light petroleum at −10 °C	*cis*-$RhCl(CO)_2(SNN(CH_3)_2)$	[10]
$[(\eta^5\text{-}C_5(CH_3)_5)RhCl_2]_2$	4 : 1	AgX (X = PF_6, ClO_4), acetone–ether	$[(\eta^5\text{-}C_5(CH_3)_5)Rh(SNN(CH_3)_2)_3]^{2+} (X^-)_2$	[8]
$(\eta^5\text{-}C_5H_5)RhI_2(CO)$	excess $S{=}NN(CH_3)_2$	$AgPF_6$, acetone–ether	$[(\eta^5\text{-}C_5H_5)Rh(SNN(CH_3)_2)_3]^{2+} (PF_6^-)_2$	[8]
$(\eta^5\text{-}C_5H_5)RhI_2(CO)$	a)	$AgClO_4$, acetone–ether, $S{=}NN(CH_3)_2$ slowly added	$[(\eta^5\text{-}C_5H_5)Rh(ClO_4)(SNN(CH_3)_2)_2]^+ ClO_4^-$	[8]
$[RhCl(1,5\text{-}C_8H_{12})]_2$	a)	$AgClO_4$, acetone–ether, $S{=}NN(CH_3)_2$ slowly added	$Rh(ClO_4)(1,5\text{-}C_8H_{12})(SNN(CH_3)_2)$	[8]
$[RhCl(1,5\text{-}C_8H_{12})]_2$	8 : 1	AgX (X = PF_6, ClO_4), acetone–ether	$[Rh(1,5\text{-}C_8H_{12})(SNN(CH_3)_2)_2]^+ X^-$	[8]
$[RhCl(C_7H_8)]_2$	8 : 1	AgX (X = PF_6, ClO_4), acetone–ether	$[Rh(C_7H_8)(SNN(CH_3)_2)_2]^+ X^-$	[8]
$[RhCl(diene)]_2$ (diene = $1,5\text{-}C_8H_{12}$, C_7H_8)		CH_2Cl_2–ether	$RhCl(diene)(SNN(CH_3)_2)$	[10]
$[Rh(1,5\text{-}C_8H_{12})(SNN(CH_3)_2)_2]^+ ClO_4^-$	4 : 1	$P(C_6H_5)_3$, acetone–ether	$[Rh(1,5\text{-}C_8H_{12})P(C_6H_5)_3(SNN(CH_3)_2)]^+ ClO_4^-$	[8]
$[Rh(1,5\text{-}C_8H_{12})(SNN(CH_3)_2)_2]^+ ClO_4^-$	4 : 1	$(4\text{-}CH_3C_6H_4)P(C_6H_5)_2$, acetone–ether	$[Rh(1,5\text{-}C_8H_{12})((4\text{-}CH_3C_6H_4)P(C_6H_5)_2)\text{-}(SNN(CH_3)_2)]^+ ClO_4^-$	[8]
cis-$RhCl(CO)_2(SNN(CH_3)_2)$		ether	*trans*-$RhCl(CO)(SNN(CH_3)_2)_2$	[10]

Table 14 (continued)

complex	mole ratio $S{=}NN(CH_3)_2$: complex	reaction conditions	main product	Ref.
trans-$RhX(CO)(E(C_6H_5)_3)_2$ (X = Cl, E = As, Sb; X = Br, NCS, E = As)	8:1	ether	*trans*-$RhX(CO)(SNN(CH_3)_2)_2$ X = Cl, Br, NCS	[10]
$IrHCl_2(P(C_6H_5)_3)_3$	1.9:1	toluene–ether, room temperature for 4 h	$IrHCl_2(P(C_6H_5)_3)_2(SNN(CH_3)_2)$	[7]
$IrHCl_2(P(C_6H_5)_3)_3$	2.4:1	1) CH_3CN, $AgBF_4$, 15 min refluxing 2) THF–ether, 15 h	*cis,trans*-$[IrHCl(P(C_6H_5)_3)_2(SNN(CH_3)_2)_2]^+$ BF_4^-	[7]
$PdCl_2$(diene) diene = 1,5-C_8H_{12}, C_7H_8	5:1	$AgPF_6$, acetone–ether	$[Pd(SNN(CH_3)_2)_4]^{2+}$ $(PF_6^-)_2$	[11]
$PdCl_2$(diene) diene = 1,5-C_8H_{12}, C_7H_8, 1,3,5,7-cyclooctatetraene, dicyclopentadiene	2.5:1	acetone or CH_2Cl_2	*cis*-$PdCl_2(SNN(CH_3)_2)_2$	[12]
$PdCl_2$(1,5-C_8H_{12})	3:1	$P(C_6H_5)_3$, acetone or CH_2Cl_2	$PdCl_2(P(C_6H_5)_3)(SNN(CH_3)_2)$	[12]
$PdCl_2$(1,5-C_8H_{12})	3:1	4-$CH_3C_6H_4P(C_6H_5)_2$, acetone or CH_2Cl_2	$PdCl_2$(4-$CH_3C_6H_4P(C_6H_5)_2)(SNN(CH_3)_2)$	[12]
$[PdCl(CH_3Odiene)]_2$ diene = 1,5-C_8H_{12}, 1,3,5,7-cyclooctatetraene, dicyclopentadiene	ca. 2:1	CH_2Cl_2	$PdCl(CH_3Odiene)(SNN(CH_3)_2)$	[12]
$PdBr_2$(1,5-C_8H_{12})	2.5:1	acetone or CH_2Cl_2	*cis*-$PdBr_2(SNN(CH_3)_2$	[11]
$PdCl_2(C_6H_5CN)_2$	1:1	benzene–ether; room temperature	$(\mu\text{-}Cl)_2\{PdCl(SNN(CH_3)_2)\}_2$	[11]

$PdCl_2(C_6H_5CN)_2$	6:1	1) benzene–ether 2) $NaB(C_6H_5)_4$, ethanol	$[Pd(SNN(CH_3)_2)_4]^{2+}\ (B(C_6H_5)_4^-)_2$	[11]
		or $HgCl_2$ in acetone	$[Pd(SNN(CH_3)_2)_4]^{2+}\ HgCl_4^{2-}$	[11]
Na_2PdCl_4	3:1	acetone–H_2O	$[Pd(SNN(CH_3)_2)_4]^{2+}\ PdCl_4^{2-}$	[11]
K_2PtCl_4	3:1	acetone–H_2O	$[Pt(SNN(CH_3)_2)_4]^{2+}\ PtCl_4^{2-}$	[11]
$PtCl_2$(diene) diene = 1,5-C_8H_{12}, C_7H_8	5:1	$AgPF_6$, acetone–ether	$[Pt(SNN(CH_3)_2)_4]^{2+}\ (PF_6^-)_2$	[11]
$PtCl_2(1,5\text{-}C_8H_{12})$	3:1	$P(C_6H_5)_3$, acetone or CH_2Cl_2	$PtCl_2(P(C_6H_5)_3)(SNN(CH_3)_2)$	[12]
$PtCl_2(1,5\text{-}C_8H_{12})$	3:1	4-$CH_3C_6H_4P(C_6H_5)_2$, acetone or CH_2Cl_2	$PtCl_2(4\text{-}CH_3C_6H_4P(C_6H_5)_2)(SNN(CH_3)_2)$	[12]
$PtCl_2(C_6H_5CN)_2$	6:1	1) acetone–ether 2) $NaB(C_6H_5)_4$, ethanol	$[Pt(SNN(CH_3)_2)_4]^{2+}(B(C_6H_5)_4^-)_2$	[11]

[a] $S{=}NN(CH_3)_2$ was not present in large excess.

References:

[1] Middleton, W. J. (J. Am. Chem. Soc. **88** [1966] 3842/4, U.S. 3344135 [1967] 3 pp.; C.A. **67** [1967] No. 116823).

[2] Penkovsky, V. V.; Shermolovitch, Yu. G.; Solovyov, A. V.; Vovna, V. I.; Borisenko, A. V. (Phosphorus Sulfur Silicon Relat. Elem. **73** [1992] 1/4).

[3] Mehlhorn, A.; Sauer, J.; Fabian, J.; Mayer, R. (Phosphorus Sulfur Relat. Elem. **11** [1981] 325/34).

[4] Seitz, G.; Overheu, W. (Chem.-Ztg. **103** [1979] 230/1).

[5] Roesky, H. W.; Emmert, R.; Isenberg, W.; Schmidt, M.; Sheldrick, G. M. (J. Chem. Soc. Dalton Trans. **1983** 183/5).

[6] Roesky, H. W.; Emmert, R.; Clegg, W.; Isenberg, W.; Sheldrick, G. M. (Angew. Chem. **93** [1981] 623/4; Angew. Chem. Int. Ed. Engl. **20** [1981] 591).

[7] Herberhold, M.; Hill, A. F. (J. Organomet. Chem. **315** [1986] 105/12).

[8] Tresoldi, G.; Sergi, S.; Lo Schiavo, S.; Piraino, P. (J. Organomet. Chem. **328** [1987] 387/91).

[9] Gieren, A.; Ruiz-Perez, C.; Hübner, T.; Herberhold, M.; Hill, A. F. (J. Chem. Soc. Dalton Trans. **1988** 1693/6).

[10] Tresoldi, G.; Sergi, S.; Lo Schiavo, S.; Piraino, P. (J. Organomet. Chem. **322** [1987] 369/76).

[11] Tresoldi, G.; Bruno, G.; Piraino, P.; Faraone, G.; Bombieri, G. (J. Organomet. Chem. **265** [1984] 311/22).

[12] Tresoldi, G.; Bruno, G.; Crucitti, F.; Piraino, P. (J. Organomet. Chem. **252** [1983] 381/7).

[13] Schmähl, D.; Krüger, F. W. (Z. Krebsforsch. **73** [1969] 191/2).

6.1.2 1-Thionitroso-piperidine, $S{=}NN(CH_2)_5$

The title compound was produced by the same methods as used for the synthesis of $S{=}NN(CH_3)_2$ (see p. 57).

A solution of $LiAlH_4$ in ether was added dropwise to a solution of $O{=}S{=}NN(CH_2)_5$ in ether at −50 °C within 15 min. After warming to room temperature and stirring for 1 h, the reaction mixture was filtered and the filtrate was concentrated by evacuation at 0.01 Torr. Crude $S{=}NN(CH_2)_5$ was obtained as a purple oil. The compound could not be obtained analytically pure. Further attempts to purify the compound by recrystallization or distillation led to decomposition.

$S{=}NN(CH_3)_2$ was also prepared by reacting sulfur with $H_2NN(CH_3)_2$.

^{1}H NMR: δ (in ppm) = 1.6 (6H, m), 3.8 (2H, br), 4.3 (2H, br).

The ^{1}H NMR spectrum indicated restricted rotation of the N–N bond.

UV (CCl_4): λ_{max} (in nm) (ε in $L \cdot mol^{-1} \cdot cm^{-1}$) = 585 (13), 318 (14500).

Reference:

Middleton, W. J. (J. Am. Chem. Soc. **88** [1966] 3842/4, U.S. 3344135 [1967] 3 pp.; C.A. **67** [1967] 116823).

6.1.3 1-Thionitroso-hexahydro-1H-azepine, $S{=}NN(CH_2)_6$

The title compound was prepared in a crude form by reacting $H_2NN(CH_2)_6$ with sulfur in ether. It was isolated as a purple oil. The compound is unstable.

UV (ether): λ_{max} (in nm) = 700, 584, 306.

Reference:

Middleton, W. J. (J. Am. Chem. Soc. **88** [1966] 3842/4, U.S. 3344135 [1967] 3 pp.; C.A. **67** [1967] 116823).

6.1.4 N-Thionitroso-diphenylamine, $S{=}NN(C_6H_5)_2$

The title compound was prepared by adding S_2Cl_2 to a solution of $(C_6H_5)_2NNH_2$ in diethyl ether at −20 °C in the presence of $N(C_2H_5)_3$ (mole ratio 1 : 1 : 2.5). After filtration and evaporating the volatiles, the residue was dissolved in petroleum ether. Filtration and evaporation gave the title compound as a viscous, dark red-brown residue. The compound is not stable enough to be purified by recrystallization or sublimation, but could be isolated as a chromium pentacarbonyl complex (see below).

IR (Nujol): ν (in cm^{-1}) = 3370 w, 1590 vs, 1500 s, 1410 w, 1305 vs, 1255 s, 1165 s, 1100 m, 1065 m, 1020 m, 995 w, 935 w, 880 m, 815 w, 745 vs, 685 vs.

$S{=}NN(C_6H_5)_2$ reacts with $[Cr(CO)_5]\cdot THF$ in THF between −20 °C and room temperature while stirring for 2 h to produce the complex $Cr(CO)_5(S{=}NN(C_6H_5)_2)$ with 36% yield.

Reference:

Roesky, H. W.; Emmert, R.; Isenberg, W.; Schmidt, M.; Sheldrick, G. M. (J. Chem. Soc. Dalton Trans. **1983** 183/5).

6.2 Transition Metal Complexes of N-Thionitroso-dimethylamine and N-Thionitroso-diphenylamine

The title complexes were prepared in general by the reaction of the uncomplexed $S{=}NN(CH_3)_2$ or $S{=}NN(C_6H_5)_2$ species with a corresponding transition metal complex via simple ligand exchange. Several other $S{=}NN(CH_3)_2$ complexes were obtained by transforming known $S{=}NN(CH_3)_2$ complexes via substitution reactions with a variety of ligands. A general method for the synthesis of cationic Rh or Ru $S{=}NN(CH_3)_2$ complexes is the reaction of in situ formed solvento species from dimeric chloride-bridged complexes and displacement of the coordinated solvent by $S{=}NN(CH_3)_2$.

In all cases the $S{=}NN(CH_3)_2$ ligand is coordinated via the S atom in a monodentate manner.

6.2.1 N-Thionitroso-dimethylamine Complex of Cr^0, $Cr(CO_5)(SNN(CH_3)_2)$

The complex was prepared by adding a solution of $[Cr(CO)_5]\cdot THF$ in THF to $S{=}NN(CH_3)_2$ at −15 °C (mole ratio 1.2 : 1). After warming to room temperature and stirring for another hour, the solvent was removed and the residue recrystallized from CH_2Cl_2. The title compound was obtained as ruby red crystals with 24% yield [1, 2].

The complex melts at ca. 110 °C with decomposition [1, 2].

The molecular structure of the complex was determined by X-ray diffraction. The crystals belong to the monoclinic space group $P2_1/c - C^5_{2h}$ (No. 14) with a = 10.422(3), b = 12.507(4), c = 9.546(3) Å, β = 110.21(3)°; V = 1167.7 Å³; Z = 4; D_x = 1.605 g/cm³; R = 0.049, R_w = 0.044 for 2388 observed reflections. The molecular structure is illustrated in **Fig. 4** including selected bond lengths and bond angles. The $S{=}NN(CH_3)_2$ ligand is end-on-coordinated via the S atom. The "CrSNNC" unit is in effect planar and lies approximately eclipsed to the carbonyl groups. The S–N distance of 1.635(2) Å corresponds to a single bond. The N–N bond (1.279(2) Å) is shortened with respect to N–N bonds of hydrazine derivatives (1.45 Å) [1, 2]. Atomic coordinates are listed in the paper [2].

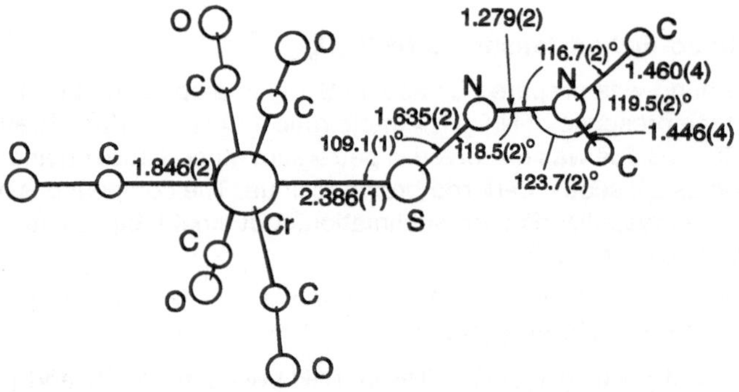

Fig. 4. Molecular structure of $Cr(CO)_5(SNN(CH_3)_2)$ [2].

^{1}H NMR ($CDCl_3$/TMS): δ (in ppm) = 0.41 (s) [1].

UV (n-hexane): λ_{max} (in nm) = 535 w, 375 w, 280 sh, 250 sh, 225 s [1].

IR (Nujol): ν (in cm^{-1}) = 2000 m, 1945 vs, 1890 vs, 1360 s, 1260 m, 1120 m, 1090 m, 1020 m, 800 m, 785 m [1].

MS (15 °C): m/e (rel. int. in %) = 282 (37) M$^+$, 254 (7), 226 (10), 198 (37), 196 (3), 192 (4), 182 (2), 170 (3), 168 (7), 154 (4), 142 (100) $[CrSNN(CH_3)_2]^+$, 140 (14), 127 (4), 126 (8), 112 (46), 108 (9), 98 (28), 90 (16), 84 (7), 80 (25), 75 (6), 58 (7), 52 (72), 46 (3), 44 (16), 43 (49) [1, 2].

The crystals are stable at room temperature [2]. The stability of the complex is comparable to that of $Cr(CO)_5(SNN(C_6H_5)_2)$; see p. 85. It is soluble in all common organic solvents, except CCl_4, without noticeable decomposition. The free ligand $S{=}NN(CH_3)_2$ was observed to form in CCl_4 [1].

References:

[1] Roesky, H. W.; Emmert, R.; Isenberg, W.; Schmidt, M.; Sheldrick, G. M. (J. Chem. Soc. Dalton Trans. **1983** 183/5).

[2] Roesky, H. W.; Emmert, R.; Clegg, W.; Isenberg, W.; Sheldrick, G. M. (Angew. Chem. **93** [1981] 623/4; Angew. Chem. Int. Ed. Engl. **20** [1981] 591).

6.2.2 N-Thionitroso-dimethylamine Complexes of Ru^{II} and Os^{II}

The complexes of Ru^{II} are compiled in Table 15 and those of Os^{II} in Table 16, p. 69.

Table 15
N-Thionitrosodimethylamine Complexes of Ru^{II}.
An asterisk preceding the compound number indicates further information at the end of the table, p. 68.

No. compound	preparation method (yield in %) properties (^1H NMR ($CDCl_3$/TMS): δ (in ppm), IR (Nujol): ν (in cm^{-1}))
neutral mononuclear Ru^{II} complexes	
1 $RuHCl(CO)(P(C_6H_5)_3)_2(SNN(CH_3)_2)$	from $RuHCl(CO)(P(C_6H_5)_3)_3$ and $S{=}NN(CH_3)_2$ (mole ratio 1 : 1.9) in toluene–ether, stirring for 4 h, adding hexane, 30 min stirring (97) bright orange crystals from CH_2Cl_2–C_2H_5OH m.p. 173 °C with decomposition the $P(C_6H_5)_3$ ligands are trans-arranged as seen by the appearance of the hydride resonance as a triplet in the ^1H NMR spectrum ^1H NMR: −8.97 (t, RuH, $^2J(P,H) = 19.8$ Hz), 3.07, 3.48 ($N(CH_3)_2$) IR: 1922 ν(CO), 1352 m, 1261 w, 1120 m, 837 w, 790 w; 1084, 1020 ($P(C_6H_5)_3$), 932 w, 898 w [1]
2 $RuCl(CO)(C_6H_5)(P(C_6H_5)_3)_2(SNN(CH_3)_2)$	from $RuCl(CO)(C_6H_5)(P(C_6H_5)_3)_2$ and $S{=}NN(CH_3)_2$ (mole ratio ca. 1 : 2) in toluene–ether, adding hexane, recrystallized from CH_2Cl_2–C_2H_5OH (94) m.p. 159 °C ^1H NMR: 2.72, 3.57 ($N(CH_3)_2$) IR: 1915 ν(CO) [1]
*3 $RuCl(CO)(4\text{-}CH_3C_6H_4)(P(C_6H_5)_3)_2(SNN(CH_3)_2)$	from $RuCl(CO)(4\text{-}CH_3C_6H_4)(P(C_6H_5)_3)_2$ and $S{=}NN(CH_3)_2$ (mole ratio ca. 1 : 2) in toluene–ether, adding hexane, recrystallized from CH_2Cl_2–C_2H_5OH (98) m.p. 160 °C ^1H NMR: 2.16 (4-$\underline{C}H_3C_6H_4$), 2.68, 3.51 ($N(CH_3)_2$) IR: 1898, 1929 ν(CO), (CH_2Cl_2: 1924 ν(CO)) [1]
4 $Ru(N_3)(CO)(4\text{-}CH_3C_6H_4)(P(C_6H_5)_3)_2(SNN(CH_3)_2)$	treating a solution of RuCl(CO)-($4\text{-}CH_3C_6H_4)(P(C_6H_5)_3)_2(SNN(CH_3)_2)$ with NaN_3 in the presence of a catalytic amount of $t\text{-}C_4H_9CN$ (good yield) [1, 2]
cationic mononuclear Ru^{II} complexes	
5 $[Ru(CO)(4\text{-}CH_3C_6H_4)(t\text{-}C_4H_9CN)(P(C_6H_5)_3)_2(SNN(CH_3)_2)]^+ PF_6^-$	from RuCl(CO)-($4\text{-}CH_3C_6H_4)(P(C_6H_5)_3)_2(SNN(CH_3)_2)$ and excess of NH_4PF_6 in the presence of $t\text{-}C_4H_9CN$ in CH_2Cl_2–H_2O, ethanol (90) [1, 2]

Table 15 (continued)

No.	compound	preparation method (yield in %) properties (^1H NMR ($CDCl_3$/TMS): δ (in ppm), IR (Nujol): ν (in cm^{-1}))
5 (continued)		yellow crystals m.p. 158 °C with decomposition ^1H NMR: 0.90 (s, t-C_4H_9), 2.23 (s, 4-$\underline{C}H_3C_6H_4$), 3.01, 3.65 ($N(CH_3)_2$) IR: 1960 ν(CO) [1]
6	*cis-trans*-$[RuH(CO)(P(C_6H_5)_3)_2(SNN(CH_3)_2)_2]^+$ SbF_6^-	from $[RuH(CO)(CH_3CN)_2(P(C_6H_5)_3)_2]^+$ SbF_6^- and S=$NN(CH_3)_2$ in THF–ether (mole ratio 2.8 : 6), stirring for 2 h (87) yellow crystals m.p. 139 °C ^1H NMR: −9.62 (t, RuH, $^2J(P,H)$ = 18.0 Hz), 3.48, 3.39, 3.29, 2.85 ($N(CH_3)_2$, equal intensity) IR: 1936 ν(CO) [1]
7	$[Ru(C_6H_6)(SNN(CH_3)_2)_3]^{2+}$ $(ClO_4^-)_2$	from $[Ru(C_6H_6)Cl_2]_2$, $AgClO_4$, and S=$NN(CH_3)_2$ in CH_3CN–ether (mole ratio 1 : 4 : 8) at room temperature (ca. 80) orange solid ^1H NMR: 4.00 (q, $N(CH_3)_2$), 4.25 (q, $N(CH_3)_2$), $^4J(H,H)$ = 0.70 Hz; 6.40 (s, C_6H_6) stable for some weeks quite insoluble in acetone and CH_3CN bi-univalent electrolyte (Λ_M = 210 to 220 $\Omega^{-1} \cdot cm^2 \cdot mol^{-1}$) [3]
8	$[Ru(C_6H_6)(SNN(CH_3)_2)_3]^{2+}$ $(PF_6^-)_2$	preparation analogous to No. 7 with $AgPF_6$ soluble in acetone–d_6 [3]

*Further information:

$RuCl(CO)(4\text{-}CH_3C_6H_4)(P(C_6H_5)_3)_2(SNN(CH_3)_2)$ (Table **15**, No. **3**). The complex reacted with t-C_4H_9CN in the presence of NH_4^+ PF_6^- in CH_2Cl_2–H_2O–C_2H_5OH to form the salt $[Ru(CO)(4\text{-}CH_3C_6H_4)(t\text{-}C_4H_9CN)(P(C_6H_5)_3)_2(SNN(CH_3)_2)]^+$ PF_6^- (Table 15, No. 5). Treatment of the complex with NaN_3 provides the azido complex $[Ru(N_3)(CO)(4\text{-}CH_3C_6H_4)(P(C_6H_5)_3)_2(SNN(CH_3)_2)]$ (Table 15, No. 4) in high yield [1, 2]. The reaction is catalyzed by t-C_4H_9CN and presumably proceeds via the cation $[Ru(CO)(4\text{-}CH_3C_6H_4)(t\text{-}C_4H_9CN)(P(C_6H_5)_3)_2(SNN(CH_3)_2)]^+$ [2].

Table 16
N-Thionitrosodimethylamine Complexes of Os^{II}.
An asterisk preceding the compound number indicates further information at the end of the table, p. 70.

No.	compound	preparation method (yield in %) properties (^1H NMR ($CDCl_3$/TMS): δ (in ppm), IR (Nujol): ν (in cm^{-1}))
neutral mononuclear Os^{II} complexes		
1	$OsHCl(CO)(P(C_6H_5)_3)_2(SNN(CH_3)_2)$	analogous to No. 1 in Table 15 from $OsHCl(CO)(P(C_6H_5)_3)_3$ and $S{=}NN(CH_3)_2$ (96) bright orange crystals m.p. 196 °C with decomposition ^1H NMR: −9.43 (t, OsH, $^2J(P,H)$ = 19.0 Hz); 3.31, 2.91 ($N(CH_3)_2$) IR: 2041 ν(OsH); 1919, 1905 ν(CO), (CH_2Cl_2: 1905 ν(CO)); 1352 m, 1261 w, 1120 m, 837 w, 790 w; 1084, 1020 ($P(C_6H_5)_3$), 932 w, 898 w [1]
*2	$OsCl(CO)(4\text{-}CH_3C_6H_4)(P(C_6H_5)_3)_2(SNN(CH_3)_2)$	1) from $OsClCO(4\text{-}CH_3C_6H_4)(P(C_6H_5)_3)_2$ and $S{=}NN(CH_3)_2$ (mole ratio ca. 1 : 2) in toluene for 20 min, addition of hexane (91) [1] 2) from $OsCl(CO)(4\text{-}CH_3C_6H_4)(P(C_6H_5)_3)_2L$ (L = SO_2, 2,1,3-benzothiadiazole, 2,1,3-benzoselenadiazole) and $S{=}NN(CH_3)_2$ [2] scarlet crystals m.p. 183 °C ^1H NMR: 2.18 ($4\text{-}\underline{CH_3}C_6H_4$), 2.61, 3.40 ($N(CH_3)_2$) IR: 1921, 1908 ν(CO), (CH_2Cl_2: 1910 ν(CO)) [1]
3	$OsCl(CS)(4\text{-}CH_3C_6H_4)(P(C_6H_5)_3)_2(SNN(CH_3)_2)$	analogous to No. 2 from OsCl(CS)-($4\text{-}CH_3C_6H_4)(P(C_6H_5)_3)_2$ and $S{=}NN(CH_3)_2$ (91) m.p. 161 °C ^1H NMR: 2.19 ($4\text{-}\underline{CH_3}C_6H_4$), 2.78, 3.49 ($N(CH_3)_2$) IR: 1279, 1268 ν(CS) [1]
cationic mononuclear Os^{II} complexes		
4	*cis-trans*-$[OsH(CO)(P(C_6H_5)_3)_2(SNN(CH_3)_2)_2]^+ BF_4^-$	from $[OsH(H_2O)(CO)(P(C_6H_5)_3)_3]^+ BF_4^-$ and $S{=}NN(CH_3)_2$ in THF−ether (mole ratio ca. 1 : 2) stirring for 2 h, adding toluene (83) yellow solid ^1H NMR: −9.63 (t, OsH, $^2J(P,H)$ = 18.0 Hz), 2.81, 3.22, 3.36, 3.41 ($N(CH_3)_2$, equal intensity) IR: 2039 ν(OsH), 1924 ν(CO) [1]
5	*cis-trans*-$[OsH(CO)(P(C_6H_5)_3)_2(SNN(CH_3)_2)_2]^+ PF_6^-$	by recrystallization of No. 4 twice from CH_2Cl_2 in the presence of a fourfold excess of NH_4PF_6 m.p. 184 °C (with decomposition) [1]

*Further information:

$OsCl(CO)(4\text{-}CH_3C_6H_4)(P(C_6H_5)_3)_2(SNN(CH_3)_2)$ (Table **16**, No. **2**). The molecular structure of the complex was determined by X-ray diffraction. The molecule crystallizes in the monoclinic space group $P2_1/n-C^5_{2h}$ (No. 14) with a = 12.235(4), b = 24.305(4), c = 14.066(2) Å, β = 103.22(2)°; V = 4070.0 Å³; Z = 4; D_m = 1.54, D_x = 1.567 g/cm³; R = 0.034 for 5508 observed reflections. The molecular structure is shown in **Fig. 5**. The osmium atom is octahedrally coordinated by its six ligands. The two phosphine ligands, the chloride and carbonyl groups, and the p-tolyl and thionitrosodimethylamine ligands occupy mutually trans coordination sites in

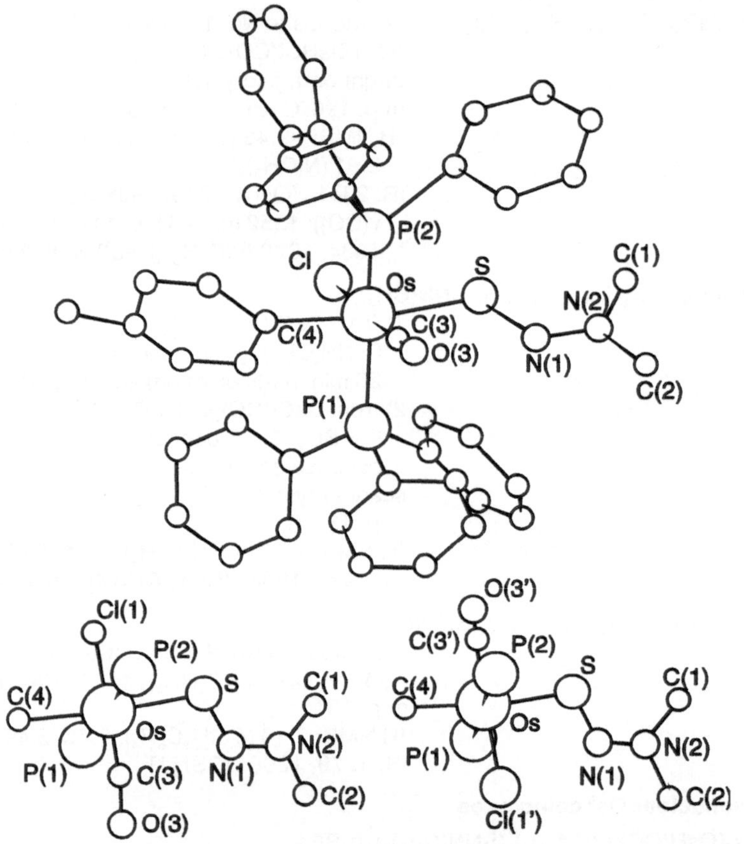

Fig. 5. Molecular structure of $OsCl(CO)(4\text{-}CH_3C_6H_4)(SNN(CH_3)_2)$-$(P(C_6H_5)_3)_2$. Two conformations (α and β) are present in the crystal [2].

the octahedron. The N-thionitroso dimethylamine ligand is planar. Its orientation towards the rest of the molecule is trans(E) with respect to the N–S bond and is characterized by an eclipsed position of this bond and the carbonyl ligand. In addition to the major molecular conformation (α, ca. 60%), a second minor conformation (β, ca. 40%) is present in the crystal in which the chloride and thionitrosoamine ligands are eclipsed. Bond lengths and bond angles are listed in Table 17 [2].

Table 17
Selected Bond Lengths and Bond Angles of α-$OsCl(CO)(4\text{-}CH_3C_6H_4)(SNN(CH_3)_2)(P(C_6H_5)_3)_2$ [2].

bond length	in Å	bond angle	in °
S–N(1)	1.616(7)	S–N(1)–N(2)	119.6(6)
N(1)–N(2)	1.293(10)	N(1)–N(2)–C(1)	123.5(7)
N(2)–C(1)	1.446(12)	C(1)–N(2)–C(2)	119.5(7)
N(2)–C(2)	1.465(13)	N(1)–N(2)–C(2)	116.9(7)
Os–S	2.411(2)	Os–S–N(1)	107.2(3)
Os–C(3)	1.798(15)	P(1)–Os–S	90.9(1)
Os–C(4)	2.117(7)	P(2)–Os–S	89.8(1)
Os–Cl	2.476(3)	S–Os–C(3)	94.5(5)
Os–P(1)	2.392(1)	S–Os–C(4)	174.7(2)
Os–P(2)	2.430(1)	C(3)–Os–C(4)	90.3(6)

$OsCl(CO)(4\text{-}CH_3C_6H_4)(P(C_6H_5)_3)_2(SNN(CH_3)_2)$ reacted with CO or 2-xylyl isocyanide to form the complex $[OsCl(CO)(4\text{-}CH_3C_6H_4)(P(C_6H_5)_3)_2L]$ (L = CO and 2-xylyl isocyanide) [2].

References:

[1] Herberhold, M.; Hill, A. F. (J. Organomet. Chem. **315** [1986] 105/12).

[2] Gieren, A.; Ruiz-Perez, C.; Huebner, T.; Herberhold, M.; Hill, A. F. (J. Chem. Soc. Dalton Trans. **1988** 1693/6).

[3] Tresoldi, G.; Sergi, S.; Lo Schiavo, S.; Piraino, P. (J. Organomet. Chem. **328** [1987] 387/91).

6.2.3 N-Thionitroso-dimethylamine Complexes of Rh^I and Rh^{III}

The method of preparation and the properties of the title complexes are compiled in Table 18.

Table 18
N-Thionitrosamine Complexes of Rh^I and Rh^{III} ($1,5\text{-}C_8H_{12}$ = 1,5-cyclooctadiene, C_7H_8 = norbornadiene).
An asterisk preceding the compound number denotes further information at the end of the table, p. 75.

No.	compound	preparation method (yield in %) properties (^1H NMR: δ (in ppm), IR (Nujol): ν (in cm^{-1}))
neutral mononuclear Rh^I complexes		
*1	*cis*-$Rh(CO)_2Cl(SNN(CH_3)_2)$	1) from $[Rh(CO)_2Cl]_2$ and $S{=}NN(CH_3)_2$ in petroleum ether at −10 °C (ca. 60) 2) from RhCl(diene)($SNN(CH_3)_2$) and CO in CH_2Cl_2 (ca. 80) (diene = $1,5\text{-}C_8H_{12}$, C_7H_8) yellow solid ^1H NMR: 3.70 (q, NCH_3), 4.0 (q, NCH_3); $^4J(H,H)$ = 0.55 Hz IR: 2075 s, 2000 s, ν(CO); 370 w ν(Rh–S), 308 m ν(Rh–Cl), 1120 ν(N–N), 780 ν(N–S) [1]

Table 18 (continued)

No.	compound	preparation method (yield in %) properties (^{1}H NMR: δ (in ppm), IR (Nujol): ν (in cm^{-1}))
*2	$RhCl(1,5\text{-}C_8H_{12})(SNN(CH_3)_2)$	from $[RhCl(1,5\text{-}C_8H_{12})]_2$ and $S{=}NN(CH_3)_2$ in CH_2Cl_2–ether (ca. 90) red needles soluble in CH_2Cl_2, $CHCl_3$ and moderately soluble in acetone ^{1}H NMR: 3.64 (q, NCH_3), 3.80 (q, NCH_3) IR: 345 w ν(Rh–S), 325 m ν(Rh–Cl) [1]
3	$Rh(ClO_4)(1,5\text{-}C_8H_{12})(SNN(CH_3)_2)$	1) from $[RhCl(1,5\text{-}C_8H_{12})]_2$, $AgClO_4$, and $S{=}NN(CH_3)_2$ in acetone–ether (ca. 60) 2) from No. 2 and $AgClO_4$ dark orange solid the complex looses the diene (slowly in the solid state, rapidly in solution) to give an unidentified solid [2]
*4	$RhCl(C_7H_8)(SNN(CH_3)_2)$	analogous to No. 2, from $[RhCl(C_7H_8)]_2$ and $S{=}NN(CH_3)_2$ (ca. 60) ^{1}H NMR: 3.52 (q, NCH_3), 3.84 (q, NCH_3); $^4J(H,H) = 0.50$ Hz IR: 340 m ν(Rh–S), 320 w ν(Rh–Cl) [1]
*5	*trans*-$Rh(CO)Cl(SNN(CH_3)_2)_2$	1) from No. 1 and $S{=}NN(CH_3)_2$ in ether (ca. 50) 2) from *trans*-$Rh(CO)Cl(As(C_6H_5)_3)_2$ (or *trans*-$Rh(CO)Cl(Sb(C_6H_5)_3)_2$) and $S{=}NN(CH_3)_2$ in ether (mole ratio 1 : 8) (ca. 80) red solid stable in the solid state for some weeks and for about 2 d in solution, moderately soluble in acetone, $CHCl_3$, and CH_2Cl_2 ^{1}H NMR: 3.65 (br, NCH_3), 3.93 (br, NCH_3); the rotation about the N–N bond is hindered IR: 1975 s ν(CO), 350 w ν(Rh–S), 300 m ν(Rh–Cl); the Cl atom is trans to the Rh–CO bond [1]
6	*trans*-$Rh(CO)Br(SNN(CH_3)_2)_2$	analogous to No. 5 from *trans*-$Rh(CO)Br(As(C_6H_5)_3)_2$ and $S{=}NN(CH_3)_2$ in ether (mole ratio 1 : 8) red solid stable in the solid state for some weeks, decomposes rapidly in acetone or chlorinated solvents IR: 1978 s ν(CO), 350 w ν(Rh–S) [1]
7	*trans*-$Rh(CO)(NCS)(SNN(CH_3)_2)_2$	analogous to No. 5 from *trans*-$Rh(CO)(NCS)(As(C_6H_5)_3)_2$ and $S{=}NN(CH_3)_2$ in ether (mole ratio 1 : 8)

Table 18 (continued)

No.	compound	preparation method (yield in %) properties (^1H NMR: δ (in ppm), IR (Nujol): ν (in cm^{-1}))
		red solid stable for some weeks in the solid state, stable in acetone solution for 1 h ^1H NMR: 3.74 (br, NCH_3), 4.04 (br, NCH_3) IR: 2080 s ν(CN), 1985 s ν(CO), 810 m ν(CS) or δ(RhCO), 460 w δ(SCN), 354 br ν(RhS) [1]
cationic mononuclear Rh^I complexes		
8	$[Rh(1,5\text{-}C_8H_{12})(P(C_6H_5)_3)(SNN(CH_3)_2)]^+ ClO_4^-$	from No. 10 ($X = ClO_4$), $P(C_6H_5)_3$, and $S{=}NN(CH_3)_2$ (mole ratio 1 : 1 : 4) in acetone–ether (ca. 60) orange solid ^1H NMR (acetone–d_6): 3.55 (q, NCH_3), 3.90 (q, (NCH_3), $^4J(H,H) = 0.66$ [2]
9	$[Rh(1,5\text{-}C_8H_{12})(4\text{-}CH_3C_6H_4P(C_6H_5)_2)(SNN(CH_3)_2)]^+ ClO_4^-$	analogous to No. 8 from No. 10 ($X = ClO_4$), $4\text{-}CH_3C_6H_4P(C_6H_5)_2$, and $S{=}NN(CH_3)_2$ (ca. 60) ^1H NMR (acetone–d_6): 2.40 (s, $4\text{-}C\underline{H}_3C_6H_4$), 3.54 (q, NCH_3), 3.92 (q, NCH_3), $^4J(H,H) = 0.68$ [2]
*10	$[Rh(1,5\text{-}C_8H_{12})(SNN(CH_3)_2)_2]^+ X^-$ $X = ClO_4$, PF_6	from $[RhCl(1,5\text{-}C_8H_{12})]_2$, AgX ($X = ClO_4$, PF_6), and $S{=}NN(CH_3)_2$ (molar ratio 1 : 2 : 8) in acetone–ether (ca. 80) orange solids ^1H NMR (acetone–d_6) ($X = ClO_4$): 3.87 (q, NCH_3), 4.15 (q, NCH_3), $^4J(H,H) = 0.66$ Hz [2]
*11	$[Rh(C_7H_8)(SNN(CH_3)_2)_2]^+ X^-$ $X = ClO_4$, PF_6	analogous to No. 10 from $[RhCl(C_7H_8)]_2$, AgX, and $S{=}NN(CH_3)_2$ (molar ratio 1 : 2 : 8) in acetone–ether (ca. 50) orange solids, stable for several weeks; soluble in acetone, acetonitrile, and DMSO ^1H NMR (acetone–d_6) ($X = ClO_4$): 3.80 (q, NCH_3), 4.16 (q, NCH_3), $^4J(H,H) = 0.66$ Hz [2]
12	*cis*-$[Rh(CO)_2(SNN(CH_3)_2)_2]^+ X^-$ $X = ClO_4$, PF_6	1) analogous to No. 10 from $[Rh(CO)_2Cl]_2$, AgX, and $S{=}NN(CH_3)_2$ (molar ratio 1 : 2 : 8) in acetone–ether (50) 2) from No. 10 or No. 11 and CO in acetone (ca. 50) orange solid stable only for a few days, decomposes rapidly in solution; it is soluble in acetone, acetonitrile, and DMSO uni-univalent electrolyte ($\Lambda_M = 160$ to 170 $\Omega^{-1} \cdot cm^2 \cdot mol^{-1}$) [2]

Table 18 (continued)

No.	compound	preparation method (yield in %) properties (^1H NMR: δ (in ppm), IR (Nujol): ν (in cm^{-1}))
neutral binuclear Rh^I complex		
13	*trans*-$(\mu\text{-}Cl)_2[Rh(CO)(SNN(CH_3)_2)]_2$	1) from No. 1 in diethyl ether at room temperature for about 1 h (ca. 60) 2) from No. 5 in solution for 2 d insoluble solid, the formula is based on analytical and IR data IR: 2050 s ν(CO), 280 m ν(Rh–Cl) [1]
neutral mononuclear Rh^{III} complexes		
14	$Rh(CO)Cl_3(SNN(CH_3)_2)_2$	from No. 5 and Cl_2 in acetone the complex decomposes as a solid and in solution too rapidly for satisfactory analysis [1]
15	$Rh(CO)ClBr_2(SNN(CH_3)_2)_2$	from No. 5 and Br_2 in acetone–ether (ca. 70) orange solid stable in the solid state, decomposition in solution IR: 2085 s ν(CO), 375 w ν(Rh–S), 310 m ν(Rh–Cl) [1]
16	$Rh(CO)ClI_2(SNN(CH_3)_2)_2$	from No. 5 and I_2 in acetone–ether (ca. 80) orange solid stable in acetone for 1 h ^1H NMR: 3.93 (q, NCH_3), 4.25 (q, NCH_3), $^4J(H,H) = 0.65$ Hz IR: 2075 s ν(CO), 375 w ν(Rh–S), 305 m ν(Rh–Cl) [1]
17	$Rh(CO)Cl(HgCl_2)(SNN(CH_3)_2)_2$	from No. 5 (in acetone) and a suspension of $HgCl_2$ in diethyl ether (mole ratio 1 : 1) (ca. 95) orange solid insoluble in common organic solvents IR: 2040 s ν(CO), 300 w ν(Rh–Cl), 240 br γ(Hg–Cl) [1]
cationic mononuclear Rh^{III} complexes		
18	$[(C_5H_5)Rh(ClO_4)(SNN(CH_3)_2)_2]^+ ClO_4^-$	from $(C_5H_5)Rh(CO)I_2$, $AgClO_4$, and $S{=}NN(CH_3)_2$ in acetone–ether; slow addition of $S{=}NN(CH_3)_2$ is necessary (in contrast to No. 19) (ca. 60) orange solid exists in solution as $[(C_5H_5)Rh(solvent)(SNN(CH_3)_2)_2]^{2+} (ClO_4^-)_2$ ^1H NMR (acetone-d_6): 3.90 (q, NCH_3), 4.18 (q, NCH_3), $^4J(H,H) = 0.65$; 5.90 (d, C_5H_5), $J(Rh,H) = 0.50$ IR: 1135 s, 1025 m ($S{=}NN(CH_3)_2$)

Table 18 (continued)

No.	compound	preparation method (yield in %) properties (^1H NMR: δ (in ppm), IR (Nujol): ν (in cm^{-1}))
		soluble in acetone, insoluble in apolar solvents bi-univalent electrolyte ($\Lambda_M = 200\ \Omega^{-1}\cdot cm^2\cdot mol^{-1}$) [2]
19	$[(C_5(CH_3)_5)Rh(SNN(CH_3)_2)_3]^{2+}(X^-)_2$ X = ClO_4, PF_6	from $[(C_5(CH_3)_5)RhCl_2]_2$, AgX, and $S{=}NN(CH_3)_2$ (molar ratio 1 : 4 : 8) in acetone–ether at room temperature (ca. 80) orange solid stable for several weeks ^1H NMR (acetone–d_6) (X = ClO_4): 1.70 (s, $C_5(CH_3)_5$); 3.85 (q, NCH_3), 4.08 (q, NCH_3), $^4J(H,H) = 0.68$ Hz bi-univalent electrolyte (Λ_M is in the range 210 to 220 $\Omega^{-1}\cdot cm^2\cdot mol^{-1}$) [2]
20	$[(C_5H_5)Rh(SNN(CH_3)_2)_3]^{2+}(PF_6^-)_2$	from $(C_5H_5)Rh(CO)I_2$, $AgPF_6$, and excess $S{=}NN(CH_3)_2$ in acetone–ether (ca. 60) orange solid stable for several weeks ^1H NMR (acetone–d_6): 4.00 (q, NCH_3), 4.20 (q, NCH_3), $^4J(H,H) = 0.70$ Hz; 6.15 (d, C_5H_5), J(Rh,H) = 0.48 Hz bi-univalent electrolyte (Λ_M is in the range 210 to 220 $\Omega^{-1}\cdot cm^2\cdot mol^{-1}$) [2]
neutral binuclear Rh^{III} complex		
21	$(\mu\text{-}Cl)_2[Rh(CO)I_2(SNN(CH_3)_2)]_2$	from No. 1 and I_2 in petroleum ether–CH_2Cl_2 (ca. 70) dark brown solid ^1H NMR: 3.80 (br, NCH_3), 4.08 (br, NCH_3) IR: 2060 s ν(CO), 285 ν(Rh–Cl) [1]

*Further information:

***cis*-$Rh(CO)_2Cl(SNN(CH_3)_2)$** (Table **18**, No. **1**). Keeping a diethyl ether solution of the title complex at room temperature yielded *trans*-$(\mu\text{-}Cl)_2[Rh(CO)(SNN(CH_3)_2)]_2$ (Table 18, No. 13) after about 1 h [1].

A solution of the title compound in CH_2Cl_2 reacted with a solution of I_2 in light petroleum ether to yield $(\mu\text{-}Cl)_2[Rh(CO)I_2(SNN(CH_3)_2)]_2$ (Table 18, No. 21) [1].

The reaction of the complex with $S{=}NN(CH_3)_2$ in diethyl ether gave *trans*-Rh(CO)Cl-$(SNN(CH_3)_2)_2$ (Table 18, No. 5) [1].

The complex reacted with $E(C_6H_5)_3$ (E = P, As, Sb) in CH_2Cl_2 yielding *trans*-$Rh(E(C_6H_5)_3)_2$-(CO)Cl [1].

The compound was stirred in CH_3I for about 1 h to give a solid with the composition $Rh(CO)Cl(CH_3I)(SNN(CH_3)_2)$ which is probably a mixture of the acetyl derivative $(\mu\text{-}Cl)_2$-$[Rh(CH_3CO)I(SNN(CH_3)_2)]_2$ and the adduct $(\mu\text{-}Cl)_2[Rh(CO)(CH_3)I(SNN(CH_3)_2)]_2$ [1].

$RhCl(1,5\text{-}C_8H_{12})(SNN(CH_3)_2)$ (Table **18**, No. **2**). When CO was bubbled through a CH_2Cl_2 solution of the title complex, *cis*-$Rh(CO)_2Cl(SNN(CH_3)_2)$ (No. 1) was formed. The reaction with $S{=}NN(CH_3)_2$ gave a brown insoluble solid which could not be isolated in a pure form [1]. No. 2 reacted with $AgClO_4$ to form $Rh(ClO_4)(1,5\text{-}C_8H_{12})(SNN(CH_3)_2)$ (No. 3) [2].

$RhCl(C_7H_8)(SNN(CH_3)_2)$ (Table **18**, No. **4**). Bubbling CO through a CH_2Cl_2 solution of No. 4 gave *cis*-$Rh(CO)_2Cl(SNN(CH_3)_2)$ (No. 1). The reaction with $S{=}NN(CH_3)_2$ gave a brown, insoluble solid which could not be isolated in a pure form [1].

***trans*-$Rh(CO)Cl(SNN(CH_3)_2)_2$** (Table **18**, No. **5**). Compound No. 5 dissolved in acetone reacted with a solution of X_2 (X = Cl, Br, I) in diethyl ether to yield $Rh(CO)ClX_2(SNN(CH_3)_2)_2$ (Table 18, Nos. 14 to 16). The chloro derivative in solid form and in solution decomposed too rapidly for satisfactory analyses [1].

Bubbling HCl through a cold acetone solution of the title complex gave a yellow solid. This material was only characterized by its IR spectrum because of its very low stability (it decomposed within a few minutes, IR: ν (in cm^{-1}) = 2310 ν(Rh–H), 2100 ν(CO), 320 ν(Rh–Cl)) [1].

Treating the complex in acetone with a suspension of $HgCl_2$ in diethyl ether gave the orange adduct $Rh(CO)Cl(HgCl_2)(SNN(CH_3)_2)_2$ (Table 18, No. 17) [1].

When a solution of the title complex in CH_2Cl_2 was treated with CH_3I, an inseparable mixture of the acetyl derivative $Rh(SNN(CH_3)_2)_2(CH_3CO)ICl$ and the adduct $Rh(SNN(CH_3)_2)_2(CO)$-$(CH_3)ICl$ was obtained [1].

After bubbling CO through a suspension of *trans*-$Rh(CO)Cl(SNN(CH_3)_2)_2$ in diethyl ether, the initially formed yellow solution of *cis*-$Rh(CO)_2Cl(SNN(CH_3)_2)$ gave the title complex after some minutes as a red solid [1].

$[Rh(1,5\text{-}C_8H_{12})(SNN(CH_3)_2)_2]^+$ X^- (Table **18**, No. **10**; X = ClO_4, PF_6). Compound No. 10 (X = ClO_4) reacted with $P(C_6H_5)_3$ in the presence of $S{=}NN(CH_3)_2$ (mole ratio 1 : 1 : 4) in acetone–diethyl ether to form $[Rh(1,5\text{-}C_8H_{12})(P(C_6H_5)_3)(SNN(CH_3)_2)]^+$ ClO_4^- (No. 8). An analogous complex was formed with $4\text{-}CH_3C_6H_4P(C_6H_5)_2$ (No. 9). Bubbling CO through an acetone solution of compound No. 10 gave $[Rh(CO)_2(SNN(CH_3)_2)_2]^+$ X^- (No. 12, X = ClO_4, PF_6) [2].

$[Rh(C_7H_8)(SNN(CH_3)_2)_2]^+$ X^- (Table **18**, No. **11**). Bubbling CO through an acetone solution of compound No. 11 gave *cis*-$[Rh(CO)_2(SNN(CH_3)_2)_2]^+$ ClO_4^- (No. 12, X = ClO_4, PF_6) [2].

References:

[1] Tresoldi, G.; Sergi, S.; Lo Schiavo, S.; Piraino, P. (J. Organomet. Chem. **322** [1987] 369/76).

[2] Tresoldi, G.; Sergi, S.; Lo Schiavo, S.; Piraino, P. (J. Organomet. Chem. **328** [1987] 387/91).

6.2.4 N-Thionitroso-dimethylamine Complexes of Ir^{III}

$IrCl_2H(P(C_6H_5)_3)_2(SNN(CH_3)_2)$. The complex was prepared with 96% yield by treating $IrCl_2H(P(C_6H_5)_3)_2$ in toluene with an ethereal solution of $S{=}NN(CH_3)_2$ (mole ratio ca. 1 : 2). The mixture was stirred for 4 h, then hexane was added to complete the separation of the product. The mixture was stirred for another 30 min, and the crystals were filtered off, washed with hexane, and recrystallized from CH_2Cl_2–C_2H_5OH.

The bright orange crystals melt at 157 °C.

IR (Nujol): ν (in cm^{-1}) = 2104 ν(Ir–H); 1352 m, 1261 w, 1120 m; 1084, 1020 ($P(C_6H_5)_3$); 932 w, 898 w, 837 w, 790 w. The trans-dichloride geometry is supported by the far-infrared spectrum ($\nu_{as}(IrCl_2)$ = 318 cm^{-1}).

^1H NMR ($CDCl_3$/TMS): δ (in ppm) = −16.93 (t, IrH, $^2J(P,H)$ = 13.6 Hz), 3.06 (NCH_3), 3.34 (NCH_3).

$[IrClH(P(C_6H_5)_3)_2(SNN(CH_3)_2)_2]^+$ BF_4^-. The complex was obtained with 52% yield by reacting $[IrClH(NCCH_3)_2(P(C_6H_5)_3)_2]^+$ with excess $S{=}NN(CH_3)_2$: A suspension of $IrCl_2H(P(C_6H_5)_3)_3$ in CH_3CN was treated with a solution of $AgBF_4$ in CH_3CN (mole ratio ca. 1 : 1), and the suspension was refluxed for 15 min. The filtrate was evaporated under reduced pressure. The residue was suspended in THF and treated with a solution of a slight excess of $S{=}NN(CH_3)_2$ in ether for 15 h. The crude product was purified by chromatography on silica gel with CH_2Cl_2 as eluent.

The yellow crystals melt at 155 °C with decomposition.

^1H NMR ($CDCl_3$/TMS): δ (in ppm) = −15.75 (t, IrH, $^2J(Ir,H)$ = 12.7 Hz); 2.53, 3.34, 3.53, 3.73 (each NCH_3, equal intensity).

Reference:

Herberhold, M.; Hill, A. F. (J. Organomet. Chem. **315** [1986] 105/12).

6.2.5 N-Thionitroso-dimethylamine Complexes of Pd^{II} and Pt^{II}

The method of preparation and the properties of the corresponding complexes are compiled in Tables 19 and 21, p. 84, respectively.

Table 19
N-Thionitroso-dimethylamine Complexes of Pd^{II} (C_7H_8 = norbornadiene, 1,5-C_8H_{12} = 1,5-cyclooctadiene, 1,3,5,7-C_8H_8 = 1,3,5,7-cyclooctatetraene, $C_{10}H_{12}$ = dicyclopentadiene).
An asterisk preceding the compound number indicates further information at the end of the table, p. 82.

No.	compound	preparation method (yield in %) properties (^1H NMR ($CDCl_3$/TMS): δ (in ppm), IR (Nujol): ν (in cm^{-1}))
neutral mononuclear Pd^{II} complexes		
*1	*cis*-$PdCl_2(P(C_6H_5)_3)(SNN(CH_3)_2)$	1) from $Pd(1,5\text{-}C_8H_{12})Cl_2$, $P(C_6H_5)_3$, and $S{=}NN(CH_3)_2$ (mole ratio 1 : 3 : 1) in acetone or CH_2Cl_2 (ca. 95) 2) from *cis*-$PdCl_2(SNN(CH_3)_2)_2$ (No. 10) and $P(C_6H_5)_3$ (mole ratio 1 : 1) in acetone, stirring overnight (ca. 70) IR: for $S{=}NN(CH_3)_2$ moiety see *cis*-$PdCl_2(SNN(CH_3)_2)_2Cl_2$ (No. 10); 325 s, 280 s, ν(Pd–Cl) yellow solid stable for some weeks in the solid state, almost insoluble in common organic solvents [1]

Table 19 (continued)

No.	compound	preparation method (yield in %) properties (^1H NMR ($CDCl_3$/TMS): δ (in ppm), IR (Nujol): ν (in cm^{-1}))
2	*cis*-$PdCl_2(4\text{-}CH_3C_6H_4P(C_6H_5)_2)(SNN(CH_3)_2)$	analogous to No. 1 yellow solid stable for some weeks in the solid state, slightly soluble in chlorinated solvents ^1H NMR: 2.35 ($C_6H_4C\underline{H}_3$-4), 3.28 (NCH_3), 3.40 (NCH_3) IR: 318 s, 282 s, ν(Pd–Cl) [1]
3	*cis*-$PdCl_2(P(C_6H_4CH_3\text{-}2)_3)(SNN(CH_3)_2)$	from *cis*-$PdCl_2(SNN(CH_3)_2)_2$ (No. 10) and $P(C_6H_4CH_3\text{-}2)_3$ (mole ratio 1 : 1) in acetone, stirring overnight (ca. 70) yellow solid stable for some weeks in the solid state, almost insoluble in common organic solvents IR: 318 s, 285 s, ν(Pd–Cl) [1]
*4	*cis*-$PdCl_2(As(C_6H_5)_3)(SNN(CH_3)_2)$	from *cis*-$PdCl_2(SNN(CH_3)_2)_2$ (No. 10) and $As(C_6H_5)_3$ (mole ratio 1 : 1) in acetone, stirring overnight (ca. 70) orange solid stable for some weeks in the solid state, almost insoluble in common organic solvents IR: 323 s, 282 s, ν(Pd–Cl) [1]
5	*cis*-$PdCl_2(Sb(C_6H_5)_3)(SNN(CH_3)_2)$	from *cis*-$PdCl_2(SNN(CH_3)_2)_2$ (No. 10) and $Sb(C_6H_5)_3$ (mole ratio 1 : 1) in acetone, stirring overnight (ca. 70) orange solid stable for some weeks in the solid state, almost insoluble in common organic solvents IR: 318 s, 270 s, ν(Pd–Cl) [1]
6	*cis*-$Pd(SCN)_2(P(C_6H_5)_3)(SNN(CH_3)_2)$	from No. 1 and KSCN (mole ratio 1 : 2) in acetone, stirring for 6 h (70) orange crystals (from $CHCl_3$–ether) stable for some weeks, slightly soluble in acetone and chlorinated solvents ^1H NMR: 3.75 (NCH_3), 4.00 (NCH_3) IR: 2105 s ν(CN), 430 m δ(SCN) [1]
7	*cis*-$PdCl(CH_3OC_8H_8)(SNN(CH_3)_2)$	from $[Pd(CH_3OC_8H_8)Cl]_2$ and $S{=}NN(CH_3)_2$ (mole ratio 1 : 2.2) in CH_2Cl_2 stirring for a few minutes (80) orange solid stable for some days at room temperature, and for some weeks at −15 °C

Table 19 (continued)

No.	compound	preparation method (yield in %) properties (^1H NMR ($CDCl_3$/TMS): δ (in ppm), IR (Nujol): ν (in cm^{-1}))
		very soluble in acetone or chlorinated solvents, insoluble in ether or saturated hydrocarbons nonconducting in acetone ^1H NMR: 3.36 (OCH_3), 3.71 (NCH_3), 4.02 (NCH_3) IR: 1095 s ν(CO), 270 m ν(Pd–Cl) [1]
8	*cis*-$PdCl(CH_3OC_8H_{12})(SNN(CH_3)_2)$	synthesis and general properties like those of No. 7 ^1H NMR: 3.22 (OCH_3), 3.77 (NCH_3), 4.02 (NCH_3) IR: 270 m, ν(Pd–Cl); 1082 s, 1062 s, ν(CO) the splitting may arise from a packing effect [1]
9	*cis*-$PdCl(CH_3OC_{10}H_{12})(SNN(CH_3)_2)$	synthesis and general properties like those of No. 7 ^1H NMR: 3.14 (OCH_3), 3.23 (NCH_3), 4.01 (NCH_3) IR: 1095 s ν(CO), 255 m ν(Pd–Cl) [1]
*10	*cis*-$PdCl_2(SNN(CH_3)_2)_2$	from $Pd(C_{10}H_{12})Cl_2$ and $S{=}NN(CH_3)_2$ (mole ratio 1 : 2.5) in acetone or CH_2Cl_2 (80); reactions with $Pd(C_7H_8)Cl_2$, $Pd(1,3,5,7\text{-}C_8H_8)Cl_2$, or $Pd(1,5\text{-}C_8H_{12})Cl_2$ proceeded similarly, but longer reaction times were required [1] the complex is also formed by decomposition of No. 14 in acetone [2] orange solid stable for some weeks in the solid state, decomposes rapidly in DMSO, very low solubility in all common organic solvents IR: bands associated with the $SNN(CH_3)_2$ ligand: 1492 s, 1375 s, 1250 w, 1132 vs ν(N–N), 1028 w, 860 m, 772 vs ν(SN), 720 w, 535 m, 415 w; 315 s and 295 s ν(Pt–Cl) [1]
*11	*cis*-$PdBr_2(SNN(CH_3)_2)_2$	from $Pd(1,5\text{-}C_8H_{12})Br_2$ and $S{=}NN(CH_3)_2$ (mole ratio 1 : 2.5) in acetone or CH_2Cl_2 (70) IR: 1132 vs ν(N–N), 782 vs ν(N–S), 390 w, 300 w, ν(Pd–$SNN(CH_3)_2$) [2]
*12	*cis*-$Pd(SCN)_2(SNN(CH_3)_2)_2$	from No. 10 and KSCN (mole ratio 1 : 2.1) in acetone, stirring for 6 h (ca. 70) orange solid stable for some weeks, sligthly soluble in acetone or chlorinated solvents ^1H NMR (acetone-d_6/TMS): 3.87, 4.12 (NCH_3) IR: 2100 s, 2090 s ν(CN), 430 m, δ(SCN) [1]

Table 19 (continued)

No.	compound	preparation method (yield in %) properties (^1H NMR ($CDCl_3$/TMS): δ (in ppm), IR (Nujol): ν (in cm^{-1}))
*13	*cis*-$Pd(SeCN)_2(SNN(CH_3)_2)_2$	from No. 10 and KSeCN (mole ratio 1: 2.1) in acetone, stirring for 6 h (ca. 70) IR: 2095 s ν(CN), 510 w ν(C–Se), 375 m δ(SeCN) [1]
cationic mononulear Pd^{II} complexes		
*14	$[Pd(SNN(CH_3)_2)_4]^{2+}$ $(Cl^-)_2$	from $Pd(C_6H_5CN)_2Cl_2$ and excess $S{=}NN(CH_3)_2$ (mole ratio 1 : 6) in ether–benzene at room temperature unstable compound, decomposes in acetone or CH_2Cl_2 slowly to give No. 10 [2]
15	$[Pd(SNN(CH_3)_2)_4]^{2+}$ $(B(C_6H_5)_4^-)_2$	from No. 14 (prepared in situ) and $NaB(C_6H_5)_4$ in ethanol (60) yellow solid stable for some days, slightly soluble in acetone IR: 1125 vs ν(N–N), 776 vs ν(N–S) bi-univalent electrolyte $\Lambda_M = 200\ \Omega^{-1} \cdot cm^2 \cdot mol^{-1}$ (5×10^{-4} M in acetone) [2]
16	$[Pd(SNN(CH_3)_2)_4]^{2+}$ $(PF_6^-)_2$	from $[Pd(diene)((CH_3)_2CO)_2]^{2+}$ $(PF_6^-)_2$ (from $Pd(diene)Cl_2$ and $AgPF_6$ in acetone) and $S{=}NN(CH_3)_2$ (mole ratio 1 : 5) in ether (ca. 60) (diene = 1,5-C_8H_{12}, C_7H_8) yellow solid stable for some days IR: 1125 vs ν(N–N), 776 vs ν(N–S) ^1H NMR: (acetone/TMS): 4.0 (s, NCH_3), 5.2 (s, NCH_3) [2]
*17	$[Pd(SNN(CH_3)_2)_4]^{2+}$ $PdCl_4^{2-}$	from Na_2PdCl_4 and $S{=}NN(CH_3)_2$ (mole ratio 1 : 3) in acetone–H_2O (ca. 60) yellow solid stable for some days, slightly soluble in acetone in which it decomposes to form No. 10 IR: 1125 vs ν(N–N), 776 vs ν(N–S), 320 s ν(Pd–Cl) [2]
18	$[Pd(SNN(CH_3)_2)_4]^{2+}$ $HgCl_4^{2-}$	from No. 14 (prepared in situ) and $HgCl_2$ in acetone (ca. 70) yellow solid stable for several days, almost insoluble in acetone IR: 1125 vs ν(N–N), 776 vs ν(N–S) [2]

Table 19 (continued)

No.	compound	preparation method (yield in %) properties (^{1}H NMR ($CDCl_3$/TMS): δ (in ppm), IR (Nujol): ν (in cm^{-1}))
neutral binuclear Pd^{II} complexes		
*19	$(\mu\text{-}Cl)_2[PdCl(SNN(CH_3)_2)]_2$	1) from $Pd(C_6H_5CN)_2$ and $S{=}NN(CH_3)_2$ (mole ratio 1 : 1) in ether–benzene (70) 2) from $Pd(C_6H_5CN)_2Cl_2$ and No. 10 in acetone (mole ratio 1 : 1), stirring for 4 h (ca. 90) yellow-brown solid stable for some weeks, almost insoluble in common organic solvents IR: 1135 vs ν(N–N), 770 vs ν(N–S), 390 m ν(Pd–$SNN(CH_3)_2$), 340 s $\nu(Pd\text{–}Cl)_{terminal}$, 295 m, 278 m $\nu(Pd\text{–}Cl)_{bridge}$; bands associated with $SNN(CH_3)_2$: 1490 s, 1375 s, 1250 m, 1135 vs, 1030 m, 860 m, 770 vs, 718 w, 540 m, 415 w bridge trans configuration [2]
20	$(\mu\text{-}Br)_2[PdCl(SNN(CH_3)_2)]_2$	from $Pd(C_6H_5CN)_2$ and No. 11 in acetone (mole ratio 1 : 1), stirring for 4 h (ca. 90) yellow-brown solid stable for several weeks IR: 1132 vs ν(N–N), 778 vs ν(N–S), 385 m ν(Pd–$SNN(CH_3)_2$), 332 s $\nu(Pd\text{–}Cl)_{terminal}$ bridge trans configuration [2]
*21	$(\mu\text{-}SCN)_2[PdCl(SNN(CH_3)_2)]_2$	1) from $Pd(C_6H_5CN)_2Cl_2$ and No. 12 in acetone (mole ratio 1 : 1), stirring for 4 h (ca. 90) 2) from No. 19 and KSCN (mole ratio 1 : 2) in acetone, 4 h stirring (80) orange solid stable solid, almost insoluble in common organic solvents IR: 2150 s ν(C–N) of bridge SCN, 1130 vs ν(N–N), 778 vs ν(N–S); 477 w, 427 w, δ(SCN); 385 m ν(Pd–$SNN(CH_3)_2$), 310 s $\nu(Pd\text{–}Cl)_{bridge}$ bridge trans configuration [2]
22	$(\mu\text{-}SeCN)_2[PdCl(SNN(CH_3)_2)]_2$	analogous to No. 21 IR: 2150 s ν(C–N) of bridge SeCN, 1130 vs ν(N–N), 778 vs ν(N–S); 412 w, 370 w, δ(SeCN); 385 m ν(Pd–$SNN(CH_3)_2$), 307 s $\nu(Pd\text{–}Cl)_{terminal}$ bridge trans configuration [2]
23	$(\mu\text{-}SCN)_2[Pd(SCN)(SNN(CH_3)_2)]_2$	1) from No. 19 and KSCN (mole ratio 1 : 4) in acetone (ca. 70) 2) from No. 21 and KSCN (mole ratio 1 : 2) in acetone (ca. 70) orange compound

Table 19 (continued)

No. compound	preparation method (yield in %) properties (^{1}H NMR ($CDCl_3$/TMS): δ (in ppm), IR (Nujol): ν (in cm^{-1}))
23 (continued)	slightly soluble in acetone IR: 2155 s ν(C–N) of bridge SCN group, 2105 s ν(C–N) of terminal SCN group, 1130 vs ν(N–N), 778 vs ν(N–S), 387 m ν(Pd–$SNNMe_2$) bridge trans configuration [2]

*Further information:

***cis*-$PdCl_2(P(C_6H_5)_3)(SNN(CH_3)_2)$** (Table **19**, No. **1**). A suspension of compound No. 1 reacted with KSCN in acetone (mole ratio 1:2.1) (stirring for 6 h) to give *cis*-$Pd(SCN)_2(P(C_6H_5)_3)(SNN(CH_3)_2)$ (Table 19, No. 6) [1].

***cis*-$PdCl_2(As(C_6H_5)_3)(SNN(CH_3)_2)$** (Table **19**, No. **4**). The molecular geometry of the compound was determined by X-ray diffraction. The crystals were found to be triclinic with the space group $P\bar{1}-C_i^1$ (No. 2). The unit cell dimensions are: a = 9.407(2), b = 10.540(2), c = 12.265(4) Å, α = 68.1(1)°, β = 78.3(1)°, γ = 86.3(1)°; V = 1104.4 Å^3; Z = 2; D_x = 1.725 g/cm^3; R = 0.029, R_w = 0.031 for 2664 independent reflections. The complex has a cis configuration in agreement with IR data. The $SNN(CH_3)_2$ ligand is bonded via the S atom to the metal. The Pd atom is nearly square-planar-coordinated. The $SNN(CH_3)_2$ moiety is nearly planar. A strong σ + π synergistic interaction of Pd with the $SNN(CH_3)_2$ ligand is assumed. The stereochemical arrangement of the complex is shown in **Fig. 6**. Selected bond lengths and bond angles are compiled in Table 20 [2].

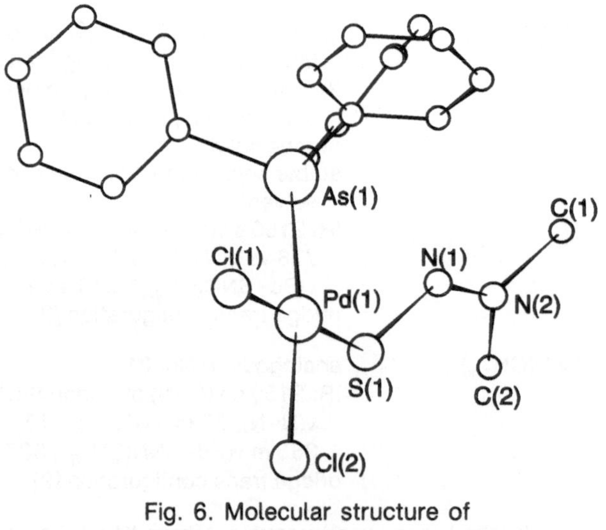

Fig. 6. Molecular structure of *cis*-$PdCl_2(As(C_6H_5)_3)(SNN(CH_3)_2)$ [2].

Table 20
Selected Bond Lengths and Bond Angles of *cis*-$PdCl_2(As(C_6H_5)_3)(SNN(CH_3)_2)$ [2].

bond length	in Å	bond angle	in °
Pd–As	2.362(1)	Cl(1)–Pd–As	84.2(1)
Pd–Cl(1)	2.313(1)	Cl(1)–Pd–S	176.6(1)
Pd–Cl(2)	2.359(1)	Cl(2)–Pd–As	175.5(1)
Pd–S	2.249(1)	Cl(1)–Pd–Cl(2)	92.0(1)
S–N(1)	1.639(4)	Cl(2)–Pd–S	84.9(1)
N(1)–N(2)	1.272(5)	S–Pd–As	98.9(1)
N(2)–C(1)	1.462(6)	Pd–S–N(1)	107.2(2)
N(2)–C(2)	1.437(7)	S–N(1)–N(2)	118.0(4)
		N(1)–N(2)–C(2)	124.5(5)
		N(1)–N(2)–C(1)	117.2(4)
		C(1)–N(2)–C(2)	118.3(5)

***cis*-$PdCl_2(SNN(CH_3)_2)_2$** (Table **19**, No. **10**). When an acetone suspension of compound No. 10 was kept together with an equimolar amount of $P(C_6H_5)_3$, 4-$CH_3C_6H_4P(C_6H_5)_2$, $P(C_6H_4CH_3\text{-}2)_3$, $As(C_6H_5)_3$, or $Sb(C_6H_5)_3$ at room temperature overnight, the complexes $PdCl_2(ER_3)(SNN(CH_3)_2)$ ($ER_3 = P(C_6H_5)_3$, 4-$CH_3C_6H_4P(C_6H_5)_2$, $P(C_6H_4CH_3\text{-}2)_3$, $As(C_6H_5)_3$, and $Sb(C_6H_5)_3$) (Table 19, Nos. 1 to 5) were formed, respectively. The title complex reacted with KXCN (X = S or Se) in acetone (mole ratio 2.1 : 1) over a period of 6 h to give the complexes $Pd(XCN)_2(SNN(CH_3)_2)_2$ (Table 19, Nos. 12 and 13) [1]. The complex reacted with $Pd(C_6H_5CN)_2$-Cl_2 in acetone to give (μ-Cl)$_2$[PdCl(SNN(CH$_3$)$_2$)]$_2$ (Table 19, No. 19) [2].

***cis*-$PdBr_2(SNN(CH_3)_2)_2$** (Table **19**, No. **11**). The compound reacted with $Pd(C_6H_5CN)_2Cl_2$ in acetone to give (μ-Br)$_2$[PdCl(SNN(CH$_3$)$_2$)]$_2$ (Table 19, No. 20) [2].

***cis*-$Pd(SCN)_2(SNN(CH_3)_2)_2$** (Table **19**, No. **12**). The compound reacted with $Pd(C_6H_5CN)_2$-Cl_2 in acetone to give (μ-SCN)$_2$[PdCl(SNN(CH$_3$)$_2$)]$_2$ (Table 19, No. 21) [2].

***cis*-$Pd(SeCN)_2(SNN(CH_3)_2)_2$** (Table **19**, No. **13**). The compound reacted with $Pd(C_6H_5CN)_2$-Cl_2 in acetone to give [(μ-SeCN)$_2$[PdCl(S=NN(CH$_3$)$_2$)]$_2$] (Table 19, No. 22) [2].

$[Pd(SNN(CH_3)_2)_4]^{2+}$ $(Cl^-)_2$ (Table **19**, No. **14**). The complex was dissolved in absolute ethanol and added to a solution of $NaB(C_6H_5)_4$ in ethanol to yield the complex $[Pd(SNN(CH_3)_2)_4]^{2+}$ $(B(C_6H_5)_4^-)_2$ (Table 19, No. 15). An analogous reaction with $HgCl_2$ in acetone gave $[Pd(SNN(CH_3)_2)_4]^{2+}$ $HgCl_4^{2-}$ (Table 19, No. 18). The unstable complex reacted with $P(C_6H_5)_3$ to yield *cis*-$PdCl_2(P(C_6H_5)_3)(SNN(CH_3)_2)$ (Table 19, No. 1) [2].

$[Pd(SNN(CH_3)_2)_4]^{2+}$ $PdCl_4^{2-}$ (Table **19**, No. **17**). The complex decomposed in acetone to form No. 10 [2].

(μ-Cl)$_2$[PdCl(SNN(CH$_3$)$_2$)]$_2$ (Table **19**, No. **19**). The title complex reacted with KXCN (X = S, Se) in acetone (mole ratio 1 : 2) to yield (μ-SCN)$_2$[PdCl(SNN(CH$_3$)$_2$)]$_2$ and (μ-SeCN)$_2$-[PdCl(SNN(CH$_3$)$_2$)]$_2$ (Table 19, Nos. 21 and 22). The corresponding reaction with KSCN in the molar ratio 1 : 4 yielded (μ-SCN)$_2$[Pd(SCN)(SNN(CH$_3$)$_2$)]$_2$ (Table 19, No. 23) [2].

(μ-SCN)$_2$[PdCl(SNN(CH$_3$)$_2$)]$_2$ (Table **19**, No. **21**). The complex reacted with KSCN (molar ratio 1 : 2) to give (μ-SCN)$_2$[Pd(SCN)(SNN(CH$_3$)$_2$)]$_2$ (Table 19, No. 23) [2].

Table 21
N-Thionitroso-dimethylamine Complexes of Pt^{II} (C_7H_8 = norbornadiene, 1,5-C_8H_{12} = 1,5-cyclooctadiene).
An asterisk preceding the compound number indicates further information at the end of the table, p. 85.

No.	compound	preparation method (yield in %) properties (^1H NMR ($CDCl_3$/TMS): δ (in ppm), IR (Nujol): ν (in cm^{-1}))
neutral mononuclear Pt^{II} complexes		
*1	*cis*-$PtCl_2(P(C_6H_5)_3)(SNN(CH_3)_2)$	from Pt(1,5-C_8H_{12})Cl_2, $S{=}NN(CH_3)_2$, and $P(C_6H_5)_3$ (mole ratio 1 : 3 : 1) in acetone or CH_2Cl_2 (ca. 95) yellow solid stable for several weeks in the solid state, almost insoluble in common organic solvents IR: 325 s, 285 s, ν(Pt–Cl) [1]
2	*cis*-$PtCl_2(4\text{-}CH_3C_6H_4P(C_6H_5)_2)(SNN(CH_3)_2)$	from Pt(1,5-C_8H_{12})Cl_2, $S{=}NN(CH_3)_2$, and 4-$CH_3C_6H_4P(C_6H_5)_3$ (mole ratio 1 : 3 : 1) in acetone or CH_2Cl_2 (ca. 95) yellow solid stable for several weeks in the solid state, slightly soluble in chlorinated solvents ^1H NMR: 2.23 ($C_6H_4C\underline{H}_3$-4), 3.10 (NCH_3), 3.25 (NCH_3) IR: 322 s, 285 s, ν(Pt–Cl) [1]
3	*cis*-$Pt(SCN)_2(P(C_6H_5)_3)(SNN(CH_3)_2)$	from No. 1 and KSCN (mole ratio 1 : 2.1) in acetone, stirring for 6 h (ca. 70) orange solid stable for several weeks, slightly soluble in acetone or chlorinated solvents IR: 2110 s, 2100 s, ν(CN); 430 m δ(SCN) [1]
cationic mononuclear Pt^{II} complexes		
*4	$[Pt(SNN(CH_3)_2)_4]^{2+}$ $(Cl^-)_2$	from $Pt(C_6H_5CN)_2Cl_2$ and $S{=}NN(CH_3)_2$ (mole ratio 1 : 6) in ether–acetone unstable, yellow-orange complex (not isolated), almost insoluble in the common organic solvents [2]
5	$[Pt(SNN(CH_3)_2)_4]^{2+}$ $(B(C_6H_5)_4^-)_2$	from No. 4 and $NaB(C_6H_5)_4$ in ethanol (60) yellow solid stable for several days, slightly soluble in acetone IR: 1130 vs ν(N–N), 780 vs ν(N–S) bi-univalent electrolyte, $\Lambda_M = 190\ \Omega^{-1}\cdot cm^2\cdot mol^{-1}$ for a 5×10^{-4} M acetone solution [2]

Table 21 (continued)

No.	compound	preparation method (yield in %) properties (^1H NMR ($CDCl_3$/TMS): δ (in ppm), IR (Nujol): ν (in cm^{-1}))
6	$[Pt(SNN(CH_3)_2)_4]^{2+}$ $(PF_6^-)_2$	from $Pt(diene)Cl_2$, $AgPF_6$, and $S{=}NN(CH_3)_2$ (mole ratio 1 : 2 : 5) in acetone–ether (ca. 60) (diene = 1,5-C_8H_{12}, C_7H_8) yellow solid IR: 1130 vs ν(N–N), 780 vs ν(S–N) [2]
7	$[Pt(SNN(CH_3)_2)_4]^{2+}$ $PtCl_4^{2-}$	from K_2PtCl_4 and $S{=}NN(CH_3)_2$ (mole ratio 1 : 3) in acetone–H_2O (ca. 60) yellow solid almost insoluble in common organic solvents IR: 1130 vs ν(N–N), 780 vs ν(S–N), 310 s ν(PtCl) of $PtCl_4^{2-}$ [2]

*Further information:

***cis*-$PtCl_2(P(C_6H_5)_3)(SNN(CH_3)_2)$** (Table **21**, No. **1**). A suspension of *cis*-$PtCl_2(P(C_6H_5)_3)$-$(SNN(CH_3)_2)$ in acetone reacted with KSCN (mole ratio 1 : 2.1) (stirring for 6 h) to give *cis*-Pt-$(SCN)_2(P(C_6H_5)_3)(SNN(CH_3)_2)$ (Table 21, No. 3) [1].

$[Pt(SNN(CH_3)_2)_4]^{2+}$ $(Cl^-)_2$ (Table **21**, No. **4**). The unstable complex, dissolved in absolute ethanol, reacted with a solution of $NaB(C_6H_5)_4$ in ethanol to yield $[Pt(SNN(CH_3)_2)_4]^{2+}$ $(B(C_6H_5)_4^-)_2$ (Table 21, No. 5) [2]. Reaction with $P(C_6H_5)_3$ yielded *cis*-$PtCl_2(P(C_6H_5)_3)(SNN(CH_3)_2)$ (Table 21, No. 1) [1].

References:

[1] Tresoldi, G.; Bruno, G.; Crucitti, F.; Piraino, P. (J. Organomet. Chem. **252** [1983] 381/7).

[2] Tresoldi, G.; Bruno, G.; Piraino, P.; Faraone, G.; Bombieri, G. (J. Organomet. Chem. **265** [1984] 311/22).

6.2.6 N-Thionitroso-diphenylamine Complex of Cr^0, $Cr(CO)_5(SNN(C_6H_5)_2)$

The complex was prepared by adding $[Cr(CO)_5]\cdot THF$ in THF to a stirred solution of $S{=}NN(C_6H_5)_2$ (mole ratio 1.56 : 1) in the same solvent at −20 °C. After warming to room temperature and stirring for another 2 h, the solvent was removed in vacuum. The deep blue oily residue was dissolved in CH_2Cl_2 and filtered through silicone-treated silica gel. The solvent was removed in vacuum and the residue recrystallized from n-heptane. The yield was 36%.

The deep blue complex melts at 118 °C with decomposition.

The molecular structure of the complex was determined by X-ray diffraction analysis. The crystals belong to the monoclinic space group $P2_1/c - C_{2h}^5$ (No. 14) with a = 7.865(1), b = 22.365(2), c = 10.546(1) Å, β = 89.44(8)°; V = 1854.9 Å³; Z = 4; D_x = 1.454 g/cm³; R = 0.079, R_w = 0.068 for 1573 observed reflections. The molecular structure is shown in **Fig. 7**, p. 86. Geometrical parameters are given in Table 22. The complex formation occurs through the S atom. The $CrSNNC_2$ unit is effectively planar, and this plane lies approximately eclipsed to the carbonyl groups. A π interaction between the trans phenyl group (which lies approximately in the $CrSNNC_2$ plane) and the nitrogen atoms is indicated by a significant lengthening of the N–N bond and shortening of the N–S and Cr–S bonds compared with the equivalent data for $Cr(CO)_5(SNN(CH_3)_2)$; see p. 66.

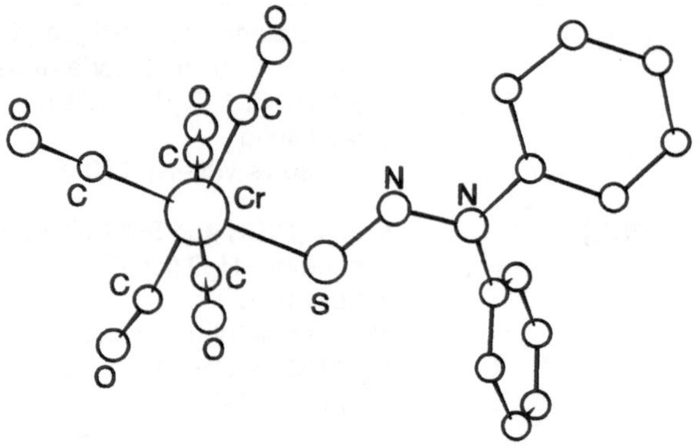

Fig. 7. Molecular structure of $Cr(CO)_5(SNN(C_6H_5)_2)$.

Table 22
Selected Bond Lengths, Bond Angles, and Torsion Angles of $Cr(CO)_5(SNN(C_6H_5)_2)$.

bond length	in Å	bond angle	in °	torsion angle	in °
Cr–C(trans)	1.848(7)	S–Cr–C(cis, mean)	89.9(10)	C(eclip.)–Cr–S–N	12.1(4)
Cr–C(cis, mean)	1.900(9)	Cr–S–N	106.4(2)	Cr–S–N–N	177.6(5)
C–O(mean)	1.139(12)	S–N–N	120.4(4)	S–N–N–C(cis)	177.7(4)
Cr–S	2.357(2)	N–N–C(cis)	116.7(5)	S–N–N–C(trans)	−3.6(9)
S–N	1.612(6)	N–N–C(trans)	120.3(5)		
N–N	1.319(7)				
N–C(cis)	1.432(8)				
N–C(trans)	1.400(9)				

^1H NMR ($CDCl_3$/TMS): δ (in ppm) = 7.08 to 7.42 (m).

IR (Nujol): ν (in cm^{-1}) = 2050 w, 1935 s, 1920 s, 1580 w, 1480 m, 1335 m, 1315 m, 1255 s, 1155 w, 1070 w, 1020 m, 1000 w, 965 w, 840 w, 800 w, 765 w, 755 m, 720 m, 690 m, 635 s.

UV (n-hexane): λ_{max} (in nm) = 620 w, 355 w, 270 sh, 235 s.

MS (100 °C): m/e (rel. int. in %) = 406 (2.6), 350 (1.1), 322 (2.6), 294 (1.7), 266 (25.9), 234 (2.8), 186 (5.4), 168 (100), 154 (7.7), 140 (6.0), 112 (6.0), 84 (20.5), 77 (22.2), 66 (9.4), 52 (6.8).

The complex is stable in moisture and air at room temperature. It is soluble in all common organic solvents, except CCl_4, without noticeable decomposition. The compound is sensitive to light and decomposes on prolonged exposure to daylight.

Reference:

Roesky, H. W.; Emmert, R.; Isenberg, W.; Schmidt, M.; Sheldrick, G. M. (J. Chem. Soc. Dalton Trans. **1983** 183/5).

7 Thiohydroxylamine and Derivatives

7.1 Thiohydroxylamine, $HSNH_2$, $DSND_2$, $HSND_2$, HSNHD

7.1.1 Formation

7.1.1.1 Formation in the Gaseous Phase

Formation from Active Nitrogen and H_2S

The $HSNH_2$ species was generated by reacting active nitrogen with H_2S in a 2450-MHz microwave discharge flow reactor. The optimal conditions were obtained by passing N_2 through the discharge at 15 mTorr (1.8 Pa) pressure and adding approximately 5 mTorr (0.6 Pa) of H_2S below the discharge. $HSNH_2$ was identified by its microwave spectrum. Deuterated species of $HSNH_2$ could be obtained by the reaction of active nitrogen with D_2S [1].

Formation from $H_2S \cdot NH_3$–N_2 Mixtures by Discharge

The production of $HSNH_2$ could be optimized when the vapor of the $H_2S \cdot NH_3$ adduct was passed with N_2 through a 2450-MHz microwave discharge flow reactor at a total pressure of about 2.5 Pa. The $H_2S \cdot NH_3$ adduct was formed by condensing H_2S and NH_3 in a liquid-nitrogen-cooled trap and then warming the trap to about −40 °C with an alcohol–dry ice bath. Similar results were obtained with the deuterated species. $HSNH_2$ and its deuterated species were observed in the microwave spectra [1].

Replacing the chemicals in the discharges by various combinations of H_2, N_2, OCS, H_2S, NH_3, and N_2H_4 showed signals in the microwave spectra which were at best equal to those obtained with only N_2 and H_2S in the discharge [1].

7.1.1.2 Formation in the Liquid Phase

The formation of $HSNH_2$ was postulated when K^+ $H_2NSSO_3^-$ was hydrolized in acidic aqueous solution [2, 3] via $H_2NSSO_3H + H_2O \rightarrow H_2NSH + H_2SO_4$.

References:

[1] Lovas, F. J.; Suenram, R. D.; Stevens, W. J. (J. Mol. Spectrosc. **100** [1983] 316/31).
[2] Meuwsen, A.; Gösl, R. (Angew. Chem. **71** [1959] 736).
[3] Gösl, R.; Meuwsen, A. (Z. Anorg. Allg. Chem. **314** [1962] 334/40).

7.1.2 Molecular Properties. Spectra

7.1.2.1 Geometrical Structure

An analysis of the microwave data revealed the existence of the cis and trans conformers of $HSNH_2$ and its deuterated species. From the three isotopic species *cis*-$HSNH_2$, *cis*-$DSND_2$, and *cis*-$HSND_2$, nine independent moments of inertia could be used to determine the values of three bond lengths and three bond angles for an r_0 structure [1].

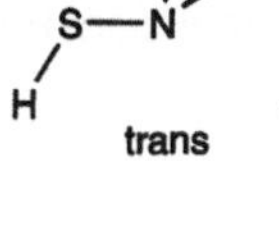

trans

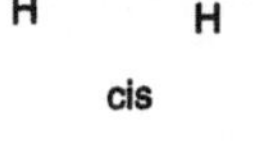

cis

***cis*-HSNH₂.** r(SN) = 1.705(3) Å, r(SH) = 1.351(35) Å, r(NH) = 1.008(13) Å, ∢(HSN) = 101.9(18)°, ∢(HNS) = 112.7(9)°, ∢(HNH) = 111.6(21)° [1].

The parameters of the r_0 structure for the trans conformer could be determined using five moments of inertia of the isotopic species *trans*-$HSNH_2$ and *trans*-$DSND_2$ and setting the H–N bond length to 1.008 Å [1].

***trans*-$HSNH_2$.** r(SN) = 1.719(4) Å, r(SH) = 1.344(29) Å, r(NH) = 1.008 Å, ∢(HSN) = 96.2(14)°, ∢(HNS) = 110.6(8)°, ∢(HNH) = 110.2(14)° [1].

The geometrical structures of *cis*-$HSNH_2$ and *trans*-$HSNH_2$ were confirmed by ab initio quantum-chemical calculations using a 4-31G basis set with a polarization function (d orbitals of the sulfur and nitrogen atoms) [1].

The energy difference between the cis and trans conformer was determined with a microwave technique. For $HSNH_2$ the intensity of the 4_{14}–3_{13} transition of *cis*-$HSNH_2$ at 113257.80 MHz compares with the intensity of the 4_{13}–3_{12} transition of *trans*-$HSNH_2$ at 113368.47 MHz. The trans conformer was found to be slightly lower in energy ($\Delta E = 87\ cm^{-1}$ = 0.161 kcal/mol). Theoretical calculations (see pp. 89/90) confirm that the two conformers are very close in energy, the trans form being lower [1].

7.1.2.2 Dipole Moment

***cis*-$HSNH_2$.** The values $\mu_a = 0.9936(4)$ D and $\mu_b = 1.742(2)$ D for the molecular electric dipole moment were determined from second-order Stark effect measurements of the 3_{13}–2_{12} rotational transition with M = 1 and 2 and the 4_{14}–3_{13} transition with M = 0, 1, and 2 [1].

***trans*-$DSND_2$.** The values $\mu_a = 0.943(3)$ D and $\mu_b = 0.19(12)$ D for the dipole moment were calculated from the M = 1, 2, and 3 Stark transitions of the 4_{14}–3_{13} rotational transition. The dipole moment of *trans*-$HSNH_2$ could not be determined, because the A rotational constant is indeterminate [1].

7.1.2.3 ^{14}N Quadrupole Coupling Constants

The ^{14}N quadrupole coupling constants eQq were derived from the splittings of a number of rotational transitions in the microwave spectrum of the cis conformer [1].

***cis*-$HSNH_2$.** $eQq_{aa} = 3.601(55)$, $eQq_{bb} = -4.772(64)$, $eQq_{cc} = 1.170(71)$ MHz [1].

***cis*-$DSND_2$.** $eQq_{aa} = 3.281(93)$, $eQq_{bb} = -4.547(56)$, $eQq_{cc} = 1.266(72)$ MHz [1].

7.1.2.4 Rotational Constants

Analysis of the microwave spectra yielded the rotational constants given in Tables 23 and 24.

Table 23
Rotational Constants A, B, C (in MHz) and Centrifugal Distortion Constants τ (in MHz) for the Isotopic Species of *cis*-$HSNH_2$ [1].

rotational parameter	*cis*-$HSNH_2$	*cis*-$DSND_2$	*cis*-$HSND_2$
A″	148316.91(32)	76182.304(83)	100435.21(16)
B″	14341.7705(134)	12047.4769(87)	12446.033(23)
C″	14097.2299(117)	11751.1396(48)	12268.472(23)
τ_1	−0.9861(16)	−0.6387(49)	−0.7519(73)
τ_2	−0.1475(13)	−0.11426(57)	−0.1088(40)
τ_3	41.16(50)	8.63(55)	5.24(90)
τ_{aaaa}	−11.42(20)	−3.49(20)	−6.00
τ_{bbbb}	−0.08046(27)	−0.04913(62)	−0.0611(24)
τ_{cccc}	−0.07733(30)	−0.04721(54)	−0.0539(21)
h_{KJ} (kHz)	6.75(48)	−0.15(8)	

Table 24
Rotational Constants A, B, C (in MHz) and Centrifugal Distortion Constants τ (in MHz) for *trans*-$HSNH_2$ and *trans*-$DSND_2$ [1].

rotational parameter	*trans*-$HSNH_2$	*trans*-$DSND_2$
A″	148000.	77349.(37)
B″	14238.032(18)	11952.322(6)
C″	13976.40(18)	11667.958(6)
τ_1	−1.7037(45)	−0.71315(55)
τ_2	−0.2084(11)	−0.12470(34)
τ_3	46.7(26)	5.67(48)
τ_{aaaa}	0	0
τ_{bbbb}	−0.0858(13)	−0.04888(39)
τ_{cccc}	−0.0869(13)	0.04658(38)

7.1.2.5 Quantum-Chemical Calculations

Several molecular properties have been the subject of quantum-chemical treatments. Table 25, p. 90, gives a survey of pertinent publications. The following abbreviations and symbols are used:

SCF	self-consistent field	E_T	total energy
MO	molecular-orbital method	ΔE	conformational energy difference
SC	single configuration	V_{min}	electrostatic potential minimum
RHF	spin-restricted Hartree-Fock method	A	relative acidity
		B	relative basicity
RMP2	second-order Møller-Plesset perturbation method	SIE	substituent interaction energy
		ν	stretching frequency
		μ	dipole moment
		NQCC	nuclear quadrupole coupling constant

Table 25
Quantum-Chemical Calculations for $HSNH_2$.

method	basis set	calculated molecular properties	Ref.
ab initio SCF-MO	STO-2G, STO-3G	E_T, geometries	[2]
ab initio SCF-MO	STO-3G	E_T, geometries, inversion barrier	[3]
ab initio SCF-MO	STO-3G	ΔE	[4]
ab initio SCF-MO	STO-3G	E_T, ^{14}N NQCC	[5]
ab initio SCF-MO	STO-3G, STO-3G*	E_T, ΔE, geometries	[6]
ab initio SCF-MO	STO-3G, 3-21G, 4-31G	E_T, geometries, B, SIE	[7]
ab initio SCF-MO	STO-3G, STO-5G	V_{min}	[8]
ab initio SCF-MO	STO-3G*, 6-31G*, 6-31G*/MP2	E_T, ΔE, geometries	[9]
ab initio SCF-MO	3-21G, 3-31G	E_T, r(S–N), ν	[10]
ab initio SCF-MO	3-21G, 4-31G*	E_T, ΔE, geometries	[11]
ab initio SCF-MO	4-31G, 6-31G	A, B, SIE	[12]
ab initio SCSCF	4-31G	ΔE, geometries, μ	[1]
ab initio SCF-MO	HF/6-31G	E_T, ΔE, geometries, inversion barrier, rotation barrier	[13]
ab initio RHF, RMP2	STO-3G, 3-21G, 4-31G, 6-31G	ΔE, E_T, geometries	[14]

7.1.2.6 Microwave Spectra

The rotational spectra were studied between 56 and 170 GHz for the normal species and several deuterated isotopic species of the cis and trans conformers, using a Stark effect-modulated microwave spectrometer. The assignments were based upon measurements of the ^{14}N quadrupole splitting, the observed Stark effect, and comparisons with values calculated from an asymmetric rotor model. The frequencies of the rotational transitions yielded the rotational constants and thus the structural parameters. Hyperfine structures of the cis conformer arising from the quadrupole effect were observed and gave the ^{14}N quadrupole coupling constants. Stark effect measurements of both conformers yielded the dipole moments [1].

***cis*-$HSNH_2$ Spectrum.** Both a- and b-type rotational transitions for J = 4–3, 3–2, and 2–1 were observed [1].

***cis*-$DSND_2$ Spectrum.** As with the normal species both a- and b-type transitions for J = 5–4, 4–3, and 3–2 were observed [1].

***cis*-$HSND_2$ Spectrum.** For this particular isotope the principal axes are switched such that a- and c-type transitions for J = 5–4 and 4–3 were observed [1].

***cis*-HSNHD Spectrum.** A series of b-type Q-branch transitions for $J_{1,J-1}$–$J_{0,J-1}$ was observed. This series was followed from J = 1 to 9. In additon, a c-type Q-branch series for $J_{1,J}$–$J_{0,J}$ was observed. Measurements were made on this series from J = 1 to J = 10. Neither series showed any splitting due to the nuclear quadrupole moment of the ^{14}N nucleus [1].

***trans*-$HSNH_2$ Spectrum.** a-Type rotational transitions for J = 6–5, 5–4, 4–3, and 3–2 could be observed [1].

***trans*-$DSND_2$ Spectrum.** a-Type R-branch transitions for J = 7–6, 5–4, 4–3, and 3–2 could be monitored. As with the normal species no b-type transitions were observed [1].

References:

[1] Lovas, F. J.; Suenram, R. D.; Stevens, W. J. (J. Mol. Spectrosc. **100** [1983] 316/31).
[2] Haddon, R. C.; Wasserman, S. R.; Wudl, F.; Williams, G. R. J. (J. Am. Chem. Soc. **102** [1980] 6687/93).
[3] Bernardi, F.; Bottoni, A.; Tonachini, G. (Croat. Chem. Acta **57** [1985] 911/25).
[4] Collins, M. P. S.; Duke, B. J. (J. Chem. Soc. Dalton Trans. **1978** 277/9).
[5] Palmer, M. H.; Simpson, I.; Findlay, R. H. (Z. Naturforsch. **36a** [1981] 34/50).
[6] Bernardi, F.; Bottoni, A.; Mangini, A.; Tonachini, G. (J. Mol. Struct. **86** [1981] 163/72 [THEOCHEM **3**]).
[7] Magnusson, E. (J. Comput. Chem. **5** [1984] 612/28).
[8] Murray, J. S.; Politzer, P. (Chem. Phys. Lett. **152** [1988] 364/70).
[9] Gimarc, B. M.; Warren, D. S. (Inorg. Chem. **30** [1991] 3276/80).
[10] Schmidt, M. W.; Truong, P. N.; Gordon, M. S. (J. Am. Chem. Soc. **109** [1987] 5217/27).

[11] Magnusson, E. (Austral. J. Chem. **39** [1986] 735/45).
[12] Magnusson, E. (Tetrahedron **41** [1985] 2939/43).
[13] Reed, A. E.; von Ragué Schleyer, P. (Inorg. Chem. **27** [1988] 3969/87).
[14] Sabio, M.; Topiol, S. (J. Mol. Struct. **206** [1990] 335/57 [THEOCHEM **65**]).

7.2 Thiohydroxylamine Radical Ion (1+), $HSNH_2^{+\bullet}$

The radical ion (1+) $HSNH_2^{+\bullet}$ was formed by electron impact at 70 eV on $ROC(=S)SNH_2$, where R = CH_3CH_2, $CH_3CH_2CH_2$, $(CH_3)_2CH$, $CH_3CH_2CH_2CH_2$, $(CH_3)_2CHCH_2$, $CH_3CH_2CH(CH_3)$, or $(CH_3)_3CCH_2$. Abundant peaks (m/e = 49, 17.1 to 7.1%) were observed in the mass spectra. Only in the mass spectrum of the neopentyl compound the abundance was very small. The formation of the radical ion was rationalized assuming the following mechanism: First, the ion $RSNH_2^{+\bullet}$ forms directly from the molecular ion by skeletal rearrangement and loss of COS. Then, $RSNH_2^{+\bullet}$ rearranges with elimination of alkene to give $HSNH_2^{+\bullet}$. $HSNH_2^{+\bullet}$ may also be generated by loss of COS from the ion $HSC(=O)SNH_2^{+\bullet}$ which is produced by a McLafferty rearrangement in the molecular ion. The process was established for the isobutyl compound by a metastable defocusing measurement.

Reference:

Holm, A.; Jensen, G. M. (Acta Chem. Scand. **26** [1972] 205/12).

7.3 Thiohydroxylamine Ion (1−), H_2NS^- Ion

The ion H_2NS^- was produced in the gaseous phase at 298 K using the flowing afterglow [1, 2] and flowing afterglow-selected ion flow tube technique [3] via the reaction $H_2N^- + OCS \rightarrow H_2NS^- + CO$. The branching ratio of the reaction was 52. H_2N^- was generated by electron impact on NH_3. Typical helium pressures and flow rates were 0.3 to 0.4 Torr and 140 to 190 STP $cm^3 \cdot s^{-1}$, respectively [1, 2].

The gas-phase proton affinity (basicity) of H_2NS^- has been bracketed by proton-transfer reactions. H_2NS^- was generated in the flowing afterglow source of the flowing afterglow-selected ion flow tube by allowing H_2N^- to react with COS. Then the ion was injected into the second flow tube where its proton transfer reactions were determined. H_2NS^- reacted with pyrrole (ΔG°_{acid} = 351 kcal/mol) and $((CH_3)_3Si)_2NH$ (ΔG°_{acid} = 353 kcal/mol), but it underwent proton transfer only sluggishly with CF_3CH_2OH (ΔG°_{acid} = 354 kcal/mol) and not at all with methyl vinyl ketone (ΔG°_{acid} = 356 kcal/mol). Therefore, the values of ΔG°_{acid} (H_2NSH) = 354 ± 3 kcal/mol and ΔH°_{acid} (H_2NSH) = 362 ± 3 kcal/mol were assigned [3]. An earlier rough estimate of the basicity of H_2NS^- gave 360 ± 3 kcal/mol. The ion is more basic than HS^- (353.4 kcal/mol), less than $CF_3CH_2O^-$ (364.4 kcal/mol), and similar to CH_3S^- (359.0 kcal/mol) [1, 2].

Hydrogen-deuterium exchange in the gaseous phase occurs readily between H_2NS^- and CF_3CH_2OD via $H_2NS^- + CF_3CH_2OD \rightleftharpoons [DH_2NS \cdot CF_3CH_2O^-] \rightarrow DHNS^- + CF_3CH_2OH$. The simultaneous reaction $CF_3CH_2OD + H_2NS^- \rightleftharpoons [H_2NSD \cdot CF_3CH_2O^-]$ cannot be excluded [1, 2].

Deuterium kinetic isotope effects in gas-phase reactions of H_2NS^- with alkyl halides were studied to distinguish S_N2 and E_2 pathways. The former showing an inverse secondary k_H/k_D ratio as small as 0.76 and the latter a normal primary k_H/k_D ratio as large as 4.7. The following rate coefficients (units are 10^{-10} $cm^3 \cdot molecule^{-1} \cdot s^{-1}$) and kinetic deuterium isotope effects (k_H/k_D) were measured: CH_3Cl 1.46 ± 0.07, CH_3CH_2Cl 3.81 ± 0.014, i-C_3H_7Cl no reaction, i-C_3D_7Cl no reaction; t-C_4H_9Cl no reaction, t-C_4D_9Cl no reaction; CH_3Br 7.04 ± 0.03, CD_3Br 6.75 ± 0.18, **k_H/k_D = 1.04 ± 0.03**; C_2H_5Br 9.05 ± 0.16, C_2D_5Br 9.07 ± 0.29, **k_H/k_D = 1.00 ± 0.04**; i-C_3H_7Br 3.49 ± 0.09, i-C_3D_7Br 3.34 ± 0.03, **k_H/k_D = 1.04 ± 0.03**; t-C_4H_9Br 1.30 ± 0.05. Methyl bromide shows no isotope effect or an inverse secondary isotope effect with H_2NS^-. The ion H_2NS^- reacts with ethyl and isopropyl bromide with no isotope effect or with a slight inverse isotope effect. The most reasonable explanation is that H_2NS^- undergoes S_N2 reactions [4].

Rate coefficients and reaction efficiencies of gas-phase S_N2 reactions of H_2NS^- with alkyl halides were determined using the flowing afterglow-selected ion flow tube technique. (Bimolecular rate coefficient in units of 10^{-10} $cm^3 \cdot molecule^{-1} \cdot s^{-1}$; given in parentheses are the reaction efficiencies defined by the ratio of the experimental rate coefficient to the collision rate coefficient (k_{exp}/k_{col}), where k_{col} is calculated from the average dipole orientation (ADO) theory): CH_3Cl 1.5 (0.085), C_2H_5Cl 0.38 (0.021), n-C_3H_7Cl 0.85 (0.043), i-C_3H_7Cl < 0.001, t-C_4H_9Cl < 0.001, neo-$C_5H_{11}Cl$ < 0.047 (< 0.0023); CH_3Br 7.0 (0.35), C_2H_5Br 9.1 (0.49), n-C_3H_7Br 11.1 (0.61), i-C_3H_7Br 3.5 (0.17), t-C_4H_9Br 1.3 (0.068), neo-$C_5H_{11}Br$ < 0.5 (< 0.02); CH_2=$CHCH_2Cl$ 7.1 (0.38), CH_3OCH_2Cl 8.5 (0.63), $CH_3C(O)CH_2Cl$ 18 (0.88), $NCCH_2Cl$ 24 (0.96) [4]; CH_3I 13 (0.88) [1, 2].

The ion showed no specially enhanced nucleophilicity due either to the presence of sulfur or to an α-hetero atom effect [2].

References:

[1] Bierbaum, V. M.; Grabowski, J. J.; DePuy, C. H. (J. Phys. Chem. **88** [1984] 1389/93).

[2] DePuy, C. H.; Bierbaum, V. M. (Tetrahedron Lett. **22** [1981] 5129/30).

[3] DePuy, C. H.; Gronert, S.; Mullin, A.; Bierbaum, V. M. (J. Am. Chem. Soc. **112** [1990] 8650/5).

[4] Gronert, S.; DePuy, C. H.; Bierbaum, V. M. (J. Am. Chem. Soc. **113** [1991] 4009/10).

7.4 N,N-Diorganyl-thiohydroxylamines and Salts of the Ions (1−)

There are only a few publications that claim the existence of this class of compounds. The unstable thiohydroxylamines were prepared in situ by protonation of the corresponding stable thiohydroxylamates at low temperatures or as intermediates in the reactions of the S^{II}–N compounds. $(CH_3)_2NS^-$ was generated by the gas-phase reaction of $(CH_3)_2N^-$ with S=C=O via sulfur atom transfer and loss of CO.

7.4.1 $HSN(CH_3)_2$

The title compound was said to be an intermediate in the decomposition reaction of $((CH_3)_2N)_2S \cdot BH_3$ leading to $(CH_3)_2NSBH_2 \cdot HN(CH_3)_2$, $((CH_3)_2N)_2S_3$, and $(CH_3)_2NH$. $HSN(CH_3)_2$ formed by solvolysis of $(CH_3)_2NSBH_2$ with $(CH_3)_2NH$. It condensed with elimination of $(CH_3)_2NH$ to give $((CH_3)_2N)_2S_3$ [1]:

$$(CH_3)_2NSBH_2 + (CH_3)_2NH \rightarrow (CH_3)_2NSH + (CH_3)_2NBH_2$$

$$3\ (CH_3)_2NSH \rightarrow (CH_3)_2NH + ((CH_3)_2N)_2S_3 + H_2$$

7.4.2 $(CH_3)_2NS^-$

$(CH_3)_2NS^-$ was produced by the gas-phase reaction of $(CH_3)_2N^-$ with S=C=O using the flowing afterglow technique: $(CH_3)_2N^- + S{=}C{=}O \rightarrow (CH_3)_2NS^- + CO$. Only a single product ion, $m/z = 76$, was observed [2].

7.4.3 $HSN(CH_2)_5$

Attempts to prepare $HSN(CH_2)_5$ by protonation of $Li^+\ (CH_2)_5NS^-$ at −70 °C with acetic acid failed. $HSN(CH_2)_5$ appears to be unstable, even at −70 °C [3].

The compound was proposed to be an intermediate in the ring-opening reaction of 3-chloro-4-nitroisothiazole with piperidine leading to the enamine $(CH_2)_5N{-}CH{=}C(NO_2)CN$ [4].

7.4.4 $Li^+\ (CH_2)_5NS^-$

$Li^+\ (CH_2)_5NS^-$ was prepared by reduction of $(CH_2)_5NSSN(CH_2)_5$ with Li (4 g atom equivalent) in liquid NH_3 at −50 °C. After evaporation of NH_3 the salt was obtained as a white solid. $Li^+\ (CH_2)_5NS^-$ was also formed by the analogous procedure with $(CH_2)_5NSN(CH_2)_5$ being the reactant [3].

When the anion was protonated with excess acetic acid at −70 °C, a pale yellow solution was obtained, but the protonated form appeared to be unstable even at −70 °C [3].

Reacting the salt (prepared from $(CH_2)_5NSSN(CH_2)_5$) with CH_3I in boiling ether for 4 h gave $(CH_2)_5NSCH_3$ and $(CH_2)_5NSSCH_3$ in the ratio 2.5 : 1 [3].

7.4.5 $HSN(CH_2)_4O$

The unstable compound formed by protonation of $Li^+\ O(CH_2)_4NS^-$ with excess acetic acid at −70 °C [3]. $HSN(CH_2)_4O$ was proposed to be an intermediate in the ring-opening reaction of 3-chloro-4-nitroisothiazole with morpholine leading to $O(CH_2)_4N{-}CH{=}C(NO_2)CN$ [4].

$HSN(CH_2)_4O$ decomposes in acetic acid solution above −40 °C [3].

When the reaction mixture of $Li^+\ O(CH_2)_4NS^-$ and excess acetic acid at −70 °C was treated with $n\text{-}C_4H_9Li$ and then with CH_3I, evolution of H_2S was detected and a trace amount of $O(CH_2)_4NSCH_3$ could be isolated, indicating the formation of $HSN(CH_2)_4O$ [3].

7.4.6 $Li^+ O(CH_2)_4NS^-$

$Li^+ O(CH_2)_4NS^-$ was prepared by reducing $O(CH_2)_4NSSN(CH_2)_4O$ with Li (4 g atom equivalent) in liquid NH_3 at −50 °C. A white salt was obtained after evaporation of NH_3. $Li^+ O(CH_2)_4NS^-$ was formed by the analogous procedure with $O(CH_2)_4NSN(CH_2)_4O$ being the reactant [3].

The salt is stable at room temperature under dry N_2, but decomposes rapidly (<1 h) in boiling THF giving sulfur and $O(CH_2)_4NH$ [3].

When the anion was protonated with 1 equivalent of acetic acid at −70 °C, a pale yellow solution of $HSN(CH_2)_4O$ was obtained. The solution darkened on warming to ambient temperature with evolution of some H_2S; workup gave sulfur (34%) and $HN(CH_2)_4O$ (64%). The anion could be protonated at −70 °C: Warming to −40 °C for 1 h, cooling to −70 °C, regenerating by adding 1 equivalent LiC_4H_9-n, and quenching with CH_3I yielded 25% $O(CH_2)_4NSCH_3$. However, if the mixture was warmed to −20 °C in a similar process, only a 2.4% yield of $O(CH_2)_4NSCH_3$ was obtained [3].

Reacting the salt (prepared from $O(CH_2)_4NSSN(CH_2)_4O$) with CH_3I at room temperature gave $O(CH_2)_4NSCH_3$ with 67% yield and $O(CH_2)_4NSSCH_3$ with 1 to 5% yield [3].

Treatment of the salt (prepared from $O(CH_2)_4NSSN(CH_2)_4O$) with CH_3CN at room temperature gave $O(CH_2)_4NSSCH_2CH_2CN$ with 33% yield and $NCCH_2CH_2SCH_2CH_2CN$ with 41% yield [3].

7.4.7 $HSN{-}C(SCF_3){=}C(SCF_3){-}C(SCF_3){=}C(SCF_3)$

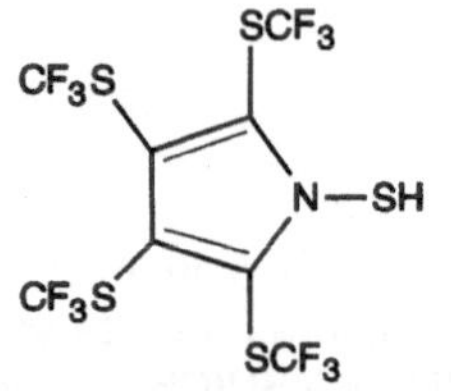

The title compound was prepared with 30% yield by treating 1-chloro-2,3,4,5-tetrakis(trifluoromethylsulfanyl)pyrrol in CCl_4 with excess NaSH (mole ratio ca. 2 : 3.5) at 45 °C for 3 h.

^1H NMR (?/TMS): δ = 5.23 ppm (NSH).

^{19}F NMR (?/$CFCl_3$): δ (in ppm) = −42.23, −43.47 (CF_3).

IR (KBr or NaCl): ν (in cm^{-1}) = 1571 m, 1491 m, 1258 m, 1170 vs, 1106 vs, 1001 s, 758 s, 690 m, 585 m.

MS (70 eV): m/e (rel. int. in %) = 498 (4) M^+, 466 (76) $M^+ - SH$, 430 (6) $M^+ - CF_3$.

The liquid compound is very unstable; it decomposed at room temperature within 2 h via [5]:

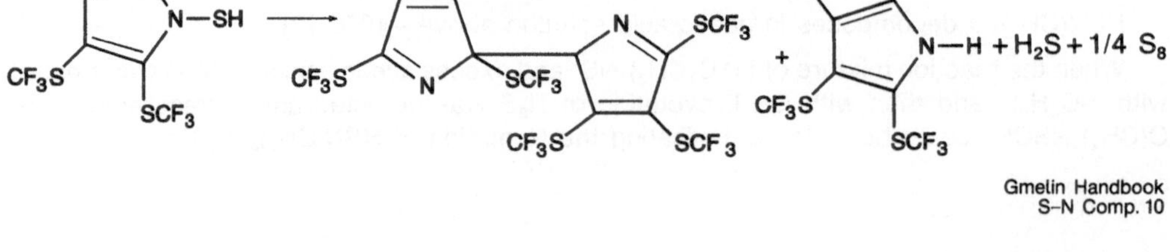

References:

[1] Nöth, H.; Mikulaschek, G. (Chem. Ber. **96** [1963] 1810/5).
[2] DePuy, C. H.; Bierbaum, V. M. (Tetrahedron Lett. **22** [1981] 5129/300).
[3] Barton, D. H. R.; Ley, S. V.; Magnus, P. D. (J. Chem. Soc. Chem. Commun. **1975** 855/6).
[4] Winn, M. (J. Org. Chem. **40** [1975] 955/6).
[5] Haas, A.; Klare, C. (Chem. Ber. **118** [1985] 4588/96).

7.5 Sodium Salt of N,N-Bis(trimethylsilyl)-thiohydroxylamine Ion (1−), $Na^+ ((CH_3)_3Si)_2NS^-$

Elemental sulfur reacted with a large excess of $((CH_3)_3Si)_2NNa$ in C_6H_6 to give the title compound which was not characterized.

Reference:

Scherer, O.; Schmidt, M. (Naturwissenschaften **50** [1963] 302).

7.6 S-Trimethylsilyl-N,N-bis(trimethylsilyl)-thiohydroxylamine, $(CH_3)_3SiSN(Si(CH_3)_3)_2$

The title compound was one of the products, when an ethereal solution of $(CH_3Si)_2NNC$ and $(CH_3)_3SiN{=}S{=}NSi(CH_3)_3$ was heated in a sealed tube at 100 °C for 12 h [1].

^{14}N NMR (solvent ?/external aqueous $NaNO_3$): $\delta = -329$ ppm (referenced to the CH_3NO_2 scale: $\delta = -332.7$ ppm) [2].

References:

[1] Wiberg, N.; Hübler, G. (Z. Naturforsch. **32b** [1977] 1003/9).
[2] Barlos, K.; Hübler, G.; Nöth, H.; Wanninger, P.; Wiberg, N.; Wrackmeyer, B. (J. Magn. Reson. **31** [1978] 363/76, 366).

7.7 S-Boranyl-N,N-dimethyl-thiohydroxylamine Dimethylamine Adduct, $(CH_3)_2NSBH_2 \cdot HN(CH_3)_2$

The title compound was formed by pyrolysis of $((CH_3)_2N)_2S \cdot BH_3$. Even though the compound was formed in all pyrolysis experiments independent of the reaction conditions, good yields were obtained only, when the decomposition was done under gentle conditions. When $((CH_3)_2N)_2S \cdot BH_3$ was slowly heated to 70 °C in vacuum (3 h), the compound was obtained by distillation at 0.5 to 1 Torr between 40 to 50 °C as an oily liquid which solidified slowly (impure material). It was also obtained, when the decomposition was performed at room temperature in vacuum for 14 h, at 70 to 90 °C at normal pressure, and in a closed tube at 120 °C for 24 h.

Boiling point: 47 to 49 °C/1 Torr; m.p. 28 °C.

IR: 25 absorptions between 3215 and 742 cm^{-1}.

Reference:

Nöth, H.; Mikulaschek, G. (Chem. Ber. **96** [1963] 1810/5).

8 Thiooximes and Derivatives

Despite the fact that oximes are one of the best known carbonyl derivatives, only a few sulfur-analog thiooximes are known because of their instability. The existence of thiooximate ions could be established by certain reactions: for example, $CH_2{=}NS^-$ was formed in the gaseous phase by an ion-molecule reaction.

8.1 $CH_2{=}NS^-$

$CH_2{=}NS^-$ was produced by the gas-phase reaction (using the flowing afterglow technique) of $CH_2{=}N^-$ with S=C=S and S=C=O: $CH_2{=}N^- + S{=}C{=}S \rightarrow CH_2{=}NS^- + CS$ and $CH_2{=}N^- + S{=}C{=}O \rightarrow CH_2{=}NS^- + CO$ [1].

$CH_2{=}N^-$ was generated from $(CH_3)_2SiCH_2N_3$. The heat of formation of $CH_2{=}NS^-$ could be calculated from this processes to be $\Delta H^\circ_{f,298} \leq 16$ kcal/mol. $CH_2{=}NS^-$ abstracts a proton from CH_3CO_2H ($\Delta H_{acid} = 348.5$ kcal/mol) but not from H_2S ($\Delta H_{acid} = 353.4$ kcal/mol); thus, its conjugated acid was assigned a value of $\Delta H_{acid} = 351 \pm 3$ kcal/mol. A resonance of the type $CH_2{=}NS^- \rightleftharpoons {}^-CH_2{=}N{=}S$ stabilizes the compound by approximately 10 kcal/mol [1].

8.2 $HSN{=}C(C_6H_5)_2$

The title compound was established by ^1H NMR spectroscopy, when a 0.5-M solution of $(C_6H_5)_2C{=}NSSi(CH_3)_3$ in $CDCl_3$ at −50 °C was quickly treated with a solution containing 1 mole equivalent of CH_3OH in $CDCl_3$. Within 15 min the ^1H NMR spectrum (external standard TMS) was recorded: In addition to the resonances associated with $(C_6H_5)_2C{=}NSSi(CH_3)_3$, a new resonance appeared at $\delta = 6$ ppm which was assigned to the SH group of $HSN{=}C(C_6H_5)$ [2].

Upon warming to room temperature the signal disappeared irreversibly, and sulfur precipitated in the NMR tube [2].

8.3 $Li^+(C_6H_5)_2C{=}NS^-$

Adding a suspension of sulfur in ether at room temperature to a stirred solution of $Li^+ (C_6H_5)_2C{=}N^-$ (mole ratio 1.4 : 1), prepared from $(C_6H_5)_2C{=}NH$ and C_4H_9Li in ether, gave within 4 h a bright orange solid which was proposed to be $Li^+ (C_6H_5)_2C{=}NS^-$. The compound was neither isolated nor characterized [2].

The salt formed with low yield by reacting $(C_6H_5)_2C{=}NSSN{=}C(C_6H_5)_2$ with an equimolar amount of n-C_4H_9Li in C_6H_6. Adding 2,4-dinitrofluorobenzene to the reacting solution to trap the anion $(C_6H_5)_2C{=}NS^-$ gave a number of products which included N-(2,4-dinitrophenylsulfanyl)-diphenylmethyleneamine, thus confirming the presence of the thiooximate anion [3].

When a mixture of $Li^+ (C_6H_5)_2C{=}N^-$ and sulfur in ether (see above) was cooled in liquid N_2 and then treated with excess $(CH_3)_3SiCl$ and warmed to room temperature, LiCl precipitated and $(C_6H_5)_2C{=}NSSi(CH_3)_3$ was obtained [2].

$Li^+ (C_6H_5)_2C{=}NS^-$ supposedly formed when $(C_6H_5)_2C{=}NSSN{=}C(C_6H_5)_2$ was reduced with lithium (3 equivalents)–ethylamine at −50 °C. Treating the resulting mixture with CH_3I gave $(C_6H_5)_2C{=}NSCH_3$ and $(C_6H_5)_2C{=}NSSCH_3$ [4].

8.4 $(CH_3)_3SiSN{=}C(C_6H_5)_2$

The title compound was prepared with 95% yield by treating a liquid-nitrogen-cooled solution of Li^+ $(C_6H_5)_2C{=}NS^-$ in ether with excess $(CH_3)_3SiCl$. Subsequently, the reaction mixture was warmed to room temperature and stirred for 3 h, the solvent was removed, and the remaining yellow oil was filtered to remove LiCl. Attempts to prepare the compound from Li^+ $(C_6H_5)_2C{=}N^-$ (obtained from $(C_6H_5)_2C{=}NSSN{=}C(C_6H_5)_2$ and $Li{-}(C_2H_5)_3N$ or $n\text{-}C_4H_9Li$) by reaction with sulfur and $(CH_3)_3SiCl$ gave the desired product contaminated with significant amounts of $(C_6H_5)_2C{=}NSi(CH_3)_3$, $(C_6H_5)_2C{=}NSC_4H_9\text{-n}$, and $(C_6H_5)_2C{=}NC_4H_9\text{-n}$.

^1H NMR (CCl_4/TMS): δ (in ppm) = 0.0 (s, $Si(CH_3)_3$), 6.8 to 7.7 (m, $(C_6H_5)_2$).

UV (hexane): λ_{max} (in nm) = 240, 290, 328 sh.

The title compound is stable for prolonged periods of time at −25 °C. It decomposes at room temperature with a half life of ~1 day with liberation of sulfur [2].

8.5 $HSN{=}\overline{C{-}C(SCH_3){=}C(C_4H_9\text{-}t){-}CH{=}C}(C_4H_9\text{-}t)$

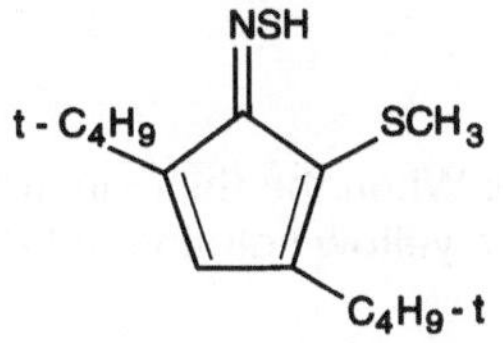

The title compound was formed by acidic hydrolysis of the thiooximate I at 25 °C. It oxidized in air to give the disulfide II with 21% yield [5].

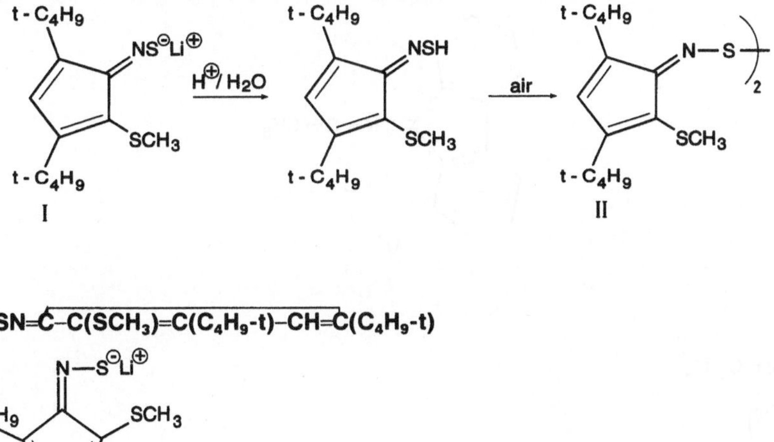

I II

8.6 $Li^+\ {}^-SN{=}\overline{C{-}C(SCH_3){=}C(C_4H_9\text{-}t){-}CH{=}C}(C_4H_9\text{-}t)$

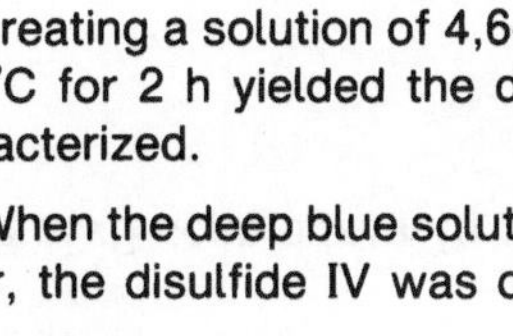

Treating a solution of 4,6-di-*tert*-butylcyclopenta-1,2,3-dithiazole (III) in THF with CH_3Li at −80 °C for 2 h yielded the deep blue thiooximate anion. The salt was neither isolated nor characterized.

When the deep blue solution was hydrolyzed with acid at 25 °C and subsequently oxidized in air, the disulfide IV was obtained [5].

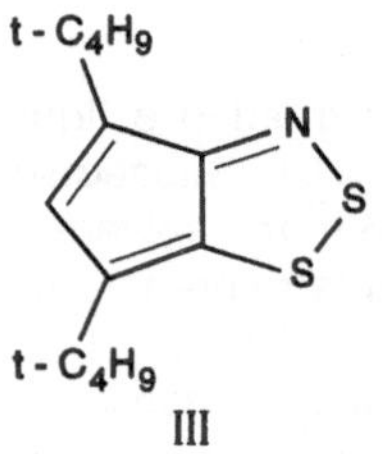
III

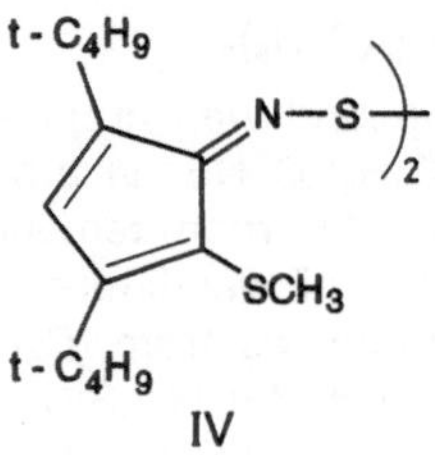
IV

8.7 9-Fluorenone Thiooxime, $HSN{=}C_{13}H_8$

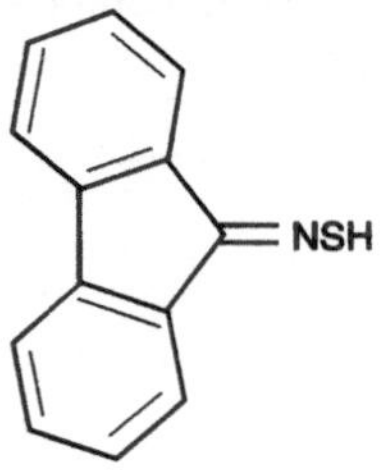

There is evidence for the existence of 9-fluorenone thiooxime: When the thiooximate salt Li^+ $^-SN{=}C_{13}H_8$ (see below) was treated with acetic acid at −70 °C, a yellow solution of 9-fluorenone thiooxime formed.

The title compound decomposes above −70 °C, $^-SN{=}C_{13}H_8$ could be regenerated by adding aqueous KOH to the yellow solution of 9-fluorenone thiooxime in acetic acid. The reaction of the yellow solution in acetic acid with CH_3N_2 gave compound V with 13% yield [4].

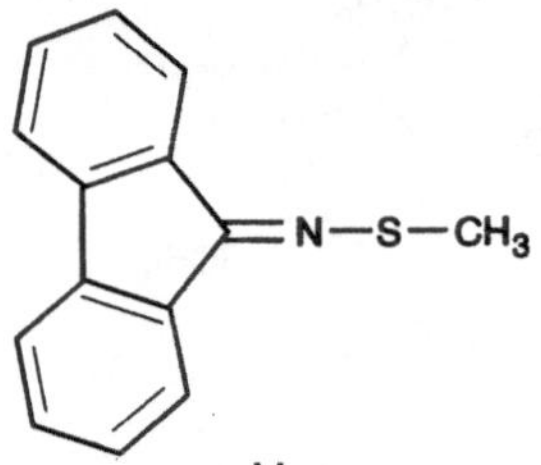
V

8.8 Li^+ $^-SN{=}C_{13}H_8$

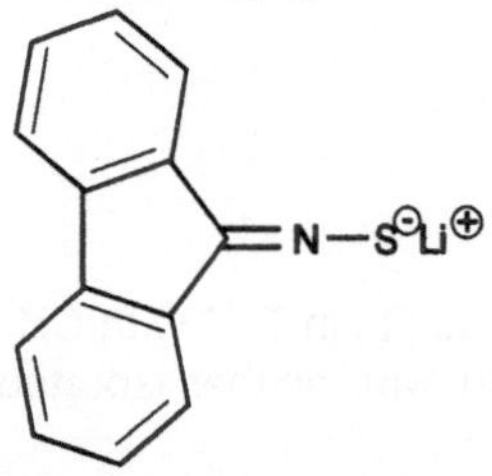

Reduction of the disulfide VI with $Li–NH_3$ at −70 °C followed by evaporation gave a red solid which was presumed to be the title compound.

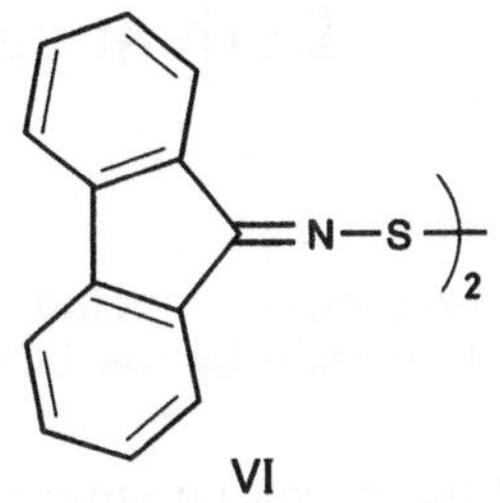

VI

UV (violet solution in THF): λ_{max} (in nm) (ε in $L \cdot mol^{-1} \cdot cm^{-1}$) = 251 (23800), 270 (3000), 575 (6200).

The salt is stable under Ar at room temperature, but it readily deteriorates in the presence of O_2. THF solutions decompose slowly into at least eight compounds: Sulfur, 9H-fluorene-9-thiol, bis(9H-fluorene-9-yl) disulfide, bifluorenylidene, 9-fluorenone, and 9-fluorenylidene-amine were detected by thin-layer chromatography. Adding acetic acid to the salt at −70°C produced a pale yellow solution of 9-fluorenone thiooxime. When the reaction was performed at −20°C H_2S and several other compounds were formed [4].

References:

[1] Kass, S. R.; Depuy, C. H. (J. Org. Chem. **50** [1985] 2874/7).
[2] Pike, S.; Walton, D. R. M. (Tetrahedron Lett. **21** [1980] 4989/90).
[3] Brown, C.; Grayson, B. T.; Hudson, R. F. (J. Chem. Soc. Chem. Commun. **1974** 1007/8).
[4] Barton, D. H. R.; Magnus, P. D.; Pennanen, S. I. (J. Chem. Soc. Chem. Commun. **1974** 1007).
[5] Hafner, K.; Stowasser, B.; Sturm, V. (Tetrahedron Lett. **26** [1985] 189/92).

9 Thiofulminic Acid and Derivatives

9.1 Thiofulminic Acid, HSN≡C

HSN≡C has not yet been prepared.

The energy of a geometry-optimized model structure of HSN≡C (and the isomers HCNS, NCSH, and HNCS) was calculated ab initio by the LCAO-MO-SCF technique giving E_T = −490.17332 hartrees.

The geometrical parameters of the optimized structure are: d(H−S) = 1.334 Å, d(S−N) = 1.678 Å, d(N−C) = 1.166 Å, ∢(H−S−N) = 96.1°, ∢(S−N−C) = 176.3°.

Calculated rotational constants A, B, C (in MHz) are 259610, 6277, and 6147, respectively.

The total dipole moment was calculated to be μ = 3.05 Debye.

Reference:

Bak, B.; Christiansen, J. J.; Nielsen, O. J.; Svanholt, H. (Acta Chem. Scand. A **31** [1977] 666/8).

9.2 C≡NS, Thiofulminic Acid Ion (1−), C≡NS⁻, Metal Complexes of C≡NS⁻

CNS and CNS^-. CNS and CNS^- have not yet been detected, but they should be observable in the gas phase as predicted by theoretical investigations.

The equilibrium geometries of CNS and CNS^-, having both $C_{\infty v}$ point group symmetry, were theoretically determined at the HF/6-31G* level (in Å). **CNS**: r(C−N) = 1.66, r(N−S) = 1.631. **CNS^-**: r(C−N) = 1.154, r(N−S) = 1.710. The CN bond lengths indicate a triple bond; the SN bond lengths correspond to a single bond.

Calculated total energies of CNS at the $^2\Pi$ state and of CNS^- at the $^1\Sigma^+$ state at HF/6-31G*-optimized geometries are (in hartrees):

E_T(CNS) = −489.7493 (6-31G*), −489.7552 (6-31 + G*), −490.1234 (MP2/6-31 + G*),
E_T(CNS^-) = −489.8730 (6-31G*), −489.8291 (6-31 + G*), −490.2296 (MP2/6-31 + G*).

Theoretically determined vibrational frequencies at the 6-31G*/6-31G* level are:
CNS: ν (in cm^{-1}) = 233 (π), 690 (σ_g), 2025 (σ_g); **CNS^-**: ν (in cm^{-1}) = 321 (π), 570 (σ_g), 2109 (σ_g).

The theoretical adiabatic electron affinity (EA) and vertical electron detachment energies (ED), respectively, of CNS were calculated at the MP2/6-31+G*//6-31G* + ZPE level: EA = 2.89 eV, ED = 3.00 eV [1].

$K_xFe(CN)_yS_z \cdot n\,H_2O$. Complexes, believed to contain the CNS^- ligand, were obtained when $K_3Fe(CN)_6$ was reacted with molten sulfur at 200 °C for 3 h yielding a dark green polymeric substance with the composition $K_{1.31}Fe_{1.00}(CN)_{4.62}S_{0.54} \cdot 0.40\,H_2O$, $K_{1.27}Fe_{1.00}(CN)_{4.45}S_{0.39} \cdot 0.38\,H_2O$, $K_{1.51}Fe_{1.00}(CN)_{4.38}S_{0.42} \cdot 0.30\,H_2O$, $K_{1.41}Fe_{1.00}(CN)_{4.34}S_{0.57} \cdot 0.36\,H_2O$, or $K_{1.35}Fe_{1.00}(CN)_{4.23}S_{0.40} \cdot 0.26\,H_2O$.

The magnetic moment at 25 °C ranges from 4.23 to 4.44 BM, indicating coordination of Fe with CNS^- ligands.

IR (KBr): ν (in cm^{-1}) = 2021 vs, 2045 sh, 2028 sh, 1215 s, 590 m, 490 s, 455 s, sh (the spectrum was tentatively assigned in the paper [3]).

The dark green substance is insoluble in H_2O [3].

$K_3Fe(CN)_4(CNS)_2$. Repeated extraction of the dark green material, $K_xFe(CN)_yS_z \cdot n\ H_2O$ (see p. 100), with H_2O at 55 °C followed by filtration and treatment of the filtrate with 1,4-dioxane gave a yellow compound which was believed to be $K_3Fe(CN)_4(CNS)_2$ [3].

The compound is paramagnetic. The magnetic moment at 25 °C ranges from 2.61 to 2.73 BM [3].

IR (KBr): ν (in cm^{-1}) = 2115 s, 2093 m, 2072 s, 2062 s, 2040 vs, 2025 vs, 1108 s, 610 m, 578 s, 383 m; no absorptions were found in the ranges 690 to 720 and 780 to 860 cm^{-1} where C–S stretching vibrations for S-bonded and N-bonded thiocyanate, respectively, are normally observed (for tentative assignments, see the paper) [3].

On heating the compound for 12 h at 100 °C and then for 24 h at 145 °C, sulfur was eliminated and $K_3Fe(CN)_6$ obtained [3].

CuCNS and CuSNC. The potential surface of the isomerization process CuCNS ⇌ CuSNC, which corresponds to the migration of a Cu^+ cation around the CNS^- anion, has been calculated by the nonempirical MO-LCAO-SCF method. It was found to contain two maxima which are separated by a considerable energy barrier and correspond to the linear CuCNS (ground state) and cyclic C–N–Cu–S (excited) structures. The linear structure CuSNC, where Cu^+ is coordinated to the S atom, lies ca. 28 kcal higher along the energy scale and is unstable towards transition to the cyclic structure C–N–Cu–S. The following geometrical parameters and total energies of CuCNS and CuSCN were calculated in the SCF approximation: CuCNS ($C_{\infty v}$): r(C–N) = 1.16 Å, r(N–S) = 1.78 Å, r(Cu–C) = 1.97 Å; E_T = –2128.5159 au. CuSNC ($C_{\infty v}$): r(C–N) = 1.16 Å, r(N–S) = 1.71 Å, r(Cu–S) = 2.24 Å; E_T = –2128.4712 au [4].

References:

[1] Koch, W.; Frenking, G. (J. Phys. Chem. **91** [1987] 49/53).

[2] Qingyun, C.; Shizheng, Z. (Sci. Sin. [Engl. Ed.] B **30** [1987] 561/71).

[3] Cole, S. B.; Kleinberg, J. (Inorg. Chim. Acta **10** [1974] 157/61).

[4] Musaev, D. G.; Charkin, O. P. (Koord. Khim. **17** [1991] 548/51; Sov. J. Coord. Chem. [Engl. Transl.] **17** [1991] 294/7).

10 Isohypothiocyanous Acid Ion (1−), $OCNS^-$

Hartree Fock- (HF) and Møller-Plesset-level 6-31G* (MP2) calculations support the existence of the noncharacterized hypothiocyanite ion $OSCN^-$, but suggest that its structure is $OCNS^-$, having the lowest energy at the HF and MP2 levels.

Calculated total energies (in au): $E_T = -564.70713$ (HF), −565.30752 (MP2).

Calculated geometrical parameters (bond lengths in Å): S–N = 1.730 (HF), 1.707 (MP2); C–N = 1.152 (HF), 1.189 (MP2); O–C = 1.195 (HF), 1.225 (MP2).

Vibrational frequencies calculated at the HF level (ν in cm^{-1}): $\nu_1 = 123$, $\nu_2 = 535$, $\nu_3 = 653$, $\nu_4 = 1496$, $\nu_5 = 2622$.

Calculations were also carried out on $SCNS^-$; see the paper.

Reference:

Pyykö, P.; Runeberg, N. (J. Chem. Soc. Chem. Commun. **1991** 547/8).

11 Thionitrous Acid and Derivatives

11.1 Thionitrous Acid, *cis*-, *trans*-HSNO and *cis*-, *trans*-DSNO

11.1.1 Preparation

cis- and *trans*-HSNO were generated together with *cis*-HOSN and SNO by photolysis of *cis*-HNSO in an Ar matrix using a medium-pressure mercury lamp [1] or 250-nm light [2]. The compounds were identified by IR spectroscopy [1, 2]. When the mixture was further irradiated with 365-nm light for several hours, the IR absorption of HSNO and HOSN steadily decreased and new IR absorption bands due to *cis*-HNSO appeared in addition to absorption bands belonging to *trans*-HNSO and HONS [2]. HSNO, DSNO, and $HS^{15}NO$ were prepared photolytically ($\lambda \approx 250$ nm) from HNSO, DNSO, and $H^{15}NSO$, respectively, in an Ar matrix [3].

Scheme 1 summarizes the various light-induced, unimolecular steps involved in the transformation process: When a freshly prepared sample of *cis*-HNSO in an Ar matrix is irradiated with 250-nm light, a mixture of *cis*-HSNO and *trans*-HSNO is formed via *cis*-HOSN. A second photolysis step at 365 nm causes *cis*- and *trans*-HSNO to disappear and *trans*-HONS to form. Simultaneously, a portion of the initial *cis*-HNSO is reformed. If this sample is then irradiated with $\lambda > 610$ nm, HONS is transformed back into HSNO but without forming HOSN. A further irradiation process at 365 nm produces HONS, a fact that establishes HSNO to be the precursor of HONS. In separate processes the cis-trans cycle of HSNO can selectively be induced by using either light in the IR region or 585-nm light [2].

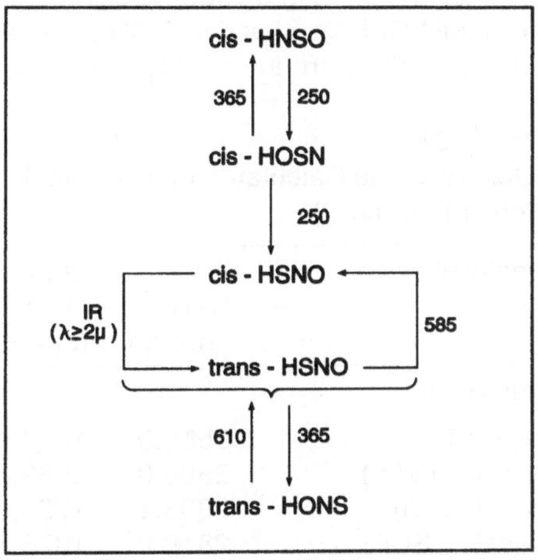

Scheme 1. Light-induced formation and transformation steps of *cis*- and *trans*-HSNO in an Ar matrix (wavelengths in nm) [2].

11.1.2 Molecular Properties. Spectra

Geometrical Sructure

The geometrical structures of *cis*- and *trans*-HSNO and the transition state (see Table 26, p. 104) were obtained from ab initio calculations at the SCF- [2, 3], CI- [3], and M3*-levels [4]; ad hoc structural parameters are given in [5].

Table 26
Geometrical Parameters of *cis*- and *trans*-HSNO and the Transition State (bond lengths in Å, bond angles in °).

	H–S	S–N	N–O	H–S–N	S–N–O	torsion	Ref.
cis-HSNO	1.337	1.752	1.170	97.6	116.3	0.0	[2]
	1.336	1.894	1.174	92.485	113.216	0.000	[3]
	1.335	1.807	1.152	95.962	114.421	0.000	[3]
	1.334	1.703	1.238	97.3	116.1		[4, 5]
trans-HSNO	1.331	1.763	1.168	93.4	113.7	180.0	[2]
	1.333	1.895	1.171	90.198	113.584	180.000	[3]
	1.381	1.815	1.150	93.041	112.894	180.000	[3]
	1.330	1.714	1.238	93.0	111.4		[4]
	1.33	1.70	1.15	95.0	115.0	180.0	[1]
transition state	1.334	1.853	1.157	92.7	113.4	86.5	[2]
gauche-HSNO	1.329	1.798	1.245	91.9	111.0		[4, 5]

Ab initio torsional potential functions are presented in [4, 5].

Molecular Vibrations

Complete sets of fundamental frequencies of both rotamers were obtained and assigned by normal-coordinate analysis using the transferable valence force field (TVFF) [3] and by ab initio MO-SCF [2, 3] and MO-CI calculations [3]. The assignments of the molecular vibrations 5(a′) and 6(a″) are interchanged in [1]. The data are summarized in Table 27.

Table 27
Observed and Calculated Fundamental Frequencies of *cis*- and *trans*-HSNO, -DSNO, and Isotopes (ν in cm^{-1}).

molecule	1(a′) S–H (stretch)	2(a′) N–O (stretch)	3(a′) H–S–N (bend)	4(a′) S–N–O (bend)	5(a′) S–N (stretch)	6(a″) torsion	Ref.
***cis*-HSNO**							
ν(obs.)	2566.0	1570.0	858.5	503.0	307.0	406.5	[3]
ν(calc.,TVFF)	2566.0	1569.3	854.6	500.7	309.7	406.5	[3]
ν(calc., CI)	2803.1	1725.4	910.7	518.4	301.6	390.0	[3]
ν(calc., SCF)	2819.2	1872.1	935.7	682.5	416.2	407.1	[2, 3]
***cis*-DSNO**							
ν(obs.)	—	1568.5	715.0	435.0	305.5	—	[3]
ν(calc., TVFF)	1842.9	1569.2	720.5	436.6	302.0	321.3	[3]
***cis*-H^{34}SNO**							
ν(calc., TVFF)	2563.7	1568.5	853.2	499.8	305.8	406.0	[3]
***cis*-HS^{15}NO**							
ν(obs.)	2566.0	1544.0	851.5	497.0	303.5	405.0	[3]
ν(calc., TVFF)	2566.0	1543.3	849.0	493.2	306.6	404.7	[3]

Table 27 (continued)

molecule	1(a′) S–H (stretch)	2(a′) N–O (stretch)	3(a′) H–S–N (bend)	4(a′) S–N–O (bend)	5(a′) S–N (stretch)	6(a″) torsion	Ref.
***cis*-HSN^{18}O**							
ν(calc., TVFF)	2566.0	1526.2	849.9	494.3	305.1	404.8	[3]
***trans*-HSNO**							
ν(obs.)	2613.0	1596.0	877.5	543.5	297.0	386.5	[3]
ν(obs.)	—	1596.8	876	542	500	298	[1]
ν(calc., TVFF)	2613.0	1596.3	879.6	546.1	297.9	386.7	[3]
ν(calc., CI)	2875.9	1751.7	967.6	591.6	373.2	369.8	[3]
ν(calc., SCF)	2891.0	1894.6	1082.9	649.7	471.8	375.9	[2, 3]
ν(calc.)	2636.9	1597.0	877.0	543.9	498.6	295.2	[1]
***trans*-DSNO**							
ν(obs.)	—	1594.5	724.0	485.5	297.0	—	[3]
ν(obs.)	—	1595.1	722.7	488	—	294	[1]
ν(calc.,TVFF)	1876.9	1594.2	721.4	282.8	296.8	302.3	[3]
ν(calc.)	1893.8	1596.7	721.3	487.0	390.7	294.8	[1]
***trans*-H^{34}SNO**							
ν(calc., TVFF)	2610.6	1595.4	878.0	345.7	293.0	386.6	[3]
***trans*-HS^{15}NO**							
ν(obs.)	2613.0	1569.0	874.3	533.0	294.0	385.0	[3]
ν(obs.)	—	1569.4	872.8	533.0	495	294.9	[1]
ν(calc., TVFF)	2613	1569.3	876.4	336.6	394.9	385.0	[3]
ν(calc.)	2336.9	1567.7	873.6	532.2	496.3	293.9	[1]
***trans*-HSN^{18}O**							
ν(calc., TVFF)	2613.0	1554.3	875.4	338.1	293.6	385.1	[3]

Force Constants

Diagonal force constants of *cis*- and *trans*-HSNO from ab initio MO-CI and MO-SCF calculations and with a transferable valence force field (TVFF) (see Table 28) were reported in [3]; interaction force constants are listed in the paper [3]. A force field using ad hoc structural parameters for *trans*-HSNO is given in [1]; attempts to fit a *cis* structure to the same data were unsuccessful [1].

Table 28
Force Constants of *cis*- and *trans*-HSNO (stretching and bending constants in mdyn/Å and mdyn·Å/rad^2, respectively).

molecule	f_{SH}	f_{NO}	f_{HSN}	f_{SNO}	f_{SN}	$f_{torsion}$	Ref.
***cis*-HSNO**							
TVFF[a)]	3.7900	11.5015	0.5851	1.6749	1.6344	0.1312	[3]
	(0.0122)	(0.5463)	(0.0131)	(0.2437)	(1.2631)	(0.0027)	
MO-CI	4.5208	13.8752	0.6578	1.7345	1.0819	0.1212	[3]
MO-SCF	4.5768	16.3909	0.8731	1.9603	1.8738	0.1273	[3]

Table 28 (continued)

molecule	f_{SH}	f_{NO}	f_{HSN}	f_{SNO}	f_{SN}	$f_{torsion}$	Ref.
***trans*-HSNO**							
TVFF[a)]	3.9299	11.7146	0.6848	1.4201	1.5441	0.1228	[3]
	(0.0124)	(0.6191)	(0.0134)	(0.1577)	(1.2481)	(0.0026)	
MO-CI	4.6188	14.1695	0.7729	1.5615	1.2516	0.1112	[3]
MO-SCF	4.6671	16.5435	0.9686	1.7633	1.8899	0.1142	[3]
tentative	4.0	11.04	0.671	0.562	1.79	0.200	[1]
fixed[a)]	(0.04)	(0.004)	(0.003)	(0.02)	(0.001)		

[a)] Numbers in parentheses represent the standard deviation.

Calculations of Excited States

The lowest electronically excited states of the more stable conformer *trans*-HSNO were calculated by CI and SCF optimizations to be: S_1 ($n\pi^*$) 2.09 eV or 2.23 eV, S_2 ($\pi\pi^*$) 4.00 or 3.86 eV; average values $S_1 \leftarrow S_0$ 580 nm, $S_2 \leftarrow S_0$ 320 nm. The theoretical results are consistent with the absorption behavior found in the photolysis experiments [2]; see Scheme 1.

IR Spectrum

Complete sets of fundamental frequencies in the IR spectra of *cis*- and *trans*-HSNO and their isotopic species observed in the Ar matrix are given in Table 27. Additional absorptions were observed in the IR spectra of the Ar matrix (ν in cm^{-1}, assignment, relative absorption in %) [3]:

***cis*-HSNO.** 1891.5 ($\nu_2+\nu_5$, 8), 1575.5 (ν_2-site, 20), 843.5 (ν_3-site, 2); *cis*-$HS^{15}NO$: 1860.5 ($\nu_2+\nu_5$), 1549 (ν_2-site), 836.5 (ν_3-site); *cis*-DSNO: 1885.5 ($\nu_2+\nu_5$) [3].

***trans*-HSNO.** 2607 (ν_1-site, <1), 1902.5 ($\nu_2+\nu_5$, 8), 1612 (ν_2-site, 18), 1590 (ν_2-site, 85), 547.5 (ν_4-site, 15), 381 (ν_6-site, 2); *trans*-$HS^{15}NO$: 2607 (ν_1-site), 1873 ($\nu_2+\nu_5$), 1583 (ν_2-site), 1563.5 (ν_3-site), 537 (ν_4-site), 368.5 (ν_6-site); *trans*-DSNO: 1898.5 ($\nu_2+\nu_5$), 1587 (ν_2-site) [3].

Similar infrared frequencies, isotopic shifts, and tentative assignments of HSNO in Ar matrix are given in [1].

11.1.3 Isomerization

Trans→cis isomerization of HSNO in an Ar matrix at 12 K was induced with UV ($\lambda = 250$ nm) and visible ($\lambda = 585$ nm) radiation. Cis→trans isomerization was induced by IR ($\lambda \geq 2$ μ) radiation of HSNO in an Ar matrix. Changes of the cis : trans ratio could be determined from IR intensity measurements [2, 3]. The isomerization processes are shown in Scheme 1, p. 103.

Ab initio MO-SCF and MO-CI calculations were performed to predict the energy geometry and barrier of internal rotation for the two HSNO rotamers. The barrier height of the cis–trans isomerization process in terms of the energy difference between the trans form and the transition state (TS) was calculated at the SCF [2] or CI level [3] to be ΔE (trans→TS) = 3200 cm^{-1} [2] or 3180 cm^{-1} [3]. The relative ground-state energy of the cis conformer is $E(cis)_{rel}$ = 70 kJ/mol [2]. The energy difference between the more stable trans isomer and the cis isomer was calculated by the ab initio MO-CI method to be E(trans) − E(cis) = −3.24 kJ/mol [3].

11.1.4 Chemical Reaction

Irradiation of HSNO in an Ar matrix with 365-nm light produced *trans*-HONS [2].

References:

[1] Tchir, P. O.; Spratley, R. D. (Can. J. Chem. **53** [1975] 2318/30).
[2] Nonella, M.; Huber, J. R.; Tae-Kyu Ha (J. Phys. Chem. **91** [1987] 5203/9).
[3] Müller, R. P.; Nonella, M.; Russegger, P.; Huber, J. R. (Chem. Phys. **87** [1984] 351/61).
[4] Cardenas-Jiron, G. I.; Cardenas-Lailhacar, C.; Torro-Labbe, A. (J. Mol. Struct. **210** [1990] 279/89 [THEOCHEM **69**]).
[5] Cardenas-Jiron, G. I.; Torro-Labbe, A. (FCTL Folia Chim. Theor. Lat. **17** [1989] 177/90).

11.2 Salt of Thionitrous Acid Ion (1−), $((C_6H_5)_3P)_2N^+$ $O{=}NS^-$

$((C_6H_5)_3P)_2N^+$ $O{=}NS^-$ was prepared by stirring a solution of 1.00 mmol $((C_6H_5)_3P)_2N^+$ $O{=}NSS^-$ and 2.02 mmol $(C_6H_5)_3P$ in 40 mL $(CH_3)_2CO$ for 24 h. The reaction was complete when the color of the solution changed from red to olive green. Subsequently, 40 mL $(C_2H_5)_2O$ were condensed onto the cooled reaction solution. On warming the solution to ambient temperature, $(C_2H_5)_2O$ diffused into the solution, and subsequently green crystals of $((C_6H_5)_3P)_2N^+$ $O{=}NS^- \cdot (CH_3)_2CO$ formed at the contact surface of the two liquids [1, 2].

The crystal structure of $((C_6H_5)_3P)_2N^+$ $O{=}NS^- \cdot (CH_3)_2CO$ at −133 °C was obtained by X-ray diffraction analysis. The crystals belong to the triclinic space group $P\bar{1}-C_i^1$ (No. 2) with the parameters a = 10.525(3) Å, b = 17.196(5) Å, c = 9.451(3) Å, α = 95.22(3)°, β = 92.32(3)°, γ = 91.65(3)°; V = 1701.1 Å³; Z = 2; D_x = 1.286 g/cm³; R = 0.054 from 6006 reflections [1].

The following geometrical parameters of $O{=}NS^-$ were determined: r(N−O) = 1.214(5) Å, r(N−S) = 1.695 Å, ∢(O−N−S) = 120.5(3)°. The geometrical data of $((C_6H_5)_3P)_2N^+$ and $(CH_3)_2CO$ are given in the paper [1]. The cation has approximately C_{2v} symmetry [1].

^{15}N NMR $((CH_3)_2CO/((C_6H_5)_3P)_2N^+\ {}^{15}NO_3^-)$: δ = 529 ppm [1, 2].

UV $((CH_3)_2CO)$: λ_{max} = 350 nm, ε = 3050 L·mol⁻¹·cm⁻¹ [1].

$((C_6H_5)_3P)_2N^+$ $O{=}NS^- \cdot (CH_3)_2CO$ liberated 1 mol of $(CH_3)_2CO$ per mole of salt when heated to 80 °C in an oil-pump vacuum. When the salt was heated to 160 °C in an oil-pump vacuum, 0.5 mol of N_2O per mol salt was evolved: $2\,((C_6H_5)_3P)_2N^+\,O{=}NS^- \rightarrow (C_6H_5)_3P + (C_6H_5)_3PS + N_2O$. The salt decomposed within several hours of exposure to air [1].

The reaction of $((C_6H_5)_3P)_2N^+$ $O{=}NS^-$ with elemental sulfur in $(CH_3)_2CO$ gave $((C_6H_5)_3P)_2N^+$ $O{=}NSS^-$ within seconds [1].

References:

[1] Seel, F.; Kuhn, R.; Simon, G.; Wagner, M.; Krebs, B.; Dartmann, M. (Z. Naturforsch. **40b** [1985] 1607/17).
[2] Seel, F.; Wagner, M. (Z. Naturforsch. **40b** [1985] 762/4).

12 Sulfur Amide Halogenides and Derivatives

12.1 Sulfur Amide Fluoride, Amino-fluoro-sulfane, $FSNH_2$

Full-geometry optimizations at the HF/6-31G* level of ab initio SCF-MO theory were carried out for the conformers and the inversion and rotation transition structures of the hypothetical molecule $FSNH_2$. Geometries and energies for the conformers are given in Table 29.

Table 29
HF/6-31G* Geometries and Energies for Conformers of $FSNH_2$.

sym.	conformn.[a]	E_T in au	E_{rel}	E_{rot} in kcal/mol	E_{inv}	R(SN) in Å	R(SF) in Å	$\Delta\theta(N)$[b] in °	$\theta(FSN)$ in °
C_s	P-pl	−552.47258	27.00			1.698	1.608	0.0	95.1
C_1	P-st	−552.48650	18.27		8.73	1.733	1.609	41.3	96.2
C_s	A-pl	−552.51327	1.47	25.53		1.615	1.625	0.0	103.8
C_s	A-st	−552.51561	0.00	18.27		1.631	1.626	11.6	103.9

[a] Notation for conformation is defined as follows: The dihedral relationship between the nitrogen lone pair and the S–F bond is designated by A (anti, ~180°), S (syn, ~0°), or P (perpendicular, ~90°). The configurations of the hydrogens at N with respect to the FS group are denoted by "st" (staggered) or "pl" (planar). – [b] The degree of pyramidalization at the nitrogen of the NH_2 group is represented by $\Delta\theta(N)$, which is equal to 360 minus the sum of the three bond angles at N.

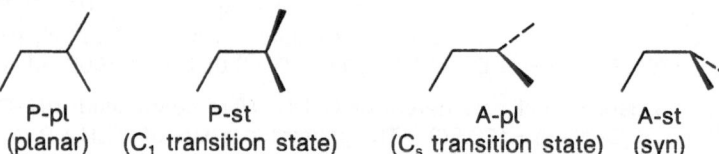

P-pl (planar) P-st (C_1 transition state) A-pl (C_s transition state) A-st (syn)

The net energetic stabilization according to the isodesmic equation $FSNH_2 + SH_2 \rightarrow FSH + HSNH_2$ for the conformers of $FSNH_2$ was calculated to be E_{stab}(P-pl) = 8.04, E_{stab}(P-st) = 8.66, E_{stab}(A-pl) = 19.48, and E_{stab}(A-st) = 18.69 kcal/mol. The stabilization energy of the most favorable A-st (syn) (18.7 kcal/mol) is relatively large. No anti conformer exists.

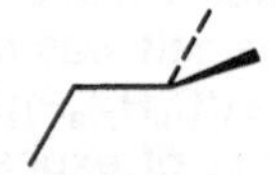

S-st (anti)

A detailed natural bond orbital (NBO) analysis of $FSNH_2$ showed that the particularly strong generalized anomeric effect in $FSNH_2$ (lengthening of the S–N bond by 0.10 Å during internal rotation associated with a barrier of 18 kcal/mol) can be quantitatively explained in terms of negative hyperconjugation, primarily of the $n_N \rightarrow \sigma^*_{SF}$- and secondarily of the $\pi_S \rightarrow \sigma^*_{NH}$-type.

Reference:

Reed, A. E.; von Ragué Schleyer, P. (Inorg. Chem. **27** [1988] 3969/87).

12.2 Substituted Amino-fluoro-sulfanes

12.2.1 N,N-Dialkyl-amino-fluoro-sulfanes, $FSNR_2$, $R = CH_3$, C_2H_5

The title compounds were formed by passing $ClSN(CH_3)_2$ or $ClSN(C_2H_5)_2$ vapor at low pressure through a column heated to 120 °C and containing AgF [1].

$FSN(CH_3)_2$

^1H NMR (neat ?/TMS, −50 °C): $\delta = 3.30$ ppm (d, CH_3), $^4J(F,H) = 13.2$ Hz [1].

^{19}F NMR (neat ?/$CFCl_3$, −50 °C): $\delta = -163.0$ ppm (sept) [1].

$FSN(C_2H_5)_2$

^1H NMR (neat ?/TMS, −50 °C): δ (in ppm) = 1.24 (t, CH_3), 3.48 (m, CH_2); $^3J(H,H) = 7.0$, $^4J(F,H) = 11.9$ Hz [1].

^{19}F NMR (neat ?/$CFCl_3$, −50 °C): $\delta = -161.5$ ppm (quint) [1].

The title compounds easily decompose, which makes it difficult to determine their properties [2].

The title compounds rapidly decompose in glass vessels via $3\ FSNR_2 + O^{2-}$ (from glass) → $FS(O)NR_2 + R_2NSSNR_2 + 2\ F^-$, $R = CH_3$, C_2H_5 [1].

References:

[1] Seel, F.; Gombler, W.; Budenz, R. (Z. Naturforsch. **27b** [1972] 78/9).
[2] Varwig, J.; Steinbeißer, H.; Mews, R.; Glemser, O. (Z. Naturforsch. **29b** [1974] 813/4).

12.2.2 N-Fluorosulfanyl-2,2,6,6-tetramethyl-piperidine, $FS\overline{N-C(CH_3)_2-(CH_2)_3-C}(CH_3)_2$

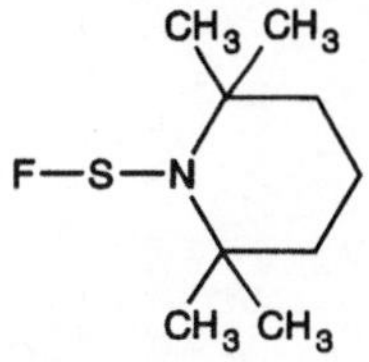

The title compound was formed by refluxing $ClS\overline{N-C(CH_3)_2-(CH_2)_3-C}(CH_3)_2$ and CsF (mole ratio 3 : 4) in toluene for 2 d. Solids were then removed by filtration, the solvent removed by distillation, and the remaining yellow liquid distilled in vacuum.

78% yield; liquid, b.p. 104 °C/12 Torr.

^1H NMR (C_6H_6/TMS): δ (in ppm) = 1.55, 1.45 (CH_3, CH_2; assignment not unambiguous).

^{19}F NMR (C_6H_6/$CFCl_3$): $\delta = -174.0$ ppm.

MS (70 eV): m/e (rel. int. in %) = M^+ (15), $M^+ - F$ (19), $M^+ - F - CH_3$ (8), $M^+ - SF - CH_3$ (53).

At room temperature the title compound is stable for a long time when stored in a dry glass vessel. In the presence of moisture it disproportionates to $FS(O)N(CH_2)_5$ and $(CH_3)_2\overline{C-(CH_2)_3-C(CH_3)_2-N}SS\overline{N-C(CH_3)_2-(CH_2)_3-C}(CH_3)_2$.

Reference:

Röschenthaler, G.-V.; Starke, R. (Z. Naturforsch. **32b** [1977] 721/2).

12.2.3 N-Hexafluoroisopropylidene-amino-fluoro-sulfane, $FSN{=}C(CF_3)_2$

$FSN{=}C(CF_3)_2$ was prepared by placing $Hg(NSF_2)_2$ in a glass bomb and then condensing onto frozen $(CF_3)_2CN_2$ (mole ratio ca. 1 : 1.8). The reaction mixture was then defrosted to react at room temperature within 24 h (occasionally exploding). The product was fractionally condensed at −80 °C in high vacuum and finally distilled under normal pressure; 75% yield [1].

Colorless liquid; b.p. 55 °C [1].

^{19}F NMR (neat ?/$CFCl_3$?): δ (in ppm) = −71.5 (SF), −67.7 (CF_{3A}), −66.1 (CF_{3B}), intensity ratio 1 : 3 : 3 [1].

IR (gas): ν (in cm^{-1}) = 1620 w ν(C=N), 1335 s, 1270 vs, 1186 ν(CF), 985 s, 941 m, 745 w, 713 s ν(SF) [1].

MS: m/e (rel. int. in %) = 215 (100) M^+, 196 (58) M^+-F, 177 (10.5) $M^+-2\,F$, 165 (4.5) CF_3CFNSF^+, 158 (1.7) $CF_2CFCNSF^+$, 150 (7.2) $(CF_3)_2C^+$, 146 (55) M^+-CF_3, 144 (11) CF_2CFCSF^+ ?, 131 (5.5) $CF_3CF_2C^+$, 127 (9.5) CF_3CNS^+, 108 (12) CF_2CNS^+, 101 (33) CF_3S^+ ?, 100 (9.5) CF_3CF^+, 96 (5.5) CF_2NS^+, 93 (2.8) CF_3CC^+, 82 (10) CF_2S^+, 81 (2.2) CF_3C^+, 76 (44.5) CF_2CN^+, 69 (67) CF_3^+, 64 (7.5) CF_2N^+, 63 (4.5) CFS^+, 58 (4) CNS^+, 51 (53) SF^+, 50 (19.5) CF_2^+, 46 (31) NS^+, 45 (4) CFN^+, 44 (11.5) CS^+, 32 (18) S^+, 31 (15.5) CF^+ [1].

$FSN{=}C(CF_3)_2$ appears rather stable compared to $FSNR_2$, where R = CH_3, C_2H_5 [2]. It could be stored in glass vessels at room temperature for weeks. It is sensitive to oxidation. Upon UV irradiation, $FSN{=}C(CF_3)_2$ decomposes forming $(CF_3)_2CFNSF_2$, $(CF_3)_2C{=}NSSN{=}C(CF_3)_2$, $(CF_3)_2$-$C{=}NSN{=}C(CF_3)_2$, and $(CF_3)_2C{=}NC(CF_3)_2NSF_2$ [1].

$FSN{=}C(CF_3)_2$ reacted with the diradical CF_3NO via cycloaddition in a reversible reaction. Structure I was tentatively assigned to the product:

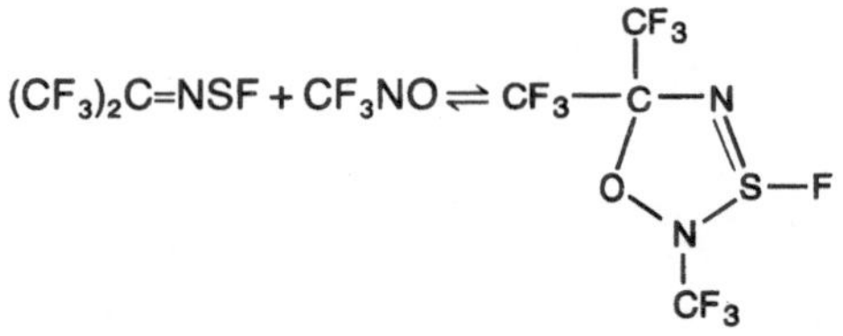

I

By the reaction with $(CF_3)_2NO^\bullet$ (mole ratio 1 : 2), $(CF_3)_2C(ON(CF_3)_2)N{=}S(F)ON(CF_3)_2$ was formed and with HgF_2 (mole ratio 1 : 2), $(CF_3)_2CFN{=}SF_2$ was obtained along with Hg_2F_2 [3].

References:

[1] Varwig, J.; Steinbeißer, H.; Mews, R.; Glemser, O. (Z. Naturforsch. **29b** [1974] 813/4).
[2] Roesky, H. W. (in: Senning, A.; Sulfur in Organic and Inorganic Chemistry, Vol. 4, Dekker, New York - Basel 1982, p. 33).
[3] Varwig, J.; Mews, R. (J. Chem. Res. Synop. **1977** 245).

12.3 Sulfur Amide Chloride, Amino-chloro-sulfane, $ClSNH_2$

Geometries and energies for the conformers and the inversion and rotation transition structures of the hypothetical molecule $ClSNH_2$ were studied by ab initio SCF-MO theory at the HF/6-31G* level. Geometries and energies are given in Table 30.

Table 30
HF/6-31G* Geometries and Energies for Conformers of $ClSNH_2$.

sym.	conformer[a)]	E_T in au	E_{rel}	E_{rot} in kcal/mol	E_{inv}	R(SN) in Å	R(SCl) in Å	$\Delta\theta(N)$[b)] in °	θ(ClSN) in °
C_s	P-pl	−912.53735	24.33			1.706	2.023	0.0	98.1
C_1	P-st	−912.55072	15.94		8.39	1.746	2.026	40.0	99.5
C_s	A-pl	−912.57327	1.79	22.64		1.622	2.077	0.0	105.6
C_s	A-st	−912.57613	0.00	15.94		1.642	2.074	13.3	105.8

a) Notation for conformation is defined as follows: The dihedral relationship between the nitrogen lone pair and the S–Cl bond is designated by A (anti, ~180°) or P (perpendicular, ~90°). The configurations of the hydrogens at N with respect to the ClS group are denoted "st" (staggered) or "pl" (planar). – b) The degree of pyramidalization at the nitrogen of the NH_2 group is represented by $\Delta\theta(N)$, equal to 360 minus the sum of the three bond angles at N.

P-pl (planar) P-st (C_1 transition state) A-pl (C_s transition state) A-st (syn)

The net energetic stabilization according to the isodesmic equation $ClSNH_2 + SH_2 \rightarrow ClSH + HSNH_2$ for the conformers of $ClSNH_2$ was calculated to be: E_{stab}(P-pl) = 4.29, E_{stab}(P-st) = 4.57, E_{stab}(A-pl) = 12.78, E_{stab}(A-st) = 12.28 kcal/mol. The most favorable A-st conformer exhibits a very large internal rotation barrier of 16 kcal/mol. This value is in the range of experimentally determined barriers for analogous aminosulfanes. No anti conformer exists. A natural bond orbital (NBO) analysis showed that the strong generalized anomeric effect (lengthening of the S–N bond by 0.10 Å during internal rotation with an associated barrier of 16 kcal/mol) can be explained quantitatively in terms of $n_N \rightarrow \sigma^*_{SCl}$ hyperconjugation. The $n_N \rightarrow \sigma^*_{SCl}$ interaction in $ClSNH_2$ has roughly the same energy as the $n_N \rightarrow \sigma^*_{SF}$ interaction in $FSNH_2$.

Reference:

Reed, A. E.; von Ragué Schleyer, P. (Inorg. Chem. **27** [1988] 3969/87).

12.4 N-Substituted Amino-chloro-sulfanes

12.4.1 $ClSNHC(O)C_6H_5$

The title compound was prepared by reacting $C_6H_5C(O)NHSi(CH_3)_3$ with freshly distilled SCl_2 (mole ratio 1 : 1.5 [1], 1 : 1.2 [2], or an excess of it [3]) in ether–pentane [1], ether [2], or THF [3] maintained at 0 °C and stirring the mixture at 0 °C for additional 6 h [1], at room temperature for 30 min [2], or at −30 °C [3].

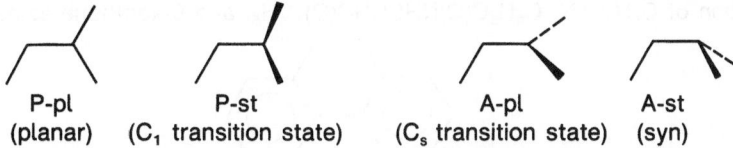

Yield: 84% [1], 80 to 87% [2], or 80 to 85% [3]; bright yellow microneedles [1], light yellow crystals [2]; m.p. 105 to 108 °C/decomposition [1, 3], m.p. 108 to 109 °C/decomposition [2].

^{1}H NMR (DMSO–d_6/TMS): δ (in ppm) = 7.50 (m, 3H), 7.97 (m, 2H), 11.66 (s, 1H) [1].

IR ($CHCl_3$): 3200 ν(NH), 1670 cm^{-1} ν(C=O) [1], see also [3].

UV (THF): λ_{max} (in nm) (ε in $L \cdot mol^{-1} \cdot cm^{-1}$) = 209 (7070), 237 (12900), 350 (140) [1].

The compound can be kept for 3 months without appreciable deterioration, as long as it is kept tightly sealed under an inert atmosphere, and it can be stored below 0 °C [1].

Heating the title compound in $CHCl_3$ at reflux temperature for 2 h until evolution of HCl ceased yielded 89% $C_6H_5C(O)N{=}S{=}NC(O)C_6H_5$. This product was also obtained with 83% yield when $ClSNHC(O)C_6H_5$ was treated with water [2]. Solutions of $ClSNHC(O)C_6H_5$ in THF reacted with a solution of morpholine or aniline in THF at −78 °C to form $C_6H_5C(O)NHSN(CH_2)_4O$ or $C_6H_5C(O)NHSNHC_6H_5$, respectively [1]. The title compound reacted with an equimolar amount of N-chlorobenzamidine, $C_6H_5C(NH_2){=}NCl$ in benzene for 4 h at 20 °C to give $C_6H_5C(O)N{=}SCl_2$, in dichloroethane at 45 to 50 °C until a homogeneous solution was formed (after about 4 h) to yield 74% $C_6H_5C(O)N{=}S(Cl)NHC(O)C_6H_5$ [2]. Solutions of $ClSNHC(O)C_6H_5$ in THF reacted with an equimolar amount of diphenyldiazomethane or diazofluorene at −78 °C [3] or −30 °C [1], or with ethyl diazoacetate at 30 °C [1] to give the corresponding α-chlorosulfenamides $RR'C(Cl)SNHC(O)C_6H_5$, $R = R' = C_6H_5$, $RR' = $ I (see below), or $R = H$, $R' = CO_2C_2H_5$, respectively. The reaction of the title compound with the more nucleophilic 9-diazoxanthone resulted in the formation of $C_6H_5C{\equiv}N$, $C_6H_5C(O)NHSNHC(O)C_6H_5$, and 9-xanthone azine [1].

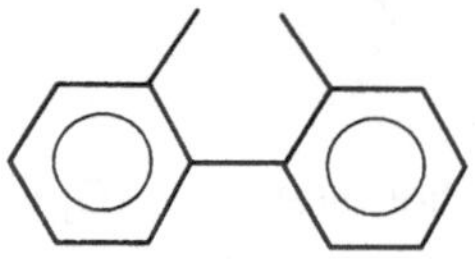

I

References:

[1] Burgess, E. M.; Penton, H. R., Jr. (J. Org. Chem. **39** [1974] 2885/92).
[2] Levchenko, E. S.; Borovikova, G. S.; Dorokhova, E. M. (Zh. Org. Khim. **13** [1977] 103/7; J. Org. Chem. USSR [Engl. Transl.] **13** [1977] 96/100).
[3] Burgess, E. M.; Penton, H. R., Jr. (J. Am. Chem. Soc. **95** [1973] 279/80).

12.4.2 $ClSNHC(O)OC_4H_9$-n

The title compound was prepared by adding a solution of n-$C_4H_9OC(O)NHCH_3$ and $(C_2H_5)_3N$ in CH_2Cl_2 to a precooled (−5 °C) solution of SCl_2 (mole ratio 1 : 1.1 : 1.1), maintaining the temperature at −15 to 0 °C, and allowing the mixture to be stirred for 1 h.

The $ClSNHC(O)OC_4H_9$-n reacted with $CH_3N{=}C{=}O$ and HF in the presence of $(C_2H_5)_3N$ in CH_2Cl_2 at 0 °C to give n-$C_4H_9OC(O)N(CH_3)SN(CH_3)C(O)F$ after letting the mixture to come to room temperature.

Reference:

Liang, W. C. (Ger. 2654313 [1976/77]; C.A. **87** [1977] No. 200836).

12.5 N,N-Disubstituted Amino-chloro-sulfanes

12.5.1 N,N-Diorganyl-Substituted Amino-chloro-sulfanes

12.5.1.1 N,N-Dialkyl-amino-chloro-sulfanes with Identical Alkyl Groups

12.5.1.1.1 $ClSN(CH_3)_2$

Preparation

The title compound can be synthesized directly from $(CH_3)_2NH$ and SCl_2 (Method I), by cleaving the sulfur–sulfur bond of $(CH_3)_2NSSN(CH_3)_2$ with a chlorinating agent (Method II), by cleaving the sulfur–nitrogen bond of $(CH_3)_2NSN(CH_3)_2$ with a chloro compound (Method III), or by some special reactions.

Method I: The best results (69% yield) were achieved by adding $(CH_3)_2NH$ to a −10 °C-precooled solution of freshly prepared SCl_2 (mole ratio 2 : 1) in $(C_2H_5)_2O$ while vigorously stirring and holding the temperature at −10 °C by cooling [1]; see also [10].

Method II: $ClSN(CH_3)_2$ was prepared with a yield of 75% by adding a solution of SO_2Cl_2 in CCl_4 to a solution of $(CH_3)_2NSSN(CH_3)_2$ (mole ratio 1.25 : 1) in CCl_4 at 60 °C and maintaining the mixture at 60 °C for another 2 h [2].

Method III: $ClSN(CH_3)_2$ was the only product of the reaction of $(CH_3)_2NSN(CH_3)_2$ with SCl_2 with ~90% yield [5]. Adding R_2BCl (R = n-C_3H_7, n-C_4H_9) at room temperature to the equimolar amount of $(CH_3)_2NSN(CH_3)_2$ and more R_2BCl (~1.1 molar amount), after the exothermic reaction ceased, gave $ClSN(CH_3)_2$ with 88% (R = n-C_3H_7) and 95% (R = n-C_4H_9) yield [3]. Benzoyl chloride at 30 °C [4] or benzenesulfenyl chloride [5] could be used instead of R_2BCl via the reaction $(CH_3)_2NSN(CH_3)_2 + C_6H_5C(O)Cl$ (or C_6H_5SCl) $\rightarrow ClSN(CH_3)_2 + C_6H_5C(O)N(CH_3)_2$ (or $C_6H_5SN(CH_3)_2$) [4, 5].

The compound was also formed with 90% yield when a solution of $ClSN{=}CCl_2$ in ether was added to a solution of $(CH_3)_2NH$ (mole ratio 1 : 2) in ether kept at ~10 °C [6]. $ClSN(CH_3)_2$ can be obtained via $(CH_3)_2NSOCH_3 + (CH_3)_3SiCl \rightarrow ClSN(CH_3)_2 + (CH_3)_3SiOCH_3$ (no details given) [7].

Properties

$ClSN(CH_3)_2$ is an orange liquid [1] or a pale green liquid with a penetrating odor [6]; b.p. 28 to 30 °C/12 Torr [3], 29 to 30 °C/13 Torr [1], 33 °C/14 Torr [8], 34 to 36 °C/15 Torr [9], 34.5 to 35.5 °C/15 Torr [6], 41 °C/16 Torr [7], 55 °C/45 Torr [2, 10]; $n_D^{24} = 1.5145$ [7].

^1H NMR (C_6H_6/TMS): $\delta = 2.68$ ppm [11]; (50% in $CDCl_3$/TMS): $\delta = 3.1$ ppm, $^1J(^{13}C,H) = 138.7 \pm 0.1$ Hz [12]; see also [2, 10, 13].

^{15}N NMR (in presence of $Cr((CH_3C(O))_2CH_2)_3/CH_3NO_2$): $\delta = -304.2$ ppm [14].

^{35}Cl NQR (77 K): $\nu = -36.000$ and -35.598 MHz [15].

$ClSN(CH_3)_2$ is a liquid that can be readily distilled in vacuum [5, 16] and visibly decomposes within 1 h even if sealed in ampules and kept in the dark [5]. It can be stored at low temperatures (−18 °C [2], −20 °C [5]) for several weeks [2] or several months [5]. It is very reactive and sensitive to hydrolysis [16]; when exposed to air it forms a white solid [6].

Chemical Reactions

Passing Cl_2 through a solution of $ClSN(CH_3)_2$ (mole ratio 2 : 1) in $CHCl_3$ at room temperature gave 70 to 80% crude $(CH_3)_2NCCl_3$. This compound was also obtained with a yield of 76% by adding a solution of $(CH_3)_2NSCl$ in $CHCl_3$ to a solution of PCl_5 (mole ratio 1 : 1) in $CHCl_3$ or

CCl_4 at room temperature within 1 h. Passing Cl_2 through a solution of $ClSN(CH_3)_2$ in $CHCl_3$ until the absorption of Cl_2 ceased gave a complex product mixture with the composition $(CH_3)_2NCCl_3Cl_2$, containing the two active chlorine atoms [17].

$ClSN(CH_3)_2$ reacted with an equimolar amount of $((CH_3)_3Si)_2NN{\equiv}C$ in CCl_4 at −20 °C to give 68% $((CH_3)_3Si)_2NN{=}C(Cl)SN(CH_3)_2$ [18]. The title compound reacted with an equimolar amount of $R_2NSi(CH_3)_3$ ($R = CH_3$, C_2H_5) to give the corresponding $R_2NSN(CH_3)_2$ and $(CH_3)_3SiCl$. The reaction of $ClSN(CH_3)_2$ with $((CH_3)_3Si)_2NR$ (mole ratio 2 : 1; $R = CH_3$, C_2H_5) led to the corresponding $(CH_3)_2NSN(R)SN(CH_3)_2$ along with $(CH_3)_3SiCl$ and the reaction with an equimolar amount of $O(CH_2)_4NSi(CH_3)_3$ to $O(CH_2)_4NSN(CH_2)_4O$, $(CH_3)_2NSN(CH_3)_2$, and $(CH_3)_3SiCl$ [5].

Adding $t\text{-}C_4H_9N{=}X{=}NSi(CH_3)_3$ (X = S, C) to a solution of $ClSN(CH_3)_2$ (mole ratio 1 : 1) in CH_2Cl_2 yielded at room temperature the corresponding $t\text{-}C_4H_9N{=}X{=}NSN(CH_3)_3$ with X = S (62%) or X = C (66%) after 2 h [19].

The title compound reacted with $O{=}S{=}NSi(CH_3)_3$ to give $O{=}S{=}NSN(CH_3)_2$ and $(CH_3)_3SiCl$ (no details); with silthianes reactions occurred according to $ClSN(CH_3)_2 + (CH_3)_3SiSR \rightarrow (CH_3)_2NSSR$ (R = alkyl, aryl, not specified) + $(CH_3)_3SiCl$ [5].

The title compound reacted with fluorophosphites, $(RO)_2PF$ (R = alkyl), to form $(RO)(F)P(O)\text{-}N(CH_3)_2$ and $(RO)_2P(S)F$ and with chlorophosphites, $(RO)_2PCl$, to form $(RO)(Cl)P(O)Cl$ and $(RO)_2P(S)N(CH_3)_2$; a reaction mechanism was proposed explaining qualitatively the different behavior of chloro- and fluorophosphites [20]. Adding a solution of $ClSN(CH_3)_2$ in benzene to $(RO)_2P(O)H$ (mole ratio 1 : 2; $R = CH_3$, C_2H_5, $n\text{-}C_3H_7$, $i\text{-}C_3H_7$) in benzene at 20 to 25 °C and stirring the mixture for 1 h at room temperature gave the corresponding $(RO)_2P(O)OP(S)(OR)_2$ with a yield of 36.5, 61.7, 65.0, and 69.2%, respectively, along with $(CH_3)_2NH \cdot HCl$ [21]. In an earlier paper a different reaction course was found in the reaction of $ClSN(CH_3)_2$ with $(C_2H_5O)_2P(O)H$ (in benzene, 3 h at 30 to 35 °C), whereby $(C_2H_5O)_2P(O)SP(O)(OC_2H_5)_2$ was the product [8]; see also [16]. Adding a solution of $(RO)_2P(S)H$ ($R = CH_3$, C_2H_5, $n\text{-}C_3H_7$, $i\text{-}C_3H_7$) in petroleum ether at 10 to 15 °C to a solution of $ClSN(CH_3)_2$ (mole ratio 5 : 2.4) in petroleum ether gave the corresponding $(RO)_2P(S)SP(S)(OR)_2$ with a yield of 21.5, 43.5, 85.0, and 82.0%, respectively, along with $(CH_3)_2NH \cdot HCl$ [21].

Adding a solution of $ClSN(CH_3)_2$ in CH_2Cl_2 to a solution of $(C_6H_5O)_2P(O)CH_2NHCH_2C(O)\text{-}OC_2H_5$ and $(C_2H_5)_3N$ (mole ratio ca. 3 : 1 : 1) in toluene precooled to 0 °C gave $(C_6H_5O)_2P(O)\text{-}CH_2N(CH_2C(O)OC_2H_5)SN(CH_3)_2$ [22].

Adding a solution of $RC{\equiv}CSLi$ ($R = CH_3$, $t\text{-}C_4H_9$) in ether to a solution of $ClSN(CH_3)_2$ (mole ratio 1 : 1) in ether precooled to −30 °C gave the corresponding $RC{\equiv}CSSN(CH_3)_2$ with 62% ($R = CH_3$) or 75% ($R = t\text{-}C_4H_9$) yield, respectively [23]. Adding $ClSN(CH_3)_2$ to a cooled solution of CH_3SNa in CH_3OH with cooling and stirring to maintain the temperature of the mixture below 20 °C gave $(CH_3)_2NSSCH_3$. Adding $ClSN(CH_3)_2$ to a to −10 °C precooled solution of $CH_2{=}CH\text{-}CH_2SNa$ (from $CH_2{=}CHCH_2SH$ and CH_3ONa at 25 °C) in C_2H_5OH while maintaining the temperature of the reaction mixture at −10 °C gave $(CH_3)_2NSSCH_2CH{=}CH_2$ [24]. Adding a solution of the title compound in ether at −50 °C to an equimolar amount of $NaN(Si(CH_3)_3)_2$ in ether and allowing the reaction mixture to come to room temperature within 8 h gave 81% $((CH_3)_3Si)_2NSN(CH_3)_2$ [25].

Adding a solution of $ClSN(CH_3)_2$ in benzene to a suspension of $(C_2H_5O)_2P(S)SNa$ in benzene raised the temperature to about 30 to 35 °C. Stirring the mixture at this temperature for 1 h gave 90% (crude) $(C_2H_5O)_2P(S)SSN(CH_3)_2$ [26].

Passing $ClSN(CH_3)_2$ at a low pressure through a preheated (100 °C) column containing HgF_2 gave $CH_3N{=}SF_2$ as the main product [27].

Reacting a solution of $Hg(SCF_3)_2$ in ether at 20 °C with a solution of the title compound (mole ratio 1 : 2) in ether and cooling the reaction mixture to −30 °C gave 88% $(CH_3)_2NSSCF_3$ and $HgCl_2$ [28].

Adding $ClSN(CH_3)_2$ under cooling to $(n\text{-}C_4H_9)_3SnOR$ (mole ratio 1:1.1; R = CH_3, C_2H_5, n-C_3H_7, n-C_4H_9) and then warming the mixture for 1 h yielded the corresponding $(CH_3)_2NSOR$ with a yield of 85, 80, 86, or 88%, respectively, along with $(n\text{-}C_4H_9)_3SnCl$ [7].

Reacting $ClSN(CH_3)_2$ with $Pb(SCN)_2$ in a 2:1 mole ratio yielded $(CH_3)_2NSSCN$ and $PbCl_2$ (no details) [4].

Passing $ClSN(CH_3)_2$ at a low pressure through a column heated to 120 °C and containing AgF gave $FSN(CH_3)_2$ [27]. Reacting $ClSN(CH_3)_2$ with an equimolar amount of AgCN yielded $(CH_3)_2NSCN$ and AgCl (no details) [4].

The reactions of $ClSN(CH_3)_2$ with organic compounds are summarized in Tables 31 and 32, pp. 115/8.

Table 31
Reactions of $ClSN(CH_3)_2$ with Organic Compounds.

$ClSN(CH_3)_2$ + organic substrate	reaction conditions	product(s) (yield in %)	Ref.
$CH_3CH{=}CH_2$	in CH_2Cl_2 with catalytic amounts of anhydrous $AlCl_3$; −20 °C	$CH_3CHClCH_2SN(CH_3)_2$ (73)	[29]
	mole ratio ca. 1:1.1; in CH_2Cl_2 at −70 °C for 2 h, −50 °C for 1 h, and warming slowly to 0 °C	$CH_3CH(SN(CH_3)_2)CH_2Cl$, $CH_3CHClCH_2SN(CH_3)_2$, in the ratio 78:22	[10]
$(CH_3)_2C{=}CH_2$	mole ratio ca. 1:1; in CH_2Cl_2 at −20 °C for 10 min	$(CH_3)_2C(SN(CH_3)_2)CH_2Cl$, $(CH_3)_2CClCH_2SN(CH_3)_2$, in the ratio 71:29	[10]
$(CH_3)_2CHCH{=}CH_2$	mole ratio ca. 1:1; in CH_2Cl_2 at −20 °C, 30 min at 0 °C	$(CH_3)_2CHCH(SN(CH_3)_2)CH_2Cl$, $(CH_3)_2CHCHClCH_2SN(CH_3)_2$, in the ratio ca. 90:10	[10]
$(CH_3)_3CCH{=}CH_2$	mole ratio ca. 1:1; in CH_2Cl_2 at −20 °C, 30 min at 0 °C, and 30 min at ambient temperature	$(CH_3)_3CCH(SN(CH_3)_2)CH_2Cl$, $(CH_3)_3CCHClCH_2SN(CH_3)_2$, in the ratio ca. 95:5	[10]
$CH_3CH_2CH{=}CH_2$	in CH_2Cl_2 with catalytic amounts of anhydrous $AlCl_3$; at −20 °C	$CH_3CH_2CHClCH_2SN(CH_3)_2$ (73)	[29]
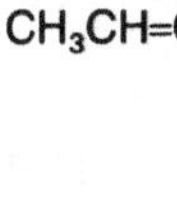	mole ratio 1:>1	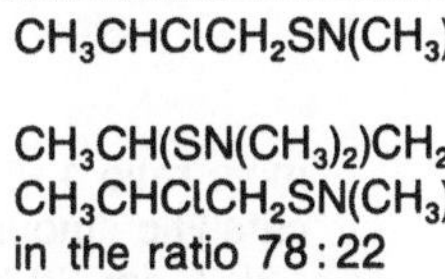 probably (18)	[6]
	in CH_2Cl_2 with catalytic amounts of anhydrous $AlCl_3$; at −20 °C	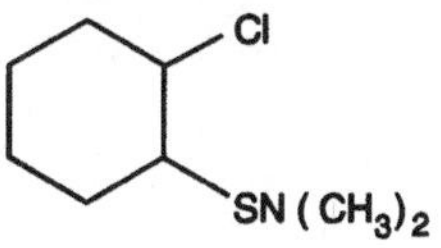	[29]

Table 31 (continued)

$ClSN(CH_3)_2$ + organic substrate	reaction conditions	product(s) (yield in %)	Ref.
	mole ratio 1:1; in CH_2Cl_2, at −20°C for 15 min, allowing to warm to −10°C	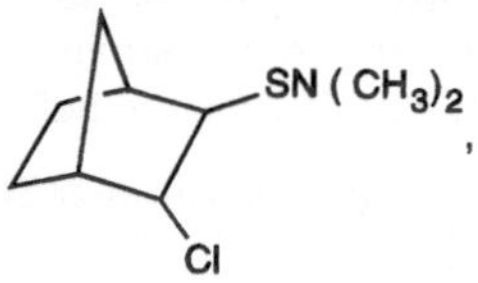 , trans (>98)	[10]
$(CF_3)_2C{=}CF_2$	in DMF in the presence of CsF; 6 h at room temperature (?)	$(CF_3)_3CSN(CH_3)_2$ (72.5)	[31]
$R^1CHOHCH{=}CHR^2$ + $(C_2H_5)_3N$ $R^1 = R^2 = CH_3$; R^1 = H, i-C_3H_7, t-C_4H_9, n-C_7H_{15}, 4-$CH_3C_6H_4$; $R^2 = CH_3$; R^1 = n-C_7H_{15}, R^2 = i-C_3H_7	in ether; −78°C to ambient temperature	$(CH_3)_2NS(O)CHR^2CH{=}CHR^1$ [a)]	[32]
$CH_3C{\equiv}CH$	mole ratio 1:5; in CH_2Cl_2 with catalytic amounts of $AlCl_3$ at −30°C for 20 h	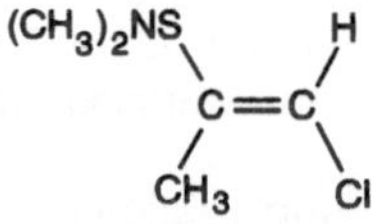 I 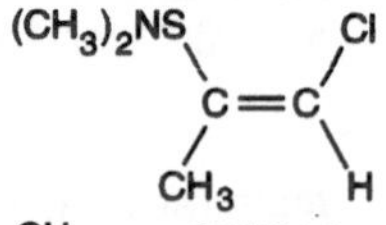 II 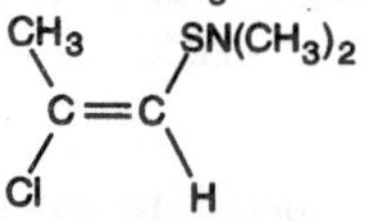 III (67; I:II:III product mixture in the ratio 84:5:10%)	[10]
i-$C_4H_9C{\equiv}CH$	mole ratio ca. 1:3; in CH_2Cl_2 at 0°C for 15 min and 5 h at ambient temperature	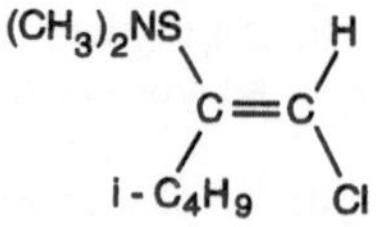 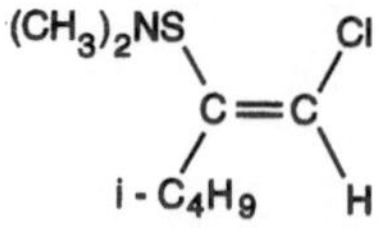 in a 9:1 mole ratio	[10]
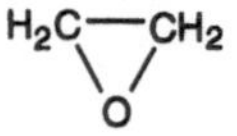		$(CH_3)_2NSOCH_2CH_2Cl$	[4]

Table 31 (continued)

$ClSN(CH_3)_2$ + organic substrate	reaction conditions	product(s) (yield in %)	Ref.
$RCHOHC{\equiv}CCH_3$	with $(C_2H_5)_3N$ (mole ratio 1:1:1); in ether at −78 °C for 15 min and 1 h at room temperature	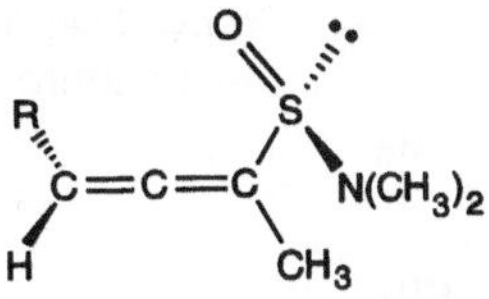 IVa 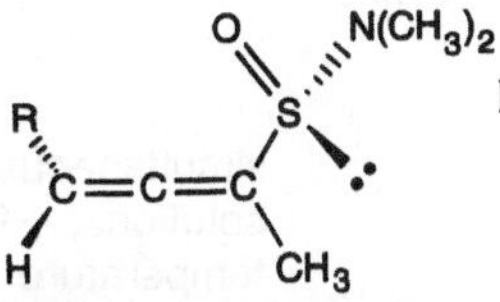 IVb	[2]
$R = CH_3$		(66)[b]; IVa:IVb ratio[c] 60:40	
$R = i\text{-}C_3H_7$		(41)[b]; IVa:IVb ratio[c] 75:25	
$R = t\text{-}C_4H_9$		(55)[b]; IVa:IVb ratio[c] 70:30	
$R = n\text{-}C_7H_{15}$		(31)[b]; IVa:IVb ratio[c] 65:35	
$(CH_3)_2NH$	mole ratio 1:2; in ether at room temperature	$(CH_3)_2NSN(CH_3)_2$ (39), $(CH_3)_2NH \cdot HCl$ (82)	[6]
$CH_3N{=}C{=}O$	with HF and $(C_2H_5)_3N$ in toluene at −50 to 10 °C and warming to room temperature within 1 h	$(CH_3)_2NSN(CH_3)C(O)F$ (50.9)	[33]
$2\text{-}CH_3\text{-}4\text{-}RC_6H_3N{=}CHNHCH_3$ ($R = Cl, CH_3$)	with $(C_2H_5)_3N$ in THF under cooling	$2\text{-}CH_3\text{-}4\text{-}RC_6H_3N{=}CHN(CH_3)\text{-}SN(CH_3)_2$	[34]
$CH_3SC(CH_3){=}NOC(O)NHCH_3$	with $(C_2H_5)_3N$ in CH_3CN; at 0 to 5 °C and warming to room temperature within 2 h	$CH_3SC(CH_3){=}NOC(O)N(CH_3)\text{-}SN(CH_3)_2$	[35]
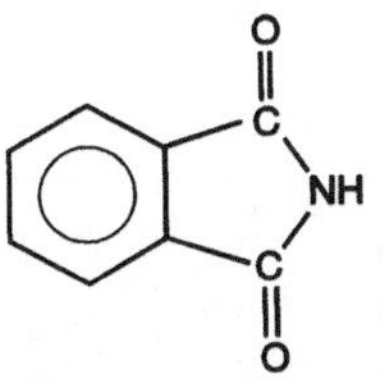	mole ratio 1:>1 and $(C_2H_5)_3N$ in toluene for 1 h	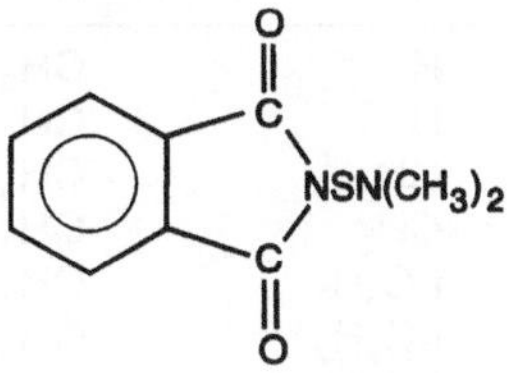 (81)	[36]
$3\text{-}i\text{-}C_3H_7C_6H_4OC(O)NHCH_3$	with $(C_2H_5)_3N$ (mole ratio 1:1:1) in THF at 10 °C, 8 d at 25 °C	$3\text{-}i\text{-}C_3H_7C_6H_4OC(O)N(CH_3)\text{-}SN(CH_3)_2$	[37]
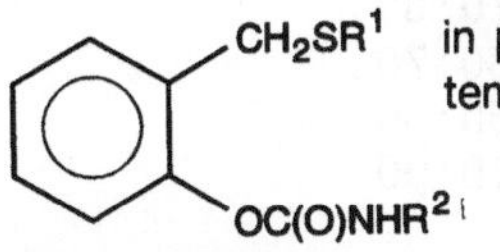 $R^1 = CH_3, C_2H_5, n\text{-}C_3H_7, i\text{-}C_3H_7$; $R^2 = CH_3$; $R^1 = R^2 = C_2H_5$	in pyridine; 18 h at room temperature	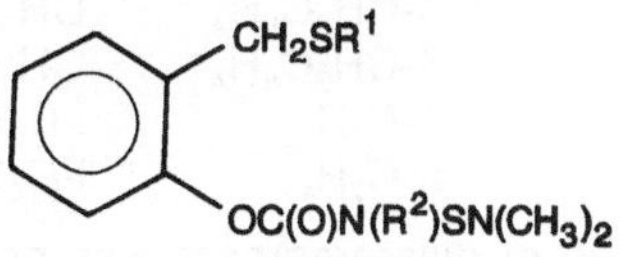	[38]

Table 31 (continued)

$ClSN(CH_3)_2$ + organic substrate	reaction conditions	product(s) (yield in %)	Ref.
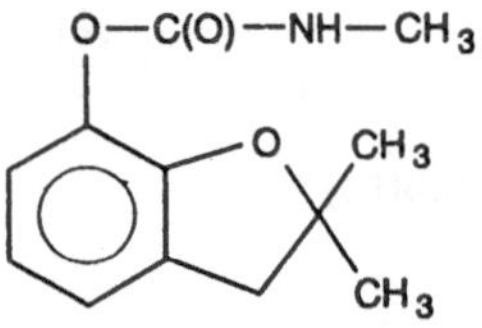	in pyridine; 18 h at room temperature	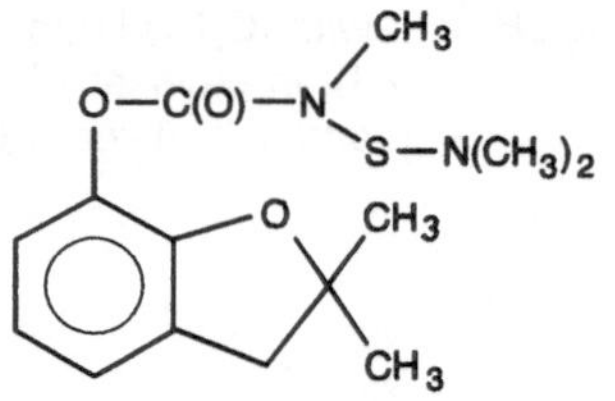	[9]
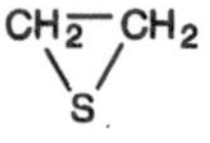	simultaneously adding CCl_4 solutions; ~12 h at room temperature	$(CH_3)_2NSSCH_2CH_2Cl$ (69)	[39]
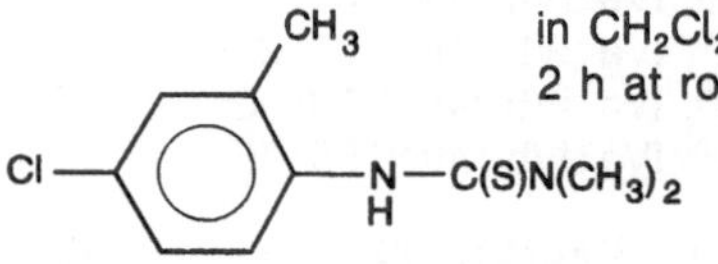 + $(C_2H_5)_3$	in CH_2Cl_2 at 0 to 10°C and 2 h at room temperature	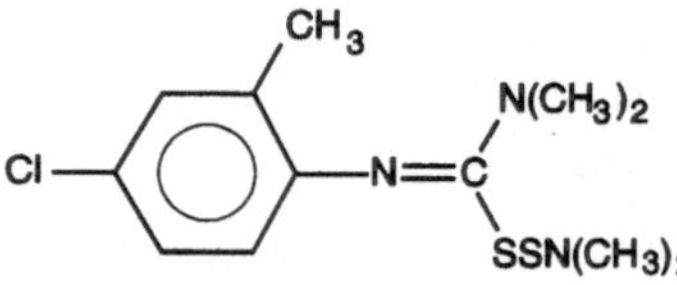	[41]

a) The crude product was left under Ar for the time listed in Table 32. – b) The crude product had to be kept at room temperature for 1 to 2 h under Ar. – c) The ratio of diastereomers was determined by ^1H NMR.

Table 32
Reactions of $ClSN(CH_3)_2$ with $R^1CHOHCH{=}CHR^2$ to $(CH_3)_2NS(O)CHR^2CH{=}CHR^1$ Diastereomers [32].
(In the presence of $(C_2H_5)_3N$ (mole ratio 1 : 1 : 2) in ether at −78°C for 15 min, and at room temperature for 1 h. The crude products were left under Ar for the time (in h) listed below.)

$R^1CHOHCH{=}CHR^2$ configuration	R^1	R^2	time (in h)	$(CH_3)_2NS(O)CHR^2CH{=}CHR^1$ yield (in %)	ratio of diastereomers[a)]
E	H	CH_3	60	37	60 : 40
Z	H	CH_3	5	41	25 : 75
E	CH_3	CH_3	4	52	60 : 40
Z	CH_3	CH_3	4	36	35 : 65
E	i-C_3H_7	CH_3	4	16	60 : 40
Z	i-C_3H_7	CH_3	4	17	30 : 70
E	t-C_4H_9	CH_3	4	33	65 : 35
Z	t-C_4H_9	CH_3	4	28	15 : 85
E	n-C_7H_{15}	CH_3	4	58	65 : 35
Z	n-C_7H_{15}	CH_3	4	52	10 : 90
E	4-$CH_3C_6H_4$	CH_3	4	11	70 : 30
Z	4-$CH_3C_6H_4$	CH_3	4	60	30 : 70
E	n-C_7H_{15}	i-C_3H_7	6	40	50 : 50
Z	n-C_7H_{15}	i-C_3H_7	4	15	10 : 90

a) The ratio of diastereomers was determined by ^1H NMR.

$ClSN(CH_3)_2$ has been found to be a useful catalyst for reactions of $SOCl_2$ with acids, especially dicarboxylic acids and their anhydrides [40].

References:

[1] Badische Anilin-Soda-Fabrik AG (Ger. 1131222 [1960/62]; C.A. **57** [1962] 13771).
[2] Baudin, J.-B.; Bkouche-Waksman, I.; Julia, S. A.; Pascard, C.; Wang, Y. (Tetrahedron **47** [1991] 3353/64).
[3] Nöth, H.; Mikulaschek, G. (Chem. Ber. **97** [1964] 709/14).
[4] Schrader, G. (private communication in [16]).
[5] Armitage, D. A.; Tso, C. C. (J. Chem. Soc. D **1971** 1413/4).
[6] Bacon, R. G. R.; Irwin, R. S. (J. Chem. Soc. **1960** 5079/87).
[7] Armitage, D. A.; Towle, I. D. H. (Phosphorus Sulfur Relat. Elem. **1** [1976] 37/9).
[8] Lorenz, W.; Schrader, G. (Ger. 820001 [1949/51]; C.A. **47** [1953] 3332).
[9] Black, L.; Fukuto, T. R. (Ger. Offen. 2433680 [1974/75]; C.A. **82** [1975] No. 156050).
[10] Müller, W. H.; Butler, P. E. (J. Org. Chem. **33** [1968] 2111/3).

[11] Nöth, H.; Mikulaschek, G.; Rambeck, W. (Z. Anorg. Allg. Chem. **344** [1966] 316/22).
[12] Jakobsen, H. J.; Senning, A.; Kaae, S. (Acta Chem. Scand. **25** [1971] 3031/6).
[13] Kühle, E. (Synthesis **1970** 561/80).
[14] Dorie, J.; Gouesnard, J.-P. (J. Chim. Phys. Phys.-Chim. Biol. **81** [1984] 15/9).
[15] Babushkina, T. A.; Levin, V. S.; Kalinkin, M. I.; Semin, G. K. (Izv. Akad. Nauk SSSR Ser. Khim. **1969** 2340; Bull. Acad. Sci. USSR Div. Chem. Sci. [Engl. Transl.] **1969** 2199).
[16] Dorlars, A. (Houben-Weyl Methoden Org. Chem. 4th Ed. **11** Pt. 2 [1958] 744/51, 746).
[17] Kuchar', V. P.; Pasternak, V. I.; Kirsanov, A. V. (Zh. Org. Khim. **7** [1971] 2084/6; J. Org. Chem. USSR [Engl. Transl.] **7** [1971] 2165/7).
[18] Wiberg, N.; Hübler, G. (Z. Naturforsch. **32b** [1977] 1003/9).
[19] Appel, R.; Montenarh, M. (Z. Naturforsch. **30b** [1975] 847/9).
[20] Gusar', N. I.; Ivanova, Z. M.; Chaus, M. P.; Gololobov, Yu. G. (Dopov. Akad. Nauk Ukr. RSR B: Geol. Khim. Biol. Nauki **1978** 520/3; C.A. **89** [1978] No. 196907).

[21] Schrader, G.; Lorenz, W.; Mühlmann, R. (Angew. Chem. **70** [1958] 690/4).
[22] Hoobler, M.; Sikorski, J. A. (Belg. 894595 [1982/83]; C.A. **99** [1983] No. 6049).
[23] Meijer, J.; Wijers, H. E.; Brandsma, L. (Recl. Trav. Chim. Pays-Bas **91** [1972] 1423/5).
[24] Williams, H. R. (U.S. 3121661 [1961/64]; C.A. **61** [1964] 6958).
[25] Rinne, D.; Blaschette, A. (Chem.-Ztg. **98** [1974] 456/7).
[26] Malz, H.; Bayer, O.; Freytag, H.; Lober, F. (U.S. 2891059 [1957/59]; C.A. **1960** 4387).
[27] Seel, F.; Gombler, W.; Budenz, R. (Z. Naturforsch. **27b** [1972] 78/9).
[28] Bur-Bur, F.; Haas, A.; Klug, W. (Chem. Ber. **108** [1975] 1365/8).
[29] Badische Anilin-Soda-Fabrik AG; Weiß, G. (Ger. 1153744 [1961/64]; C.A. **60** [1964] 1711).
[30] Müller, W. H. (Angew. Chem. **81** [1969] 475/84; Angew. Chem. Int. Ed. Engl. **8** [1969] 482/92).

[31] Zeifman, Yu. V.; Lantseva, L. T.; Knunyants, I. L. (Izv. Akad. Nauk SSSR Ser. Khim. **1978** 2640/3; Bull. Acad. Sci. USSR Div. Chem. Sci. [Engl. Transl.] **27** [1978] 2362/6).
[32] Baudin, J.-B.; Bkouche-Waksman, I.; Hareau, G.; Julia, S. A.; Lorne, R.; Pascard, C. (Tetrahedron **47** [1991] 6655/72).
[33] Thurman, D. E. (Ger. 2628574 [1975/77]; C.A. **86** [1977] No. 155209, Ger. 2628575 [1975/77]; C.A. **86** [1977] No. 155665).
[34] Drabek, J.; Böger, M. (Ger. 2621077 [1974/76]; C.A. **86** [1977] No. 89430).
[35] Gemrich, E. G., II; Lee, B. L.; Nelson, S. J.; Rizzo, V. L. (J. Agric. Food Chem. **26** [1978] 391/5).
[36] Lawrence, J. P. (Rubber Chem. Technol. **49** [1976] 330/40).

[37] Rizzo, V. L. (Ger. 2655212 [1976/77]; C.A. **88** [1978] No. 6344).
[38] Hoffmann, H.; Hammann, I.; Homeyer, B.; Stendel, W. (Ger. 2737606 [1977/79]; C.A. **91** [1979] No. 56482).
[39] Gorodilov, V. N.; Smirnova, V. N.; Ivin, S. Z. (U.S.S.R. 175957 [1964/65]; C.A. **64** [1966] 14093).
[40] Serednitskii, Ya. A.; Shibarov, V. V.; Tolopko, D. K. (Izv. Vyssh. Uchebn. Zaved. Khim. Khim. Tekhnol. **17** [1974] 397/400; C.A. **81** [1974] No. 13060).

[41] Böger, M.; Drabek, J. (Ger. 2654080 [1972/77]; C.A. **87** [1977] No. 102078).

12.5.1.1.2 $ClSN(CF_3)_2$

The title compound was formed by heating a mixture of $(CF_3)_2NCl$ and sulfur (mole ratio 1:1.1) in the absence of moisture and air under autogenous pressure in a "Hastelloy" C bomb for 1 h at 150 °C and for 1 h at 225 °C; 21% yield, along with $(CF_3)_2NSN(CF_3)_2$ (20%), $(CF_3)_2NSSN(CF_3)_2$, and $CF_3N{=}CF_2$. Heating the reaction mixture for 1 h at 150 °C and 2 h at 225 °C gave $ClSN(CF_3)_2$ with a yield of 5% along with $(CF_3)_2NSN(CF_3)_2$, $(CF_3)_2NSSN(CF_3)_2$, $(CF_3)_2NSSCl$, very small amounts of $(CF_3)_2NS_3Cl$, $(CF_3)_2NS_xN(CF_3)_2$ (x = 3, 4), and the volatiles $CF_3N{=}CF_2$, $(CF_3)_2NH$, and SOF_2. A better yield (~37%) was obtained by heating the mixture for 1 h at 100 °C, 1 h at 200 °C, 1 h at 275 °C, and 1 h at 325 °C [1]. The title compound was also obtained by sealing $(CF_3)_2NCl$ and S_2Cl_2 (mole ratio ~1:2) together in a silica ampule and irradiating it with unfiltered light from a mercury-arc lamp for 4 d. $ClSN(CF_3)_2$ could not be completely purified, but its mixture with SCl_2 corresponded to a yield of 99% [2].

$ClSN(CF_3)_2$ was formed directly by heating a mixture of SCl_2, $(CF_3)_2NH$, and anhydrous KF (mole ratio 1.9:1:6.6) in the absence of moisture and air under autogenous pressure in a "Hastelloy" C bomb for 1 h at 150 °C and for 4 h at 275 °C; in addition, $(CF_3)_2NSN(CF_3)_2$ and $CF_3N{=}CFN(CF_3)_2$ were identified [1]. $ClSN(CF_3)_2$ was formed by sealing SCl_2 with $((CF_3)_2N)_2Hg$ (mole ratio 2:1) in a glass ampule and setting it aside overnight at room temperature; 49% yield. Other spectroscopically identified products were obtained with the following estimated yields: $CF_3N{=}CF_2$ 12%, $(CF_3)_2NCl$ 19%, and sulfur chlorides [2].

The title compound is a pale [2] yellow [1, 2] liquid, b.p. 49.5 to 51 °C, of high density and strong acrid odor [1]. Its vapor pressure in the range −28 to 45 °C is represented by $\log p\,(\text{Torr}) = -1458/T\,(K) + 7.43$, giving an extrapolated boiling point of 47.3 °C and a Trouton constant of $\Delta S_v = 20.8\ \text{cal}\cdot\text{mol}^{-1}\cdot\text{K}^{-1}$ [2].

IR (gas): ν (in μ) = 7.55, 7.95, 8.10, and 14.10 [1]. The IR spectrum of the compound confirmed the presence of the bis(trifluoromethyl)amino group. A shoulder at 530 cm^{-1} is probably associated with an S–Cl stretching vibration; 14 other principle bands appear between 1358 and 425 cm^{-1} [2].

There is a very intense absorption in the UV spectrum at about 200 nm; in addition, an absorption at 313 nm ($\varepsilon = 28.9$) may be attributed to the S–Cl bond [2].

In the absence of air $ClSN(CF_3)_2$ is stable when stored and can be heated to about 300 °C for short periods of time without decomposition [1].

$ClSN(CF_3)_2$ is insoluble in water, but is slowly hydrolyzed by water at room temperature [1]. Hydrolysis in cold water at room temperature gave sulfur, CO_2, $(CF_3)_2NH$, and $(CF_3)_2$-$NSN(CF_3)_2$ [2]. It also was readily hydrolyzed by aqueous alkalies at 25 °C [1] and rapidly by 15% aqueous NaOH [2].

Condensing the title compound onto a large excess of AgCN (mole ratio ~1 : 25) and leaving the mixture at room temperature for 1.5 h gave $(CF_3)_2NSN(CF_3)_2$ as the main product along with some $CF_3N{=}CF_2$, SOF_2, and AgCN [3].

Sealing $ClSN(CF_3)_2$ and an equimolar amount of ethylene in a Pyrex ampule and heating the mixture to 70 °C for 5 h gave $(CF_3)_2NSN(CF_3)_2$, $(CF_3)_2NH$, $CF_3N{=}CF_2$, $(ClCH_2CH_2)_2S$, and SiF_4 [3]. The title compound reacted with acyclic and alicyclic hydrocarbon olefins to addition compounds. The additions are trans-stereospecific. All reactions were carried out in vacuum in sealed tubes at −78 °C in the dark (5 to 7 d) and they also gave $(CF_3)_2NH$ (1 to 11.5%) [4]. For details, see Table 33.

Table 33
Reaction of $ClSN(CF_3)_2$ with Hydrocarbon Alkenes [4].

alkene (substrate)	mole ratio $ClSN(CF_3)_2$: substrate	alkene recovered (in %)	products (yield in %)	
$CH_2{=}C(CH_3)_2$	1 : 1.94	48.5	$(CF_3)_2NSCH_2C(CH_3)_2Cl$ (81),	
			$(CF_3)_2NSCH_2C(CH_3){=}CH_2$, $(CF_3)_2NSCH{=}C(CH_3)_2$	(6.5)
$CH_2{=}CHC_2H_5$	1 : 2.82	60	$(CF_3)_2NSCH_2CH(C_2H_5)Cl$, $(CF_3)_2NSCH(C_2H_5)CH_2Cl$	(94)[a)]
$CH_2{=}CH(n\text{-}C_3H_7)$	1 : 1.31	37.5	$(CF_3)_2NSCH_2CH(n\text{-}C_3H_7)Cl$, $(CF_3)_2NSCH(n\text{-}C_3H_7)CH_2Cl$	(82.5)[a)]
$CH_2{=}CHC(CH_3)_3$	1.06 : 1	6.5	$(CF_3)_2NSCH_2CHClC(CH_3)_3$ (82)	
$CH_3CH{=}C(CH_3)_2$	1.02 : 1	9.5	$(CF_3)_2NSCH(CH_3)C(CH_3)_2Cl$ (88)	
cis-$CH_3CH{=}CHCH_3$	1 : 2.02	43.5	*threo*-$(CF_3)_2NSCH(CH_3)CH(CH_3)Cl$ (89)	
trans-$CH_3CH{=}CHCH_3$	1 : 2.31	57	*erythro*-$(CF_3)_2NSCH(CH_3)CH(CH_3)Cl$ (92)	
cyclopentene	1 : 1.05	14	*trans*-$(CF_3)_2NSCH(CH_2)_3CHCl$ (89)	
cyclohexene	1 : 2.49	66	*trans*-$(CF_3)_2NSCH(CH_2)_4CHCl$ (82)	

a) The adduct mixtures could not be separated by GLC.

The reaction of $ClSN(CF_3)_2$ with unsymmetrical fluoro olefins in daylight or under photochemical conditions produced adducts with a ca. 1 : 1 ratio. Addition to octafluoro-2-butene or hexafluoro-2-butyne gave mixtures of the *syn* and *anti* adducts. All reactions were carried out in vacuum in sealed Pyrex ampules [5]. For the reaction conditions, see Table 34.

Table 34
Reactions of $ClSN(CF_3)_2$ with Fluoro-olefins, Tetrachloroethene, and Hexafluoro-2-butyne [5].

alkene or alkyne (substrate)	mole ratio $(CF_3)_2NSCl$: substrate	conditions	products (yield in %)
$CH_2{=}CHF$	1 : 1.28	UV, 22 h	$(CF_3)_2NSCH_2CHFCl$ (26), $(CF_3)_2NSCHFCH_2Cl$ (25), $(CF_3)_2NSN(CF_3)_2$ (16), $(CF_3)_2NSSN(CF_3)_2$ (3), $(CF_3)_2NSSSN(CF_3)_2$ (1), $CH_2ClCHFCl$ (8), $(CF_3)_2NCH_2CHFSN(CF_3)_2$ (1)

Table 34 (continued)

alkene or alkyne (substrate)	mole ratio $(CF_3)_2NSCl$: substrate	conditions	products (yield in %)
$CH_2{=}CF_2$	1:1.055	UV, 21 h	$(CF_3)_2NSCH_2CF_2Cl$ (36), $(CF_3)_2NSCF_2CH_2Cl$ (38), $(CF_3)_2NSN(CF_3)_2$ (13), $(CF_3)_2NSSN(CF_3)_2$ (8), CH_2ClCF_2Cl (23), $(CF_3)_2NCH_2CF_2SN(CF_3)_2$ (5)
$CHF{=}CF_2$	1.02:1	light, 4 d	$(CF_3)_2NSCF_2CHFCl$ (35), $(CF_3)_2NSCHFCF_2Cl$ (26), $(CF_3)_2NSN(CF_3)_2$ (1), $(CF_3)_2NSSN(CF_3)_2$ (10.5), $CHFClCF_2Cl$ (40), $(CF_3)_2NSSCHFCF_2Cl$ (1.5), $(CF_3)_2NSSCF_2CHFCl$ (1.5), $(CF_3)_2NS(C_2HF_3)_2Cl$ (3.5)
CF_2CFCl[a]	1:1.04	light, 22 d	$(CF_3)_2NSCF_2CFCl_2$ (31), $(CF_3)_2NSCFClCF_2Cl$ (31), $(CF_3)_2NSN(CF_3)_2$ (15), $(CF_3)_2NSSN(CF_3)_2$ (10.5), $CF_2ClCFCl_2$ (16), $(CF_3)_2NCF_2CFClSN(CF_3)_2$ (4), $CF_2ClCFClCF_2CFCl_2$ (6.5), $S{-}CFCl{-}CF_2{-}CFCl{-}CF_2$ (11), $(CF_3)_2NSCl$ (2)
$CF_2{=}CFCl$[a]	1:1.57	UV, 22 h	$(CF_3)_2NSCF_2CFCl_2$ (26), $(CF_3)_2NSCFClCF_2Cl$ (26), $(CF_3)_2NSN(CF_3)_2$ (11.5), $(CF_3)_2NSSN(CF_3)_2$ (12), $(CF_3)_2NSSSN(CF_3)_2$ (3), $CF_2ClCFCl_2$ (3), $(CF_3)_2NSSCFClCF_2Cl$ (3), $(CF_3)_2NS(C_2F_3Cl)_2Cl$ (3.5), $(CF_3)_2NSS(C_2F_3Cl)_2Cl$ (1)
$CF_2{=}CHCF_3$	1:1.15	UV, 30 h	$(CF_3)_2NSCH(CF_3)CF_2Cl$ (35), $(CF_3)_2NSCF_2CHClCF_3$ (22), $(CF_3)_2NSN(CF_3)_2$ (11), $(CF_3)_2NSSN(CF_3)_2$ (7), $CF_2ClC(CF_3)FCl$ (7), $(CF_3)_2NSS(C_3HF_5)_2Cl$ (3), $(CF_3)_2NSCl$ (1)
$CF_2{=}CFCF_3$[a]	1.18:1	UV, 21 h	$(CF_3)_2NSCF_2CFClCF_3$ (33), $(CF_3)_2NSCF(CF_3)CF_2Cl$ (33), $(CF_3)_2NSN(CF_3)_2$ (20), $(CF_3)_2NSS(CF_3)_2$ (10), $CF_2ClC(CF_3)FCl$ (3), $(CF_3)_2NNSCl$ (9)
$CF_3CF{=}CFCF_3$[b]	1:1.25	UV, 21 h	$(CF_3)_2NSCF(CF_3)CFClCF_3$ (59)[c], $(CF_3)_2NSN(CF_3)_2$ (8), $(CF_3)_2NSSN(CF_3)_2$ (22), $(CF_3)_2NSSSN(CF_3)_2$ (11), $CF_3CFClC(CF_3)FCl$ (36), $(CF_3)_2NSSC(CF_3)FC(CF_3)FCl$ (5)
$CCl_2{=}CCl_2$	~1:1	UV, 21 h	$(CF_3)_2NSCCl_2CCl_3$ (32), $(CF_3)_2NSN(CF_3)_2$ (15), $(CF_3)_2NSSN(CF_3)_2$ (48), CCl_3CCl_3 (9.5), $(CF_3)_2NSSCCl_2CCl_3$ (3)
$CF_3C{\equiv}CCF_3$	~1:1	light, 7 d	(*E*)-$(CF_3)_2NSC(CF_3){=}CClCF_3$ (42), (*Z*)-$(CF_3)_2NSC(CF_3){=}CClCF_3$ (21), $(CF_3)_2NSN(CF_3)_2$ (4), $(CF_3)_2NSSN(CF_3)_2$ (10.5), $CF_3CCl{=}CClCF_3$ (11), $[CF_3CCl{=}C(CF_3)]_2$ (4), $S{-}CCF_3{=}CCF_3{-}CCF_3{=}CCF_3$ (2), $(CF_3)_2NSSC(CF_3){=}CClCF_3$ (10), $[CF_3CCl{=}C(CF_3)]_2S$ (4.5)

[a] 1 : 1 adducts could not be separated by GLC. – [b] Mixture of *cis* and *trans* isomers (1 : 3 ratio). – [c] Mixture of *erythro* and *threo* adducts (1 : 1 ratio).

The reaction of the compound with anisole (20 °C, dark, 48 h) yielded $(CF_3)_2NH$ (68%), HCl (66%), $(CF_3)_2NSN(CF_3)_2$ (21%), and $(4\text{-}CH_3OC_6H_4)_2S$ (41%). The reaction with phenol (20 °C, dark, 24 h) in ether gave $(CF_3)_2NSN(CF_3)_2$ (95%) and $4\text{-}ClC_6H_4OH$ (97%) [4].

The compound may be used as a solvent for fluorinated polymers, e.g., it readily dissolves low-molecular-weight polytetrafluoroethylene giving a solution of 5 to 10% polymer by weight. The solution thus obtained can be applied to fibrous materials, such as paper fabrics, wood, etc., to make them fire-retarding and water-repelling [1].

References:

[1] Tullock, C. W. (U.S. 3121112 [1959/64]; C.A. **60** [1964] 13143).
[2] Emeléus, H. J.; Tattershall, B. W. (J. Chem. Soc. **1964** 5892/4).
[3] Emeléus, H. J.; Tattershall, B. W. (J. Inorg. Nucl. Chem. **28** [1966] 1823/7).
[4] Service, C. F.; Tipping, A. E. (J. Fluorine Chem. **19** [1981] 91/5).
[5] Service, C. F.; Tipping, A. E. (J. Fluorine Chem. **20** [1982] 135/40).

12.5.1.1.3 $ClSN(C_2H_5)_2$

Preparation

Directly from $(C_2H_5)_2NH$ and SCl_2 or S_2Cl_2. A solution of SCl_2 in CH_2Cl_2 precooled to −5 °C was treated with a solution of pyridine in CH_2Cl_2 at −5 to 0 °C and after 15 min with a solution of $(C_2H_5)_2NH$ in CH_2Cl_2 at −5 to 0 °C. The resulting solution was stirred for 2 h while the temperature was allowed to rise to ambient temperature [1]. The compound was also formed with a yield of 69% by adding $(C_2H_5)_2NH$ at 20 °C to a solution of S_2Cl_2 in petroleum ether (b.p. 40 to 56 °C) [2].

Cleaving the Sulfur–Sulfur Bond of $(C_2H_5)_2NSSN(C_2H_5)_2$ with a Chlorinating Agent. $ClSN(C_2H_5)_2$ was formed with 70% yield by passing dry Cl_2 at 0 to 5 °C through a solution of $(C_2H_5)_2NSSN(C_2H_5)_2$ in CCl_4 until there was a slight excess (of Cl_2). The excess Cl_2 was then removed at 0 to 5 °C in vacuum or by purging with an inert gas. The resulting solution was then cleared with animal charcoal, and after filtration the solvent was evaporated under reduced pressure at 30 to 50 °C to give the crude compound which was distilled in vacuum [3]; see also [4 to 6]. $ClSN(C_2H_5)_2$ was also synthesized with 65% yield by adding a solution of SO_2Cl_2 in CCl_4 without cooling to a solution of $(C_2H_5)_2NSSN(C_2H_5)_2$ (mole ratio 1 : 1) in CCl_4. The reaction is slightly exothermic with evolution of SO_2, causing the temperature to rise to about 40 °C. The mixture was then heated for 10 min to 70 to 80 °C, allowed to cool, and was cleared with animal charcoal [3]; see also [4 to 7]. The reaction could also be performed at −78 °C with an excess of SO_2Cl_2 (mole ratio ca. 2 : 3) without solvent [8].

Cleaving the Sulfur–Nitrogen Bond of $(C_2H_5)_2NSN(C_2H_5)_2$ with a Chloro Compound. $ClSN(C_2H_5)_2$ was obtained with ca. 90% yield as the only product of the exothermic reaction of $(C_2H_5)_2NSN(C_2H_5)_2$ with sulfur dichloride via $(C_2H_5)_2NSN(C_2H_5)_2 + SCl_2 \rightarrow 2\ ClSN(C_2H_5)_2$ [9]. Instead of SCl_2 benzenesulfenyl chloride was also used according to $(C_2H_5)_2NSN(C_2H_5)_2 + C_6H_5SCl \rightarrow ClSN(C_2H_5)_2 + C_6H_5SN(C_2H_5)_2$ [9].

Special Procedures. $ClSN(C_2H_5)_2$ was obtained by reacting $(C_2H_5)_2NSOCH_3$ with RC(O)Cl ($R = CH_3$, C_6H_5) at 20 to 25 °C in an inert atmosphere [10], by the reaction $(C_2H_5)_2NSOCH_3 + (CH_3)_3SiCl \rightarrow ClSN(C_2H_5)_2 + (CH_3)_3SiOCH_3$ (no details given) [11], or with 50% yield by adding a solution of $ClSN{=}CCl_2$ in ether to a solution of $(C_2H_5)_2NH$ (mole ratio 1 : 2) in ether kept at ~10 °C [12].

Properties

$ClSN(C_2H_5)_2$ is an orange-yellow [2] or orange liquid [3, 9,12] (see also [4 to 6]) of pungent [9], unpleasant odor [12]; b.p. 38 °C/3 Torr [11], 50 °C/7 Torr [2], 56 °C/12 Torr [12], 60 to 62 °C/11 Torr [1], 61 to 62 °C/12 Torr [7], 62 to 64 °C/13 Torr [3]; see also [4 to 6]. $n_D^{24} = 1.5073$ [11].

The ^1H NMR spectrum of $ClSN(C_2H_5)_2$ in $CD_3C_6D_5$ (0.68 M) showed at 35 °C sharp triplet and quartet ethyl signals. The methylene protons of $(C_2H_5)_2NSCl$ in $CD_3C_6H_5$ became diastereotopic on the ^1H NMR time scale in the temperature range 0 to 30 °C with a coalescence temperature of 28 ± 2 °C for 0.17, 0.45, and 0.93 M solutions in $CD_3C_6D_5$ due to bond rotation. However, in CH_2Cl_2 solution the compound did not show any signal splitting down to −60 °C; possibly chlorine exchange is more rapid in this solvent [14].

Solutions of similar concentrations of $(C_2H_5)_2NSCl$ and $(C_2H_5)_2NSBr$ in $CD_3C_6D_5$ showed sharp signals at δ values intermediate of those of the pure compounds at temperatures above 50 °C indicative of rapid halogen exchange. Below −40 °C distinct signals were observed, attributable to each of the separate components [14].

^{15}N NMR (neat/CH_3NO_2): δ = −274.7 ppm [14].

$ClSN(C_2H_5)_2$ is a liquid that can be readily distilled in vacuum [6, 9] and is visibly decomposed within 1 h, even if sealed in ampules and kept in the dark [9]. It is very reactive and sensitive to hydrolysis [6].

Chemical Reactions

$ClSN(C_2H_5)_2$ reacted with the equimolar amount of $R_2NSi(CH_3)_3$ (R = CH_3, n-C_3H_7) to give only symmetrical products and $(CH_3)_3SiCl$: 2 $ClSN(C_2H_5)_2$ + 2 $R_2NSi(CH_3)_3 \rightarrow R_2NSNR_2$ + $(C_2H_5)_2NSN(C_2H_5)_2$ + $(CH_3)_3SiCl$ (R = CH_3, n-C_3H_7) [9]. Adding t-C_4H_9N=X=$NSi(CH_3)_3$ (X = S, C) to a solution of the title compound (mole ratio 1 : 1) in CH_2Cl_2 gave after 2 h at room temperature the corresponding t-C_4H_9N=X=$NSN(C_2H_5)_2$ in 52% (X = S) or 69% (X = C) yield [15].

$ClSN(C_2H_5)_2$ reacted with O=S=$NSi(CH_3)_3$ to give O=S=$NSN(C_2H_5)_2$ and $(CH_3)_3SiCl$ (no details); with silthianes reactions occur via $ClSN(C_2H_5)_2$ + $(CH_3)_3SiSR \rightarrow (C_2H_5)_2NSSR$ (R = alkyl, aryl, not specified) + $(CH_3)_3SiCl$ [9].

Adding the title compound under cooling and stirring to the trialkyl phosphites $(RO)_3P$ (mole ratio 1 : 1; R = C_2H_5, n-C_3H_7, n-C_4H_9), while maintaining the temperature of the mixture below 10 °C, gave the corresponding $(RO)_2P(O)SN(C_2H_5)_2$ with a yield of 74, 57, or 70%, respectively, along with RCl [16]; see also [17].

$ClSN(C_2H_5)_2$ was reacted with $(C_2H_5O)_2PF$ under cooling for 1 h and at room temperature for 1 h to give 79% $(C_2H_5)_2N(C_2H_5O)P(O)F$ and 64.6% $(C_2H_5O)_2P(S)F$, respectively [18]. Adding $(C_2H_5O)_2PCl$ to the title compound (mole ratio 2 : 1) at 25 °C and at 55 to 70 °C produced within 2 to 3 h 95% $(C_2H_5O)_2P(S)N(C_2H_5)_2$, 65% $C_2H_5OP(O)Cl_2$, and 85% C_2H_5Cl [19]. A reaction mechanism was proposed explaining qualitatively the different behavior of fluoro- and chlorophosphates [20].

Adding the title compound at room temperature to an equimolar amount of $(CH_3O)_2P$-$N(CH_3)_2$ gave 45% $(C_2H_5)_2NSP(O)(OCH_3)N(CH_3)_2$ along with CH_3Cl [21]. Adding $ClSN(C_2H_5)_2$ at 20 to 65 °C to an equimolar amount of $(C_2H_5O)_2PN(CH_3)C(O)OC_2H_5$ gave 43% $C_2H_5OC(O)$-$N(CH_3)P(O)(OC_2H_5)SN(C_2H_5)_2$ along with C_2H_5Cl [22].

$ClSN(C_2H_5)_2$ reacted with $C_2H_5OPF_2$ (mole ratio 1 : 2; 1 h under cooling, 1 h at room temperature, and heating at 80 °C) to give 65% $(C_2H_5)_2NP(O)F_2$ and 60% $C_2H_5OP(S)F_2$ [18, 23].

The reaction of the compound with n-$C_3H_7OPF_2$ proceeds analogously yielding 68% $(C_2H_5)_2N$-$P(O)F_2$ [18]. The title compound reacted with $C_2H_5OPCl_2$ (mole ratio 1 : 2) under cooling, for 1 h at 20 °C, and for 30 min at 95 to 100 °C to yield 96% $C_2H_5OP(S)(Cl)N(C_2H_5)_2$ and 90% $POCl_3$ and C_2H_5Cl [19].

$ClSN(C_2H_5)_2$ reacted as a soft electrophile with $C_2H_5OP(N(C_2H_5)_2)N(C_2H_5)C(O)OC_2H_5$ without solvent while keeping the temperature at 40 °C yielding 63% $(C_2H_5)_2NSP(O)(N(C_2H_5)_2)$-$N(C_2H_5)C(O)OC_2H_5$ and 90% C_2H_5Cl, with $C_2H_5OP(N(C_2H_5)C(O)OC_2H_5)_2$ without solvent under cooling yielding 20% $(C_2H_5)_2NSP(O)(N(C_2H_5)C(O)OC_2H_5)_2$ and 90% C_2H_5Cl [24]. Adding a solution of $C_2H_5P(OC_2H_5)_2$ in benzene at 0 to 10 °C to a solution of the compound (mole ratio 1 : 1) in benzene gave $C_2H_5OP(O)(C_2H_5)SN(C_2H_5)_2$ [17].

Adding a solution of $(C_2H_5O)_2P(O)H$ in benzene to a solution of the title compound and $(C_2H_5)_3N$ (mole ratio 1 : 1 : 1) at 0 °C and allowing the mixture to react for 2 h at room temperature gave $(C_2H_5O)_2P(O)SN(C_2H_5)_2$ with a yield of 35% [17].

$ClSN(C_2H_5)_2$ reacted with the equimolar amount of $(C_2H_5O)_2P(O)NHNHC_6H_4R$ (R = H, 3-CF_3) and excess $(C_2H_5)_3N$ in ether at 0 to 2 °C for 0.5 h to give the corresponding $(C_2H_5O)P$-$(O)NHN(C_6H_4R)SN(C_2H_5)_2$ with 88 and 68.7% yield, respectively [25]; see also [26].

Adding $(C_2H_5O)_2P(S)H$ to a −5 °C-cooled solution of the compound (mole ratio 2 : 1) in CCl_4 such that the temperature did not rise above 0 °C gave $(C_2H_5O)_2P(S)SP(S)(OC_2H_5)_2$ along with $(C_2H_5)_2NH \cdot HCl$ [27].

$ClSN(C_2H_5)_2$ reacted with $(C_2H_5O)_2P(S)NHNHC_6H_4R$ (R = H, 3-CF_3) (mole ratio 1 : 1 each) and a slight excess of $(C_2H_5)_3N$ in ether at −5 to +5 °C for 0.5 to 2.5 h to give the corresponding $(C_2H_5O)_2P(S)N(SN(C_2H_5)_2)NHC_6H_4R$ with 95 to 100% (R = H) or ~100% (R = 3-CF_3) yield [25]; see also [26].

Addition of the title compound to cooled solutions (−5 to −10 °C) of RSNa (R = CH_3, CH_2=$CHCH_2$, CH_2=$C(CH_3)CH_2$) in C_2H_5OH (R = CH_3), CH_3OH (R = CH_2=$CHCH_2$), or toluene (R = CH_2=$C(CH_3)CH_2$) gave the corresponding $(C_2H_5)_2NR$ [29].

A solution of the title compound in benzene was added to a suspension of $(C_2H_5O)_2P(S)$-SNa in benzene while the temperature rose to about 30 to 35 °C; stirring the mixture at this temperature for 1 h gave 75 to 80% $(C_2H_5O)_2P(S)SSN(C_2H_5)_2$ [30].

When a solution of $ClSN(C_2H_5)_2$ in ether at −50 °C was added to an equimolar amount of $NaN(Si(CH_3)_3)_2$ in ether and the mixture was allowed to come to room temperature within 8 h, $((CH_3)_3Si)_2NSN(C_2H_5)_2$ was obtained with 68% yield [31].

Passing $ClSN(C_2H_5)_2$ at low pressures through a preheated column (100 °C) containing HgF_2 gave C_2H_5N=SF_2 as the main product [32]. Adding a solution of $Hg(SCF_3)_2$ in anhydrous ether at 20 °C to a solution of the title compound (mole ratio 1 : 2) in anhydrous ether and cooling the reaction mixture to −30 °C yielded 85% $(C_2H_5)_2NSSCF_3$ and $HgCl_2$ [33]. The title compound reacted with $Hg(SC_6F_5)_2$ (mole ratio 2 : 1) in dry ether to give $(C_2H_5)_2NSSC_6F_5$ [34].

Adding $ClSN(C_2H_5)_2$ under cooling to $(n\text{-}C_4H_9)_3SnOCH_3$ (mole ratio 1 : 1.1) and warming the mixture then for 1 h gave 65% $(C_2H_5)_2NSOCH_3$ along with $(n\text{-}C_4H_9)_3SnCl$ [11].

Adding a solution of the title compound in CH_3CN to an equimolar amount of $AgN(SO_2CH_3)_2$ in CH_3CN and allowing the mixture to react for 6 h at 0 °C gave $(C_2H_5)_2NSN(SO_2CH_3)_2$ [35]. Passing $ClSN(C_2H_5)_2$ at low pressure through a preheated column (120 °C) containing AgF gave $FSN(C_2H_5)_2$ [32].

$ClSN(C_2H_5)_2$ reacted with 1-alkynyllithium derivatives $LiC{\equiv}CR$ (R = H, n-C_4H_9, C_6H_5, $(CH_3)_3Si$) to yield the corresponding N,N-diethyl-1-alkynesulfenamides $(C_2H_5)_2NSC{\equiv}CR$ [28], see next page.

substrate LiC≡CR R	reaction conditions	product(s) $(C_2H_5)_2NSC{\equiv}CR$ R	yield[a)] in %
H	0°C, 1 h	H	(25)
		$SN(C_2H_5)_2$	(75)
	−78 to +20°C, 1 h	H	(75), 43
		$SN(C_2H_5)_2$	(30)
n-C_4H_9	−78 to +20°C, 2 h	n-C_4H_9	90
C_6H_5	−78 to +20°C, 2 h	C_6H_5	90
$Si(CH_3)_3$	−78 to +20°C, 2 h	$Si(CH_3)_3$	95

a) The yield of crude product (given in parentheses) was estimated from ^1H NMR data.

Reactions of $ClSN(C_2H_5)_2$ with organic compounds are summarized in Table 35.

References:

[1] Anderson, M. (U.S. 4623658 [1986]; C.A. **107** [1987] No. 23100).
[2] Badische Anilin-Soda-Fabrik AG (Ger. 1131222 [1960/62]; C.A. **57** [1962] 13771).
[3] Farbenfabriken Bayer AG (Br. 790021 [1954/58]; C.A. **1958** 13806).
[4] Kühle, E. (The Chemistry of the Sulfenic Acids, Thieme, Stuttgart 1973, p. 10).
[5] Kühle, B. (Synthesis **1970** 561/80, 566).
[6] Dorlars, A. (Houben-Weyl Methoden Org. Chem. 4th Ed. **10** Pt. 2 [1958] 744/51, 745/6).
[7] Berger, H.; Gall, R.; Kampe, W.; Bicker, U.; Kuhn, R. (Ger. 2948832 [1979/81]; C.A. **95** [1981] No. 115257).
[8] Hatch, C. E., III (J. Org. Chem. **43** [1978] 3953/7).
[9] Armitage, D. A.; Tso, C. C. (J. Chem. Soc. D **1971** 1413/4).
[10] Musin, B. M.; Ivanov, V. B.; Ivanov, B. E. (Izv. Akad. Nauk SSSR Ser. Khim. **1988** 1693; Bull. Acad. Sci. USSR Div. Chem. Sci. [Engl. Transl.] **1988** 1509).

[11] Armitage, D. A.; Towle, I. D. H. (Phosphorus Sulfur Relat. Elem. **1** [1976] 37/9).
[12] Bacon, R. G. R.; Irwin, R. S. (J. Chem. Soc. **1960** 5079/87).
[13] Jackson, W. R.; Kee, T. G.; Spratt, R.; Jennings, W. B. (Tetrahedron Lett. **1973** 3581/4).
[14] Dorie, J.; Gouesnard, J.-P. (J. Chim. Phys. Phys.-Chim. Biol. **81** [1984] 15/9).
[15] Appel, R.; Montenarh, M. (Z. Naturforsch. **30b** [1975] 847/9).
[16] Michalski, J.; Pliska, B. (Dokl. Akad. Nauk SSSR **147** [1962] 111/2; Dokl. Chem. [Engl. Transl.] **142/147** [1962] 961/2).
[17] Michalski, J.; Pliszka-Krawiecka, B. (J. Chem. Soc. C **1966** 2249/52).
[18] Gusar, N. I.; Ivanova, Z. M.; Chaus, M. P.; Gololobov, Yu. G. (Zh. Obshch. Khim. **46** [1976] 1981/6; J. Gen. Chem. USSR [Engl. Transl.] **46** [1976] 1910/4).
[19] Ivanova, Z. M.; Gusar, N. I.; Gololobov, Yu. G. (Zh. Vses. Khim. O-va. im. D. I. Mendeleeva **18** [1973] 349/50; C.A. **79** [1973] No. 104677).
[20] Gusar, N. I.; Ivanova, Z. M.; Chaus, M. P.; Gololobov, Yu. G. (Dopov. Akad. Nauk Ukr. RSR B: Geol. Khim. Biol. Nauki **1978** 520/3; C.A. **89** [1978] No. 196907).

[21] Alimov, P. I.; Antokhina, L. A. (Izv. Akad. Nauk SSSR Ser. Khim. **1966** 1486/8; Bull. Acad. Sci. USSR Div. Chem. Sci. [Engl. Transl.] **1966** 1432/4).
[22] Antokhina, L. A.; Alimov, P. I. (Izv. Akad. Nauk SSSR Ser. Khim. **1971** 806/10; Bull. Acad. Sci. USSR Div. Chem. Sci. [Engl. Transl.] **1971** 723/6).
[23] Ivanova, Zh. M.; Gusar', N. I.; Miroshnichenko, V. V.; Gololobov, Yu. G. (Zh. Obshch. Khim. **42** [1972] 2115; J. Gen. Chem. USSR [Engl. Transl.] **42** [1972] 2110).

(continued on p. 129)

Table 35
Reactions of $ClSN(C_2H_5)_2$ with Organic Compounds.

$ClSN(C_2H_5)_2$ + organic substrate	reaction conditions	product(s) (yield in %)	Ref.
$CF_3CF{=}CF_2 + CsF$	in abs. DMF, 6 h at room temperature	$(CF_3)_2CFSN(C_2H_5)_2$ (25)	[36]
$(CF_3)_2C{=}CF_2 + CsF$	in abs. DMF, 6 h at room temperature	$(CF_3)_3CSN(C_2H_5)_2$ (69.5)	[36]
$CF_2{=}C(CF_3)C(O)OCH_3$	in monoglyme, 3.5 h at 45 to 50°C	$(C_2H_5)_2NSC(CF_3)_2C(O)OCH_3$ (52)	[36]
$CF_3CH_2C(O)OCH_3$	no reaction		[37]
$(CF_3)_2CHC(O)OCH_3 + (C_2H_5)_3N$	in ether at 0 to 20°C	$(C_2H_5)_2NSC(CF_3)_2C(O)OCH_3$ (70 to 80)	[38]
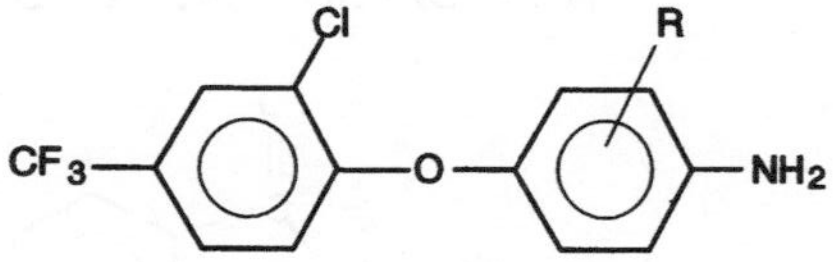		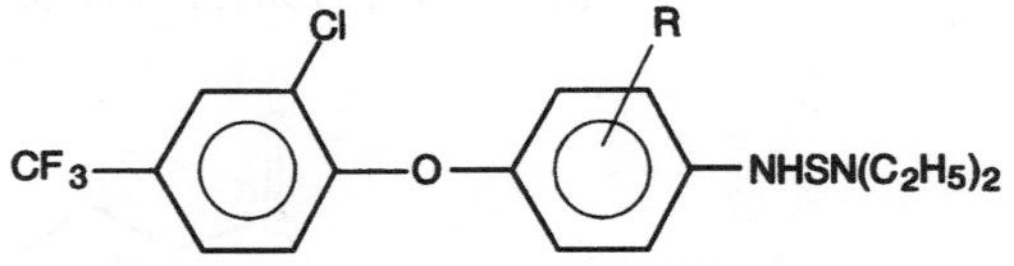	[1]
$(CH_3)_2NH$	mole ratio 1:2, in ether at room temperature	$(CH_3)_2NSN(CH_3)_2$, $(C_2H_5)_2NSN(CH_3)_2$, $(CH_3)_2NH \cdot HCl$	[12]
$(C_2H_5)_2NH$	mole ratio 1:2, in ether at room temperature	$(C_2H_5)_2NSN(C_2H_5)_2$ (69)	[12]
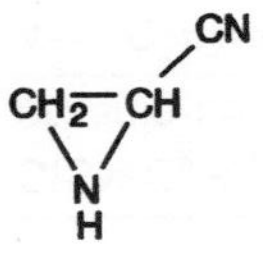	in ether	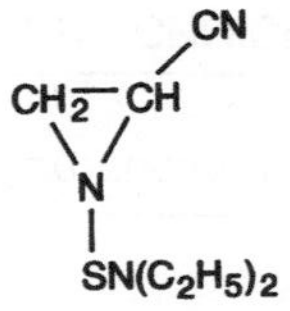	[7]

Table 35 (continued)

$ClSN(C_2H_5)_2$ + organic substrate	reaction conditions	product(s) (yield in %)	Ref.
$ArN{=}CHNHCH_3 + (C_2H_5)_3N$ (Ar = 2-CH_3-4-ClC_6H_3, 2,4-$(CH_3)_2C_6H_3$, 2,4,5-$Cl_3C_6H_2$)	in THF, with cooling	$ArN{=}CHN(CH_3)SN(C_2H_5)_2$	[39]
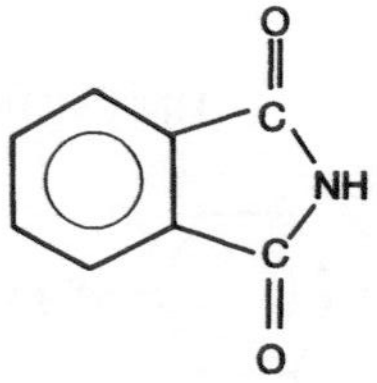 + $(C_2H_5)_3N$	in THF, at 0 to 10 °C for 0.5 to 1 h	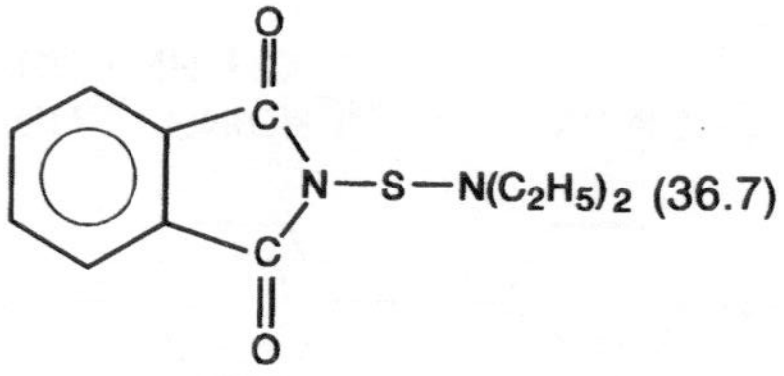 (36.7)	[40 to 43]
$CH_3N{=}C{=}O + HF + (C_2H_5)_3N$	in CH_2Cl_2, at −10 to 0 °C, and 1 h at 5 °C	$(C_2H_5)_2NSN(CH_3)C(O)F$ (62.9)	[44, 45], see also [8, 46]
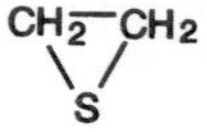	simultaneous addition of CCl_4 solutions, ~12 h at room temperature	$(C_2H_5)_2NSSCH_2CH_2Cl$ (51.4)	[47]
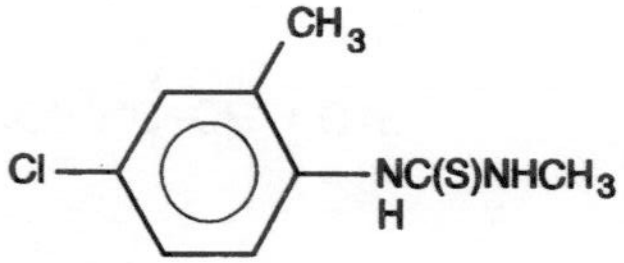 + $(C_2H_5)_3N$	in CH_2Cl_2, at 0 to 10 °C and 2 h at room temperature	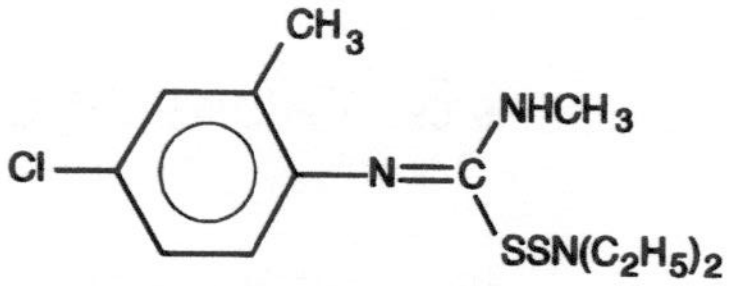	[48]

(continued from p. 126)

[24] Antokhina, L. A.; Alimov, P. I. (Izv. Akad. Nauk SSSR Ser. Khim. **1974** 401/3; Bull. Acad. Sci. USSR Div. Chem. Sci. [Engl. Transl.] **1974** 367/9).
[25] Gusar', N. I.; Randina, L. V.; Shurubura, A. K. (Zh. Obshch. Khim. **59** [1989] 548/56; J. Gen. Chem. USSR [Engl. Transl.] **59** [1989] 486/92).
[26] Gusar, N. I.; Gololobov, Yu. G. (Heteratom Chem. **3** [1992] 407/14).
[27] Almasi, L.; Hantz, A. (Chem. Ber. **97** [1964] 661/6).
[28] Baudin, J.-B.; Julia, S. A.; Lorne, R. (Bull. Soc. Chim. Fr. **1987** No. 1, pp. 181/8).
[29] Williams, H. R. (U.S. 3121661 [1961/64]; C.A. **61** [1964] 6958).
[30] Malz, H.; Bayer, O.; Freytag, H.; Lober, F. (U.S. 2891059 [1957/59]; C.A. **1960** 4387).

[31] Rinne, D.; Blaschette, A. (Chem.-Ztg. **98** [1974] 456/7).
[32] Seel, F.; Gombler, W.; Budenz, R. (Z. Naturforsch. **27b** [1972] 78/9).
[33] Bur-Bur, F.; Haas, A.; Klug, W. (Chem. Ber. **108** [1975] 1365/8).
[34] Livingston, M. J.; Peach, M. E. (J. Fluorine Chem. **9** [1977] 85/8).
[35] Blaschette, A.; Näveke, M. (Chem.-Ztg. **115** [1991] 61/4).
[36] Zeifman, Yu. V.; Lantseva, L. T.; Knunyants, I. L. (Izv. Akad. Nauk SSSR Ser. Khim. **1978** 2640/3; Bull. Acad. Sci. USSR Div. Chem. Sci. [Engl. Transl.] **27** [1978] 2362/6).
[37] Zeifman, Yu. V.; Lantseva, L. T. (Izv. Akad. Nauk SSSR Ser. Khim. **1980** 1102/6; Bull. Acad. Sci. USSR Div. Chem. Sci. [Engl. Transl.] **29** [1980] 809/13).
[38] Zeifman, Yu. V.; Lantseva, L. T.; Knunyants, I. L. (Izv. Akad. Nauk SSSR Ser. Khim. **1978** 1229; Bull. Acad. Sci. USSR Div. Chem. Sci. [Engl. Transl.] **27** [1978] 1073).
[39] Drabek, J.; Böger, M. (Ger. 2621077 [1974/76]; C.A. **86** [1977] No. 89430).
[40] Lawrence, J. P. (Fr. 2190819 [1970/74]; C.A. **81** [1974] No. 122404).

[41] Lawrence, J. P. (Ger. 2329431 [1970/74]; C.A. **81** [1974] No. 50818).
[42] Lawrence, J. P. (U.S. 3928340 [1972/75]; C.A. **84** [1976] No. 166003).
[43] Lawrence, J. P. (U.S. 3944552 [1972/76]; C.A. **85** [1976] No. 22609).
[44] Thurman, D. E. (Ger. 2628574 [1975/77]; C.A. **86** [1977] No. 155209).
[45] Thurman, D. E. (Ger. 2628575 [1975/77]; C.A. **86** [1977] No. 155665).
[46] Durden, J. A., Jr. (U.S. 4071627 [1976/78]; C.A. **89** [1978] No. 43450).
[47] Gorodilov, V. N.; Smirnova, V. N.; Ivin, S. Z. (U.S.S.R. 175957 [1964/65]; C.A. **64** [1966] 14093).
[48] Böger, M.; Drabek, J. (Ger. 2654080 [1972/77]; C.A. **87** [1977] No. 102078).

12.5.1.1.4 $ClSN(C_3H_7\text{-}n)_2$

The title compound was formed by treating a precooled (−5 °C) solution of SCl_2 in CH_2Cl_2 with a solution of pyridine in CH_2Cl_2 at −5 to 0 °C and after 15 min with a solution of $(n\text{-}C_3H_7)_2NH$ in CH_2Cl_2 at −5 to 0 °C. The resulting mixture was stirred for 2 h while the temperature was allowed to rise to ambient [1]. $ClSN(C_3H_7\text{-}n)_2$ was also formed by the reaction of $(n\text{-}C_3H_7)_2NSSN(C_3H_7\text{-}n)_2$ with SO_2Cl_2 (mole ratio ~2 : 3) at −78 °C without a solvent [2]. The title compound can be synthesized with ca. 90% yield as the only product of the exothermic reaction $(n\text{-}C_3H_7)_2NSN(C_3H_7\text{-}n)_2 + SCl_2 \rightarrow 2\ ClSN(C_3H_7\text{-}n)_2$ [3]. $ClSN(C_3H_7\text{-}n)_2$ can be obtained according to $(n\text{-}C_3H_7)_2NSOCH_3 + (CH_3)_3SiCl \rightarrow (CH_3)_3SiOCH_3 + ClSN(C_3H_7\text{-}n)_2$ (no details given) [4].

Pungent, orange-colored liquid [3], b.p. 45 °C/0.01 Torr [4], 94 to 95 °C/15 Torr [1], n_D^{24} = 1.4956 [4].

The title compound can be readily distilled under vacuum [3, 5], but is visibly decomposed within 1 h, even when sealed in ampules and kept in the dark [3]. It is very reactive and sensitive to hydrolysis [5].

Adding $ClSN(C_3H_7\text{-n})_2$ under cooling to $(n\text{-}C_4H_9)_3SnOCH_3$ (mole ratio 1 : 1.1) and warming the mixture for 1 h gave 92% $(n\text{-}C_3H_7)_2NSOCH_3$ along with $(n\text{-}C_4H_9)_3SnCl$ [4].

Reactions of $ClSN(C_3H_7\text{-n})_2$ with some organic compounds are tabulated on p. 131.

References:

[1] Anderson, M. (U.S. 4623658 [1986]; C.A. **107** [1987] No. 23100).
[2] Hatch, C. E., III (J. Org. Chem. **43** [1978] 3953/7).
[3] Armitage, D. A.; Tso, C. C. (J. Chem. Soc. D **1971** 1413/4).
[4] Armitage, D. A.; Towle, I. D. H. (Phosphorus Sulfur **1** [1976] 37/9).
[5] Dorlars, A. (Houben-Weyl Methoden Org. Chem. 4th Ed. **11** Pt. 2 [1958] 744/51).
[6] Drabek, J.; Böger, M. (Ger. 2621077 [1974/76]; C.A. **86** [1977] No. 89430).
[7] Schneider, R. (Ger. 2340925 [1972/74]; C.A. **81** [1974] No. 3949).
[8] Black, L.; Fukuto, T. R. (Ger. 2433680 [1972/75]; C.A. **82** [1975] No. 156050).
[9] Böger, M.; Drabek, J. (Ger. 2654080 [1972/77]; C.A. **87** [1977] No. 102078).

12.5.1.1.5 $ClSN(C_3H_7\text{-i})_2$

The title compound was formed by treating a precooled (−5 °C) solution of SCl_2 in CH_2Cl_2 with a solution of pyridine in CH_2Cl_2 at −5 to 0 °C and after 15 min with a solution of $(i\text{-}C_3H_7)_2NH$ in CH_2Cl_2 at −5 to 0 °C. The resulting solution was stirred for 2 h while the temperature was allowed to rise to ambient [1]. $ClSN(C_3H_7\text{-i})_2$ was also formed by cleaving the sulfur–sulfur bond of $(i\text{-}C_3H_7)_2NSSN(C_3H_7\text{-i})_2$ by introducing dry chlorine at 0 to 5 °C into a solution of $(i\text{-}C_3H_7)_2NSSN(C_3H_7\text{-i})_2$ (mole ratio 1.25 : 1) in CCl_4 maintained at 60 °C. The reaction mixture was then heated to 60 °C for 2 h. The solvent and unreacted SO_2Cl_2 were removed in vacuum and the residue was distilled to give 81% of the product [3]. The title compound was obtained with ca. 90% yield as the only product of the exothermic reaction $(i\text{-}C_3H_7)_2NSN(C_3H_7\text{-i})_2 + SCl_2 \rightarrow 2\ ClSN(C_3H_7\text{-i})_2$ [4]. By a special procedure $ClSN(C_3H_7\text{-i})_2$ was formed along with $ClC{\equiv}N$ by decomposing of $(i\text{-}C_3H_7)_2NSN{=}CCl_2$ at 50 °C [5].

$ClSN(C_3H_7\text{-i})_2$ is a pungent, orange [4] or pungent, brown, viscous liquid [5]; b.p. 39 to 40 °C/0.01 Torr [3], 92 to 94 °C/20 Torr [1].

^1H NMR ($CDCl_3$/TMS): δ (in ppm) = 1.2 (d, 4 CH_3, J = 7 Hz), 4.05 (sept, 2 CH, J = 7 Hz) [3]. The geminal methyl groups of $(i\text{-}C_3H_7)_2NSCl$ in $CD_3C_6D_5$ become diastereotopic on the ^1H NMR time-scale in the temperature range 0 to 30 °C. ΔG^* was determined (from $k_c = \pi\Delta\nu/\sqrt{2}$ and the Eyring equation ($\Delta\nu = 9.6$ Hz at 60 MHz)) to be 60.8 kJ/mol at the coalescence temperature of 5 °C. The coalescence temperature was similar for 0.09 and 0.48 M solutions. Based on the notion that halogen exchange is a bimolecular process, the concentration independence suggests that the observed rate process is not sulfur inversion. In CH_2Cl_2 solution the compound did not show any signal splitting down to −60 °C (possibly chlorine exchange is more rapidly in this solvent) [6].

^{15}N NMR (neat/CH_3NO_2): δ = −250.7 ppm; (in the presence of $Cr(acac)_3/CH_3NO_2$): δ = −249.7 ppm [7].

$ClSN(C_3H_7\text{-i})_2$ is a readily vacuum-distillable liquid [4, 8]. It is stable at room temperature [4] and can be stored at −18 °C for several weeks [3]. $ClSN(C_3H_7\text{-i})_2$ becomes gum-like when heated above 50 °C [5]. It is very reactive and sensitive to hydrolysis [8].

Reactions of $ClSN(C_3H_7\text{-}n)_2$ with Organic Compounds.

reactant	reaction conditions	product(s) (yield in %)	Ref.
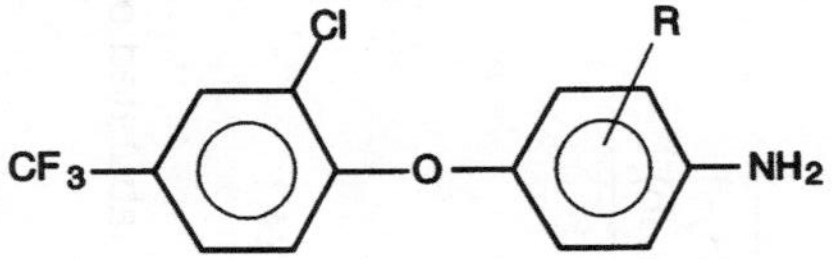		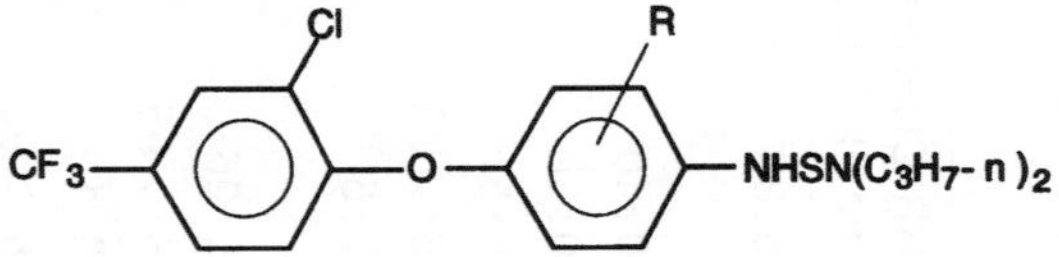	[1]
2-CH_3-4-RC_6H_3N=$CHNHCH_3$ (R = Cl, CH_3)	in THF with cooling	2-CH_3-4-RC_6H_3N=$CHN(CH_3)SN(C_3H_7\text{-}n)_2$	[6]
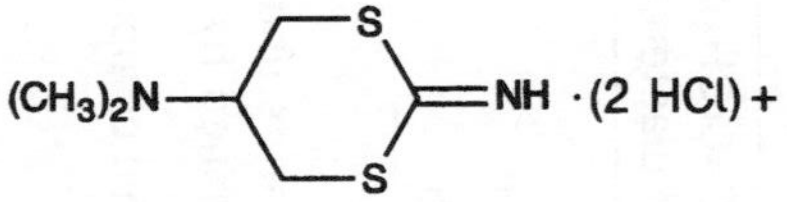 $(C_2H_5)_3N$	in CH_3CN, 3 h	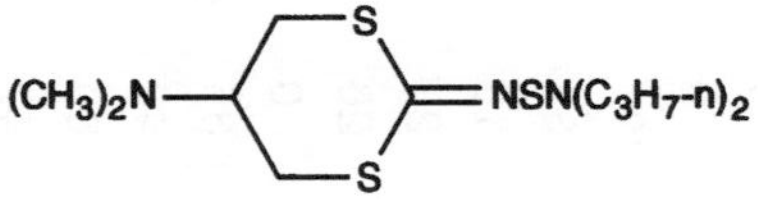	[7]
CH_3N=C=O + HF + $(C_2H_5)_3N$	in CH_2Cl_2, cooling	$(n\text{-}C_3H_7)_2NSN(CH_3)C(O)F$ (53)	[2]
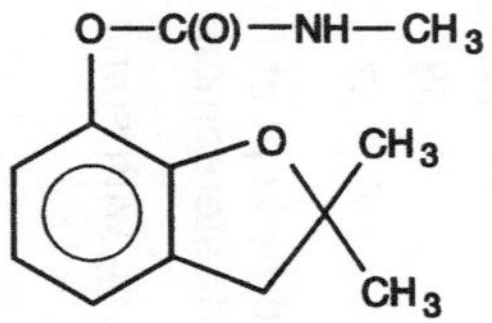	in pyridine, 18 h at room temperature	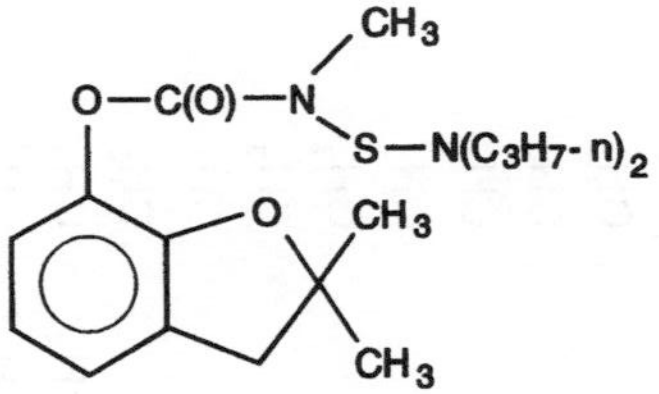	[8]
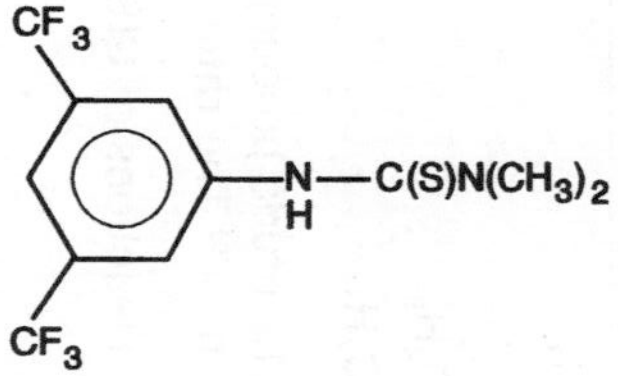	in CH_2Cl_2 at 0 to 10 °C, 2 h at room temperature	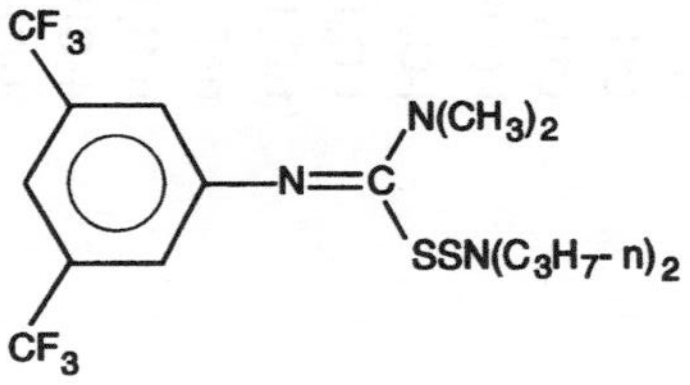	[9]

The corresponding $(i\text{-}C_3H_7)_2NS(O)CHR^2CH{=}CHR^1$ diastereomers were formed by adding a solution of the title compound in ether at −78 °C to a mixture of the allylic alcohol $R^1CHOHCH{=}CHR^2$ in ether and $LiC_4H_9\text{-}n$ (mole ratio 1 : 1 : 1) in hexane cooled to 0 °C and stirring the mixture at −78 °C for 15 min, heating it to room temperature, and leaving it stand for the time indicated in the table below. The ratios of the diastereomers were determined by ^1H NMR [9].

$ClSN(C_3H_7\text{-}i)_2 + R^1CHOHCH{=}CHR^2$ configuration	R^1	R^2	time (in h)	$(i\text{-}C_3H_7)_2NS(O)CHR^2CH{=}CHR^1$ yield (in %)	ratio of diastereomers
E	H	CH_3	24	4	65:35
Z	H	CH_3	120	0	—
E	CH_3	CH_3	60	21	60:40
Z	CH_3	CH_3	48	37	10:90
E	$i\text{-}C_3H_7$	CH_3	120	18	95:5
Z	$i\text{-}C_3H_7$	CH_3	48	25	<5:>95
E	$t\text{-}C_4H_9$	CH_3	24	26	95:5
Z	$t\text{-}C_4H_9$	CH_3	120	0	—
E	$n\text{-}C_7H_{15}$	CH_3	120	35	75:25
Z	$n\text{-}C_7H_{15}$	CH_3	48	42	<5:>95
E	$4\text{-}CH_3C_6H_4$	CH_3	18	60	90:10
Z	$4\text{-}CH_3C_6H_4$	CH_3	72	47	<5:>95
E	$n\text{-}C_7H_{15}$	$i\text{-}C_3H_7$	72	26	95:5
Z	$n\text{-}C_7H_{15}$	$i\text{-}C_3H_7$	48	11	<5:>95

Adding a solution of the compound in ether to a solution of the propargyl alcohol $RCHOHC{\equiv}CCH_3$ (for R see the table below) and $(C_2H_5)_3N$ (mole ratio 1 : 1 : 1) in ether precooled to −78 °C, stirring at −78 °C for 15 min, warming to room temperature, and stirring for 1 h at room temperature gave the corresponding α-allenic sulfinamide diastereomers Ia and Ib [3].

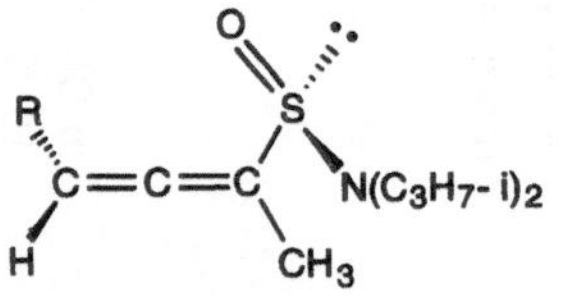

Ia

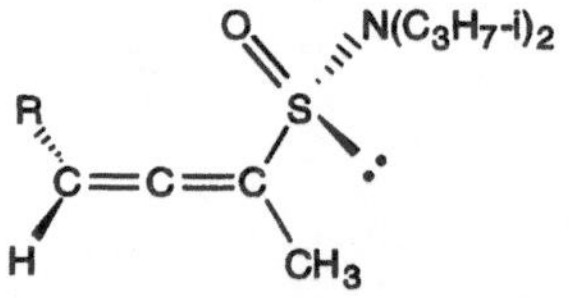

Ib

$ClSN(C_3H_7\text{-}i)_2 + RCHOHC{\equiv}CCH_3$ R	yield[a] (in %)	$RCH{=}C{=}C(CH_3)S(O)N(C_3H_7\text{-}i)_2$ ratio of diastereomers (Ia : Ib)[b]
CH_3	72	70:30
$i\text{-}C_3H_7$	72	>95:<5
$t\text{-}C_4H_9$	48	>95:<5
$n\text{-}C_7H_{15}$	57	80:20

[a] The crude products had to be kept at room temperature under Ar for 24 to 48 h. – [b] The ratio of the diastereomers was determined by ^1H NMR.

Reactions of $ClSN(C_3H_7\text{-}i)_2$ with some other organic compounds are tabulated on p. 133.

Reactions of $ClSN(C_3H_7\text{-}i)_2$ with Organic Compounds.

reactant	reaction conditions	product(s) (yield in %)	Ref.
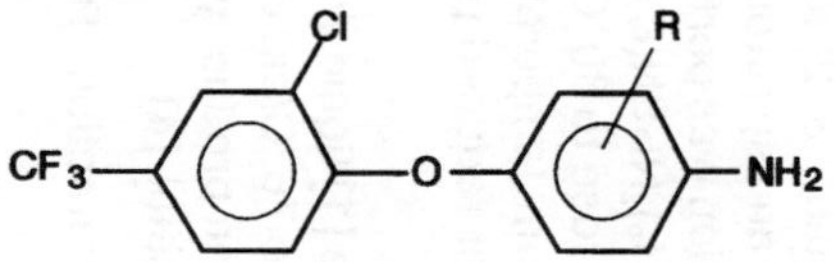		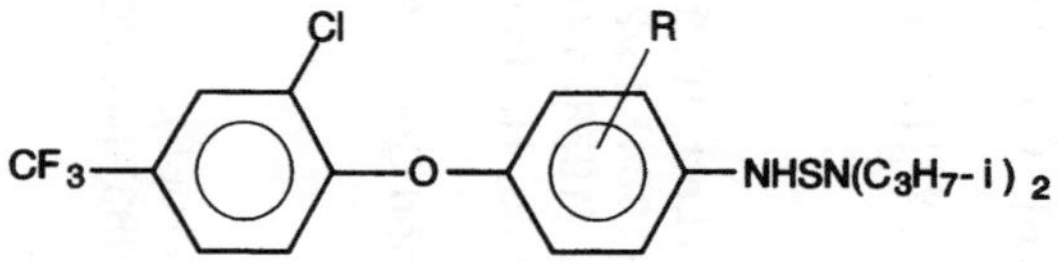	[1]
$(i\text{-}C_3H_7)_2NH$	mole ratio 1:2; in ether	$(i\text{-}C_3H_7)_2NSN(C_3H_7\text{-}i)_2$ (7)	[5]
$CH_3SC(CH_3){=}NOC(O)NHCH_3$ + $(C_2H_5)_3N$	in CH_3CN; at 0 to 5 °C, allowing to warm to room temperature within 2 h	$CH_3SC(CH_3){=}NOC(O)N(CH_3)SN(C_3H_7\text{-}i)_2$	[10]
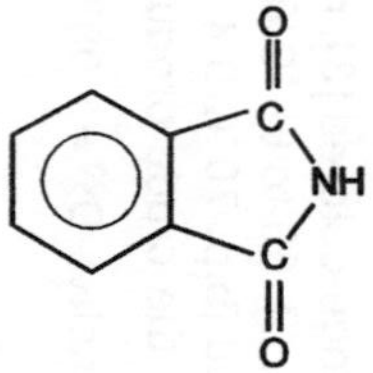 + $(C_2H_5)_3N$ (excess)	in THF; at 0 to 10 °C for 1 h	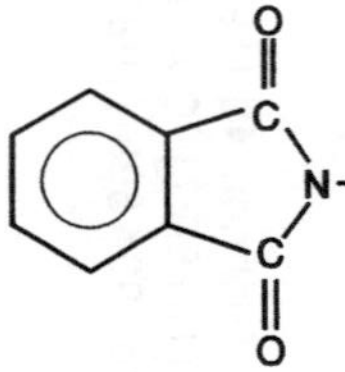 $-S-N(C_3H_7\text{-}i)_2$ (45.6) [2] (62) [11]	[2, 11]
$CH_3N{=}C{=}O + HF + (C_2H_5)_3N$	in CH_2Cl_2 or toluene; at −50 °C to room temperature	$(i\text{-}C_3H_7)_2NSN(CH_3)C(O)F$	[12]
$(C_6H_5O)_2P(O)CH_2NHCH_2C(O)OCH_3$ + $(C_2H_5)_3N$	mole ratio ~3:1:1; in CH_2Cl_2–toluene, at 0 °C	$(C_6H_5O)_2P(O)CH_2N(CH_2C(O)OCH_3)SN(C_3H_7\text{-}i)_2$	[13]

References:

[1] Anderson, M. (U.S. 4623658 [1986]; C.A. **107** [1987] No. 23100).

[2] Lawrence, J. P. (Fr. 2190819 [1970/74]; C.A. **81** [1974] No. 122404, Ger. 2329431 [1970/74]; C.A. **81** [1974] No. 50818, U.S. 3944552 [1972/76]; C.A. **85** [1976] No. 22609).

[3] Baudin, J.-B.; Bkouche-Waksman, I.; Julia, S. A.; Pascard, C.; Wang, Y. (Tetrahedron **47** [1991] 3353/64).

[4] Armitage, D. A.; Tso, C. C. (J. Chem. Soc. D **1971** 1413/4).

[5] Bacon, R. G. R.; Irwin, R. S. (J. Chem. Soc. **1960** 5079/87).

[6] Jackson, W. R.; Kee, T. G.; Spratt, R.; Jennings, W. B. (Tetrahedron Lett. **1973** 3581/4).

[7] Dorie, J.; Gouesnard, J.-P. (J. Chim. Phys. Phys.-Chim. Biol. **81** [1984] 15/9).

[8] Dorlars, A. (Houben-Weyl Methoden Org. Chem. 4th Ed. **11** Pt. 2 [1958] 744/51).

[9] Baudin, J.-B.; Bkouche-Waksman, I.; Hareau, G.; Julia, S. A.; Lorne, R.; Pascard, C. (Tetrahedron **47** [1991] 6655/72).

[10] Gemrich, E. G., II; Lee, B. L.; Nelson, S. J.; Rizzo, V. L. (J. Agric. Food Chem. **26** [1978] 391/5).

[11] Lawrence, J. P. (Rubber Chem. Technol. **49** [1976] 333/40).

[12] Thurman, D. E. (Ger. 2628575 [1975/77]; C.A. **86** [1977] No. 155665).

[13] Hoobler, M.; Sikorski, J. A. (Belg. 894595 [1982/83]; C.A. **99** [1983] No. 6049).

12.5.1.1.6 $ClSN(C_4H_9\text{-n})_2$

The title compound was formed by adding a solution of $(n\text{-}C_4H_9)_2NH$ in petroleum ether at −78 °C to excess SCl_2. The deep orange product was distilled at 60 °C in high vacuum and condensed in a trap at 0 °C [1]. The title compound was also formed by adding S_2Cl_2 to a solution of $(n\text{-}C_4H_9)_2NH$ in hexane at 0 to 5 °C and stirring the mixture for 0.5 h at 0 to 5 °C and then for 1 h at room temperature. The hydrochloride was removed by filtration, and Cl_2 was bubbled through the CCl_4 solution of the concentrated filtrate [2]. $ClSN(C_4H_9\text{-n})_2$ was formed by introducing (dry) Cl_2 at 0 to 5 °C into a solution of $(n\text{-}C_4H_9)_2NSSN(C_4H_9\text{-n})_2$ in CCl_4 until there was a slight excess (of Cl_2). Excess Cl_2 was removed at 0 to 5 °C in vacuum or by purging with inert gas. The resulting mixture was cleared with animal charcoal, and after filtration the solvent was evaporated under reduced pressure at 30 to 50 °C to give the crude compound with a yield of 82% which was distilled in vacuum [3]. Instead of Cl_2, SO_2Cl_2 or PCl_5 can be used: all SO_2Cl_2 was added at once to $(n\text{-}C_4H_9)_2NSSN(C_4H_9\text{-n})_2$ (mole ratio ~3 : 2) at −78 °C, and the mixture was stirred at 0 to 5 °C for 2 to 3 h and then subjected to an aspirator vacuum followed by high vacuum yielding 99% product [4]; alternatively, the reaction was performed in a reactor at 20 °C in solution (hexane) [5] or by adding a solution of $(n\text{-}C_4H_9)_2NSSN(C_4H_9\text{-n})_2$ in CCl_4 without cooling to a solution of PCl_5 in CCl_4. The inside temperature rose to 30 °C and the mixture was kept for another 5 min at 40 to 50 °C. After cooling to room temperature, the solution was cleared with animal charcoal, the solvent evaporated under reduced pressure, and the residue distilled in vacuum [3]; see also [6].

$ClSN(C_4H_9\text{-n})_2$ is a yellow [4] or orange-colored [3], moisture-sensitive [1] liquid with a foul smell [4]; b.p. 61 to 63 °C/0.08 Torr [3, 6], 66 to 68 °C/0.18 Torr [2], 75 to 76 °C/ca. 0.45 Torr [3]. The material could be flash-distilled (b.p. 70 °C/0.4 Torr); however, the product yield was low. Batch distillation of a sample of the crude product was unsuccessful [4].

Refractive index: $n_D^{20} = 1.4920$; density: $D_4^{20} = 1.0039$ g/cm³; mole refraction: $R_D^{20} = 56.2$ (calculated 55.7) [1].

Reactions of $ClSN(C_4H_9\text{-}n)_2$ with Organic Compounds.

reactant	reaction conditions	product(s) (yield in %)	Ref.
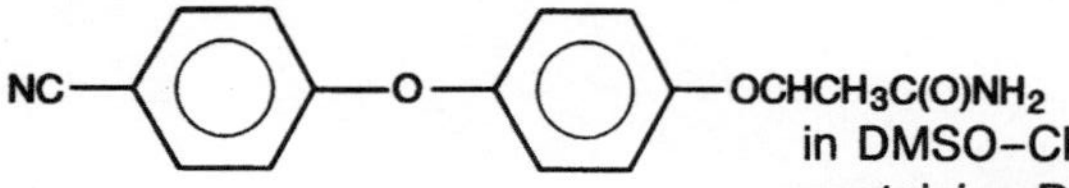	in DMSO–CH_2Cl_2 containing DBU at room temperature	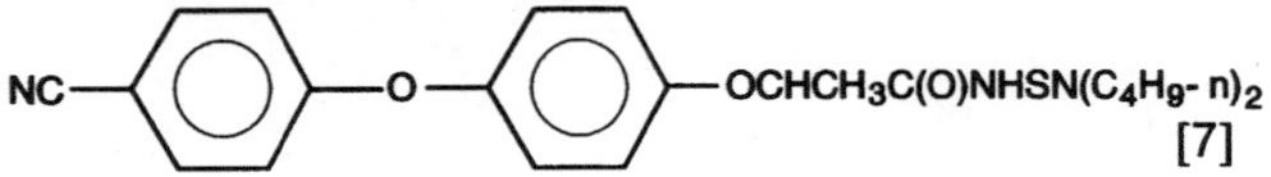	[7]
$2\text{-}CH_3\text{-}4\text{-}RC_6H_3N{=}CHNHCH_3$ R = Cl, CH_3	in THF with cooling	$2\text{-}CH_3\text{-}4\text{-}RC_6H_3N{=}CHN(CH_3)SN(C_4H_9\text{-}n)_2$	[8]
$C_6H_5C(O)NHN(C_4H_9\text{-}t)\text{-}C(O)C_6H_5 + NaH$	in THF–mineral oil; 5 min at room temperature	$C_6H_5C(O)N(C_4H_9\text{-}t)N(C(O)C_6H_5)SN(C_4H_9\text{-}n)_2$	[9]
$CH_3SC(CH_3){=}NOC(O)\text{-}NHCH_3$ (methomyl) $+ (C_2H_5)_3N$	in CH_3CN; at 0 to 5 °C and then warming to room temperature within 2 h	$CH_3SC(CH_3){=}NOC(O)N(CH_3)SN(C_4H_9\text{-}n)_2$	[10]
$CH_3SC(CH_3)_2CH{=}NOC(O)\text{-}NHCH_3$ (aldicarb)	in pyridine at 5 °C for 1 h and 16 h at room temperature	$CH_3SC(CH_3)_2CH{=}NOC(O)N(CH_3)SN(C_4H_9\text{-}n)_2$	[2]
$(C_6H_5O)_2P(O)CH_2NH\text{-}CH_2C(O)OC_2H_5 + (C_2H_5)_3N$	mole ratio ca. 3 : 1 : 1; in CH_2Cl_2–toluene, at 0 °C	$(C_6H_5O)_2P(O)CH_2N(CH_2C(O)OC_2H_5)SN(C_4H_9\text{-}n)_2$	[11]
$CH_3N{=}C{=}O + HF + (C_2H_5)_3N$	in toluene or CH_2Cl_2; at −50 °C to room temperature	$(n\text{-}C_4H_9)_2NSN(CH_3)C(O)F$ (60)	[12]
$CH_3N{=}C{=}O + HF$	in pyridine; at room temperature	$(n\text{-}C_4H_9)_2NSN(CH_3)C(O)F$	[4]
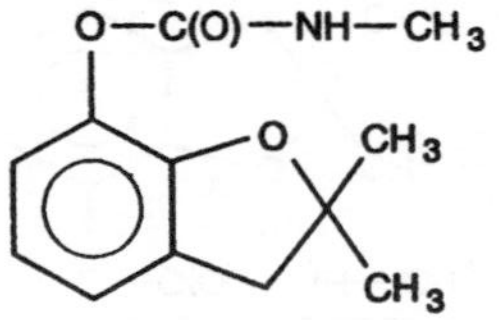	in pyridine; 18 h at room temperature	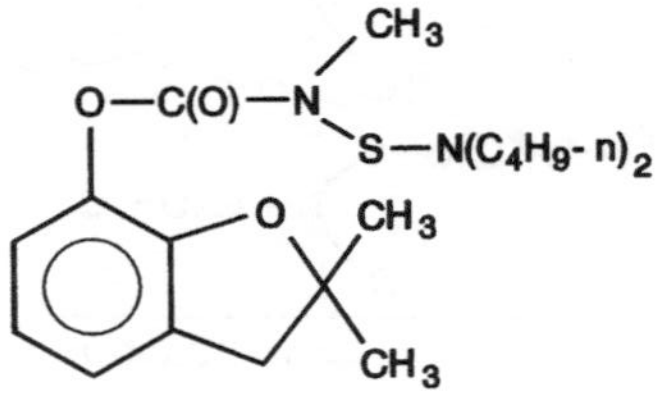	[13]

reactant	reaction conditions	product(s) (yield in %)	Ref.
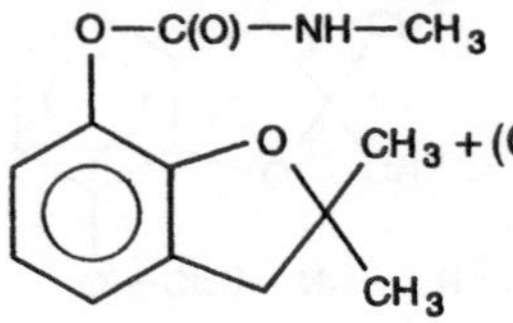 + $(C_2H_5)_3N$	in hexane; at 20 to 35 °C; the reaction was examined by modifying the reaction parameters	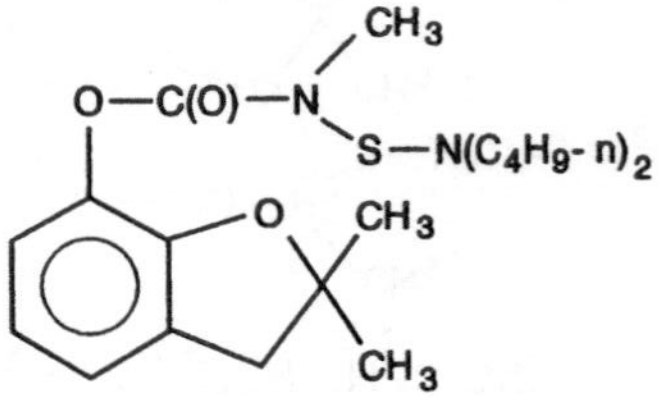 (90)	[5]
2-CH_3-4-$ClC_6H_3NHC(S)$-$N(CH_3)_2$ + $(C_2H_5)_3N$	in CH_2Cl_2; at 0 to 10 °C, and 2 h at room temperature	2-CH_3-4-$ClC_6H_3N{=}C(N(CH_3)_2)SSN(C_4H_9\text{-}n)_2$	[14]

^1H NMR ($CDCl_3$/TMS): δ (in ppm) = 0.80 to 2.00 (m, $(CH_2)_2CH_3$), 3.25 (t, NCH_2) [4].

Raman: ν (in cm^{-1}; intensities in parentheses) = 2024 (3), 1422 (5), 1138 (1), 876 (0), 835 (0), 364 (10), 248 (2), 142 (1) [1].

Reactions of $ClSN(C_4H_9\text{-n})_2$ with some organic compounds are tabulated on pp. 135/6.

References:

[1] Fehér, F.; Kruse, W. (Chem. Ber. **91** [1958] 2528/31).
[2] Fukuto, T. R.; Black, A. L. (U.S. 4108991 [1976/78]; C.A. **90** [1979] No. 151581).
[3] Farbenfabriken Bayer AG (Br. 790021 [1954/58]; C.A. **1958** 13806).
[4] Hatch, C. E., III (J. Org. Chem. **43** [1978] 3953/7).
[5] Ager, J. W.; Harding, M. J. C.; Hatch, C. E., III (Eur. Appl. 51273 [1981/82]; C.A. **97** [1982] No. 109573).
[6] Dorlars, A. (Houben-Weyl Methoden Org. Chem. 4th Ed. **11** Pt. 2 [1958] 744/51).
[7] Akahira, R.; Someya, S.; Nakanishi, A.; Amon, S.; Ito, M.; Nonaka, Y. (Jpn. Kokai Tokkyo Koho 62-12755 [1985/87]; C.A. **107** [1987] No. 134039).
[8] Drabek, J.; Böger, M. (Ger. 2621077 [1974/76]; C.A. **86** [1977] No. 89430).
[9] Drabek, J. (Eur. Appl. 395581 [1990]; C.A. **114** [1991] No. 185026).
[10] Gemrich, E. G., III; Lee, B. L.; Nelson, S. J.; Rizzo, V. L. (J. Agric. Food Chem. **26** [1978] 391/5).

[11] Hoobler, M.; Sikorski, J. A. (Belg. 894595 [1982/83]; C.A. **99** [1983] No. 6049).
[12] Thurman, D. E. (Ger. 2628574 [1975/77]; C.A. **86** [1977] No. 155209, Ger. 2628575 [1975/77]; C.A. **86** [1977] No. 155665).
[13] Black, L.; Fukuto, T. R. (Ger. 2433680 [1972/75]; C.A. **82** [1975] No. 156050).
[14] Böger, M.; Drabek, J. (Ger. 2654080 [1972/77]; C.A. **87** [1977] No. 102078).

12.5.1.1.7 $ClSN(C_4H_9\text{-i})_2$

The title compound was formed by introducing (dry) Cl_2 at 0 to 5 °C into a solution of $(i\text{-}C_4H_9)_2NSSN(C_4H_9\text{-i})_2$ in CCl_4 until there was a slight excess of Cl_2, which was then removed at 0 to 5 °C in vacuum or by purging with an inert gas. The resulting solution was then cleared with animal charcoal, the solvent evaporated under reduced pressure at 30 to 50 °C to give a crude product with 82% yield which was purified by distillation to give pure $ClSN(C_4H_9\text{-i})_2$ [1].

Orange-colored liquid, b.p. 61 to 64 °C/0.45 Torr [1]; see also [2].

Adding a solution of $ClSN(C_4H_9\text{-i})_2$ in benzene to a suspension of $(C_2H_5O)_2P(S)SNa$ in benzene caused the temperature to rise to about 30 to 35 °C, after which the mixture was stirred at this temperature for 1 h to give $(C_2H_5O)_2P(S)SSN(C_4H_9\text{-i})_2$ [3]. A solution of the title compound in ether reacted with $(c\text{-}C_6H_{11})_2NH$ (mole ratio 1 : 2) at room temperature to yield $(c\text{-}C_6H_{11})_2NH \cdot HCl$ (90%), $(c\text{-}C_6H_{11})_2NSN(C_6H_{11}\text{-c})_2$ (24%), and the unsymmetrical compound $(c\text{-}C_6H_{11})_2NSN(C_4H_9\text{-i})_2$ [4]. $ClSN(C_4H_9\text{-i})_2$ reacted under cooling with a solution of ArN=CH-$NHCH_3$ (Ar = 2-$C_2H_5OC_6H_4$, 2-CNC_6H_4, 2-CH_3-4-ClC_6H_3, 2-CH_3-4-BrC_6H_3, 3,5-$Cl_2C_6H_3$) and $(C_2H_5)_3N$ in THF to give the corresponding ArN=CHN(CH_3)SN(C_4H_9-i)$_2$ [5]. Reacting a mixture of $ClSN(C_4H_9\text{-i})_2$ and carbofuran (I) in pyridine at room temperature for 18 h gave the carbamate (II) [6].

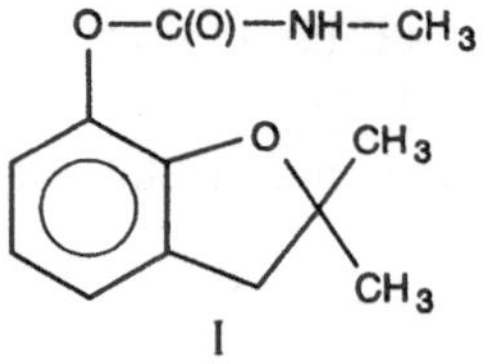

I

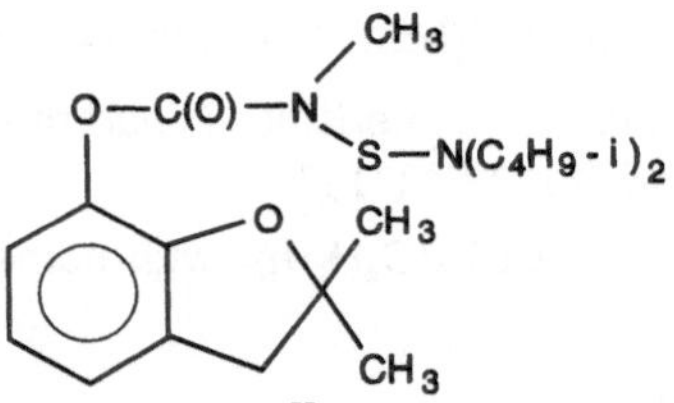

II

References:

[1] Farbenfabriken Bayer AG (Br. 790021 [1954/58]; C.A. **1958** 13806).
[2] Dorlars, A. (Houben-Weyl Methoden Org. Chem. 4th Ed. **11** Pt. 2 [1958] 744/51).
[3] Malz, H.; Bayer, O.; Freytag, H.; Lober, F. (U.S. 2891059 [1957/59]; C.A. **1960** 4387).
[4] Bacon, R. G. R.; Irwin, R. S. (J. Chem. Soc. **1960** 5079/87).
[5] Drabek, J.; Böger, M. (Ger. 2621077 [1974/76]; C.A. **86** [1977] No. 89430).
[6] Black, L.; Fukuto, T. R. (Ger. 2433680 [1972/75]; C.A. **82** [1975] No. 156050).

12.5.1.1.8 $ClSN(C_4H_9\text{-}s)_2$

The title compound was formed by adding a solution of $ClSN{=}CCl_2$ in ether to a solution of $(s\text{-}C_4H_9)_2NH$ (mole ratio 1:2) in ether kept at ~10 °C. It could not be obtained in a pure form. Decomposition occurred when isolated.

$ClSN(C_4H_9\text{-}s)_2$ reacted with $(s\text{-}C_4H_9)_2NH$ (mole ratio 1:2) in ether to give $(s\text{-}C_4H_9)_2N\text{-}SN(C_4H_9\text{-}s)_2$ (not distillable at 10^{-4} Torr without resinification).

Reference:

Bacon, R. G. R.; Irwin, R. S. (J. Chem. Soc. **1960** 5079/87).

12.5.1.1.9 $ClSN(C_6H_{13}\text{-}n)_2$

The preparation of the title compound was not described.

Reacting a mixture of $ClSN(C_6H_{13}\text{-}n)_2$ and carbofuran (I) in pyridine at room temperature for 18 h gave the carbamate (II).

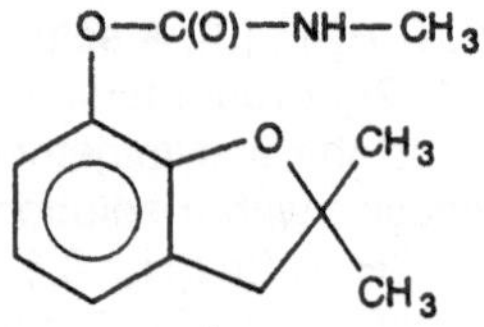

I

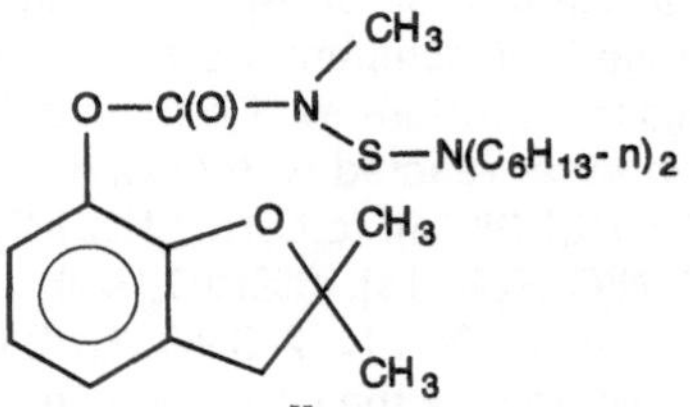

II

Reference:

Black, L.; Fukuto, T. R. (Ger. 2433680 [1972/75]; C.A. **82** [1975] No. 156050).

12.5.1.1.10 $ClSN(CH_2CH_2Cl)_2$

The title compound was formed by introducing dry Cl_2 at 0 to 5 °C into a solution of $(ClCH_2CH_2)_2NSSN(CH_2CH_2Cl)_2$ in CCl_4. The excess Cl_2 was then removed at 0 to 5 °C in vacuum or by purging with an inert gas. The resulting solution was then cleared with animal charcoal, the solvent was evaporated under reduced pressure, and the residue was distilled in vacuum to give $ClSN(CH_2CH_2Cl)_2$ as a yellow-brown liquid, b.p. 88 to 90 °C/0.1 Torr.

Reference:

Farbenfabriken Bayer AG (Br. 790 021 [1954/58]; C.A. **1958** 13 806).

12.5.1.1.11 $ClSN(CH_2C(O)OC_2H_5)_2$

The title compound was formed by adding pyridine at 0 °C and successively diethyl iminodiacetate, $HN(CH_2C(O)OC_2H_5)_2$, at 10 to 20 °C to a solution of SCl_2 (mole ratio 1 : 1 : 1) in CCl_4 and reacting the mixture at 10 to 20 °C for 1 h. It was not isolated [1]; see also [2]. $ClSN(CH_2C(O)OC_2H_5)_2$ reacted with carbofuran and excess pyridine in CH_2Cl_2 to give compound I with 85% yield [1].

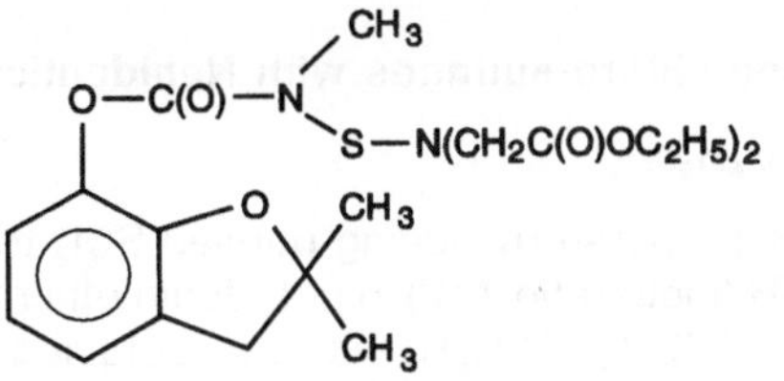

I

References:

[1] Goto, T.; Yasudomi, N.; Tanaka, A. K.; Osaki, N.; Takao, H.; Kawata, M.; Imada, J.; Endo, Y.; Umetsu, N. (Nippon Noyaku Gakkaishi **13** [1988] 38/47; C.A. **110** [1989] No. 7982).
[2] Asai, N.; Soeda, T.; Tanaka, A.; Goto, T. (Belg. 892 670 [1981/82]; C.A. **98** [1983] No. 34 224).

12.5.1.1.12 $ClSN(CH_2C_6H_5)_2$

The title compound (erroneously quoted in the paper) was formed as an orange-colored liquid by the reaction of $(C_6H_5CH_2)_2NSSN(CH_2C_6H_5)_2$ in CCl_4 with Cl_2.

Reference:

Farbenfabriken Bayer AG (Br. 790 021 [1954/58]; C.A. **1958** 13 806).

12.5.1.1.13 $ClSN(C_6H_{11}\text{-}c)_2$

The title compound was formed by decomposition (1 h on the water pump at 40 °C) of $(c\text{-}C_6H_{11})_2NSN{=}CCl_2$ [1].

The brown liquid decomposed upon heating [1].

The crude title compound reacted with $(c\text{-}C_6H_{11})_2NH$ (mole ratio 1 : 2) in ether to give 87% $(c\text{-}C_6H_{11})_2NSN(C_6H_{11}\text{-}c)_2$ [1]. Adding $ClSN(C_6H_{11}\text{-}c)_2$ in toluene to a mixture of (excess) phthalimide and $(C_2H_5)_3N$ in toluene gave compound I after 1 h [2].

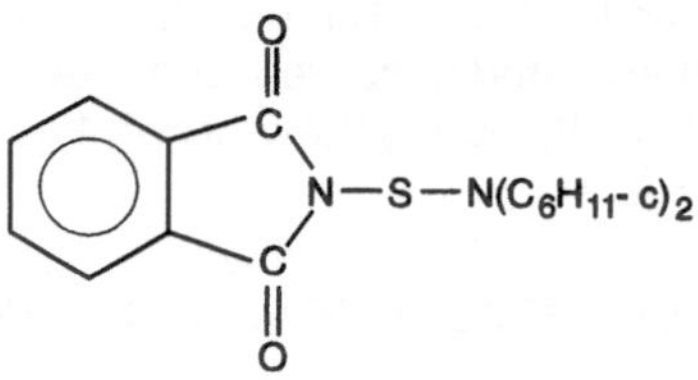

I

References:

[1] Bacon, R. G. R.; Irwin, R. S. (J. Chem. Soc. **1960** 5079/87).
[2] Lawrence, J. P. (Rubber Chem. Technol. **49** [1976] 333/40).

12.5.1.2 N,N-Dialkyl-amino-chloro-sulfanes with Nonidentical Alkyl Groups

12.5.1.2.1 $ClSN(CH_3)CH_2C_6H_5$

The title compound was prepared by adding purified SCl_2 to a stirred, −78 °C-precooled solution of $CH_3(C_6H_5CH_2)NH$ (mole ratio 1 : 2) in petroleum ether and letting the mixture react for 1 h; 80% yield, oil [1]. $ClSN(CH_3)CH_2C_6H_5$ was also obtained by reacting $C_6H_5CH_2N(CH_3)$-$SSN(CH_3)CH_2C_6H_5$ with Cl_2 in CCl_4 [2] or with SO_2Cl_2 (mole ratio ca. 2 : 3) at −78 °C [3].

^1H NMR ($CDCl_3$/TMS): δ (in ppm) = 3.03 (s, CH_3), 4.38 (s, CH_2), 7.33 (s, C_6H_5) [1].

The low-temperature ^1H NMR spectrum (displayed in the paper) of $ClSN(CH_3)CH_2C_6H_5$ exhibits chemical shift nonequivalence (AB quartet) of diastereotopic benzyl methylene protons in $CDCl_3$ or $CD_3C_6D_5$. The chemical shift nonequivalence is due to the molecular chirality in the dimer I which results from slow rotation about the S–N bond. At higher temperatures the signals from diastereotopic groups coalesce. The influence of solvent ($CDCl_3$ and $CD_3C_6D_5$) and addends ($(CH_3)_4NCl$ and $(C_2H_5)_4NClO_4$) on chemical shifts, coalescence temperature, and free energy of activation is shown below [1].

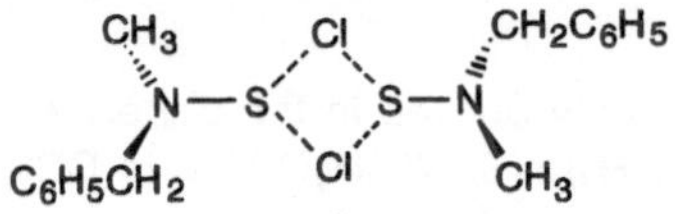

I

solvent	addend	Δν (in Hz)	J_{AB} (in Hz)	T_c (in °C)	ΔG* (in kcal/mol)
$CD_3C_6D_5$		20.8	13.8	39	15.5
$CDCl_3$		1.9	13.5	−43	11.3
$CD_3C_6D_5$	$(CH_3)_4NCl$ (1×10^{-5} M)	21	14.0	22	14.6
$CDCl_3$	$(CH_3)_4NCl$ (4.6×10^{-5} M)	12	13.5	−72	10.0
$CDCl_3$	$(C_2H_5)_4NClO_4$ (7×10^{-5} M)	12	13.5	−58	10.8

^{15}N NMR (in presence of $Cr(acac)_3/CH_3NO_2$): δ = −292.8 ppm [4].

$ClSN(CH_3)CH_2C_6H_5$ reacted with $LiN(Si(CH_3)_3)_2$ to $C_6H_5CH_2N(CH_3)SN(Si(CH_3)_3)_2$ [5]. A solution of $ClSN(CH_3)CH_2C_6H_5$ in THF was reacted with $NaOC_3H_7$-i (mole ratio 1 : 1) for 3 h at room temperature to give $C_6H_5CH_2N(CH_3)SOC_3H_7$-i [1].

The title compound reacted with N-methylcarbamoyl fluoride, $CH_3NHC(O)F$, to give 47% $C_6H_5CH_2N(CH_3)SN(CH_3)C(O)F$ [3]. It reacted with the product of the reaction of II and with NaH suspended in THF and DMSO between −65 °C and room temperature to give III [6].

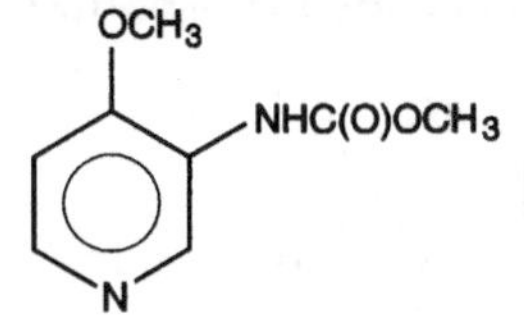
II

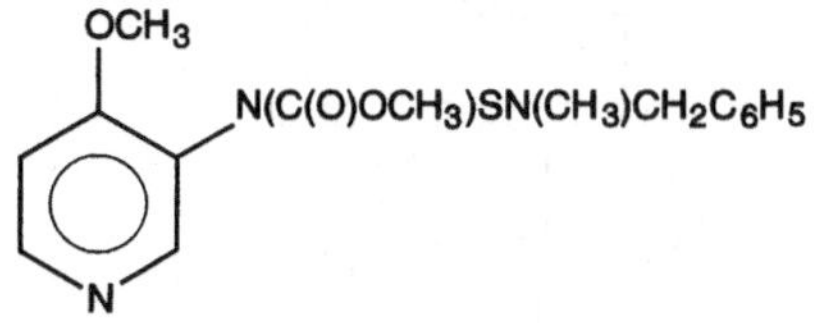
III

$ClSN(CH_3)CH_2C_6H_5$ reacted with $CH_3SC(CH_3)_2CH{=}NOC(O)NHCH_3$ ("aldicarb") and pyridine within 1 h at 5 °C and 17 h at room temperature to give $CH_3SC(CH_3)_2CH{=}NOC(O)N(CH_3)S$-$N(CH_3)CH_2C_6H_5$ [2].

References:

[1] Raban, M.; Noyd, D. A.; Bermann, L. (J. Org. Chem. **40** [1975] 752/5).
[2] Fukuto, T. R.; Black, A. L. (U.S. 4108991 [1976/78]; C.A. **90** [1979] No. 151581).
[3] Hatch, C. E., III (J. Org. Chem. **43** [1978] 3953/7).
[4] Dorie, J.; Gouesnard, J.-P. (J. Chim. Phys. Phys.-Chim. Biol. **81** [1984] 15/9).
[5] Neidlein, R.; Lenhard, T. (Chem. Ber. **116** [1983] 3133/40).
[6] Baker, D. R.; Walker, F. H.; Brownell, K. H. (U.S. 4800205 [1987/89]; C.A. **111** [1989] No. 7238).

12.5.1.2.2 $ClSN(CH_3)C_4H_9$-n

A preparation of the title compound was not reported.

$ClSN(CH_3)C_4H_9$-n reacted under cooling with a solution of 2-CH_3-4-$RC_6H_3N{=}CHNHCH_3$ (R = Cl, CH_3) and $(C_2H_5)_3N$ in THF to give the corresponding 2-CH_3-4-$RC_6H_3N{=}CHN(CH_3)S$-$N(CH_3)C_4H_9$-n [1]. It reacted with $CH_3NHC(O)ON{=}C(CH_3)SR$ (R = CH_3, C_2H_5) and pyridine in CH_2Cl_2 at 5 to 10 °C within 16 h to give the corresponding n-$C_4H_9N(CH_3)SN(CH_3)C(O)O$-$N{=}C(CH_3)SR$ [2].

References:

[1] Drabek, J.; Böger, M. (Ger. 2621077 [1974/76]; C.A. **86** [1977] No. 89430).
[2] Drabek, J.; Böger, M. (Ger. 2727614 [1976/78]; C.A. **88** [1978] No. 190124).

12.5.1.2.3 $ClSN(CH_3)C_6H_{11}$-c

The title compound was prepared by adding a solution of SCl_2 in benzene at 15 °C to CH_3-(c-C_6H_{11})NH while holding the temperature constant; 91% yield [1].

Orange liquid; b.p. 73 to 74 °C/0.5 Torr [1].

Adding $(C_2H_5O)_2P(S)H$ to a precooled solution (−5 °C) of $ClSN(CH_3)C_6H_{11}$-c (mole ratio 2 : 1) in CCl_4 while holding the temperature below 0 °C gave $(C_2H_5O)_2P(S)SP(S)(OC_2H_5)_2$ along with $(c\text{-}C_6H_{11})NHCH_3 \cdot HCl$ [2]. $ClSN(CH_3)C_6H_{11}$-c reacted with a solution of 2-CH_3-4-$RC_6H_3N{=}CH$-$NHCH_3$ (R = Cl, CH_3) and $(C_2H_5)_3N$ in THF under cooling to give the corresponding 2-CH_3-4-$RC_6H_3N{=}CHN(CH_3)SN(CH_3)C_6H_{11}$-c [3]. Reacting a mixture of $ClSN(CH_3)C_6H_{11}$-c with I ("carbofuran") in pyridine at room temperature for 18 h yielded II [4].

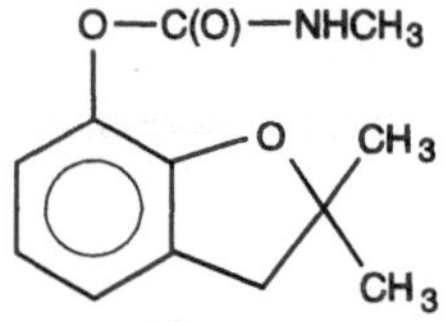

I

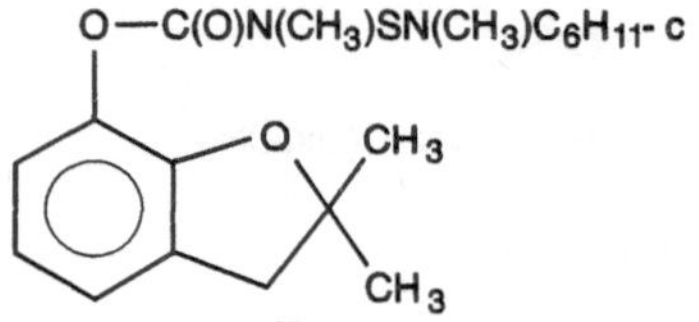

II

References:

[1] Badische Anilin-Soda-Fabrik AG; Weiß, G.; Schulze, G. (Ger. 1131222 [1960/62]; C.A. **57** [1962] 13771).
[2] Almasi, L.; Hantz, A. (Chem. Ber. **97** [1964] 661/6).
[3] Drabek, J.; Böger, M. (Ger. 2621077 [1974/76]; C.A. **86** [1977] No. 89430).
[4] Black, L.; Fukuto, T. R. (Ger. 2433680 [1972/75]; C.A. **82** [1975] No. 156050).

12.5.1.2.4 $ClSN(CH_3)CF_3$

The title compound was obtained by stirring $(CH_3)(CF_3)NSF_2^+ SbF_6^-$ with a ca. 10% excess of NOCl in SO_2 at −10 °C for 1 h and separating it from the by-products $(CF_3(CH_3)N)_2S_x$ (x = 2, 3) by fractional condensation at 10^{-3} Torr (−60, −80, −196 °C). The crude product (−80 °C trap) was purified by another fractional condensation to give $ClSN(CH_3)CF_3$ with a yield of 52% as a yellow liquid, b.p. 82.5 °C.

^{1}H NMR (SO_2/TMS): $\delta = 3.22$ ppm, $^4J(H,F) = 1.1$ Hz.

^{19}F NMR (SO_2/C_6F_6): $\delta = -63.24$ ppm.

IR (gas): 19 absorption bands between 3030 and 450 cm^{-1}.

MS: m/e (rel. int. in %) = 166, 164 (11.8, 34.1) $CF_3CH_2NSCl^+$, 162 (9.6), 130 (45.9), 128 (96.3), 69 (100.0) CF_3^+.

Reference:

Henle, H.; Mews, R. (Chem. Ber. **115** [1982] 3547/54).

12.5.1.2.5 $ClSN(C_2H_5)CH_2C_6H_5$

The title compound was formed by reacting $C_6H_5CH_2N(C_2H_5)SSN(C_2H_5)CH_2C_6H_5$ with SO_2Cl_2 (mole ratio ca. 2 : 3) at −78 °C [1].

$ClSN(C_2H_5)CH_2C_6H_5$ reacted with N-methylcarbamoyl fluoride, $CH_3NHC(O)F$, to give $C_6H_5CH_2N(C_2H_5)SN(CH_3)C(O)F$ (60%) [1]. Adding a solution of the title compound in toluene to a mixture of excess phthalimide and $(C_2H_5)_3N$ in toluene yielded 80% I after 1 h [2].

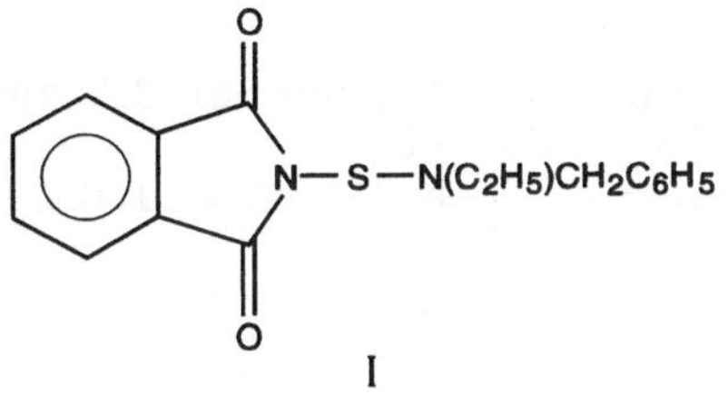

I

Reacting a mixture of the title compound with "carbofuran" in pyridine at room temperature for 18 h gave II [3].

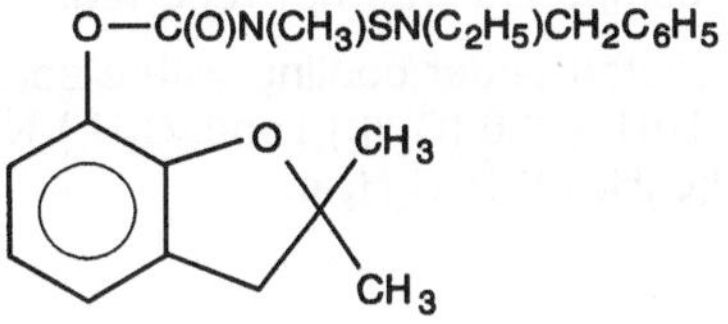

II

References:

[1] Hatch, C. E., III (J. Org. Chem. **43** [1978] 3953/7).
[2] Lawrence, J. P. (Rubber Chem. Technol. **49** [1976] 333/40).
[3] Black, L.; Fukuto, T. R. (Ger. 2433680 [1972/75]; C.A. **82** [1975] No. 156050).

12.5.1.2.6 $ClSN(C_2H_5)C_4H_9$-n

A formation of $ClSN(C_2H_5)C_4H_9$-n was not reported.

It reacted with $CH_3N{=}C{=}O$ and HF in the presence of $(C_2H_5)_3N$ in toluene or CH_2Cl_2 between −50 °C and room temperature to give n-$C_4H_9(C_2H_5)NSN(CH_3)C(O)F$.

Reference:

Thurman, D. E. (Ger. 2628575 [1975/77]; C.A. **86** [1977] No. 155665).

12.5.1.2.7 $ClSN(C_2H_5)C_6H_{11}$-c

The title compound was formed by introducing dry Cl_2 at 0 to 5 °C into a solution of c-C_6H_{11}-$N(C_2H_5)SSN(C_2H_5)C_6H_{11}$-c in CCl_4. The excess Cl_2 was then removed at 0 to 5 °C in vacuum or by purging the mixture with an inert gas. The resulting solution was cleared with animal charcoal, and the solvent was evaporated under reduced pressure. Subsequently, the residue was distilled in vacuum to give $ClSN(C_2H_5)C_6H_{11}$-c as an orange-colored liquid; b.p. 90 to 92 °C/0.4 Torr [1]; see also [2].

Adding a solution of $ClSN(C_2H_5)C_6H_{11}$-c in benzene to a suspension of $(C_2H_5O)_2P(S)SNa$ in benzene, which caused the temperature to rise to about 30 to 35 °C, and stirring the mixture in this temperature range for 1 h produced c-$C_6H_{11}(C_2H_5)NSSP(S)(OC_2H_5)_2$ [3].

References:

[1] Farbenfabriken Bayer AG (Br. 790021 [1954/58]; C.A. **1958** 13806).
[2] Kühle, E. (Synthesis **1970** 561/80, 566).
[3] Malz, H.; Bayer, O.; Freytag, H.; Lober, F. (U.S. 2891059 [1957/59]; C.A. **1960** 4387).

12.5.1.2.8 $ClSN(C_3H_7\text{-n})CH_2C_3H_5\text{-c}$

A preparation of the title compound was not reported.

$ClSN(C_3H_7\text{-n})CH_2C_3H_5\text{-c}$ reacted under cooling with a solution of $R^1R^2C_6H_3N{=}CHNHCH_3$ ($R^1 = 2\text{-}CH_3$, $R^2 = 4\text{-}Cl$, $4\text{-}CH_3$; $R^1R^2 = 2,6\text{-}(C_2H_5)_2$) and $(C_2H_5)_3N$ in THF to give the corresponding $R^1R^2C_6H_3N{=}CHN(CH_3)SN(C_3H_7\text{-n})CH_2C_3H_5\text{-c}$.

Reference:

Drabek, J.; Böger, M. (Ger. 2621077 [1974/76]; C.A. **86** [1977] No. 89430).

12.5.1.2.9 $ClSN(C_3H_7\text{-n})C_8H_{17}\text{-t}$

The title compound was prepared by passing dry Cl_2 at 0 to 5 °C through a solution of $t\text{-}C_8H_{17}(n\text{-}C_3H_7)NSSN(C_8H_{17}\text{-t})C_3H_7\text{-n}$ in CCl_4. Excess Cl_2 was removed between 0 and 5 °C in vacuum or with the help of inert gas. The resulting solution was cleared with animal charcoal, the solvent evaporated under reduced pressure, and the residue distilled in vacuum to give the title compound as a reddish brown liquid; b.p. 86 to 88 °C/0.1 Torr.

Reference:

Farbenfabriken Bayer AG (Br. 790021 [1954/58]; C.A. **1958** 13806).

12.5.1.2.10 $ClSN(C_3H_7\text{-i})CH_2C_6H_5$

The title compound was prepared by reacting SCl_2 with $(C_6H_5CH_2)(i\text{-}C_3H_7)NH$ in a mole ratio of 1:2 [1].

^1H NMR: At ambient temperature the title compound exhibits chemical shift equivalence of benzylmethylene protons and isopropyl methyl groups. Nonequivalence of the chemical shifts have been found at low temperatures; referring to the coalescence of the benzyl AB quartet, the coalescence temperature T_c in $CD_3C_6D_5$ is 31 °C, corresponding to $\Delta v = 15$ Hz and $J_{AB} = 15$ Hz. The free energy of activation $\Delta G_c^{\ddagger}$ was calculated from the coalescence rate using the Eyring equation to be 15.1 kcal/mol. The low-temperature (−40 °C) ^1H NMR spectrum (in $CD_3C_6H_5$) is displayed in the paper [1].

However, in CH_2Cl_2 solution $ClSN(C_3H_7\text{-i})CH_2C_6H_5$ did not show any signal splitting down to −60 °C, since chlorine exchange is perhaps more rapid in this solvent [2].

References:

[1] Raban, M.; Cho, T.-M. (Int. J. Sulfur Chem. A **1** [1971] 269/71).
[2] Jackson, W. R.; Kee, T. G.; Spratt, R.; Jennings, W. B. (Tetrahedron Lett. **1973** 3581/4).

12.5.1.2.11 $ClSN(C_3H_7\text{-}i)CH_2CH_2C(O)OC_2H_5$

The title compound was formed by adding SO_2Cl_2 to a solution of $(C_2H_5OC(O)(CH_2)_2\text{-}N(C_3H_7\text{-}i)S)_2$ in $ClCH_2CH_2Cl$ and stirring the mixture for 1 h at 0 to 10 °C (the title compound was not isolated) [1]. Cl_2 was introduced instead of SO_2Cl_2 at 5 °C, and the mixture was reacted for 3 h at −30 to 0 °C [2].

Reacting the title compound with "carbofuran" and pyridine or $(C_2H_5)_3N$ with precooled (to 0 °C) $(CH_2Cl)_2$ at 0 to 10 °C for 2 to 3 h gave I (92%) [1 to 3].

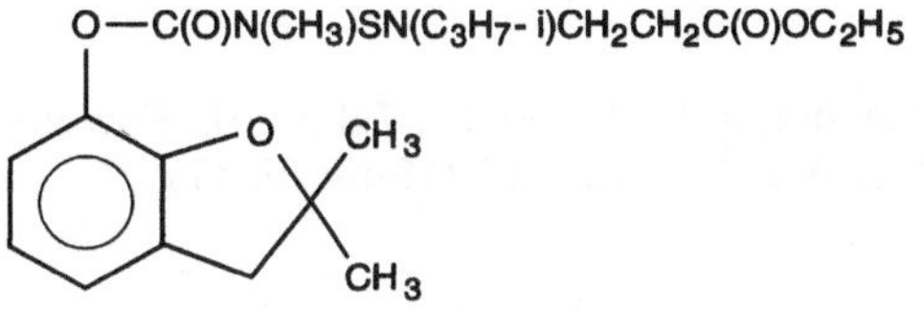

I

The title compound was reacted with a solution of $CH_3NHC(O)ON{=}C(CH_3)SCH_3$ in CH_2Cl_2 and $(C_2H_5)_3N$ at 0 to 5 °C for 2 h to give $C_2H_5OC(O)CH_2CH_2N(C_3H_7\text{-}i)SN(CH_3)C(O)ON{=}C\text{-}(CH_3)SCH_3$ [4].

References:

[1] Goto, T.; Yasudomi, N.; Tanaka, A. K.; Osaki, N.; Takao, H.; Kawata, M.; Imada, J.; Endo, Y.; Umetsu, N. (Nippon Noyaku Gakkaishi **13** [1988] 39/47).

[2] Otsuka Chemical Co., Ltd. (Jpn. Kokai Tokkyo Koho 59-161 353 [1983/84]; C.A. **102** [1985] No. 112 897).

[3] Tanaka, A. K.; Umetsu, N.; Fukuto, T. R. (J. Agric. Food Chem. **33** [1985] 1049/55; C.A. **103** [1985] No. 208 826).

[4] Goto, T.; Takao, H.; Yasudomi, N.; Osaki, N.; Murata, T. (Belg. 892 302 [1981/82]; C.A. **97** [1982] No. 144 415).

12.5.1.2.12 $ClSN(CH_2C_6H_5)CH_2CH_2C(O)OC_2H_5$

A preparation of the title compound was not reported.

It reacted with $CH_3NHC(O)ON{=}C(CH_3)SCH_3$ and $(C_2H_5)_3N$ in the presence of SO_2 (catalyst) in $C_2H_4Cl_2$ or hexane at 10 °C and 30 min at room temperature to $CH_3SC(CH_3){=}NOC(O)NCH_3S\text{-}N(CH_2C_6H_5)CH_2CH_2C(O)OC_2H_5$ with 93.3% yield and 94.5% purity (using $C_2H_4Cl_2$ as solvent), or with 74.1% yield and 90.1% purity (using hexane as solvent).

Reference:

Tada, I. (Jpn. Kokai Tokkyo Koho 03-31 252 [1989/91] 4 pp.; C.A. **115** [1991] No. 8109).

12.5.1.2.13 $ClSN(CH_2CN)CH_2P(O)(OC_2H_5)_2$

The title compound was formed by adding pyridine at 0 °C and then $(C_2H_5O)_2P(O)CH_2\text{-}NHCH_2CN$ at 10 to 20 °C to a solution of SCl_2 (mole ratio 1 : 1 : 1) in CCl_4.

Dissolving the title compound, "carbofuran", and excess pyridine in CH_2Cl_2 gave I (81%).

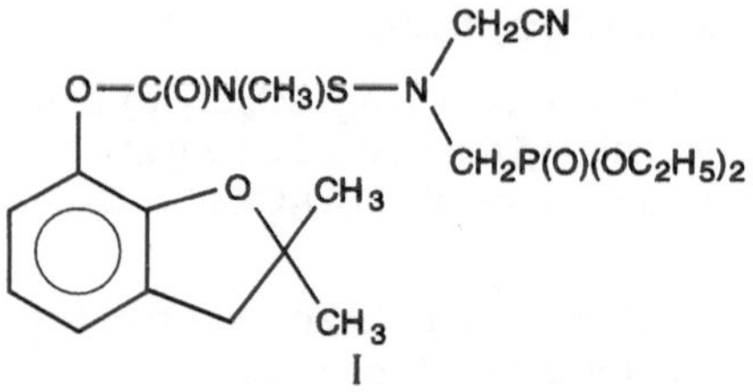

I

Reference:

Goto, T.; Yasudomi, N.; Tanaka, A. K.; Osaki, N.; Takao, H.; Kawata, M.; Imada, J.; Endo, Y.; Umetsu, N. (Nippon Noyaku Gakkaishi **13** [1988] 39/47).

12.5.1.2.14 $ClSN(CH_2CN)CH_2P(O)(OC_6H_5)_2$

The title compound was prepared by adding a solution of $(C_6H_5O)_2P(O)CH_2NHCH_2CN$ and $(C_2H_5)_3N$ in toluene to a precooled solution of SCl_2 (mole ratio 1 : 1 : 1) in pentane (0 °C) [1 to 3] or toluene (−20 °C) [4], while maintaining the temperature below 10 °C [1 to 3] or −10 °C [4], and stirring the mixture at 0 °C for 45 min [2] or at −20 °C for 3 h [4]. 79% yield; yellow oil [1 to 4].

^{31}P NMR: δ = −11.15 ppm [2, 3].

The title compound reacts with secondary amines R_2NH (R = CH_3, i-C_3H_7, n-C_4H_9, c-C_6H_{11}, C_6H_5; R_2 = C_5H_{10}, $S(CH_2)_4$) or R_2NK (R_2 = phthalimido) and $(C_2H_5)_3N$ to give the corresponding $(C_6H_5O)_2P(O)CH_2N(CH_2CN)SNR_2$ [1].

Adding a solution of the title compound in toluene at 0 °C to a solution of a mercaptan RSH (R = CH_3, i-C_3H_7, $(CH_3)_2CHCH_2CH_2$, n-C_8H_{17}, $C_6H_5CH_2CH_2$, 3-$CF_3C_6H_4CH_2$, c-C_6H_{11}, 4-CH_3O-C_6H_4, 4-ClC_6H_4, 3-$CF_3C_6H_4$, 2-naphthyl) [2] (see also [3]) or at −20 °C (R = t-C_4H_9) [4] and $(C_2H_5)_3N$ (mole ratio 1 : 1 : 1) in toluene and letting the mixture react for 2 h at 10 °C yielded the corresponding $(C_6H_5O)_2P(O)CH_2N(SSR)CH_2CN$ [2]; see also [3].

References:

[1] Sikorski, J. A.; Hoobler, M. A. (U.S. 4433996 [1981/84]; C.A. **101** [1984] No. 23718).
[2] Sikorski, J. A.; Curtis, T. G. (Belg. 894590 [1982/83]; C.A. **99** [1983] No. 53970).
[3] Hoobler, M. A.; Sikorski, J. A. (Belg. 894591 [1982/83]; C.A. **99** [1983] No. 6051).
[4] Sikorski, J. A.; Curtis, T. G. (U.S. 4395276 [1981/83]; C.A. **99** [1983] No. 135541).

12.5.1.2.15 $ClSN(CH_2CN)CH_2P(O)(OC_6H_3R^1R^2)_2$, R^1 = H, R^2 = 2-CH_3O, 4-CH_3O; R^1 = 4-Cl, R^2 = 3-CH_3

A preparation of the compounds was not reported.

The title compounds were reacted with $(C_2H_5)_2NH$ and $(C_2H_5)_3N$ to give the corresponding $(R^1R^2C_6H_3O)_2P(O)CH_2N(CH_2CN)SN(C_2H_5)_2$.

Reference:

Sikorski, J. A.; Hoobler, M. A. (U.S. 4433996 [1981/84]; C.A. **101** [1984] No. 23718).

12.5.1.2.16 $ClSN(CH_2CH_2CH_2OCH_3)CH_2CH_2C(O)OCH_3$

A preparation of the title compound was not reported.

The title compound was reacted with "carbofuran" and $(C_2H_5)_3N$ in $(CH_2Cl)_2$ for 2 h at 10 °C to give I.

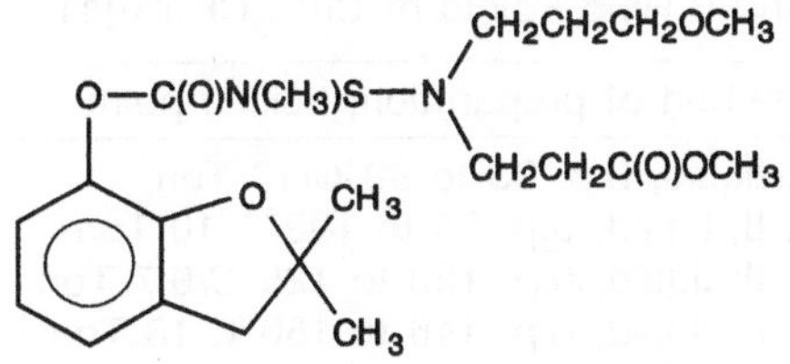

I

Reference:

Otsuka Pharmaceutical Co., Ltd. (Jpn. Kokai Tokkyo Koho 59-88461 [1982/84]; C.A. **101** [1984] No. 191672).

12.5.1.3 N-Aryl-N-alkyl-amino-chloro-sulfanes

12.5.1.3.1 $ClSN(CH_3)C_6H_5$

The title compound was apparently formed along with $C_6H_5NHCH_3 \cdot HCl$ by the reaction of $ClSN{=}CCl_2$ with $C_6H_5NHCH_3$ (mole ratio 1:2) in ether at ~20 °C for 1 h [1].

Yellow liquid which darkened and resinified [1].

Adding a solution of $ClSN(CH_3)C_6H_5$ in toluene to a mixture of excess phthalimide and $(C_2H_5)_3N$ in toluene yielded 82% I after 1 h [2].

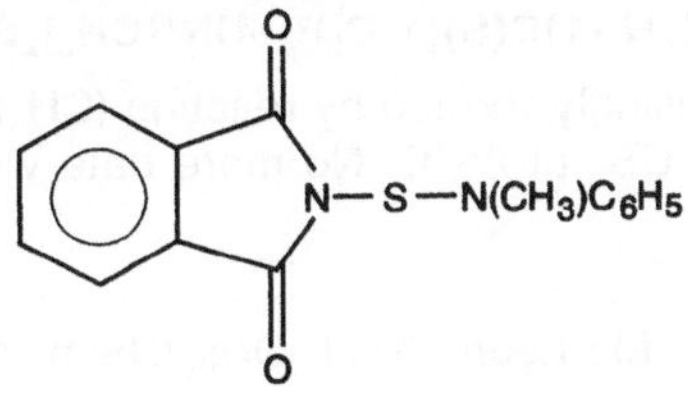

I

References:

[1] Bacon, R. G. R.; Irwin, R. S. (J. Chem. Soc. **1960** 5079/87).

[2] Lawrence, J. P. (Rubber Chem. Technol. **49** [1976] 333/40).

12.5.1.3.2 $ClSN(CF_3)R$, R = aryl

The title compounds (see Table 36, p. 148) were prepared as follows:

Method I: Adding S_2Cl_2 under cooling (staying below 25 °C) to a precooled (15 to 20 °C) mixture of the respective CF_3NHR and $(C_2H_5)_3N$ (mole ratio ca. 1:2:2) in CCl_4, removing the hydrochloride by filtration, and introducing Cl_2 under ice cooling at 5 to 10 °C to the remaining clear filtrate [1].

Method II: Treating a solution of the respective disulfide $CF_3N(R)SSN(R)CF_3$ in benzene with Cl_2 at 15 to 20 °C [1]; see also [2].

Table 36
Methods of Preparation and Boiling Points of $ClSN(CF_3)R$ [1].

R	method of preparation; boiling point
C_6H_4Cl-2	I; liquid, b.p. 93 to 99 °C/13 Torr
C_6H_4Cl-4	I, II; liquid, b.p. 95 to 100 °C/10 Torr
$C_6H_4NO_2$-4	I, II; liquid, b.p. 120 to 125 °C/0.7 Torr
$C_6H_4C(O)OC_2H_5$-4	I, II; liquid, b.p. 146 to 150 °C/13 Torr
$C_6H_3Cl_2$-2,3	I; liquid, b.p. 114 to 122 °C/10 Torr
$C_6H_3Cl_2$-2,4	I (20 to 25 °C, introducing Cl_2 at 15 to 20 °C); II (87% yield); liquid, b.p. 110 to 115 °C/10 Torr
$C_6H_3Cl_2$-2,5	I (< 25 °C, introducing Cl_2 at < 10 °C); II (instead of Cl_2, sulfuryl chloride can be used, but the yield is lower); liquid, b.p. 114 or 115 to 122 °C/12 Torr, or 115 to 123 °C/13 Torr
$C_6H_3Cl_2$-3,4	II; liquid, b.p. 119 to 122 °C/15 Torr
$C_6H_2Cl_3$-2,4,6	I; liquid, b.p. 138 to 145 °C/13 Torr

References:

[1] Farbenfabriken Bayer AG; Kühle, E.; Klauke, E. (Ger. 1187627 [1962/65]; C.A. **62** [1965] 16118).
[2] Kühle, E. (Synthesis **1970** 561/80, 567; The Chemistry of the Sulfenic Acids, Thieme, Stuttgart 1973).

12.5.1.3.3 $ClSN^+(CH_3)_2(4\text{-}C_6H_4\text{-}1)C(S)(1\text{-}C_6H_4\text{-}4)N^+(CH_3)_2SCl\ 2\ Cl^-$

The title compound surprisingly formed by reacting $(CH_3)_2N$–(4-C_6H_4-1)–C(S)–(1-C_6H_4-4)–$N(CH_3)_2$ with excess SCl_2 in CS_2 at 25 °C. No more data were given.

Reference:

Still, I. W. J.; Kutney, G. W.; Mc Lean, D. (J. Org. Chem. **47** [1982] 555/60).

12.5.1.3.4 N-(4-Chloro-6-ethylamino-1,3,5-triazin-2-yl)-N-ethyl-amino-chloro-sulfane

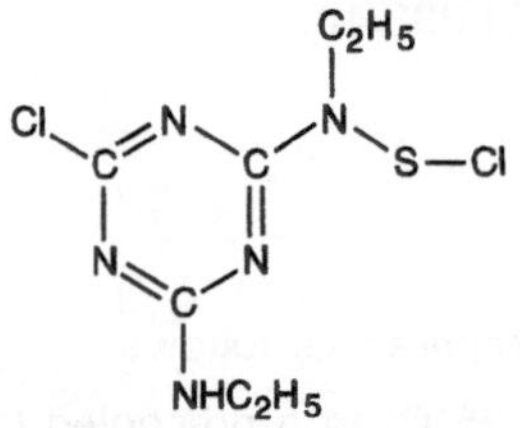

The title compound was formed by adding SCl_2 at room temperature and then pyridine between 25 to 30 °C to a solution of 4,6-bis(ethylamino)-2-chloro-s-triazine (mole ratio ca. 1 : 1 : 1) in CH_2Cl_2 and letting the mixture react for 30 min at room temperature. Then the hydrochloride was removed by filtration to get the title compound (not isolated) in CH_2Cl_2 solution [1 to 3].

^1H NMR (CH_2Cl_2/TMS): $\delta = 3.95$ ppm (q, CH_2NSCl) [1 to 3].

The title compound reacted (in situ) with CH_3OH and pyridine to give the 2-substituted methoxysulfanyl derivative I (X = OCH_3) [1]; see also [2]. The reaction with CH_3SH or C_6H_5SH and pyridine gave the corresponding 2-substituted disulfanyl derivative I, where X = SCH_3 or SC_6H_5 [1]; see also [2, 3].

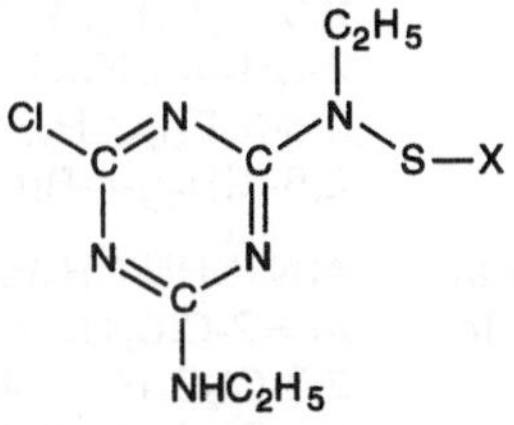

I

References:

[1] Cleveland, J. D. (U.S. 3864342 [1972/75]; C.A. **82** [1975] No. 171090).
[2] Cleveland, J. D. (U.S. 3909237 [1972/75]; C.A. **83** [1975] No. 206335).
[3] Cleveland, J. D. (U.S. 3796712 [1972/74]; C.A. **80** [1974] No. 133486).

12.5.1.4 N,N-Diphenyl-amino-chloro-sulfane, $ClSN(C_6H_5)_2$

The title compound was apparently formed with 97% yield along with 96% $(C_6H_5)_2NH \cdot HCl$ by reacting $ClSN{=}CCl_2$ with $(C_6H_5)_2NH$ (mole ratio 1 : 2) in ether for 12 h at ~20 °C [1].

The dark green liquid suddenly decomposed upon warming to 30 °C with evolution of HCl and production of a resin [1].

Adding a solution of $ClSN(C_6H_5)_2$ in CH_2Cl_2 to a precooled (0 °C) solution of $(C_6H_5O)_2P(O)$-$CH_2NHCH_2C(O)OCH_2C_6H_5$ and $(C_2H_5)_3N$ (mole ratio ca. 3 : 1 : 1) in toluene yielded $(C_6H_5O)_2P$-$(O)CH_2N(CH_2C(O)OCH_2C_6H_5)SN(C_6H_5)_2$ [2].

References:

[1] Bacon, R. G. R.; Irwin, R. S. (J. Chem. Soc. **1960** 5079/87).
[2] Hoobler, M.; Sikorski, J. A. (Belg. 894595 [1982/83]; C.A. **99** [1983] No. 6049).

12.5.1.5 N-Formyl-N-alkyl-amino-chloro-sulfanes

12.5.1.5.1 $ClSN(CH_3)C(O)H$

The title compound was prepared by adding a solution of CH_3NHCHO and $(C_2H_5)_3N$ in CH_2Cl_2 to a mixture of SCl_2 and CH_2Cl_2 cooled in an ice bath [1].

Liquid, b.p. 39 °C/0.5 Torr [1].

Reactions of $ClSN(CH_3)C(O)H$ are tabulated in Table 37, p. 150.

Table 37
Chemical Reactions of $ClSN(CH_3)C(O)H$.

substrate	reaction conditions	product(s)	Ref.
$ArN{=}CHNHCH_3$ + $(C_2H_5)_3N$	in THF	$ArN{=}CHN(CH_3)SN(CH_3)C(O)H$ Ar = $2\text{-}ClC_6H_4$, $4\text{-}NO_2C_6H_4$, $2,3\text{-}Cl_2C_6H_3$, $2,4\text{-}Cl_2C_6H_3$, $3,5\text{-}Cl_2C_6H_3$, $2\text{-}CH_3\text{-}4\text{-}ClC_6H_3$, $2,4\text{-}(CH_3)_2C_6H_3$, $3,5\text{-}(CF_3)_2C_6H_3$, $2,4,5\text{-}Cl_3C_6H_2$, $2,5\text{-}(CH_3)_2\text{-}4\text{-}BrC_6H_2$	[2]
$ArN{=}CHNHCH_3$	in pyridine at 0 to 10°C for 20 min	$ArN{=}CHN(CH_3)SN(CH_3)C(O)H$ Ar = $2\text{-}ClC_6H_4$, $4\text{-}NO_2C_6H_4$, $2,3\text{-}Cl_2C_6H_3$, $2,4\text{-}Cl_2C_6H_3$, $2,6\text{-}Cl_2C_6H_3$, $2\text{-}CH_3\text{-}3\text{-}ClC_6H_3$, $2,4\text{-}(CH_3)_2C_6H_3$, $3,5\text{-}(CF_3)_2C_6H_3$, $2,4,5\text{-}Cl_3C_6H_2$, $2,5\text{-}(CH_3)_2\text{-}4\text{-}BrC_6H_2$	[3]
$2\text{-}CH_3\text{-}4\text{-}ClC_6H_3\text{-}N{=}CHNHC_4H_9\text{-}n$	in pyridine at 0 to 10°C for 20 min	$2\text{-}CH_3\text{-}4\text{-}ClC_6H_3N{=}CHN(C_4H_9\text{-}n)SN(CH_3)\text{-}C(O)H$	[3]
$NH_2C(O)NHC_6H_5$ + pyridine	in CH_2Cl_2	$C_6H_5NHC(O)NHSN(CH_3)C(O)H$	[4]
$CH_3NHC(O)N\text{-}(CH_3)C_6H_5$	in CH_2Cl_2	$HC(O)N(CH_3)SN(CH_3)C(O)N(CH_3)C_6H_5$	[4, 5]
$CH_3NHC(O)NHAr$ + pyridine	in CH_2Cl_2, 1 h at room temperature	$HC(O)N(CH_3)SN(CH_3)C(O)NHAr$ Ar = RC_6H_4, R = H [5], 2-F [4 to 6], 2-Cl [4, 5], $4\text{-}NO_2$ [4, 5]; Ar = $RR'C_6H_3$, RR' = $3,4\text{-}Cl_2$, 3-Cl-4-Br [4, 5]	[4 to 6]
$C_2H_5NHC(O)NH\text{-}C_6H_3Cl_2\text{-}3,4$	in $CHCl_2$	$HC(O)N(CH_3)SN(C_2H_5)C(O)NHC_6H_3Cl_2\text{-}3,4$	[4, 5]
$CH_3N{=}C{=}O$ + HF + $(C_2H_5)_3N$	in toluene at 0 to 8°C	$HC(O)N(CH_3)SN(CH_3)C(O)F$	[7]
$RSC(CH_3){=}NO\text{-}C(O)NHCH_3$ + pyridine	in CH_2Cl_2 at 5 to 10°C for 16 h	$RSC(CH_3){=}NOC(O)N(CH_3)SN(CH_3)C(O)H$	[8, 9]

References:

[1] Kohn, G. K. (Ger. 2128672 [1970/71]; C.A. **76** [1972] No. 45777).
[2] Böger, M.; Drabek, J. (Ger. 2600987 [1975/76]; C.A. **85** [1976] No. 123638).
[3] Böger, M.; Drabek, J. (Swiss 604500 [1975/78]; C.A. **90** [1979] No. 1703).
[4] Brown, M. S.; Kohn, G. K. (U.S. 3891424 [1970/75]; C.A. **83** [1975] No. 78931).
[5] Brown, M. S. (U.S. 3824281 [1970/74]; C.A. **81** [1974] No. 77698).
[6] Brown, M. S. (Ger. 2155391 [1971/72]; C.A. **77** [1972] No. 75032).
[7] D'Silva, T. D. J. (Ger. 2654282 [1976/77]; C.A. **87** [1977] No. 200835).
[8] Drabek, J.; Böger, M. (U.S. 4364959 [1978/82]; C.A. **98** [1983] No. 215201).
[9] Drabek, J.; Böger, M. (Ger. 2727614 [1976/78]; C.A. **88** [1978] No. 190124).

12.5.1.5.2 $ClSN(C_2H_5)C(O)H$

The title compound was prepared by adding a solution of $C_2H_5NHC(O)H$ and pyridine in CH_2Cl_2 at 10 °C to a solution of freshly distilled SCl_2 in CH_2Cl_2 and stirring the mixture for 1 h [1].

Liquid, b.p. 25 to 30 °C/0.5 Torr [1].

$ClSN(C_2H_5)C(O)H$ was reacted with a solution of the formamidine 2-CH_3-4-RC_6H_3N=$CHNHCH_3$ (R = Cl, Br, CH_3) [2, 3] in pyridine at 0 to 10 °C for 20 min [2] or with $(C_2H_5)_3N$ in THF [3] to give the corresponding 2-CH_3-4-RC_6H_3N=$CHN(CH_3)SN(C_2H_5)C(O)H$ [2, 3].

Adding a solution of $ClSN(C_2H_5)C(O)H$ in CH_2Cl_2 to a mixture of $CH_3NHC(O)NHC_6H_4R$ (R = 2-F, 4-NO_2) and pyridine in CH_2Cl_2 yielded the corresponding $HC(O)N(C_2H_5)SN(CH_3)C(O)$-$NHC_6H_4R$ [4]. Adding the title compound to a mixture of 2-CH_3-4-$ClC_6H_3NHC(S)N(CH_3)_2$ and $(C_2H_5)_3N$ in CH_2Cl_2 at 0 to 10 °C and reacting the mixture for 2 h at room temperature gave 2-CH_3-4-ClC_6H_3N=$C(N(CH_3)_2)SSN(C_2H_5)C(O)H$ [5].

References:

[1] Kohn, G. K. (Ger. 2128672 [1970/71]; C.A. **76** [1972] No. 45777).
[2] Böger, M.; Drabek, J. (Swiss 604500 [1975/78]; C.A. **90** [1979] No. 1703).
[3] Böger, M.; Drabek, J. (Ger. 2600987 [1975/76]; C.A. **85** [1976] No. 123638).
[4] Brown, M. S.; Kohn, G. K. (U.S. 3891424 [1970/75]; C.A. **83** [1975] No. 78931), Brown, M. S. (U.S. 3824281 [1970/74]; C.A. **81** [1974] No. 77698).
[5] Böger, M.; Drabek, J. (Ger. 2654080 [1972/77]; C.A. **87** [1977] No. 102078).

12.5.1.5.3 $ClSN(CH_2CH_2OCH_3)C(O)H$

The preparation of the title compound was not described.

$ClSN(CH_2CH_2OCH_3)C(O)H$ reacted with the formamidines $R^1R^2C_6H_3N$=$CHNHCH_3$ (R^1R^2 = 3,5-Cl_2; 2-CH_3-4-Cl (or Br, CH_3)) and $(C_2H_5)_3N$ in THF to give the corresponding $R^1R^2C_6H_3N$=$CHN(CH_3)SN(CH_2CH_2OCH_3)C(O)H$.

Reference:

Böger, M.; Drabek, J. (Ger. 2600987 [1975/76]; C.A. **85** [1976] No. 123638).

12.5.1.5.4 $ClSN(R)C(O)H$, R = n-C_3H_7, i-C_3H_7, i-C_4H_9

The preparation of the title compounds was not described.

$ClSN(R)C(O)H$ was reacted with a solution of the formamidines 2-CH_3-4-$R'C_6H_3N$=$CHNHCH_3$ (R′ = Cl [1], CH_3 [1, 2]) in pyridine at 0 to 10 °C for 20 min [1] or with $(C_2H_5)_3N$ in THF [2] to give the corresponding 2-CH_3-4-$R'C_6H_3N$=$CHN(CH_3)SN(R)C(O)H$ [1, 2].

References:

[1] Böger, M.; Drabek, J. (Swiss 604500 [1975/78]; C.A. **90** [1979] No. 1703).
[2] Böger, M.; Drabek, J. (Ger. 2600987 [1975/76]; C.A. **85** [1976] No. 123638).

12.5.1.5.5 $ClSN(CH_2CH(C_2H_5)C_4H_9\text{-n})C(O)H$

The preparation of the title compound was not described.

The title compound reacted with a solution of 2-CH_3-4-$RC_6H_3N{=}CHNHCH_3$ (R = Cl, CH_3) and $(C_2H_5)_3N$ in THF to produce the corresponding 2-CH_3-4-$RC_6H_3N{=}CHN(CH_3)SN(CH_2CH$-$(C_2H_5)C_4H_9$-n)C(O)H [1].

The title compound was reacted with $CH_3NHC(O)ON{=}C(CH_3)SR$ (R = CH_3, C_2H_5) and pyridine in CH_2Cl_2 at 5 to 10°C for 16 h yielding the corresponding HC(O)(n-$C_4H_9(C_2H_5)$-$CHCH_2)NSN(CH_3)C(O)ON{=}C(CH_3)SR$ (R = CH_3, C_2H_5) [2].

References:

[1] Böger, M.; Drabek, J. (Ger. 2600987 [1975/76]; C.A. **85** [1976] No. 123638).
[2] Drabek, J.; Böger, M. (Ger. 2727614 [1976/78]; C.A. **88** [1978] No. 190124, U.S. 4364959 [1978/82]; C.A. **98** [1983] No. 215201).

12.5.1.5.6 $ClSN(CH_2C_3H_5\text{-c})C(O)H$

The preparation of the title compound was not described.

$ClSN(CH_2C_3H_5$-c)C(O)H reacted with a mixture of $R^1R^2C_6H_3N{=}CHNHCH_3$ (R^1 = 2-CH_3, R^2 = 4-Cl, 4-CH_3; R^1R^2 = 3,5-$(CF_3)_2$) and $(C_2H_5)_3N$ in THF to the corresponding $R^1R^2C_6H_3N{=}CHN$-$(CH_3)SN(CH_2C_3H_5$-c)C(O)H.

Reference:

Böger, M.; Drabek, J. (Ger. 2600987 [1975/76]; C.A. **85** [1976] No. 123638).

12.5.1.5.7 $ClSN(C_3H_5\text{-c})C(O)H$

The preparation of the title compound was not described.

$ClSN(C_3H_5$-c)C(O)H reacted with 2-R^1-4-R^2-$C_6H_3N{=}CHNHCH_3$ (R^1 = H, R^2 = NO_2; R^1 = CH_3, R^2 = Cl, CH_3) and $(C_2H_5)_3N$ in THF to form the corresponding 2-R^1-4-$R^2C_6H_3N{=}CHN(CH_3)SN$-$(C_3H_5$-c)C(O)H [1]. It reacted with $CH_3NHC(O)ON{=}C(CH_3)SR$ (R = CH_3, C_2H_5) and pyridine in CH_2Cl_2 at 5 to 10°C for 16 h to give the corresponding HC(O)N(C_3H_5-c)$SN(CH_3)C(O)ON{=}$ $C(CH_3)SR$ [2].

References:

[1] Böger, M.; Drabek, J. (Ger. 2600987 [1975/76]; C.A. **85** [1976] No. 123638).
[2] Drabek, J.; Böger, M. (U.S. 4364959 [1978/82]; C.A. **98** [1983] No. 215201, Ger. 2727614 [1976/78]; C.A. **88** [1978] No. 190124).

12.5.1.5.8 $ClSN(C_6H_{11}\text{-c})C(O)H$

The title compound was prepared by mixing a solution of Cl_2 in CCl_4 with a solution of HC(O)N(C_6H_{11}-c)SSN(C_6H_{11}-c)C(O)H in CH_2Cl_2 at room temperature overnight, yielding 99% of the yellow, liquid product [1].

Instead of Cl_2, SO_2Cl_2 can also be used (no details) [2].

$ClSN(C_6H_{11}$-c)C(O)H reacted with olefines R^1R^2C=CR^3R^4 to produce the addition compounds $R^1R^2C(Cl)C(SN(C_6H_{11}$-c)C(O)H)$R^3R^4$ and $R^1R^2C(SN(C_6H_{11}$-c)C(O)H)C(Cl)R^3R^4 [2]. When reacted with cyclohexene at room temperature overnight, N-(2-chlorocyclohexylsulfanyl)-N-cyclohexyl-formamide was obtained with 97.6% yield [1]. The title compound reacted with a solution of 2-CH_3-4-RC_6H_3N=$CHNHCH_3$ (R = Cl, CH_3) and $(C_2H_5)_3N$ in THF to the corresponding 2-CH_3-4-RC_6H_3N=$CHN(CH_3)SN(C_6H_{11}$-c)C(O)H [3].

$ClSN(C_6H_{11}$-c)C(O)H was reacted with $CH_3NHC(O)ON$=$C(CH_3)SCH_3$ and pyridine in CH_2Cl_2 at 5 to 10 °C for 16 h to give $HC(O)N(C_6H_{11}$-c)$SN(CH_3)C(O)ON$=$C(CH_3)SCH_3$ [4].

References:

[1] Hopper, R. J. (U.S. 3997605 [1975/76]; C.A. **86** [1977] No. 120883).
[2] Goodyear Tire and Rubber Co. (Res. Disl. No. 143 [1976] 28; C.A. **84** [1976] No. 179730).
[3] Böger, M.; Drabek, J. (Ger. 2600987 [1975/76]; C.A. **85** [1976] No. 123638).
[4] Drabek, J.; Böger, M. (U.S. 4364959 [1978/82]; C.A. **98** [1983] No. 215201, Ger. 2727614 [1976/78]; C.A. **88** [1978] No. 190124).

12.5.1.5.9 $ClSN(C_8H_{15}$-c)C(O)H

The preparation of the title compound was not described.

$ClSN(C_8H_{15}$-c)C(O)H reacted with $CH_3NHC(O)ON$=$C(CH_3)SR$ (R = CH_3, C_2H_5) and pyridine in CH_2Cl_2 at 5 to 10 °C to give the corresponding $HC(O)N(C_8H_{15}$-c)$SN(CH_3)C(O)ON$=$C(CH_3)SR$ after 16 h.

Reference:

Drabek, J.; Böger, M. (U.S. 4364959 [1978/82]; C.A. **98** [1983] No. 215201, Ger. 2727614 [1976/78]; C.A. **88** [1978] No. 190124).

12.5.1.6 N-Halogenocarbonyl-N-alkyl(or phenyl)-amino-chloro-sulfanes

12.5.1.6.1 $ClSN(CH_3)C(O)F$

The title compound was synthesized with 69.8% yield by adding 2,4-lutidine, 2,4-$(CH_3)_2C_5H_3N$, at 5 to 10 °C to a precooled (−15 °C) mixture of SCl_2 and $CH_3NHC(O)F$ (mole ratio 1 : 2 : 1) and stirring it at 15 to 20 °C for 80 min. The crude product was dissolved in ether, and Cl_2 was passed through the solution to give $ClSN(CH_3)C(O)F$. The title compound was also obtained by adding a solution of 2,4-lutidine, 2,4-$(CH_3)_2C_5H_3N$, in ether at 5 to 10 °C to a mixture of SCl_2 and $CH_3NHC(O)F$ (mole ratio 1 : 2 : 1) and stirring the mixture for 30 min [1]. The title compound was formed by adding a solution of pyridine in CH_2Cl_2 to a solution of $CH_3NHC(O)F$ and freshly distilled SCl_2 (mole ratio ca. 1 : 1 : 1) in CH_2Cl_2 cooled to −10 °C; after completing the addition, the reaction mixture was warmed to room temperature and stirred for 20 h [2]. Similarly, a mixture of $CH_3NHC(O)F$ and SCl_2 in ether was reacted with a solution of $(C_2H_5)_3N$ (mole ratio 1 : 1.1 : 1) in ether at −5 to 0 °C for 90 min [3].

The title compound was also formed by treating a solution of $FC(O)N(CH_3)SSN(CH_3)C(O)F$ in CH_2Cl_2 with excess chlorine at 0 to 5 °C and leaving the reaction mixture at this temperature for a while. Then the solvent was removed in vacuum, and the remaining crude product was distilled [4].

Yellow liquid [1] or yellow oil [2], b.p. 30 to 32 °C/12 Torr [4], 30 to 35 °C/7 Torr [2], 54 to 56 °C/40 Torr [1].

Refractive index: $n_{20}^{D} = 1.4722$ [4].

^1H NMR ($CDCl_3$/TMS): $\delta = 3.50$ ppm [2].

Reactions of $ClSN(CH_3)C(O)F$ are summarized in the following table:

Table 38
Chemical Reactions of $ClSN(CH_3)C(O)F$.

substrate	reaction conditions	product(s) (yield in %)	Ref.
$ClCH_2C(CH_3)=CH_2$	in CH_2Cl_2, 45 min at reflux temp.	$(ClCH_2)_2C(CH_3)SN(CH_3)C(O)F$	[5]
$(CH_3)_2CHC(O)CH(CH_3)_2$	in CH_2Cl_2, 1 h at room temperature	$(CH_3)_2CHC(O)C(CH_3)_2SN(CH_3)C(O)F$	[6]
2-methyl-cyclopentanone	in CH_2Cl_2, 2 h at room temperature	$\overline{CH_2CH_2CH_2C(O)C}(CH_3)SN(CH_3)C(O)F$	[6]
$R^1C(OR^2)=NH$ or their hydrochlorides ($R^1 = i\text{-}C_4H_9$, $R^2 = n\text{-}C_4H_9$; $R^1 = C_6H_5$, $2,6\text{-}Cl_2C_6H_3$, $R^2 = C_2H_5$; $R^1 = C_6H_5$, $R^2 = CH_3$, $n\text{-}C_4H_9$)		$R^1C(OR^2)=NSN(CH_3)C(O)F$	[7]
$C_6H_5C(SC_6H_5)=NH$		$C_6H_5SC(C_6H_5)=NSN(CH_3)C(O)F$	[7]
$(CH_3)_2CHC{\equiv}N + HCl$	30 min (HCl) insertion, 72 h at room temperature	$(CH_3)_2C(C{\equiv}N)SN(CH_3)C(O)F$	[5]
$4\text{-}RC_6H_4SH$ ($R = H$, $t\text{-}C_4H_9$)	in pentane	$4\text{-}RC_6H_4SSN(CH_3)C(O)F$	[8]
$(C_2H_5O)_2P(O)NHC_3H_7\text{-}i + (C_2H_5)_3N$	mole ratio 1 : 1 : 1; in THF at 0 °C and 1 1/2 h at 0 °C	$(C_2H_5O)_2P(O)N(C_3H_7\text{-}i)SN(CH_3)C(O)F$	[9]

Table 38 (continued)

substrate	reaction conditions	product(s) (yield in %)	Ref.
$CH_3SC(CH_3)$=NOH + 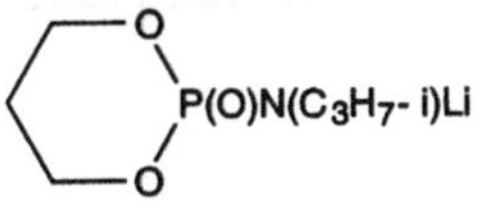	in THF–n-hexane at −5 °C	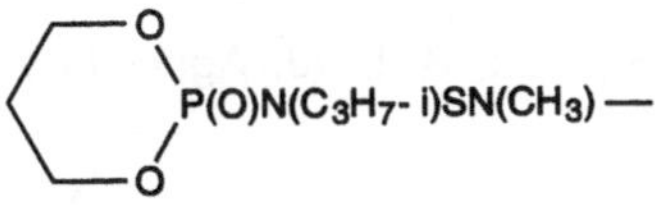 C(O)ON=$C(CH_3)SCH_3$	[10]
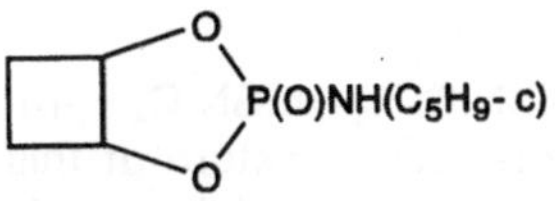 + $(C_2H_5)_3N$	mole ratio 1 : 1 : 1; in THF at 0 °C, and 1 1/2 h at 0 °C	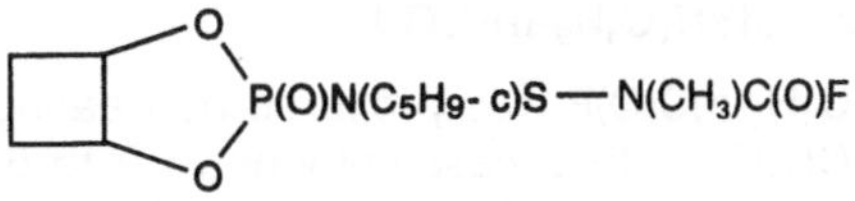	[9]
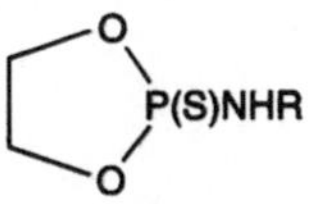 R = i-C_3H_7, t-C_4H_9, + $(C_2H_5)_3N$	mole ratio 1 : 1 : 1; in THF at 0 °C and 1 1/2 h at 0 °C	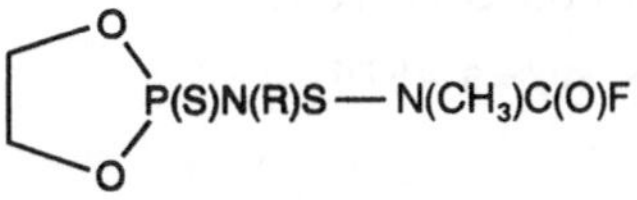 R = C_3H_7-i [3, 9], R = C_4H_9-t [9]	[3, 9]
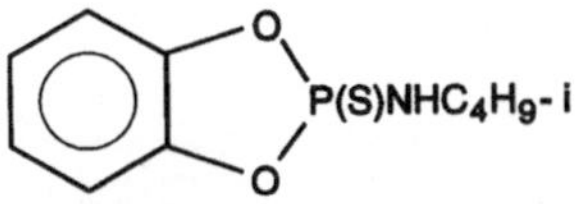 + $(C_2H_5)_3N$	mole ratio 1 : 1 : 1; in THF at 0 °C and 1 1/2 h at 0 °C	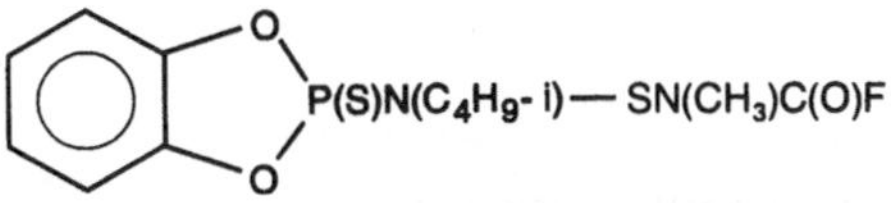	[9]
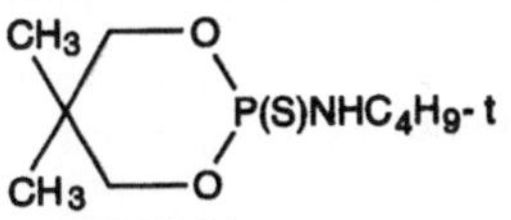 + $(C_2H_5)_3N$	in THF at −5 °C [2] or in CH_2Cl_2 [11]	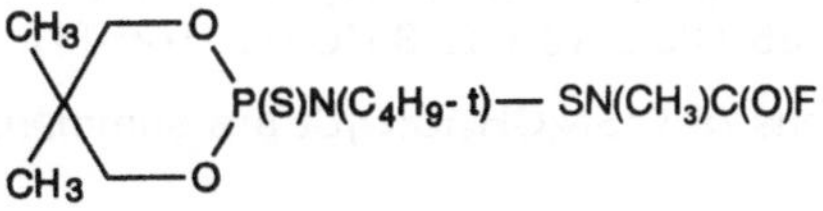	[2, 11]

References:

[1] Nelson, S. J. (U.S. 4333883 [1981/82]; C.A. **97** [1982] No. 109574).
[2] Dutton, F. E.; Gemrich, E. G., II; Lee, L.; Nelson, S. J.; Parham, P. H.; Seaman, W. J. (J. Agric. Food Chem. **29** [1981] 1114/8; C.A. **95** [1981] No. 163819).
[3] Upjohn Co. (Neth. 8100193 [1980/81]; C.A. **95** [1981] No. 204011).
[4] Kühle, E.; Klauke, E. (Ger. 2033079 [1970/71]; C.A. **76** [1972] No. 45780, Ger. 1931054 [1969/71]; C.A. **74** [1971] No. 63910).
[5] D'Silva, T. D. J. (U.S. 4485113 [1979/84]; C.A. **102** [1985] No. 148750, U.S. 4058549 [1976/77]; C.A. **88** [1978] No. 50329).
[6] D'Silva, T. D. J. (U.S. 4066689 [1976/78]; C.A. **88** [1978] No. 104715).
[7] Dutton, F. E.; Nelson, S. J. (U.S. 4473580 [1982/84]; C.A. **102** [1985] No. 24467).
[8] D'Silva, T. D. J. (U.S. 4091016 [1974/78]; C.A. **89** [1978] No. 129291, Ger. 2530278 [1972/76]; C.A. **84** [1976] No. 135347).

[9] Nelson, S. J. (Ger. 3019634 [1979/80]; C.A. **95** [1981] No. 132314, Ger. 3019590 [1978/80]; C.A. **94** [1981] No. 191729).
[10] Nelson, S. J. (U.S. 4297285 [1980/81]; C.A. **96** [1982] No. 85599, U.S. 4292256 [1980/81]; C.A. **96** [1982] No. 20128).

[11] Johnson, D. B.; Cox, B. L. (J. Agric. Food Chem. **33** [1985] 255/9; C.A. **102** [1985] No. 108165).

12.5.1.6.2 $ClSN(C_4H_9\text{-}n)C(O)F$

$ClSN(C_4H_9\text{-}n)C(O)F$ was prepared by treating a solution of $FC(O)N(C_4H_9\text{-}n)SSN(C_4H_9\text{-}n)$-$C(O)F$ in CH_2Cl_2 with excess chlorine at 0 to 5 °C and leaving the reaction mixture at this temperature for a while. Then the solvent was removed in vacuum, and the remaining crude product was distilled.

Liquid, b.p. 81 to 83 °C/17 Torr.

Refractive index: $n_D^{20} = 1.4575$.

Reference:

Kühle, E.; Klauke, E. (Ger. 2023079 [1970/71]; C.A. **76** [1972] No. 45780, Ger. 1931054 [1969/71]; C.A. **74** [1971] No. 63910).

12.5.1.6.3 $ClSN(CH_3)C(O)Cl$

The title compound was prepared by adding SCl_2 to a solution of $CH_3N{=}C{=}O$ and catalytic amounts of $(C_2H_5)_4NCl$ in CH_2Cl_2. A standard workup gave the title compound as a liquid; b.p. 32 to 35 °C/0.2 Torr or 35 °C/0.5 Torr [1].

Reactions of $ClSN(CH_3)C(O)Cl$ are summarized in the following table:

Table 39
Chemical Reactions of $ClSN(CH_3)C(O)Cl$.

substrate	reaction conditions	product(s)	Ref.
$(CH_3)_2C{=}CH_2$	in benzene at 5 to 10 °C and 1 h at ambient temperature	$(CH_3)_2C(CH_2Cl)SN(CH_3)C(O)Cl$	[2]
$ClCH_2C(CH_3){=}CH_2$	in CH_2Cl_2	$(ClCH_2)_2C(CH_3)SN(CH_3)C(O)Cl$	[2]
cyclohexene	mole ratio 1 : >1; at 0 °C, several hours	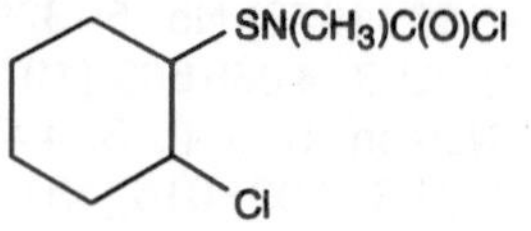	[3]

Table 39 (continued)

substrate	reaction conditions	product(s)	Ref.
norbornene	mole ratio 1 : >1; at 0 °C, several hours	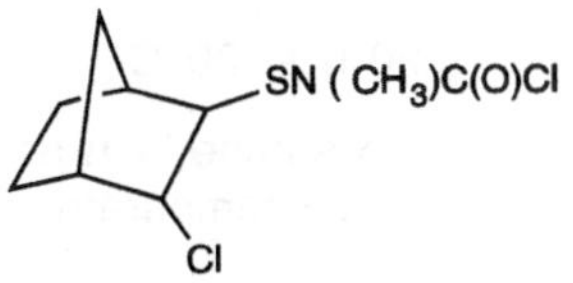	[3]
$(CH_3)_2CHC{\equiv}N$ + HCl	20 min HCl-insertion, 26 h at room temperature	$(CH_3)_2C(C{\equiv}N)SN(CH_3)C(O)Cl$	[2]
$R^1R^2CHC(O)H$ ($R^1 = CH_3$, $R^2 = CH_3$, CH_2Cl; $R^1 = C_2H_5$, $R^2 = C_2H_5$, n-C_4H_9)	in CH_2Cl_2; $R^1 = R^2 = CH_3$: mole ratio 1 : 1, 0.5 h at 35 °C [4], see also [5]; $R^1 = CH_3$, $R^2 = CH_2Cl$: mole ratio 1 : 1, without solvent at 5 to 10 °C, and 1 h at room temperature [4, 5]; $R^1 = R^2 = C_2H_5$: mole ratio 1 : 2, 2 h at 40 °C [4], see also [5]; $R^1 = C_2H_5$, $R^2 =$ n-C_4H_9: mole ratio 1 : 1, room temperature, and 1 h at room temperature [4], see also [5]	$ClC(O)N(CH_3)SCR^1R^2C(O)H$	[4, 5]
i-$C_3H_7C(O)R$ (R = i-C_3H_7, C_6H_5)	without solvent; R = i-C_3H_7: 1 h at room temperature [4, 5]; $R = C_6H_5$: mole ratio 1 : 1, heating at 30 °C, and 0.5 h at 40 °C [4]; see also [5]	$ClC(O)N(CH_3)SC(CH_3)_2C(O)R$	[4, 5]
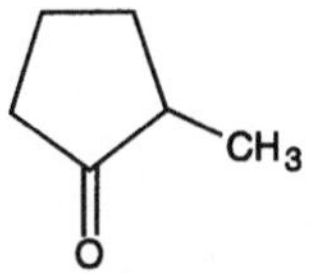	in CH_2Cl_2 cooled to 10 °C and 0.5 h at 15 °C	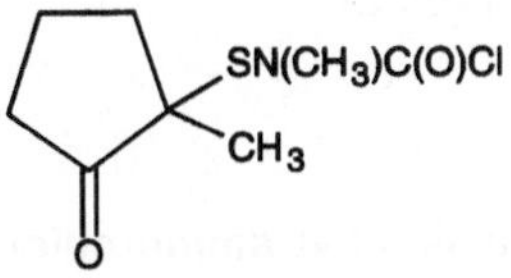	[4, 5]
$C_2H_5OC(O)CHN_2$	mole ratio 1 : 1; in $(C_2H_5)_2O$ cooled to −10 °C and 45 min [5] to 1 h [4] at room temperature	$C_2H_5OC(O)CH(Cl)SN(CH_3)C(O)Cl$	[4, 5]

Table 39 (continued)

substrate	reaction conditions	product(s)	Ref.
$(C_2H_5OC(O))_2CN_2$	20 h at 30 °C	$(C_2H_5OC(O))_2C(Cl)SN(CH_3)C(O)Cl$	[5]
4-t-$C_4H_9C_6H_4SH$	in pentane, 1 h at reflux temperature	4-t-$C_4H_9C_6H_4SSN(CH_3)C(O)Cl$	[6]

References:

[1] Kohn, G. K. (Ger. 2128672 [1970/71]; C.A. **76** [1972] No. 45777).
[2] D'Silva, T. D. J. (U.S. 4485113 [1979/84]; C.A. **102** [1985] No. 148750, U.S. 4058549 [1976/77] No. 50329).
[3] Kobzina, J. W.; Moore, J. F.; Kohn, G. K. (U.S. 3826644 [1970/74]; C.A. **82** [1975] No. 52660, U.S. 3711530 [1970/73]; C.A. **78** [1973] No. 110695).
[4] D'Silva, T. D. J. (U.S. 4081550 [1976/78]; C.A. **89** [1978] No. 75268).
[5] D'Silva, T. D. J. (U.S. 4066689 [1976/78]; C.A. **88** [1978] No. 104715).
[6] D'Silva, T. D. J. (U.S. 4091016 [1974/78]; C.A. **89** [1978] No. 129291).

12.5.1.6.4 $ClSN(C_4H_9\text{-}n)C(O)Cl$

The title compound was prepared by reacting n-$C_4H_9N{=}C{=}O$ and SCl_2 without a catalyst in an ampule in the presence of sunlight, in a sealed pressure vessel in a boiling water bath for 6 h, or with $(C_2H_5)_4NCl$ catalyst in CH_2Cl_2 solution.

Reference:

Kohn, G. K. (Ger. 2128672 [1970/71]; C.A. **76** [1972] No. 45777).

12.5.1.6.5 $ClSN(C_6H_5)C(O)Cl$

The preparation of the title compound by reacting $C_6H_5N{=}C{=}O$ with SCl_2 in CH_2Cl_2 at room temperature for 3 d was indicated by a strong IR absorption band at 1765 cm^{-1}. But isolation by distillation failed, showing the thermal instability of the 1:1 adduct.

Reference:

Komatsu, M.; Harada, N.; Kashiwagi, H.; Ohshiro, Y.; Agawa, T. (Phosphorus Sulfur Relat. Elem. **16** [1983] 119/33).

12.5.1.7 N-Alkanoyl-N-alkyl-amino-chloro-sulfanes

12.5.1.7.1 $ClSN(CH_3)C(O)CH_3$

The title compound was prepared by adding a solution of $CH_3C(O)NHCH_3$ and pyridine in CH_2Cl_2 to a precooled (−10 °C) solution of freshly distilled SCl_2 and CH_2Cl_2 while holding the temperature between −10 and 0 °C. The reaction mixture was then stirred for 1 h at 0 °C and worked up as usual to give the liquid title compound [1].

$ClSN(CH_3)C(O)CH_3$ reacted with 2-CH_3-4-RC_6H_3N=CHNHCH$_3$ (R = Br, CH_3: with $(C_2H_5)_3N$ in THF [2]; R = Cl, Br, CH_3: in pyridine at 0 to 10 °C for 20 min [3]; see also [2]) to give the corresponding 2-CH_3-4-RC_6H_3N=CHN(CH_3)SN(CH_3)C(O)CH_3. Adding a solution of $ClSN(CH_3)C(O)CH_3$ in CH_2Cl_2 to a mixture of $CH_3NHC(O)NHAr$ (Ar = RC_6H_4, R = 2-F, 4-Cl, 4-NO_2; 3,4-$Cl_2C_6H_3$) and pyridine in CH_2Cl_2 yielded the corresponding $CH_3C(O)N(CH_3)SN(CH_3)C(O)NHAr$ [4]. Adding $ClSN(CH_3)C(O)CH_3$ and then $(C_2H_5)_3N$ at −10 °C [5] or 0 to 5 °C [6] to the reaction product of CH_3N=C=O and HF in toluene and letting the mixture react at room temperature for 24 h and at 50 to 60 °C for 8 h gave $CH_3C(O)N(CH_3)SN(CH_3)C(O)F$ [5, 6].

References:

[1] Kohn, G. K. (Ger. 2128672 [1970/71]; C.A. **76** [1972] No. 45777).

[2] Böger, M.; Drabek, J. (Ger. 2600987 [1975/76]; C.A. **85** [1976] No. 123638).

[3] Böger, M.; Drabek, J. (Swiss 604500 [1975/78]; C.A. **90** [1979] No. 1703).

[4] Brown, M. S.; Kohn, G. K. (U.S. 3891424 [1970/75]; C.A. **83** [1975] No. 78931), Brown, M. S. (U.S. 3824281 [1970/74]; C.A. **81** [1974] No. 77698).

[5] Drabek, J. (Eur. Appl. 90776 [1983]; C.A. **100** [1984] No. 68158).

[6] D'Silva, T. D. J. (Ger. 2654282 [1976/77]; C.A. **87** [1977] No. 200835).

12.5.1.7.2 $ClSN(CH_3)C(O)CH_2Cl$

The title compound was prepared by adding a mixture of $ClCH_2C(O)NHCH_3$ and pyridine in CH_2Cl_2 to a precooled (−10 °C) solution of freshly distilled SCl_2 in CH_2Cl_2 while holding the temperature between −10 and 0 °C and letting the reaction mixture come to room temperature within about 2 h. A standard workup furnished liquid $ClSN(CH_3)C(O)CH_2Cl$; b.p. 68 °C/1.5 Torr.

Reference:

Kohn, G. K. (Ger. 2128672 [1970/71]; C.A. **76** [1972] No. 45777).

12.5.1.7.3 $ClSN(CH_3)C(O)C_2H_5$

The preparation of the title compound was not described.

Adding a solution of $ClSN(CH_3)C(O)C_2H_5$ in CH_2Cl_2 to a mixture of $CH_3NHC(O)NHC_6H_4F$-2 and pyridine in CH_2Cl_2 gave $C_2H_5C(O)N(CH_3)SN(CH_3)C(O)NHC_6H_4F$-2 [1]. Adding $ClSN(CH_3)C(O)C_2H_5$ and then $(C_2H_5)_3N$ at −10 °C to the reaction product of CH_3N=C=O and HF in toluene and reacting the mixture at room temperature for 24 h and at 50 to 60 °C for 8 h yielded $C_2H_5C(O)N(CH_3)SN(CH_3)C(O)F$ [2].

References:

[1] Brown, M. S.; Kohn, G. K. (U.S. 3891424 [1970/75]; C.A. **83** [1975] No. 78931), Brown, M. S. (U.S. 3824281 [1970/74]; C.A. **81** [1974] No. 77698).

[2] Drabek, J. (Eur. Appl. 90776 [1983]; C.A. **100** [1984] No. 68158).

12.5.1.7.4 $ClSN(CH_2C(O)OC_2H_5)C(O)C_2H_5$

The title compound was prepared by adding pyridine at 0 °C and then $C_2H_5OC(O)CH_2NHC(O)C_2H_5$ at 10 to 20 °C to a solution of SCl_2 (mole ratio 1 : 1 : 1) in CCl_4.

Reacting a solution of $ClSN(CH_2C(O)OC_2H_5)C(O)C_2H_5$ in CCl_4 with "carbofuran", CH_3-NHC(O)OR (R = I), and excess pyridine in CH_2Cl_2 gave $ROC(O)N(CH_3)SN(CH_2C(O)OC_2H_5)C$-$(O)C_2H_5$ with 70% yield.

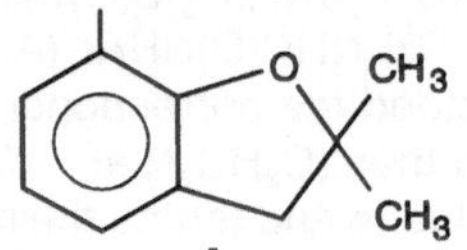

I

Reference:

Goto, T.; Yasudomi, N.; Tanaka, A. K.; Osaki, N.; Takao, H.; Kawata, M.; Imada, J.; Endo, Y.; Umetsu, N. (Nippon Noyaku Gakkaishi **13** [1988] 39/47).

12.5.1.7.5 $ClSN(CH_3)C(O)C_3H_7$-n

The title compound was prepared by adding a solution of SCl_2 in CH_2Cl_2 to a solution of n-$C_3H_7C(O)NHCH_3$ in CH_2Cl_2 while maintaining the temperature at 10 °C, adding a solution of pyridine in CH_2Cl_2 after 30 min, and letting the mixture react for 2 h at room temperature. Standard workup yielded the compound as an oil; b.p. 82 to 84 °C/13 Torr [1].

Adding a solution of $ClSN(CH_3)C(O)C_3H_7$-n in CH_2Cl_2 to a mixture of $CH_3NHC(O)NHC_6H_4R$ (R = 2-F, 3-NO_2, 4-NO_2) and pyridine in CH_2Cl_2 gave the corresponding n-$C_3H_7C(O)N(CH_3)SN$-$(CH_3)C(O)NHC_6H_4R$ [2]. Adding $ClSN(CH_3)C(O)C_3H_7$-n and then $(C_2H_5)_3N$ at −10 °C to the reaction product of $CH_3N{=}C{=}O$ and HF in toluene and reacting the mixture at room temperature for 24 h and then at 50 to 60 °C for 8 h produced n-$C_3H_7C(O)N(CH_3)SN(CH_3)C(O)F$ [3].

A solution of the compound in $(C_2H_5)_2O$ reacted with a mixture of 2-fluoro-4-[2-chloro-4-(trifluoromethyl)phenoxy]aniline and $(C_2H_5)_3N$ in $(C_2H_5)_2O$ to give II [1].

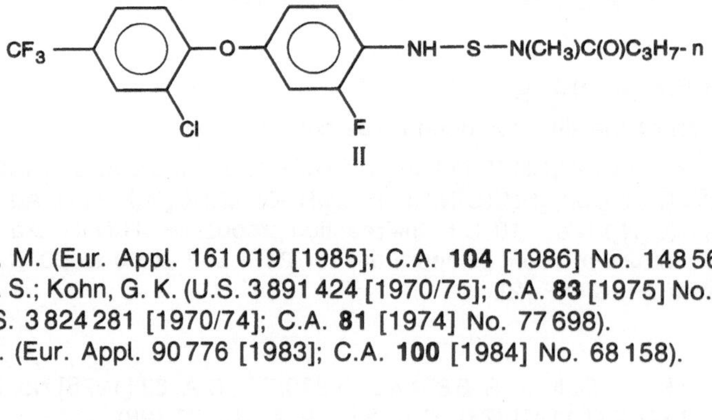

II

References:

[1] Anderson, M. (Eur. Appl. 161019 [1985]; C.A. **104** [1986] No. 148560).
[2] Brown, M. S.; Kohn, G. K. (U.S. 3891424 [1970/75]; C.A. **83** [1975] No. 78941), Brown, M. S. (U.S. 3824281 [1970/74]; C.A. **81** [1974] No. 77698).
[3] Drabek, J. (Eur. Appl. 90776 [1983]; C.A. **100** [1984] No. 68158).

12.5.1.7.6 $ClSN(CH_3)C(O)C_3H_7$-i

The preparation of the title compound was not described.

Adding $ClSN(CH_3)C(O)C_3H_7$-i and then $(C_2H_5)_3N$ at −10 °C to the reaction product of $CH_3N{=}C{=}O$ and HF in toluene and reacting the mixture at room temperature for 24 h and then at 50 to 60 °C for 8 h gave i-$C_3H_7C(O)N(CH_3)SN(CH_3)C(O)F$.

Reference:

Drabek, J. (Eur. Appl. 90776 [1983]; C.A. **100** [1984] No. 68158).

12.5.1.7.7 $ClSN(CH_3)C(O)C_4H_9$-n

The preparation of the title compound was not described.

$ClSN(CH_3)C(O)C_4H_9$-n reacted with a cold solution (0 °C) of 3,5-$(CF_3)_2C_6H_3$–NH–CH-(CCl_3)–C_6H_4Cl-4 and $(C_2H_5)_3N$ to give (3,5-$(CF_3)_2C_6H_3$)4-$ClC_6H_4CH(CCl_3)NSN(CH_3)C(O)C_4H_9$-n.

Reference:

Fahmy, M. A. B. H. (PCT Int. Appl. 9112228 [1991]; C.A. **115** [1991] No. 255782).

12.5.1.7.8 $ClSN(CH_3)C(O)C_6H_{13}$-n

The preparation of the title compound was not described.

$ClSN(CH_3)C(O)C_6H_{13}$-n reacted with a solution of 2-CH_3-4-RC_6H_3N=$CHNHCH_3$ (R = Cl, CH_3) in THF and $(C_2H_5)_3N$ to give the corresponding 2-CH_3-4-RC_6H_3N=$CHN(CH_3)SN(CH_3)C$-$(O)C_6H_{13}$-n.

Reference:

Böger, M.; Drabek, J. (Ger. 2600987 [1975/76]; C.A. **85** [1976] No. 123638).

12.5.1.7.9 $ClSN(CH_3)C(O)C_3H_5$-c

The preparation of the title compound was not described.

$ClSN(CH_3)C(O)C_3H_5$-c reacted with a solution of 2-CH_3-4-RC_6H_3N=$CHNHCH_3$ (R = Cl, CH_3) in THF and $(C_2H_5)_3N$ yielding the corresponding 2-CH_3-4-RC_6H_3N=$CHN(CH_3)SN(CH_3)C(O)$-C_3H_5-c [1]. It reacted with $CH_3NHC(O)ON$=$C(CH_3)SR$ (R = CH_3, C_2H_5) and pyridine in CH_2Cl_2 at 5 to 10 °C for 16 h to give the corresponding c-$C_3H_5C(O)N(CH_3)SN(CH_3)C(O)ON$=$C(CH_3)SR$ [2].

References:

[1] Böger, M.; Drabek, J. (Ger. 2600987 [1975/76]; C.A. **85** [1976] No. 123638).
[2] Drabek, J.; Böger, M. (U.S. 4364959 [1978/82]; C.A. **98** [1983] No. 215201, Ger. 2727614 [1976/78]; C.A. **88** [1978] No. 190124).

12.5.1.7.10 $ClSN(CH_3)C(O)C_6H_{11}$-c

The preparation of the title compound was not described.

Adding a solution of $ClSN(CH_3)C(O)C_6H_{11}$-c in CH_2Cl_2 to a mixture of $CH_3NHC(O)NHC_6H_4R$ (R = 2-F, 4-NO_2) and pyridine in CH_2Cl_2 gave the corresponding c-$C_6H_{11}C(O)N(CH_3)SN(CH_3)C$-$(O)NHC_6H_4R$.

Reference:

Brown, M. S.; Kohn, G. K. (U.S. 3891424 [1970/75]; C.A. **83** [1975] No. 78931, U.S. 3824281 [1970/74]; C.A. **81** [1974] No. 77698).

12.5.1.8 N-Aroyl-N-alkyl-amino-chloro-sulfanes

12.5.1.8.1 $ClSN(CH_3)C(O)Ar$, Ar = C_6H_5, 2-ClC_6H_4, 3-ClC_6H_4, 4-ClC_6H_4, 4-$CH_3OC_6H_4$, 4-$CH_3C_6H_4$, $R^1R^2C_6H_3$ (R^1R^2 = 2,4-Cl_2, 2,6-Cl_2, 3,4-Cl_2, 3-CF_3-6-Cl)

The preparation of the title compounds was not described.

Ar = C_6H_5. The title compound reacted with $CH_3N{=}C{=}O$, anhydrous HF, and $(C_2H_5)_3N$ to give $C_6H_5C(O)N(CH_3)SN(CH_3)C(O)F$ [1] and with $CH_3NHC(O)OR$ (R = I, 3,5-$(CH_3)_2$-4-R′C_6H_2 (R′ = NH_2, $NHCH_2CN$, $N(CH_2C{\equiv}CH)_2$, SCH_3)) and $(C_2H_5)_3N$ in toluene for 18 h at 20 °C and 6 h at 60 °C to give the corresponding $C_6H_5C(O)N(CH_3)SN(CH_3)C(O)OR$ [2].

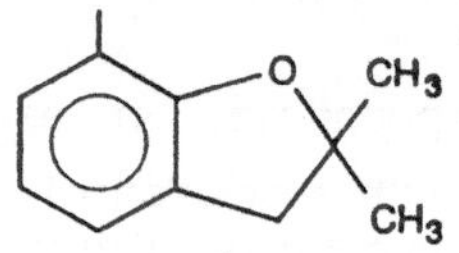

I

Ar = 2-ClC_6H_4. The title compound reacted with $CH_3N{=}C{=}O$, anhydrous HF, and $(C_2H_5)_3N$ to give 2-$ClC_6H_4C(O)N(CH_3)SN(CH_3)C(O)F$ [1] and with $CH_3NHC(O)OR$ (R = I) and $(C_2H_5)_3N$ in toluene for 16 h at 20 °C and 6 h at 60 °C to give 2-$ClC_6H_4C(O)N(CH_3)SN(CH_3)C(O)OR$ [2].

Ar = 3-ClC_6H_4. The title compound was reacted with $CH_3NHC(O)OR$ (R = I) and $(C_2H_5)_3N$ in toluene for 16 h at 20 °C and 6 h at 60 °C to give 3-$ClC_6H_4C(O)N(CH_3)SN(CH_3)C(O)OR$ [2].

Ar = 4-ClC_6H_4. The title compound was reacted with $CH_3N{=}C{=}O$, anhydrous HF, and $(C_2H_5)_3N$ to give 4-$ClC_6H_4C(O)N(CH_3)SN(CH_3)C(O)F$ [1] and with $CH_3NHC(O)OR$ (R = I, C_6H_4R' (R′ = 3-i-C_3H_7, 2-i-C_3H_7O)) and $(C_2H_5)_3N$ in toluene for 16 h at 20 °C and 6 h at 60 °C to give the corresponding 4-$ClC_6H_4C(O)N(CH_3)SN(CH_3)C(O)OR$ [2].

Ar = 4-$CH_3OC_6H_4$. The title compound was reacted with $CH_3NHC(O)OR$ (R = I) and $(C_2H_5)_3N$ in toluene for 16 h at 20 °C and 6 h at 60 °C to give 4-$CH_3OC_6H_4C(O)N(CH_3)SN(CH_3)C(O)OR$ [2].

Ar = 4-$CH_3C_6H_4$. The title compound reacted with $CH_3N{=}C{=}O$, anhydrous HF, and $(C_2H_5)_3N$ to give 4-$CH_3C_6H_4C(O)N(CH_3)SN(CH_3)C(O)F$ [1].

Ar = 2,4-$Cl_2C_6H_3$. The title compound was reacted with $CH_3NHC(O)OR$ (R = I, 2-R′C_6H_4 (R′ = i-C_3H_7O, i-C_3H_7)) and $(C_2H_5)_3N$ in toluene for 16 h at 20 °C and 6 h at 60 °C to give the corresponding 2,4-$Cl_2C_6H_3C(O)N(CH_3)SN(CH_3)C(O)OR$ [2].

Ar = $R^1R^2C_6H_3$ (R^1R^2 = 2,6-Cl_2; 3,4-Cl_2, 3-CF_3-6-Cl). The title compounds were reacted with $CH_3NHC(O)OR$ (R = I) and $(C_2H_5)_3N$ in toluene for 16 h at 20 °C and 6 h at 60 °C to give the corresponding $R^1R^2C_6H_3C(O)N(CH_3)SN(CH_3)C(O)OR$ [2].

References:

[1] Drabek, J.; Böger, M. (Eur. Appl. 49684 [1981/82]; C.A. **97** [1982] No. 91977).
[2] Drabek, J.; Böger, M. (Ger. Offen. 3235109 [1982/86]).

12.5.1.8.2 $ClSN(C_3H_7\text{-i})C(O)C_6H_4CH_3\text{-4}$

The preparation of the title compound was not described.

$ClSN(C_3H_7\text{-i})C(O)C_6H_4CH_3\text{-4}$ reacted with $CH_3NHC(O)OR$ (R = I) and $(C_2H_5)_3N$ in toluene (16 h at 20 °C and 6 h at 60 °C) to give $4\text{-}CH_3C_6H_4C(O)N(C_3H_7\text{-i})SN(CH_3)C(O)OR$.

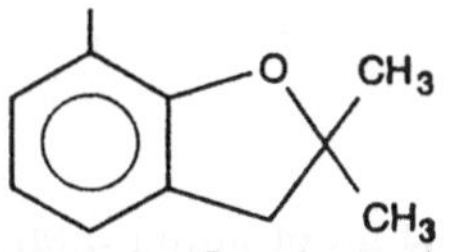

I

Reference:

Drabek, J.; Böger, M. (Ger. Offen. 3235109 [1982/86]).

12.5.1.8.3 $ClSN(R)C(O)C_6H_5$, R = n-C_4H_9, n-C_6H_{13}, n-$C_{10}H_{21}$

The preparation of the title compounds was not described.

$ClSN(R)C(O)C_6H_5$ reacted with $CH_3NHC(O)OR'$ (R' = I) and $(C_2H_5)_3N$ in toluene (16 h at 20 °C and 6 h at 60 °C) to give $C_6H_5C(O)N(R)SN(CH_3)C(O)OR'$.

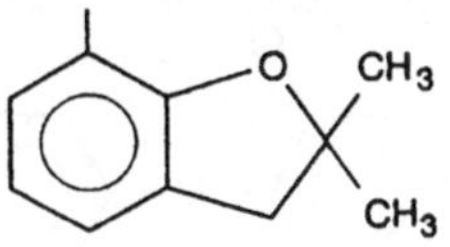

I

Reference:

Drabek, J.; Böger, M. (Ger. Offen. 3235109 [1982/86]).

12.5.1.9 N-Thiobenzoyl-N-methyl-amino-chloro-sulfane, $ClSN(CH_3)C(S)C_6H_5$

The title compound was probably formed by the reaction of $C_6H_5C(S)NHCH_3$ with SCl_2 (excess) in CCl_4 or benzene at 20 to 25 °C [1, 2], but structure I was favored for the substance obtained [1].

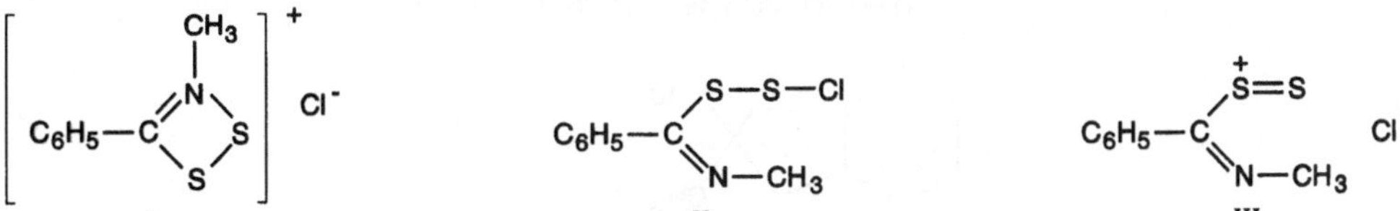

I II III

In a later reexamination [2], structures II and III were also suggested, but could not be confirmed.

References:

[1] El'tsov, A. V.; Lopatin, V. E.; Mikhel'son, M. G. (Zh. Org. Khim. **6** [1970] 402; J. Org. Chem. USSR [Engl. Transl.] **6** [1970] 394).

[2] El'tsov, A. V.; Lopatin, V. E. (Zh. Org. Khim. **7** [1971] 1279/84; J. Org. Chem. USSR [Engl. Transl.] **7** [1971] 1319/23).

12.5.1.10 N-Organyloxycarbonyl-N-alkyl-amino-chloro-sulfanes

12.5.1.10.1 $ClSN(CH_3)C(O)OCH_3$

The title compound was prepared by adding SCl_2 and then pyridine at 25 to 30 °C to a solution of $HN(CH_3)C(O)OCH_3$ (mole ratio 1.3 : 1.1 : 1) in CH_2Cl_2 [1], see also [2, 3].

Liquid, b.p. 68 to 70 °C/9 Torr; $n_D^{25} = 1.5038$ [3].

The ^1H NMR spectrum of $ClSN(CH_3)C(O)OCH_3$ (not isolated) showed a sharp singlet of the N–CH_3 group at 3.45 ppm (in CH_2Cl_2, relative to TMS) [1, 2].

The title compound reacted with solutions of the formamidines $ArN{=}CHNHCH_3$ (Ar = 2,3-$Cl_2C_6H_3$, 2,6-$Cl_2C_6H_3$, 3,5-$Cl_2C_6H_3$, 2-CH_3-4-ClC_6H_3, 2,4-$(CH_3)_2C_6H_3$ [4, 5], 2,4,5-$Cl_3C_6H_2$ [5]) in pyridine at 0 to 10 °C for 20 min [4] or in THF and $(C_2H_5)_3N$ [5] to give the corresponding $ArN{=}CHN(CH_3)SN(CH_3)C(O)OCH_3$ [4, 5]. A solution of $ClSN(CH_3)C(O)OCH_3$ reacted with the amine I in the presence of $(C_2H_5)_3N$ to give the compound II [6].

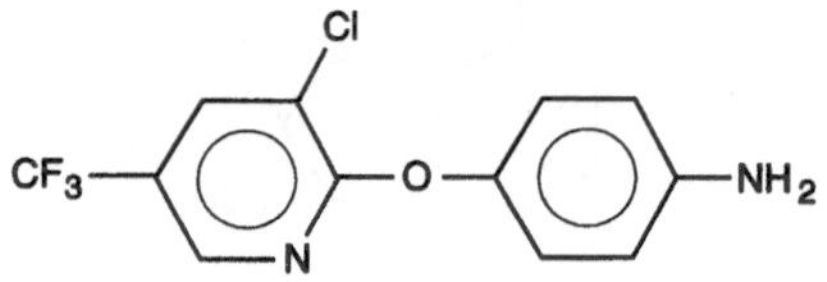

I

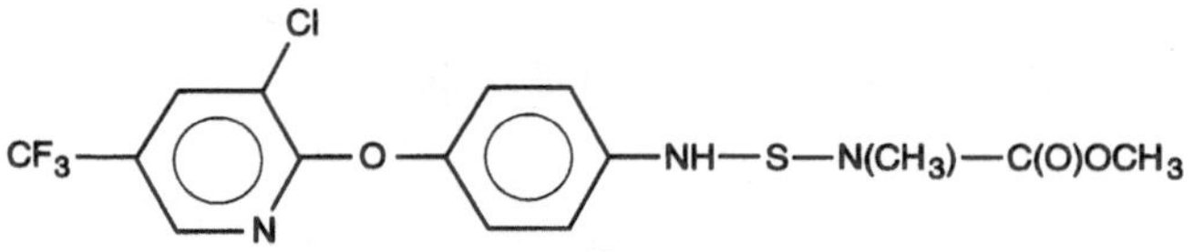

II

$ClSN(CH_3)C(O)OCH_3$ was reacted with 3-i-$C_3H_7C_6H_4OC(O)NHCH_3$ and pyridine at 5 °C and at room temperature overnight to give 3-i-$C_3H_7C_6H_4OC(O)N(CH_3)SN(CH_3)C(O)OCH_3$ [3]. With "carbofuran" (2,3-dihydro-2,2-dimethyl-benzofuran-7-yl-methyl carbamate) analogously [3] or in CH_2Cl_2 at 5 to 10 °C and at room temperature [7] III was obtained after 12 h [3, 7].

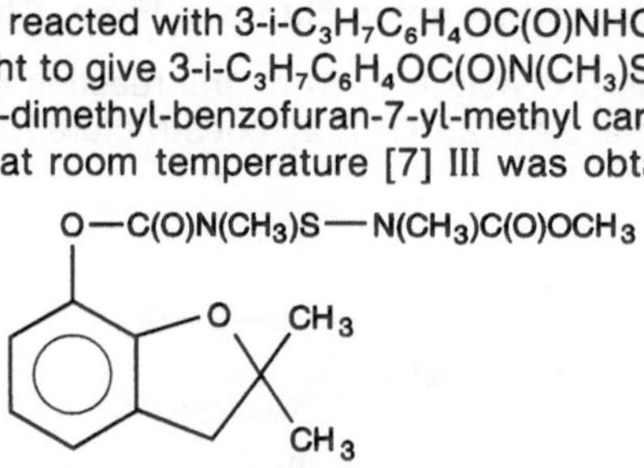

III

The title compound was reacted with $CH_3NHC(O)ON{=}C(R)SCH_3$ (R = CH_3 [3, 8], $(CH_3)_2NC(O)$ [3]) and pyridine at 5 °C and at room temperature overnight [3] or in CH_2Cl_2 at 5 to 10 °C for 16 h [8] to give the corresponding $CH_3OC(O)N(CH_3)SN(CH_3)C(O)ON{=}C(R)SCH_3$ [3, 8].

Adding $ClSN(CH_3)C(O)OCH_3$ to a mixture of $(CH_3)_2NC(S)NHC_6H_3CH_3$-2-Cl-4 and $(C_2H_5)_3N$ in CH_2Cl_2 at 0 to 10 °C and letting the mixture react for 2 h at room temperature gave 2-CH_3-4-$ClC_6H_3N{=}C(N(CH_3)_2)SSN(CH_3)C(O)OCH_3$ [9]. Adding C_6H_5SH and pyridine at 0 °C to a solution of $ClSN(CH_3)C(O)OCH_3$ in CH_2Cl_2 gave $CH_3OC(O)N(CH_3)SSC_6H_5$ [1]; see also [2].

References:

[1] Brown, M. S. (U.S. 3914259 [1971/75]; C.A. **84** [1976] No. 30721), Brown, M. S.; Kohn, G. K. (U.S. 3843689 [1972/74]).
[2] Cleveland, J. D. (U.S. 3843688 [1972/74]; C.A. **82** [1975] No. 43175).
[3] Fahmy, M. A. H.; Mallipudi, N. M.; Fukuto, T. R. (J. Agric. Food Chem. **26** [1978] 550/7).
[4] Böger, M.; Drabek, J. (Swiss 604500 [1975/78]; C.A. **90** [1979] No. 1703).
[5] Böger, M.; Drabek, J. (Ger. 2600987 [1975/76]; C.A. **85** [1976] No. 123638).
[6] Nanjo, K.; Kariya, A.; Katsurayama, T.; Tsuji, A. (Jpn. Kokai Tokkyo Koho 63-091367 [1986/88]; C.A. **109** [1988] No. 210904).
[7] Drabek, J.; Böger, M. (Ger. 2812622 [1977/78]; C.A. **90** [1979] No. 23028).
[8] Drabek, J.; Böger, M. (U.S. 4364959 [1978/82]; C.A. **98** [1983] No. 215201, Ger. 2727614 [1976/78]; C.A. **88** [1978] No. 190124).
[9] Böger, M.; Drabek, J. (Ger. 2654080 [1972/77]; C.A. **87** [1977] No. 102078).

12.5.1.10.2 $ClSN(CH_3)C(O)OC_2H_5$

The title compound was formed by adding $C_2H_5OC(O)NHCH_3$ and $(C_2H_5)_3N$ in CH_2Cl_2 to a to 0 °C cooled solution of SCl_2 (mole ratio 1 : 1 : 1.4) in CH_2Cl_2 and stirring the mixture for 1 h at 0 to 5 °C [1]. Instead of $(C_2H_5)_3N$, pyridine could be used as acid acceptor [2].

Amber-colored oil [1], b.p. 54 °C/2.5 Torr; $n_D^{25} = 1.4782$ [2].

$ClSN(CH_3)C(O)OC_2H_5$ was reacted with a solution of the formamidines $ArN{=}CHNHCH_3$ (Ar = 2-CH_3-4-$R'C_6H_3$ (R' = Cl, Br, CH_3), 2,4,5-$Cl_3C_6H_2$) in pyridine at 0 to 10 °C for 20 min [3] or in THF in the presence of $(C_2H_5)_3N$ [4] to give the corresponding $ArN{=}CHN(CH_3)SN(CH_3)$-$C(O)OC_2H_5$ [3, 4]. The title compound was reacted with "carbofuran" and pyridine at 5 °C and at room temperature overnight to give I.

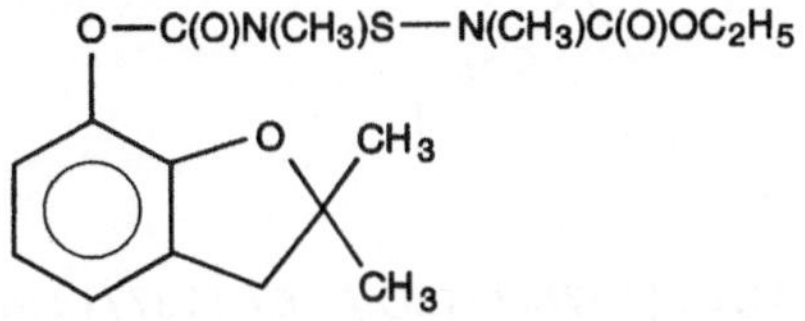

I

With $CH_3NHC(O)ON{=}C(R)SCH_3$ (R = CH_3, $(CH_3)_2NC(O)$) and pyridine analogously the corresponding $C_2H_5OC(O)N(CH_3)SN(CH_3)C(O)ON{=}C(R)SCH_3$ were obtained [2]. The title compound reacted with $CH_3N{=}C{=}O$ and HF in the presence of $(C_2H_5)_3N$ in CH_2Cl_2 at 0 to 5 °C to give $C_2H_5OC(O)N(CH_3)SN(CH_3)C(O)F$ after letting the mixture to come to room temperature [1].

References:

[1] Liang, W. C. (Ger. 2654313 [1976/77]; C.A. **87** [1977] No. 200836).
[2] Fahmy, M. A. H.; Mallipudi, M. N.; Fukuto, T. R. (J. Agric. Food Chem. **26** [1978] 550/7).
[3] Böger, M.; Drabek, J. (Swiss 604500 [1975/78]; C.A. **90** [1979] No. 1703).
[4] Böger, M.; Drabek, J. (Ger. 2600987 [1975/76]; C.A. **85** [1976] No. 123638).

12.5.1.10.3 $ClSN(CH_3)C(O)OCH_2C_6H_5$

The title compound was formed as an oil by adding a solution of $C_6H_5CH_2OC(O)NHCH_3$ and $(C_2H_5)_3N$ in CH_2Cl_2 to a to 0 °C cooled solution of SCl_2 (mole ratio 1 : 1.1 : 1.4) in CH_2Cl_2 at −6 to 5 °C and reacting the mixture for 1 h at 0 °C.

$ClSN(CH_3)C(O)OCH_2C_6H_5$ reacted with $CH_3N{=}C{=}O$ and HF in the presence of $(C_2H_5)_3N$ in CH_2Cl_2 at 0 °C to give 42% $C_6H_5CH_2OC(O)N(CH_3)SN(CH_3)C(O)F$ after letting the mixture to come to room temperature.

Reference:

Liang, W. C. (Ger. 2654313 [1976/77]; C.A. **87** [1977] No. 200836).

12.5.1.10.4 $ClSN(CH_3)C(O)OCH_2CH_2OCH_3$

The title compound was prepared by adding a solution of $CH_3OCH_2CH_2OC(O)NHCH_3$ and $(C_2H_5)_3N$ in CH_2Cl_2 to a to 0 °C cooled solution of SCl_2 (mole ratio 1 : 1.1 : 1.4) in CH_2Cl_2 at −10 to 4 °C and stirring the mixture at 5 °C for 1 h [1].

$ClSN(CH_3)C(O)OCH_2CH_2OCH_3$ reacted with a solution of 2-CH_3-4-$RC_6H_3N{=}CHNHCH_3$ (R = Cl, CH_3) in THF and $(C_2H_5)_3N$ to give the corresponding 2-CH_3-4-$RC_6H_3N{=}CHN(CH_3)SN$-$(CH_3)C(O)OCH_2CH_2OCH_3$ [2]. It reacted with $CH_3N{=}C{=}O$ and HF in the presence of $(C_2H_5)_3N$ in CH_2Cl_2 at 0 °C for 1 h to give 61% $CH_3OCH_2CH_2OC(O)N(CH_3)SN(CH_3)C(O)F$ after letting the mixture to come to room temperature [1]. Adding the title compound to a mixture of 2,4-$(CH_3)_2C_6H_3NHC(S)NHCH_3$ and $(C_2H_5)_3N$ in CH_2Cl_2 at 0 to 10 °C and reacting the mixture for 2 h at room temperature gave 2,4-$(CH_3)_2C_6H_3N{=}C(NHCH_3)SSN(CH_3)C(O)OCH_2CH_2OCH_3$ [3]. The title compound reacted with "carbofuran" and pyridine in CH_2Cl_2 over 12 h at room temperature to give I [4].

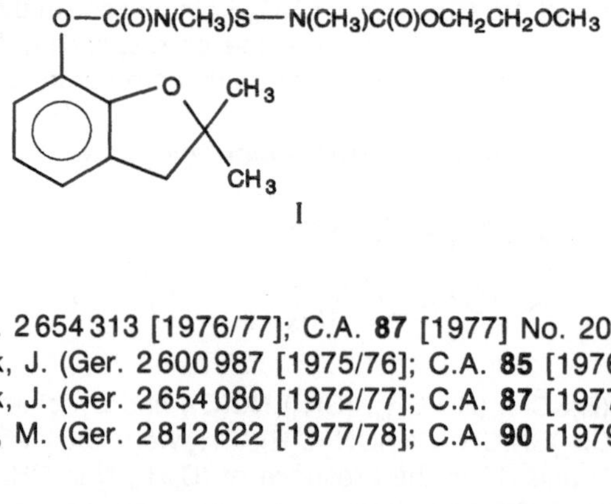

I

References:

[1] Liang, W. C. (Ger. 2654313 [1976/77]; C.A. **87** [1977] No. 200836).
[2] Böger, M.; Drabek, J. (Ger. 2600987 [1975/76]; C.A. **85** [1976] No. 123638).
[3] Böger, M.; Drabek, J. (Ger. 2654080 [1972/77]; C.A. **87** [1977] No. 102078).
[4] Drabek, J.; Böger, M. (Ger. 2812622 [1977/78]; C.A. **90** [1979] No. 23028).

12.5.1.10.5 $ClSN(CH_3)C(O)OCH_2CH_2OCH_2CH_2OCH_3$

The title compound was formed by adding a solution of $CH_3OCH_2CH_2OCH_2CH_2OC(O)$-$NHCH_3$ and $(C_2H_5)_3N$ in CH_2Cl_2 to a to 0 °C cooled solution of SCl_2 (mole ratio 1 : 1.1 : 1.5) in CH_2Cl_2 at −2 to 4 °C and stirring the mixture at 5 °C for 1 h.

The title compound reacted with $CH_3N{=}C{=}O$ and HF in the presence of $(C_2H_5)_3N$ in CH_2Cl_2 at 0 °C within 1 h to give 90% $CH_3OCH_2CH_2OCH_2CH_2OC(O)N(CH_3)SN(CH_3)C(O)F$ after letting the mixture to come to room temperature.

Reference:

Liang, W. C. (Ger. 2654313 [1976/77]; C.A. **87** [1977] No. 200836).

12.5.1.10.6 $ClSN(CH_3)C(O)OC_3H_7$-x, x = n, i

The title compounds were formed from the reaction of x-$C_3H_7OC(O)NHCH_3$ (x = n, i) with SCl_2 in CH_2Cl_2 using pyridine as acid acceptor [1].

$ClSN(CH_3)C(O)OC_3H_7$-n. Liquid, b.p. 60 °C/1.4 Torr; $n_D^{25} = 1.4821$ [1].

$ClSN(CH_3)C(O)OC_3H_7$-i. Liquid, b.p. 64 °C/5 Torr; $n_D^{25} = 1.4738$ [1].

The title compounds reacted with $CH_3SC(R){=}NOC(O)NHCH_3$ (x = n, i; R = CH_3; x = i, R = $(CH_3)_2NC(O)$), 3-i-$C_3H_7C_6H_4OC(O)NHCH_3$, or "carbofuran" and pyridine at 5 °C and at room temperature overnight to give the corresponding $CH_3SC(R){=}NOC(O)N(CH_3)SN(CH_3)C(O)$-$OC_3H_7$-x, 3-i-$C_3H_7C_6H_4OC(O)N(CH_3)SN(CH_3)C(O)OC_3H_7$-x, or I, respectively [1].

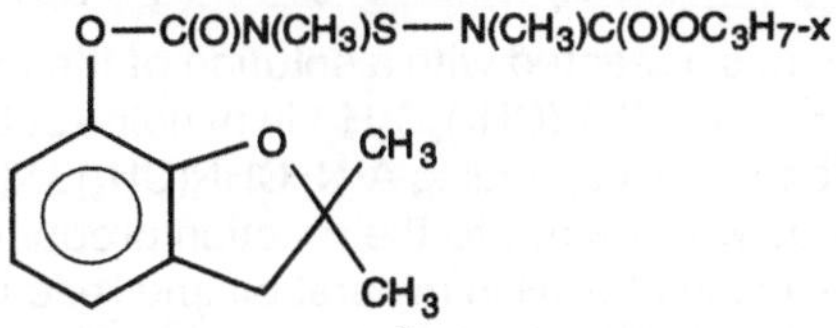

I

Reacting a solution of the compound with x = n in $(C_2H_5)_2O$ [2] or CH_2Cl_2 [3] with 2-R-4-$CF_3C_6H_3OC_6H_4NH_2$ (R = F [2], Cl [3]) and $(C_2H_5)_3N$, in $(C_2H_5)_2O$ at 15 to 20°C [2] or in CH_2Cl_2 at 0 to 5 °C and allowing the mixture to stand for 2 h at ambient temperature [3] gave the corresponding 2-R-4-$CF_3C_6H_3OC_6H_4NHSN(CH_3)C(O)OC_3H_7$-n [2, 3].

References:

[1] Fahmy, M. A. H.; Mallipudi, N. M.; Fukuto, T. R. (J. Agric. Food Chem. **26** [1978] 550/7).
[2] Anderson, M. (Eur. Appl. 161019 [1985]; C.A. **104** [1986] No. 148560).
[3] Anderson, M. (U.S. 4623658 [1986]; C.A. **107** [1987] No. 23100).

12.5.1.10.7 $ClSN(CH_3)C(O)OCH(CH_3)CH_2OCH_3$

The title compound was prepared by adding a solution of $CH_3OCH_2CH(CH_3)OC(O)NHCH_3$ and $(C_2H_5)_3N$ in CH_2Cl_2 to a to 0 °C cooled solution of SCl_2 (mole ratio 1 : 1.1 : 1.5) at −8 to 0 °C and stirring the mixture at 0 to 5 °C for 1 h.

The title compound reacted with $CH_3N{=}C{=}O$ and HF in the presence of $(C_2H_5)_3N$ in CH_2Cl_2 at 0 °C within 1 h to give $CH_3OCH_2CH(CH_3)OC(O)N(CH_3)SN(CH_3)C(O)F$ after letting the mixture to come to room temperature.

Reference:

Liang, W. C. (Ger. 2654313 [1976/77]; C.A. **87** [1977] No. 200836).

12.5.1.10.8 **$ClSN(CH_3)C(O)OC_4H_9$-x**, x = n, s, i, t

The title compounds with x = n, s, t were prepared by reaction of x-$C_4H_9OC(O)NHCH_3$ (x = n, s, t) with SCl_2 in CH_2Cl_2 using pyridine (x = n [1], x = s [2]) or $(C_2H_5)_3N$ (mole ratio 1 : 1.1 : 1.5, x = t [3]) as acid acceptor. The preparation of $ClSN(CH_3)C(O)OC_4H_9$-i was not described.

$ClSN(CH_3)C(O)OC_4H_9$-n was also formed by adding at 30 °C n-$C_4H_9OC(O)NHCH_3$ in controlled amounts from a feed vessel to SCl_2 (mole ratio 1 : 1.8). After completion of the addition, the internal temperature was raised stepwise to 40 °C and held there for 3 h. Then excess SCl_2 was distilled off and the residue was purified by distillation under high vacuum to give the product in a yield of 85% [4], see also [5]. It was formed by adding SCl_2 (?) to n-$C_4H_9OC(O)NHCH_3$ at 45 to 50 °C for 1 h, and at 60 °C for ~2 h, then treating the reaction mixture with Cl_2 at 5 to 10 °C for ~1.7 h followed by bubbling N_2 through the reaction mixture at 50 °C for 2 h; 93.3% yield of 96.6% purity [6], see also [7]; without Cl_2 the yield was 20.5%, purity 22.1% [7].

$ClSN(CH_3)C(O)OC_4H_9$-n. Liquid, b.p. 74 to 76 °C/0.5 bar [4].

$ClSN(CH_3)C(O)OC_4H_9$-s. The compound was not isolated. The ^1H NMR (CH_2Cl_2/TMS) showed a singlet for the N–CH_3 group at $\delta = 3.4$ ppm [2].

$ClSN(CH_3)C(O)OC_4H_9$-i, $ClSN(CH_3)C(O)OC_4H_9$-t. No properties reported.

The compounds with x = n, s, i reacted with a solution of the formamidines ArN=CHNHCH$_3$ (Ar = 3,5-$Cl_2C_6H_3$, 2-CH_3-4-ClC_6H_3, 2,4-$(CH_3)_2C_6H_3$) in pyridine at 0 to 10 °C for 20 min [8], or in THF and $(C_2H_5)_3N$ [9] to give the corresponding ArN=CHN(CH_3)SN(CH_3)C(O)OC$_4$H$_9$-x (x = n, s, i) [8, 9]. Adding the compounds with x = n, t to the reaction product of $C_6H_5C(O)NHN(C_4H_9$-t)C-(O)C_6H_5 in THF with a suspension of NaH in mineral oil and reacting the mixture at room temperature for 5 h gave $C_6H_5C(O)N(C_4H_9$-t)N(C(O)C_6H_5)SN(CH_3)C(O)OC$_4$H$_9$-x (x = n, t) [10]. $ClSN(CH_3)C(O)OC_4H_9$-n reacted with 3-i-$C_3H_7C_6H_4OC(O)NHCH_3$ and pyridine at 5 °C and at room temperature overnight to give 3-i-$C_3H_7C_6H_4OC(O)N(CH_3)SN(CH_3)C(O)OC_4H_9$-n [1]. It reacted with 4-$FC_6H_4OC_6H_4OCH_2CH_2NHC(O)OC_2H_5$ in pyridine at 5 to 10 °C and another 16 h at room temperature to give 4-$FC_6H_4OC_6H_4OCH_2CH_2N(C(O)OC_2H_5)SN(CH_3)C(O)OC_4H_9$-n [11]. It reacted with 2,3-dihydro-2,2-dimethyl-benzofuran-7-yl-methyl carbamate ("carbofuran") in CH_2Cl_2 and pyridine at 5 to 10 °C and additionally 12 h at room temperature to give I [12], see also [1, 6].

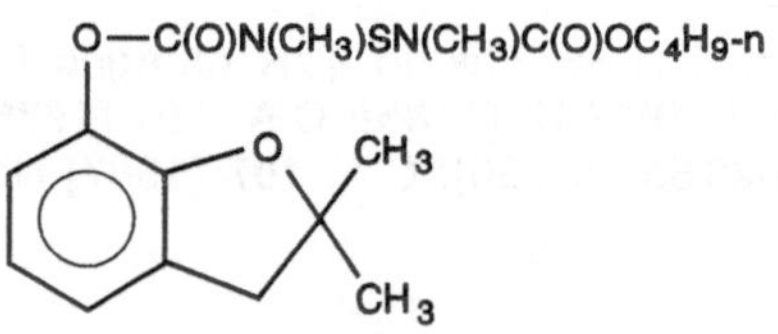

I

Adding $ClSN(CH_3)C(O)OC_4H_9$-n to a mixture of 3,5-$(CF_3)_2C_6H_3NHC(O)N(CH_3)_2$ and $(C_2H_5)_3N$ in CH_2Cl_2 at 0 to 10 °C and reacting the mixture for 2 h at room temperature gave 3,5-$(CF_3)_2C_6H_3N$=C(N(CH_3)$_2$)SSN(CH_3)C(O)OC$_4$H$_9$-n [13]. Adding $ClSN(CH_3)C(O)OC_4H_9$-n and then pyridine at 5 to 10 °C to a solution of $CH_3NHC(O)ON$=C(CH_3)SR (R = CH_3, C_2H_5) in CH_2Cl_2 and stirring the mixture for 16 h at 5 to 10 °C gave the corresponding n-$C_4H_9OC(O)$-N(CH_3)SN(CH_3)C(O)ON=C(CH_3)SR [14]. $ClSN(CH_3)C(O)OC_4H_9$-n reacts with II and $(C_2H_5)_3N$ to give III [15].

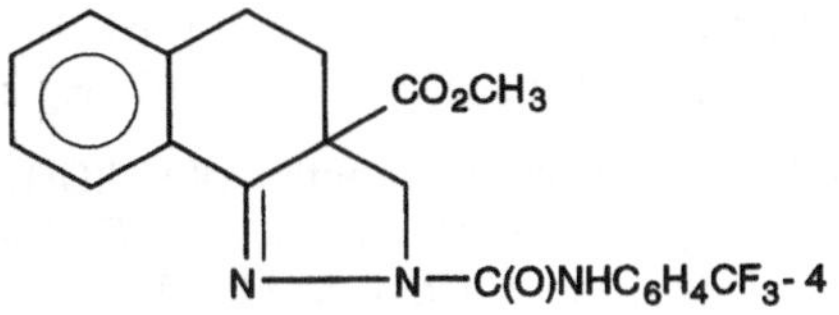

II

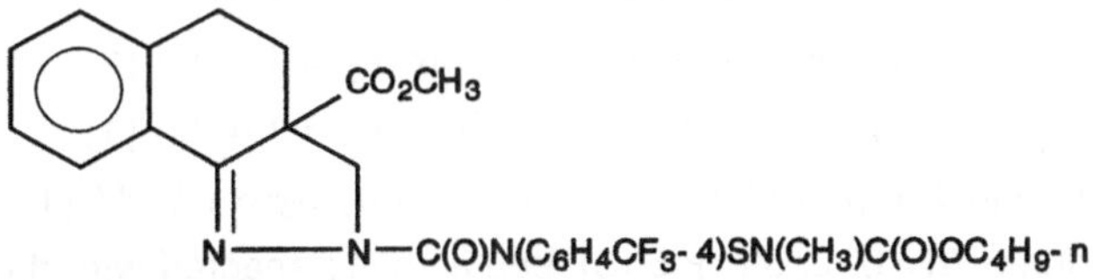

III

Adding $ClSN(CH_3)C(O)OC_4H_9$-n to a solution of 2-thiabicyclo[2.2.2]octan-3-one oxime (IV) and pyridine (mole ratio ca. 1 : 1 : 1) in CH_2Cl_2 at 0 °C overnight and allowing the mixture to come to room temperature gave the 2-thiabicyclo[2.2.2]octan-3-one oxime V [16].

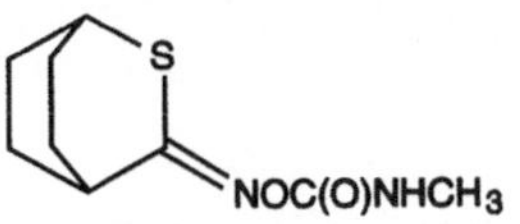

IV

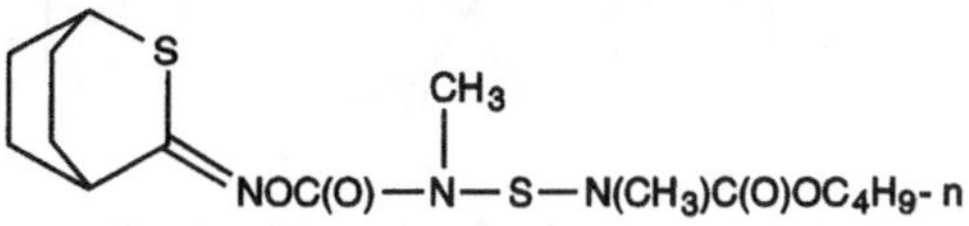

V

$ClSN(CH_3)C(O)OC_4H_9$-t reacted with $CH_3N{=}C{=}O$ and HF in the presence of $(C_2H_5)_3N$ in CH_2Cl_2 at 0 °C to give t-$C_4H_9OC(O)N(CH_3)SN(CH_3)C(O)F$ after letting the mixture to come to room temperature [3].

References:

[1] Fahmy, M. A. H.; Mallipudi, N. M.; Fukuto, T. R. (J. Agric. Food Chem. **26** [1978] 550/7).
[2] Cleveland, J. D. (U.S. 3843688 [1972/74]; C.A. **82** [1975] No. 43175).
[3] Liang, W. C. (Ger. 2654313 [1976/77]; C.A. **87** [1977] No. 200836).
[4] Fäh, H. (U.S. 4587353 [1983/86]; C.A. **105** [1986] No. 60311, Br. 2161802 [1984/86]; C.A. **105** [1986] No. 152554).
[5] Hansuya, K. F. (Jpn. 61-134368 [1984/86]; C.A. **106** [1987] No. 119288).
[6] Koja, I.; Nakao, T.; Sakaguchi, H. (Jpn. 01-180872 [1988/89]; C.A. **112** [1990] No. 7052).
[7] Koja, I.; Nakao, T.; Sakaguchi, H. (Jpn. Kokai Tokkyo Koho 02-184666 [1989/90]; C.A. **113** [1990] No. 211393).
[8] Böger, M.; Drabek, J. (Swiss 604500 [1975/78]; C.A. **90** [1979] No. 1703).
[9] Böger, M.; Drabek, J. (Ger. 2600987 [1975/76]; C.A. **85** [1976] No. 123638).
[10] Drabek, J. (Eur. Appl. 395581 [1990]; C.A. **114** [1991] No. 185026).

[11] Karrer, F. (Ger. 4010325 [1990]; C.A. **115** [1991] No. 28914).
[12] Drabek, J.; Böger, M. (Ger. 2812622 [1977/78]; C.A. **90** [1979] No. 23028).
[13] Böger, M.; Drabek, J. (Ger. 2654080 [1972/77]; C.A. **87** [1977] No. 102078).
[14] Drabek, J.; Böger, M. (U.S. 4364959 [1978/82]; C.A. **98** [1983] No. 215201).
[15] Fahmy, M. A. H.; Harrison, C. R.; Lahm, G. P.; Stevenson, T. M. (PCT Int. Appl. 9003369 [1989/90]; C.A. **113** [1990] No. 172010).
[16] Magee, T. A.; Battershell, R. D.; Limpel, L. E.; Ho, A. W.-W.; Friedman, A. J.; Corkins, H. G.; Brand, W. W.; Buchman, R.; Storace, L. (Eur. Appl. 51990 [1981/82]; C.A. **98** [1983] No. 89174, U.S. 4424213 [1981/84]; C.A. **100** [1984] No. 138983).

12.5.1.10.9 $ClSN(CH_3)C(O)OC_{5+n}H_{11+2n}$, n = 0 ($C_5H_{11}$-n), n = 2 ($C_7H_{15}$-n), n = 3 ($C_8H_{17}$-n, $CH_2CHC_2H_5C_4H_9$-n), n = 5 ($C_{10}H_{21}$-n), n = 7 ($C_{12}H_{25}$-n)

The title compounds were formed from the reaction of $HN(CH_3)C(O)OC_{5+n}H_{11+2n}$ (n = 0, 2, 3, 5, 7) with SCl_2 in CH_2Cl_2 using pyridine (C_5H_{11}-n, C_7H_{15}-n, C_8H_{17}-n, $C_{10}H_{21}$-n [1]) or $(C_2H_5)_3N$ ($CH_2CHC_2H_5C_4H_9$-n, $C_{12}H_{25}$-n [2]) as acid acceptor [1, 2].

n = 0 (C_5H_{11}-n). Liquid, b.p. 76 to 78 °C/1 Torr; n_D^{25} = 1.4780 [1].

n = 2 (C_7H_{15}-n). Liquid, b.p. 85 to 86 °C/0.15 Torr; n_D^{25} = 1.4752 [1].

n = 3 (C_8H_{17}-n). Liquid, b.p. 92 to 95 °C/0.05 Torr; n_D^{25} = 1.4750 [1].

n = 5 ($C_{10}H_{21}$-n). Liquid, b.p. 124 to 126 °C/0.05 Torr; n_D^{25} = 1.4737 [1].

The title compounds (n = 0, 2, 3 [1], 5 [3], see also [1]) reacted with the carbofuran I and pyridine at 5 °C and room temperature overnight to give the corresponding compounds II [1].

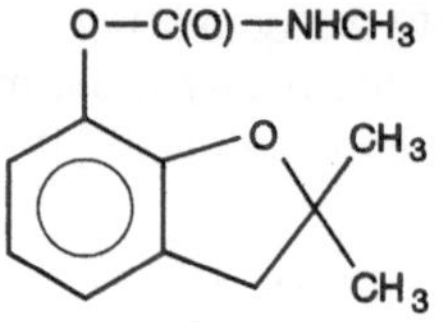

I

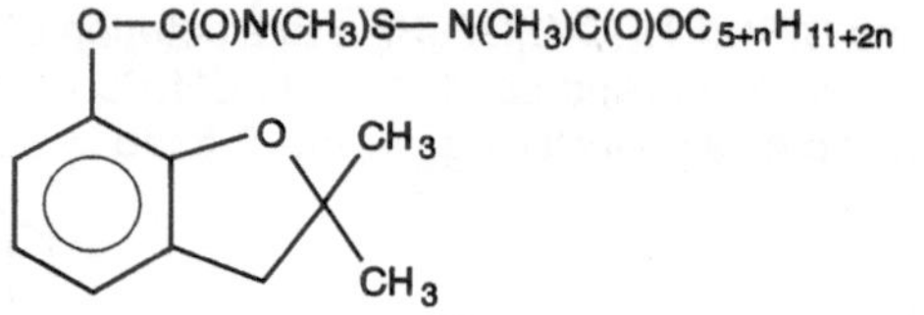

II

They reacted with $ArOC(O)NHCH_3$ (n = 0, 2, 3 (C_5H_{11}-n, C_7H_{15}-n, C_8H_{17}-n): Ar = 2-$(CH_3)_2$-$CHOC_6H_4$; n = 2 (C_7H_{15}-n): Ar = 3-$(CH_3)_2CHC_6H_4$; n = 3 (C_8H_{17}-n): Ar = 1-naphthyl (1-$C_{10}H_7$)) under similar conditions to give the corresponding compounds $ArOC(O)N(CH_3)SN(CH_3)C(O)O$-$C_{5+n}H_{11+2n}$ [1]. The title compounds (n = 3, 5 (C_8H_{17}-n, $C_{10}H_{21}$-n)) reacted with $ArN{=}CHNHCH_3$ (Ar = 2-CH_3-4-ClC_6H_3, 2,4-$(CH_3)_2C_6H_3$) in THF to give $ArN{=}CHN(CH_3)SN(CH_3)C(O)OC_{5+n}H_{11+2n}$ [4]. $ClSN(CH_3)C(O)OC_8H_{17}$-n reacted with $ArO(CH_2)_2NHC(O)OC_2H_5$ (Ar = 4-$C_6H_5OC_6H_4$) [5] and $ArNHC(S)NHCH_3$ (Ar = 2,4-$(CH_3)_2C_6H_3$, 3,5-$(CF_3)_2C_6H_3$) [6] to give the corresponding $ArO(CH_2)_2N(C(O)OC_2H_5)SN(CH_3)C(O)OC_8H_{17}$-n and $ArN{=}C(NHCH_3)SSN(CH_3)C(O)OC_8H_{17}$-n, respectively. The title compounds (n = 3 ($CH_2CHC_2H_5C_4H_9$-n); n = 7 ($C_{12}H_{25}$-n)) reacted with $CH_3N{=}C{=}O$ and HF in the presence of $(C_2H_5)_3N$ in CH_2Cl_2 at 0 °C within 1 h to give $C_{5+n}H_{11+2n}O$-$C(O)N(CH_3)SN(CH_3)C(O)F$ after letting the mixture to come to room temperature [2].

References:

[1] Fahmy, M. A. H.; Mallipudi, N. M.; Fukuto, T. R. (J. Agric. Food Chem. **26** [1978] 550/7).
[2] Liang, W. C. (Ger. 2654313 [1976/77]; C.A. **87** [1977] No. 200836).
[3] Drabek, J.; Böger, M. (Ger. 2812622 [1977/78]; C.A. **90** [1979] No. 23028).
[4] Böger, M.; Drabek, J. (Ger. 2600987 [1975/76]; C.A. **85** [1976] No. 123638).
[5] Innofinance Altalanos Innovacios Pénzintézet (Neth. 8700074 [1987/88]; C.A. **110** [1989] No. 187820, Belg. 1000110 [1987/88]; C.A. **109** [1988] No. 68867).
[6] Böger, M.; Drabek, J. (Ger. 2654080 [1972/77]; C.A. **87** [1977] No. 102078).

12.5.1.10.10 $ClSN(CH_3)C(O)OC_6H_{11}$-c

The preparation of the title compound was not described.

$ClSN(CH_3)C(O)OC_6H_{11}$-c reacted with a solution of formamidines $ArN{=}CHNHCH_3$ (Ar = 2-CH_3-4-ClC_6H_3, 2,4-$(CH_3)_2C_6H_3$, 2,4,5-$Cl_3C_6H_2$) in THF and $(C_2H_5)_3N$ to give the corresponding $ArN{=}CHN(CH_3)SN(CH_3)C(O)OC_6H_{11}$-c [1]. It reacted with 2,3-dihydro-2,2-dimethyl-benzofuran-7-yl-methyl carbamate (carbofuran) in CH_2Cl_2 and pyridine at 5 to 10 °C and over 12 h at room temperature to give I [2].

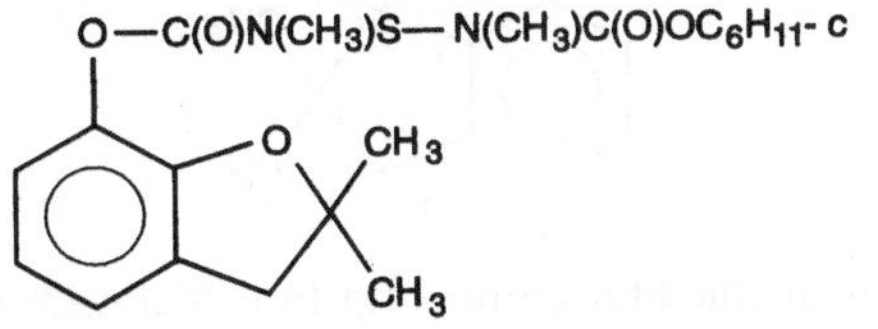

I

References:

[1] Böger, M.; Drabek, J. (Ger. 2600987 [1975/76]; C.A. **85** [1976] No. 123638).
[2] Drabek, J.; Böger, M. (Ger. 2812622 [1977/78]; C.A. **90** [1979] No. 23028).

12.5.1.10.11 ClSN(CH$_3$)C(O)OAr, Ar = $C_6H_4(C_4H_9$-s)-3, 1-naphthyl ($C_{10}H_7$-1)

The title compounds were formed by adding SCl_2 and then pyridine at 25 to 30 °C to a solution of $CH_3NHC(O)OAr$ (Ar = $C_6H_4(C_4H_9$-s)-3, 1-naphthyl ($C_{10}H_7$-1)) (mole ratio 1.1 : 1.2 : 1) [1, 2].

The ^{1}H NMR of the title compounds (not isolated) in CH_2Cl_2 showed a sharp singlet at 3.4 and 3.6 ppm, respectively, (relative to TMS) for the N–CH_3 group [1, 2].

Adding a solution of RSH (R = CH_3, C_6H_5) and pyridine in CH_2Cl_2 to a solution of ClSN(CH$_3$)C(O)OC$_6$H$_4$(C$_4$H$_9$-s)-3 in CH_2Cl_2 at 0 °C gave RSSN(CH$_3$)C(O)OC$_6$H$_4$(C$_4$H$_9$-s)-3 [1, 2], see also [3].

References:

[1] Brown, M. S. (U.S. 3914259 [1971/75]; C.A. **84** [1976] No. 30721).
[2] Brown, M. S.; Kohn, G. K. (U.S. 3843689 [1972/74]).
[3] Cleveland, J. D. (U.S. 3843688 [1972/74]; C.A. **82** [1975] No. 43175).

12.5.1.10.12 ClSN(CH$_3$)C(O)OC$_{10}$H$_{11}$O

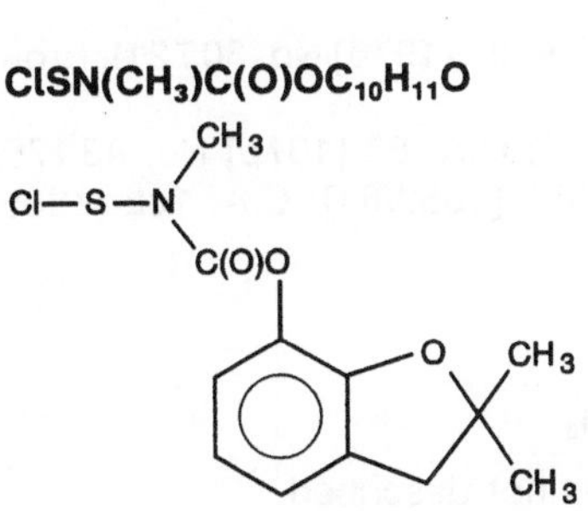

The title compound was prepared by adding a solution of SCl_2 in CH_2Cl_2 to a stirred solution of "carbofuran" (I) (mole ratio ca. 1 : 1) in CH_2Cl_2 at −5 °C and adding a solution of a slight excess of $(C_2H_5)_3N$ in CH_2Cl_2 to the resulting reaction mixture at 0 to 5 °C. The mixture was allowed to come to room temperature and then stirred for 3 h [1]. Previously, pyridine was used as acid acceptor at 25 to 30 °C [2, 3].

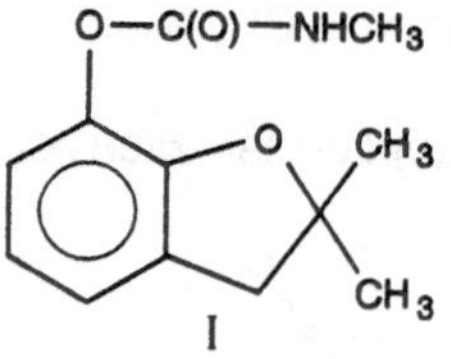

I

The ¹H NMR spectrum of the title compound (not isolated) showed a sharp singlet at 3.7 ppm (in CH_2Cl_2, relative to TMS) for the N–CH_3 group [2, 3].

Adding to the title compound in CH_2Cl_2 a solution of ROH (R = CH_3, C_2H_5, i-C_3H_7, n-C_4H_9, t-C_4H_9, $CH(CH_2CH_3)_2$, n-C_6H_{13}, n-C_8H_{17}, n-$C_{10}H_{21}$, n-$C_{12}H_{25}$, n-$C_{18}H_{37}$, c-C_6H_{11}, $CH_2C_6H_5$, $CH_2C_6H_4OCH_3$-4, $CH_2C_6H_4C_4H_9$-t-4) and $(C_2H_5)_3N$ in CH_2Cl_2 at 0 °C for 3 h gave the corresponding compound II [1].

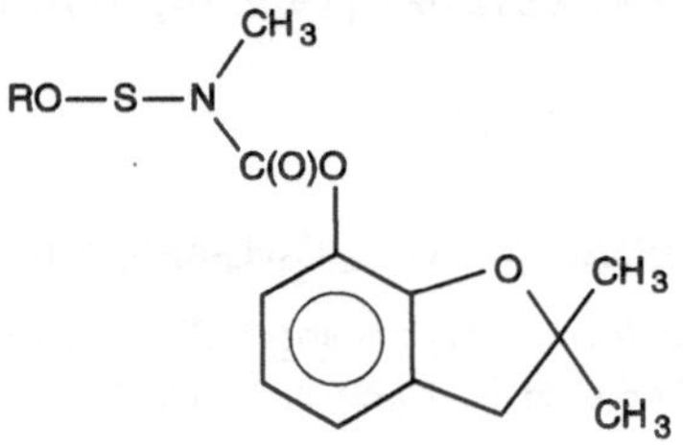

II

A solution of the title compound in CH_2Cl_2 reacted with a mixture of $C_6H_5C(OC_3H_7\text{-i})$=N-H·HCl and $(C_2H_5)_3N$ in CH_2Cl_2 for 2 h at 0 °C to give the benzimidate III [4].

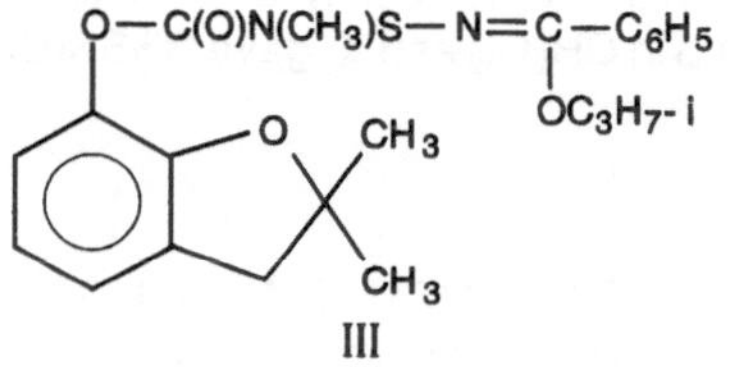

III

References:

[1] Kawata, M.; Umetsu, N.; Goto, T.; Fukuto, T. R. (Nippon Noyaku Gakkaishi **13** [1988] 595/603; C.A. **110** [1989] No. 212521).

[2] Brown, M. S. (U.S. 3914259 [1971/75]; C.A. **84** [1976] No. 30721), Brown, M. S.; Kohn, G. K. (U.S. 3843689 [1972/74]).

[3] Cleveland, J. D. (U.S. 3843688 [1972/74]; C.A. **82** [1975] No. 43175).

[4] Dutton, F. E.; Nelson, S. J. (U.S. 4473580 [1982/84]; C.A. **102** [1985] No. 24467).

12.5.1.10.13 ClSN(CH_3)C(O)ON=C(CH_3)SCH_3

The preparation of the title compound was not described.

Adding the title compound (in situ) in ether to a cooled (0 °C) solution of the phosphorus compounds $R^2R^3P(S)NHR^1$, $(CH_3O)_2P(S)NHR^1$ (R^1 = CH_3, i-C_3H_7, n-C_4H_9), $(C_2H_5O)_2P(S)NHR^1$ (R^1 = i-C_3H_7, C_6H_5), $(CH_3O)(CH_3S)P(S)NHCH_3$, $(CH_3O)P(R^2)(S)NHR^1$ (R^1 = CH_3, i-C_3H_7; R^2 = CH_3, C_6H_5), $(CH_3)(C_6H_5O)P(S)NHR^1$ (R^1 = CH_3, i-C_3H_7), $(CH_3)(\text{i-}C_3H_7O)P(S)NHR^1$ (R^1 = i-C_3H_7, C_6H_5), $(C_2H_5)(C_6H_5O)P(S)NHR^1$ (R^1 = n-C_4H_9, c-C_6H_{11}, C_6H_5, $CH_2C_6H_5$, 4-ClC_6H_4), (i-C_3H_7O)-

$(C_6H_5)P(S)NHC_3H_7$-i, $(C_2H_5)(4\text{-}ClC_6H_4O)P(S)NHR^1$ ($R^1 = CH_3$, C_6H_5), $(C_2H_5O)(C_6H_5)P(S)NHCH_2$-$C_6H_5$), $(C_2H_5)(2\text{-}ClC_6H_4O)P(S)NHC_6H_5$ or $(C_6H_5)(C_6H_5O)P(S)NHC_2H_5$) in DMF and reacting the mixture for 4 h at 25 °C gave the corresponding $(CH_3S)(CH_3)C{=}NOC(O)N(CH_3)SNR^1P(S)R^2R^3$.

Reference:

Nelson, S. J. (U.S. 4081536 [1977/78]; C.A. **89** [1978] No. 43787).

12.5.1.10.14 $ClSN(C_2H_5)C(O)OR$, $R = CH_3$, C_2H_5, C_3H_7-i, C_4H_9-n, C_6H_{13}-n

The title compounds were prepared by reacting $ROC(O)NHC_2H_5$ ($R = CH_3$, C_2H_5, C_3H_7-i, C_4H_9-n, C_6H_{13}-n) with SCl_2 in CH_2Cl_2 using pyridine as acid acceptor [1]. The preparation of $ClSN(C_2H_5)C(O)OC_4H_9$-n was not described.

$R = CH_3$. Liquid, b.p. 52 to 54 °C/2.8 Torr; $n_D^{25} = 1.4920$ [1].

$R = C_2H_5$. Liquid, b.p. 82 to 84 °C/10 Torr; $n_D^{25} = 1.4810$ [1].

$R = C_3H_7$-i. Liquid, b.p. 44 to 46 °C/1.25 Torr; $n_D^{25} = 1.4706$ [1].

$R = C_6H_{13}$-n. Liquid, b.p. 82 to 84 °C/0.35 Torr; $n_D^{25} = 1.4731$ [1].

Adding $ClSN(C_2H_5)C(O)OC_4H_9$-n to a stirred solution of $4\text{-}ClC_6H_4NH_2$ (mole ratio 5 : 1) in ether at room temperature and allowing the mixture to stand for 2 h gave 90% $4\text{-}ClC_6H_4$-$NHSN(C_2H_5)C(O)OC_4H_9$-n [5].

$ClSN(C_2H_5)C(O)OCH_3$ reacted with a solution of the formamidines $2\text{-}CH_3\text{-}4\text{-}RC_6H_3N{=}CH$-$NHCH_3$ in pyridine at 0 to 10 °C for 20 min ($R = Cl$, Br, CH_3) [2], or in THF and $(C_2H_5)_3N$ ($R = Br$, CH_3) [3] to give the corresponding $2\text{-}CH_3\text{-}4\text{-}RC_6H_3N{=}CHN(CH_3)SN(C_2H_5)C(O)OCH_3$ [2, 3]. It reacted with $2\text{-}CH_3\text{-}4\text{-}ClC_6H_3NHC(S)NR'CH_3$ ($R' = H$ or CH_3) and $(C_2H_5)_3N$ in CH_2Cl_2 at 0 to 10 °C and then at room temperature for 2 h to give $2\text{-}CH_3\text{-}4\text{-}ClC_6H_3N{=}C(NR'CH_3)SSN(C_2H_5)C$-$(O)OCH_3$ [4]. The title compounds with $R = i\text{-}C_3H_7$ and $n\text{-}C_6H_{13}$ reacted with $CH_3SC(R'){=}NOC$-$(O)NHCH_3$ ($R' = CH_3$, $(CH_3)_2NC(O)$) and pyridine at 5 °C and at room temperature overnight to give the corresponding $CH_3SC(R'){=}NOC(O)N(CH_3)SN(C_2H_5)C(O)OR$ [1]. A similar reaction of the title compounds with $R = C_2H_5$ and i-C_3H_7 with $ArOC(O)NHCH_3$ ($R = C_2H_5$, $Ar = 2\text{-}(CH_3)_2CH$-OC_6H_4; $R = i\text{-}C_3H_7$, $Ar = 2\text{-}i\text{-}C_3H_7C_6H_4$) gave the corresponding $ArOC(O)N(CH_3)SN(C_2H_5)C$-$(O)OR$ [1]. The title compounds with $R = CH_3$ and C_2H_5 reacted with "carbofuran" and pyridine at 5 °C and then at room temperature overnight to give the corresponding compounds I [1].

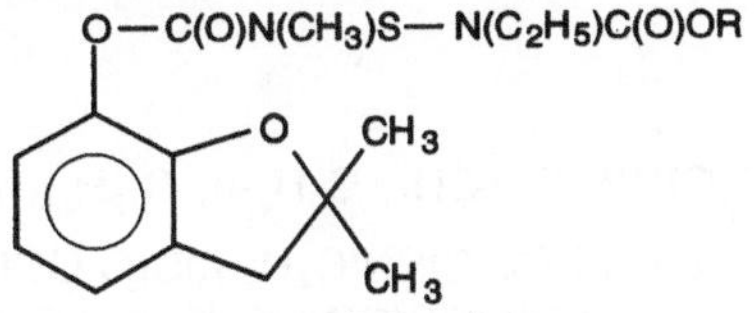

I: $R = CH_3$, C_2H_5

References:

[1] Fahmy, M. A. H.; Mallipudi, N. M.; Fukuto, T. R. (J. Agric. Food Chem. **26** [1978] 550/7).
[2] Böger, M.; Drabek, J. (Swiss 604500 [1975/78]; C.A. **90** [1979] No. 1703).
[3] Böger, M.; Drabek, J. (Ger. 2600987 [1975/76]; C.A. **85** [1976] No. 123638).
[4] Böger, M.; Drabek, J. (Ger. 2654080 [1972/77]; C.A. **87** [1977] No. 102078).
[5] Wie, S. I.; Sylwester, A. P.; Wing, K. D.; Hammock, B. D. (J. Agric. Food Chem. **30** [1982] 943/8; C.A. **97** [1982] No. 105489).

12.5.1.10.15 $ClSN(CH_2CH_2Cl)C(O)OCH_3$

The preparation of the title compound was not described.

$ClSN(CH_2CH_2Cl)C(O)OCH_3$ reacted with a solution of the formamidines $ArN{=}CHNHCH_3$ ($Ar = 2\text{-}CH_3\text{-}4\text{-}ClC_6H_3$, $2,4\text{-}(CH_3)_2C_6H_3$, $2,4,5\text{-}Cl_3C_6H_2$ [1, 2], $2,6\text{-}(C_2H_5)_2C_6H_3$ [2]) in pyridine at 0 to 10 °C for 20 min [1], or in THF and $(C_2H_5)_3N$ [2] to give the corresponding $ArN{=}CHN(CH_3)SN(CH_2CH_2Cl)C(O)OCH_3$ [1, 2].

References:

[1] Böger, M.; Drabek, J. (Swiss 604500 [1975/78]; C.A. **90** [1979] No. 1703).
[2] Böger, M.; Drabek, J. (Ger. 2600987 [1975/76]; C.A. **85** [1976] No. 123638).

12.5.1.10.16 $ClSN(C_3H_7\text{-}i)C(O)OC_2H_5$

The title compound was formed from the reaction of $C_2H_5OC(O)NHC_3H_7\text{-}i$ with SCl_2 in CH_2Cl_2 using pyridine as acid acceptor [1].

Liquid, b.p. 62 to 65 °C/3.5 Torr; $n_D^{25} = 1.4735$ [1].

$ClSN(C_3H_7\text{-}i)C(O)OC_2H_5$ reacted with a mixture of the ethyl carbamate $4\text{-}C_6H_5O\text{-}C_6H_4OCH_2CH_2NHC(O)OC_2H_5$ and pyridine (mole ratio 1 : 1 : 1) under ice cooling and 18 h at ambient temperature to give $4\text{-}C_6H_5OC_6H_4OCH_2CH_2N(C(O)OC_2H_5)SN(C_3H_7\text{-}i)C(O)OC_2H_5$ [2]. It reacted with "carbofuran" and pyridine at 5 °C and at room temperature overnight to give I [1].

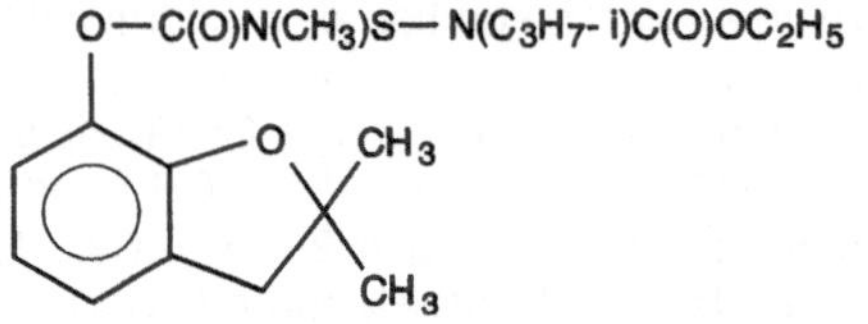

I

References:

[1] Fahmy, M. A. H.; Mallipudi, N. M.; Fukuto, T. R. (J. Agric. Food Chem. **26** [1978] 550/7).
[2] Innofinance Altalanos Innovacios Pénzintézet (Belg. 1000110 [1987/88]; C.A. **109** [1988] No. 68867, Neth. 8700074 [1987/88]; C.A. **110** [1989] No. 187820).

12.5.1.10.17 $ClSN(C_4H_9\text{-}n)C(O)OR$, $R = CH_3$, $C_4H_9\text{-}n$, $C_{12}H_{25}\text{-}n$, $C_6H_{11}\text{-}c$

A preparation was only described for $ClSN(C_4H_9\text{-}n)C(O)OCH_3$. It was formed by adding SCl_2 and then $(C_2H_5)_3N$ to a solution of $CH_3OC(O)NHC_4H_9\text{-}n$ in CH_2Cl_2 at 0 to 5 °C (not isolated) [1].

Adding $ClSN(C_4H_9\text{-}n)C(O)OCH_3$ to the reaction product of $C_6H_5C(O)NHN(C_4H_9\text{-}t)C(O)C_6H_5$ in THF with a suspension of NaH in mineral oil and reacting the mixture at room temperature for 5 h yielded $C_6H_5C(O)N(C_4H_9\text{-}t)N(C(O)C_6H_5)SN(C_4H_9\text{-}n)C(O)OCH_3$ [4].

The title compounds with $R = CH_3$ and $C_4H_9\text{-}n$ reacted with a solution of the formamidines $ArN{=}CHNHCH_3$ ($Ar = 2\text{-}CH_3\text{-}4\text{-}ClC_6H_3$, $2,4\text{-}(CH_3)_2C_6H_3$ [2], $2,6\text{-}(C_2H_5)_2C_6H_3$ [2, 3]) in pyridine at 0 to 10 °C for 20 min [2], or in a mixture of THF and $(C_2H_5)_3N$ [3] to give the corresponding $ArN{=}CHN(CH_3)SN(C_4H_9\text{-}n)C(O)OR$ [2, 3].

The title compounds with R = CH_3 and C_4H_9-n reacted with $CH_3NHC(O)ON{=}C(CH_3)SR'$ (R' = CH_3, C_2H_5) and pyridine in CH_2Cl_2 at 5 to 10 °C for 16 h to give the corresponding $ROC(O)N(C_4H_9\text{-n})SN(CH_3)C(O)ON{=}C(CH_3)SR'$ [5].

The title compounds with R = CH_3, C_4H_9-n, and C_6H_{11}-c reacted with a solution of 2,3-dihydro-2,2-dimethyl-benzofuran-7-yl-methyl carbamate ("carbofuran") in CH_2Cl_2 and pyridine at 5 to 10 °C and then 12 h at room temperature to give the corresponding compounds I [1].

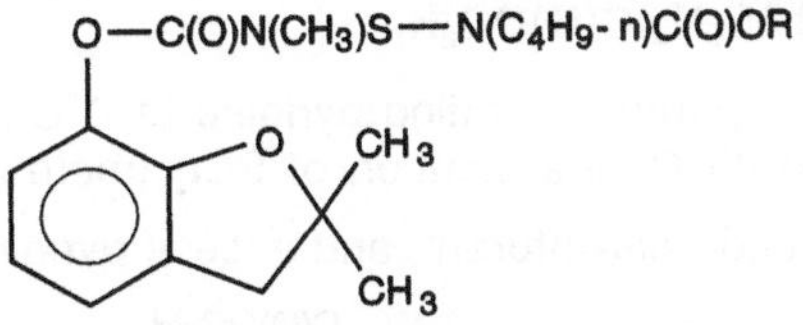

I: R = CH_3, C_4H_9-n, C_6H_{11}-c

Adding $ClSN(C_4H_9\text{-n})C(O)OC_4H_9\text{-n}$ to a mixture of 3,5-$(CF_3)_2C_6H_3NHC(S)NHCH_3$ and $(C_2H_5)_3N$ in CH_2Cl_2 at 0 to 10 °C and reacting the mixture for 2 h at room temperature gave 3,5-$(CF_3)_2C_6H_3N{=}C(NHCH_3)SSN(C_4H_9\text{-n})C(O)OC_4H_9\text{-n}$ [6].

$ClSN(C_4H_9\text{-n})C(O)OC_{12}H_{25}\text{-n}$ reacted with a mixture of 4-$C_6H_5OC_6H_4OCH_2CH_2NHC(O)$-$OC_2H_5$ and pyridine (mole ratio 1 : 1 : 1) under ice cooling and 18 h at ambient temperature to give II [7].

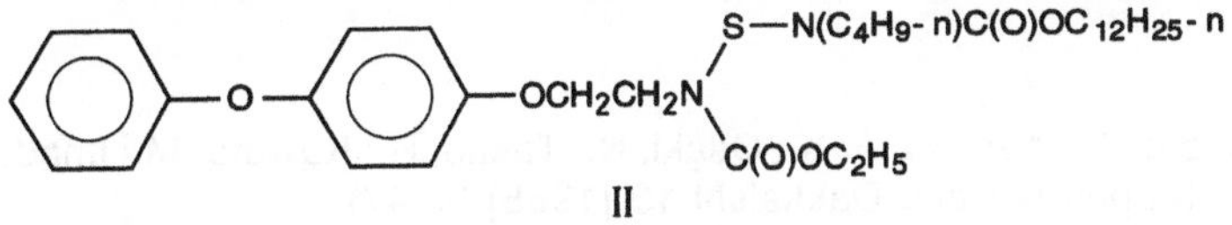

II

References:

[1] Drabek, J.; Böger, M. (Ger. 2812622 [1977/78]; C.A. **90** [1979] No. 23028).
[2] Böger, M.; Drabek, J. (Swiss 604500 [1975/78]; C.A. **90** [1979] No. 1703).
[3] Böger, M.; Drabek, J. (Ger. 2600987 [1975/76]; C.A. **85** [1976] No. 123638).
[4] Drabek, J. (Eur. Appl. 395581 [1990]; C.A. **114** [1991] No. 185026).
[5] Drabek, J.; Böger, M. (U.S. 4364959 [1978/82]; C.A. **98** [1983] No. 215201, Ger. 2727614 [1976/78]; C.A. **88** [1978] No. 190124).
[6] Böger, M.; Drabek, J. (Ger. 2654080 [1972/77]; C.A. **87** [1977] No. 102078).
[7] Innofinance Altalanos Innovacios Pénzintézet (Neth. 8700074 [1987/88]; C.A. **110** [1989] No. 187820).

12.5.1.10.18 $ClSN(C_8H_{17}\text{-n})C(O)OCH_3$

The preparation was not described.

Reacting a solution of $(C_2H_5)_3N$ in CH_2Cl_2 with a mixture of the title compound and 4-(5-bromo-2-pyrimidinyloxy)-3-chloroaniline in CH_2Cl_2 for 3 h at room temperature yielded III.

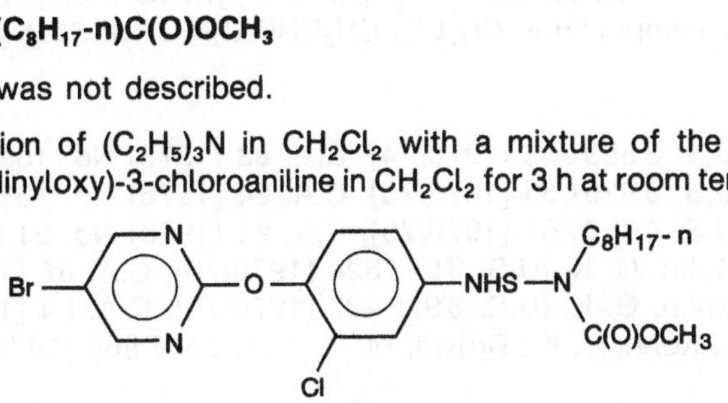

III

Reference:

Haga, T.; Yamada, N.; Sugi, H.; Koyanagi, T.; Okada, H. (Eur. Appl. 335 408 [1989]; C.A. **112** [1990] No. 138 771).

12.5.1.10.19 $ClSN(CH_2C(O)OC_2H_5)C(O)OC_2H_5$

The title compound was prepared by adding pyridine at 0 °C and successively $C_2H_5OC(O)CH_2NHC(O)OC_2H_5$ at 10 to 20 °C to a solution of SCl_2 (mole ratio 1 : 1 : 1) in CCl_4.

Dissolving the title compound, "carbofuran", and excess pyridine in CH_2Cl_2 gave 85% IV.

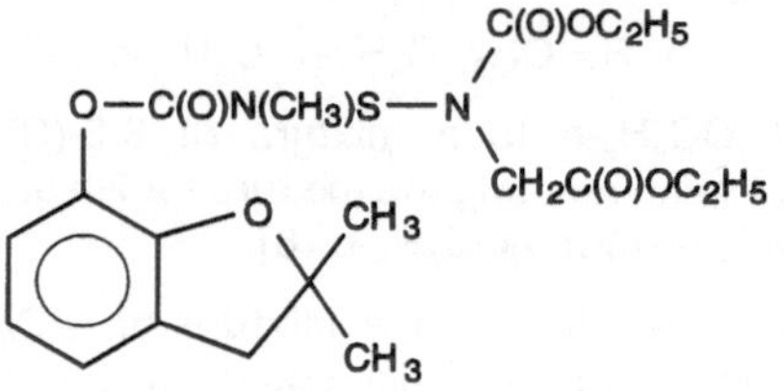

IV

Reference:

Goto, T.; Yasudomi, N.; Tanaka, A. K.; Osaki, N.; Takao, H.; Kawata, M.; Imada, J.; Endo, Y.; Umetsu, N. (Nippon Noyaku Gakkaishi **13** [1988] 39/47).

12.5.1.11 N-Chlorosulfanyl-N-alkyl(or aryl)-N'-alkyl(or aryl)-ureas

12.5.1.11.1 $ClSN(CH_3)C(O)NHCH_3$

The title compound was prepared by adding pyridine to a solution of $CH_3NHC(O)NHCH_3$ and SCl_2 (mole ratio 1.2 : 1 : 1.1) in CH_2Cl_2 at 25 to 30 °C. Removal of the hydrochloride by filtration left the product in CH_2Cl_2 solution (it was not isolated) [1 to 13].

The ^1H NMR spectrum (CH_2Cl_2/TMS) showed a doublet at 2.95 ppm for the CH_3NH group and a singlet at 3.5 ppm for the CH_3NS group [1 to 13].

$ClSN(CH_3)C(O)NHCH_3$ reacted with formamidines 2-CH_3-4-RC_6H_3N=$CHNHCH_3$ (R = Cl, CH_3) and $(C_2H_5)_3N$ in THF at 0 °C to give the corresponding 2-CH_3-4-RC_6H_3N=$CHN(CH_3)SN(CH_3)C(O)NHCH_3$ [14]. When a solution of C_6H_5SH and pyridine in CH_2Cl_2 was added at 0 °C to a solution of the title compound in CH_2Cl_2, $CH_3NHC(O)N(CH_3)SSC_6H_5$ was obtained [15].

References:

[1] Brown, M. S. (U.S. 3 853 966 [1970/74]; C.A. **82** [1975] No. 139 799).
[2] Brown, M. S. (U.S. 3 925 056 [1970/75]; C.A. **84** [1976] No. 73 950).
[3] Brown, M. S. (U.S. 4 159 281 [1970/79]; C.A. **91** [1979] No. 91 196).
[4] Brown, M. S.; Kohn, G. K. (U.S. 3 997 324 [1970/76]; C.A. **86** [1977] No. 139 655).
[5] Brown, M. S.; Kohn, G. K. (U.S. 3 928 407 [1970/75]; C.A. **84** [1976] No. 135 344).
[6] Cleveland, J. D.; Kohn, G. K.; Brown, M. S. (Ger. 2 407 560 [1974/75]; C.A. **84** [1976] No. 4704).
[7] Cleveland, J. D. (U.S. 3 897 481 [1972/75]; C.A. **83** [1975] No. 163 866).

[8] Cleveland, J. D. (U.S. 3946062 [1972/76]; C.A. **85** [1976] No. 46055).
[9] Cleveland, J. D. (U.S. 3857883 [1972/74]; C.A. **82** [1975] No. 155429).
[10] Magee, T. A. (U.S. 3899318 [1973/75]; C.A. **84** [1976] No. 5014).

[11] Magee, T. A. (U.S. 3853853 [1973/74]; C.A. **82** [1975] No. 140217).
[12] Chevron Research Co. (Fr. 2263237 [1974/75]; C.A. **84** [1976] No. 73951).
[13] Chevron Research Co. (Fr. 2290424 [1974/76]; C.A. **86** [1977] No. 106204).
[14] Böger, M.; Drabek, J. (Ger. 2632692 [1972/77]; C.A. **86** [1977] No. 151527).
[15] Cleveland, J. D.; Kohn, G. K.; Brown, M. S. (Br. 1416123 [1974/75]; C.A. **84** [1976] No. 89856).

12.5.1.11.2 $ClSN(CH_2C_3H_5\text{-c})C(O)NHCH_3$

The preparation of the title compound was not described.

$ClSN(CH_2C_3H_5\text{-c})C(O)NHCH_3$ reacted with formamidines 2-CH_3-4-$RC_6H_3N{=}CHNHCH_3$ (R = Cl, Br, CH_3) and $(C_2H_5)_3N$ in THF at 0 °C to give the corresponding 2-CH_3-4-RC_6H_3-$N{=}CHN(CH_3)SN(CH_2C_3H_5\text{-c})C(O)NHCH_3$.

Reference:

Böger, M.; Drabek, J. (Ger. 2632692 [1972/77]; C.A. **86** [1977] No. 151527).

12.5.1.11.3 $ClSN(C_6H_3Cl_2\text{-}3,4)C(O)NHCH_3$

The title compound was formed by reacting SCl_2 with $CH_3NHC(O)NHC_6H_3Cl_2$-3,4 and pyridine (mole ratio 1.1 : 1 : 1.2) in CH_2Cl_2 at 25 to 30 °C. It was not purified [1 to 12].

Clear yellow oil. The ^1H NMR spectrum (in CH_2Cl_2 ?/TMS) showed a singlet at 3.5 ppm for the CH_3–N group [1 to 12].

Adding a solution of the alcohol ROH (R = CH_3, C_2H_5 [3, 5], n-C_3H_7 [1, 3, 5], i-C_3H_7, n-C_8H_{17}, norbornyl [3, 5]) and pyridine in CH_2Cl_2 at 0 °C to a solution of the title compound in CH_2Cl_2 formed the corresponding $ROSN(C_6H_3Cl_2\text{-}3,4)C(O)NHCH_3$ [1, 3, 5].

Adding 3,4-$Cl_2C_6H_3NH_2$ and pyridine in CH_2Cl_2 to a solution of the title compound in CH_2Cl_2 at 0 °C yielded 3,4-$Cl_2C_6H_3NHSN(C_6H_3Cl_2\text{-}3,4)C(O)NHCH_3$ [1, 4]. Adding 3,4-$Cl_2C_6H_3$(n-C_3H_7)-NH and pyridine to the title compound and stirring the mixture overnight furnished 3,4-$Cl_2C_6H_3$-(n-C_3H_7)$NSN(C_6H_3Cl_2\text{-}3,4)C(O)NHCH_3$ [1, 4].

Adding 1-naphthyl-N-methyl carbamate, (1-$C_{10}H_7$)$OC(O)NHCH_3$, and then pyridine to a solution of the title compound in CH_2Cl_2 and storing the mixture overnight generated (1-$C_{10}H_7$)-$OC(O)N(CH_3)SN(C_6H_3Cl_2\text{-}3,4)C(O)NHCH_3$ [1, 2].

Adding a solution of the title compound to a slurry of $(CH_3)_2NC(O)NHC_6H_3Cl_2$-3,4 and pyridine at about 25 °C and letting the mixture react at 25 °C for 2.5 h produced $(CH_3)_2NC(O)$-(3,4-$Cl_2C_6H_3$)$NSN(C_6H_3Cl_2\text{-}3,4)C(O)NHCH_3$ [1, 2].

When a solution of the mercaptane RSH (R = CH_3 [9] (see also [4]), C_2H_5, n-C_3H_7, i-C_3H_7, n-C_4H_9, c-C_6H_{11} [9]) and pyridine in CH_2Cl_2 was added at 0 °C to a CH_2Cl_2 solution of the title compound, the corresponding disulfides $RSSN(C_6H_3Cl_2\text{-}3,4)C(O)NHCH_3$ were formed [4, 9].

A solution of the title compound in CH_2Cl_2 reacted with $ROC(O)CH_2SH$ (R = CH_3, C_2H_5) and pyridine in CH_2Cl_2 at 0 °C to give the corresponding urea $ROC(O)CH_2SSN(C_6H_3Cl_2$-3,4)C(O)-$NHCH_3$ [12]; see also [1, 4, 8, 10].

Mixing the title compound with an equimolar amount of pyridine at about 25 °C in CH_2Cl_2 and warming the mixture to 45 °C over a 15-min period gave 1,5,2,4,6,8-dithiatetrazocine-3,7-dione (I, R = 3,4-$Cl_2C_6H_3$) [13, 14].

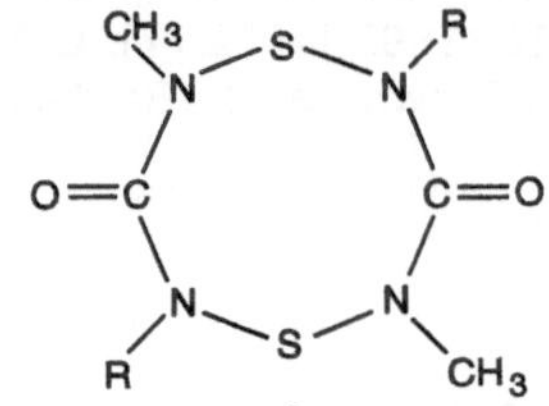

I

References:

[1] Brown, M. S. (U.S. 3853966 [1970/74]; C.A. **82** [1975] No. 139799).
[2] Brown, M. S. (U.S. 3925056 [1970/75]; C.A. **84** [1976] No. 73950).
[3] Brown, M. S.; Kohn, G. K. (U.S. 3997324 [1970/76]; C.A. **86** [1977] No. 139655).
[4] Brown, M. S. (U.S. 4159281 [1970/79]; C.A. **91** [1979] No. 91196).
[5] Brown, M. S.; Kohn, G. K. (U.S. 3928407 [1970/75]; C.A. **84** [1976] No. 135344).
[6] Cleveland, J. D. (U.S. 3857883 [1972/74]; C.A. **82** [1975] No. 155429).
[7] Cleveland, J. D. (U.S. 3946062 [1972/76]; C.A. **85** [1976] No. 46055).
[8] Cleveland, J. D. (U.S. 3897481 [1972/75]; C.A. **83** [1975] No. 163866).
[9] Cleveland, J. D.; Kohn, G. K.; Brown, M. S. (Br. 1416123 [1974/75]; C.A. **84** [1976] No. 89856).
[10] Cleveland, J. D.; Kohn, G. K.; Brown, M. S. (Ger. 2407560 [1974/75]; C.A. **84** [1976] No. 4704).

[11] Chevron Research Co. (Fr. 2263237 [1974/75]; C.A. **84** [1976] No. 73951).
[12] Chevron Research Co. (Fr. 2290424 [1974/76]; C.A. **86** [1977] No. 106204).
[13] Magee, T. A. (U.S. 3899318 [1973/75]; C.A. **84** [1976] No. 5014).
[14] Magee, T. A. (U.S. 3853853 [1973/74]; C.A. **82** [1975] No. 140217).

12.5.1.11.4 $ClSN(CH_3)C(O)NHC_8H_{15}$-c

The preparation of the title compound was not described.

Mixing $ClSN(CH_3)C(O)NHC_8H_{15}$-c with an equimolar amount of pyridine at about 25 °C in CH_2Cl_2 and warming the mixture to 45 °C over a 15-min period yielded 1,5,2,4,6,8-dithiatetrazocine-3,7-dione (I, R = C_8H_{15}-c; see above) [1, 2].

References:

[1] Magee, T. A. (U.S. 3853853 [1973/74]; C.A. **82** [1975] No. 140217).
[2] Magee, T. A. (U.S. 3899318 [1973/75]; C.A. **84** [1976] No. 5014).

12.5.1.11.5 $ClSN(CH_3)C(O)NHC_6H_4F$-2

The title compound was prepared by reacting SCl_2 with 2-$FC_6H_4NHC(O)NHCH_3$ and pyridine (mole ratio 1.1 : 1 : 1.2) in CH_2Cl_2 cooled in an ice bath [1 to 16].

The clear red oil was not purified. The ^1H NMR of the oil showed the N–CH_3 singlet at 3.5 ppm relative to TMS [1 to 16].

Adding a solution of NaI in water at about 0 °C to a solution of $ClSN(CH_3)C(O)NHC_6H_4F$-2 in CH_2Cl_2 gave 2-$FC_6H_4NHC(O)N(CH_3)SSN(CH_3)C(O)NHC_6H_4F$-2 [1, 4].

Adding cyclohexene to a solution of $ClSN(CH_3)C(O)NHC_6H_4F$-2 in CH_2Cl_2 cooled to 0 °C produced N-(2-chlorocyclohexylsulfanyl)-N-methyl-N′-(2-fluorophenyl)-urea [1].

Adding a solution of the alcohol ROH (R = CH_3, C_2H_5, n-C_3H_7, i-C_3H_7, n-C_8H_{17}, c-C_6H_{11}, norbornyl) and pyridine in CH_2Cl_2 at 0 °C to a solution of $ClSN(CH_3)C(O)NHC_6H_4F$-2 in CH_2Cl_2 gave the corresponding $ROSN(CH_3)C(O)NHC_6H_4F$-2 [3, 5]. Adding a solution of the mercaptane RSH (R = c-C_6H_{11}, C_6H_5) and pyridine in CH_2Cl_2 to a precooled (0 °C) solution of the title compound in CH_2Cl_2 yielded the corresponding disulfide $RSSN(CH_3)C(O)NHC_6H_4F$-2 [8]; see also [10, 11].

Adding a solution of $HSCH_2CH_2SH$ and pyridine in CH_2Cl_2 at 0 °C to the title compound gave the disulfide (2-$FC_6H_4NHC(O)N(CH_3)SSCH_2)_2$ [1].

Adding a solution of $CH_3OC(O)SH$ and pyridine in CH_2Cl_2 at 0 °C to the title compound yielded the corresponding $CH_3OC(O)SSN(CH_3)C(O)NHC_6H_4F$-2 [1, 4, 6, 13]. Adding a solution of the title compound in CH_2Cl_2 to a mixture of $(CH_3)_2C(O)NHC_6H_3Cl_2$-3,4 and pyridine in CH_2Cl_2 and stirring the mixture at about 25 °C for 3 h gave $(CH_3)_2C(O)N(C_6H_3Cl_2$-3,4)SN-$(CH_3)C(O)NHC_6H_4F$-2 [1, 2].

Mixing the title compound with an equimolar amount of pyridine at about 25 °C in CH_2Cl_2 and warming to 45 °C over a 15-min period gave 1,5,2,4,6,8-dithiatetrazocine-3,7-dione (I, R = 2-FC_6H_4; see p. 178) [15, 16].

References:

[1] Brown, M. S. (U.S. 3853966 [1970/74]; C.A. **82** [1975] No. 139799).
[2] Brown, M. S. (U.S. 3925056 [1970/75]; C.A. **84** [1976] No. 73950).
[3] Brown, M. S.; Kohn, G. K. (U.S. 3997324 [1970/76]; C.A. **86** [1977] No. 139655).
[4] Brown, M. S. (U.S. 4159281 [1970/79]; C.A. **91** [1979] No. 91196).
[5] Brown, M. S.; Kohn, G. K. (U.S. 3928407 [1970/75]; C.A. **84** [1976] No. 135344).
[6] Cleveland, J. D. (U.S. 3897481 [1972/75]; C.A. **83** [1975] No. 163866).
[7] Cleveland, J. D.; Kohn, G. K.; Brown, M. S. (Ger. 2407560 [1974/75]; C.A. **84** [1976] No. 4704).
[8] Cleveland, J. D.; Kohn, G. K.; Brown, M. S. (Br. 1416123 [1974/75]; C.A. **84** [1976] No. 89856).
[9] Cleveland, J. D. (U.S. 3946062 [1972/76]; C.A. **85** [1976] No. 46055).
[10] Brown, M. S. (U.S. 3960943 [1970/76]; C.A. **85** [1976] No. 94113).

[11] Brown, M. S.; Kohn, G. K. (Can. 1030553 [1975/78]; C.A. **89** [1978] No. 108757).
[12] Chevron Research Co. (Fr. 2263237 [1974/75]; C.A. **84** [1976] No. 73951).
[13] Chevron Research Co. (Fr. 2290424 [1974/76]; C.A. **86** [1977] No. 106204).
[14] Chevron Research Co. (Jpn. Kokai 75101328 [1974/75]; C.A. **85** [1976] No. 77699).
[15] Magee, T. A. (U.S. 3899318 [1973/75]; C.A. **84** [1976] No. 5014).
[16] Magee, T. A. (U.S. 3853853 [1973/74]; C.A. **82** [1975] No. 140217).

12.5.1.11.6 ClSN(CH$_3$)C(O)NHR, R = 4-ClC$_6$H$_4$, 3-CF$_3$C$_6$H$_4$, 2,5-Cl$_2$C$_6$H$_3$

The preparation of the title compounds was not described.

Mixing the title compounds with an equimolar amount of pyridine at about 25 °C in CH$_2$Cl$_2$ and warming the mixture to 45 °C over a 15-min period furnished 1,5,2,4,6,8-dithiatetrazocine-3,7-dione (I, R = 4-ClC$_6$H$_4$, 3-CF$_3$C$_6$H$_4$, 2,5-Cl$_2$C$_6$H$_3$; see p. 178) [1, 2].

References:

[1] Magee, T. A. (U.S. 3853853 [1973/74]; C.A. **82** [1975] No. 140217).
[2] Magee, T. A. (U.S. 3899318 [1973/75]; C.A. **84** [1976] No. 5014).

12.5.1.11.7 ClSN(CH$_3$)C(O)NHC$_6$H$_3$Cl$_2$-3,4

The preparation of the title compound was not described.

The title compound reacted with cycloheptene or cyclooctene (?) (mole ratio 1 : 1) in CH$_2$Cl$_2$ at 0 °C to yield II, where R = 3,4-Cl$_2$C$_6$H$_3$NHC(O)N(CH$_3$)–, n = 3, 4.

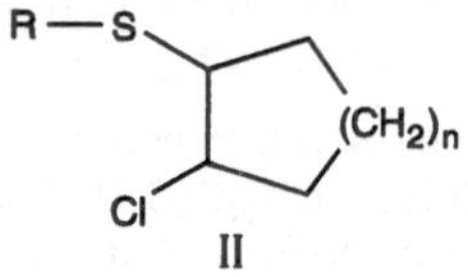

II

Reference:

Cleveland, J. D. (U.S. 3946062 [1972/76]; C.A. **85** [1976] No. 46055).

12.5.1.11.8 ClSN(CH$_3$)C(O)NHC$_6$H$_3$Br-4-Cl-3

The preparation of the title compound was not described; presumably it was prepared by reacting 3-Cl-4-BrC$_6$H$_3$NHC(O)NHCH$_3$, pyridine, and SCl$_2$ (mole ratio 1 : 1.2 : 1.1) in CH$_2$Cl$_2$ [1 to 4].

A solution of the title compound in CH$_2$Cl$_2$ reacted with ROC(O)CH$_2$SH (R = CH$_3$, C$_2$H$_5$) and pyridine at 0 °C to give the corresponding 3-Cl-4-BrC$_6$H$_3$NHC(O)N(CH$_3$)SSCH$_2$C(O)OR [1 to 4].

References:

[1] Brown, M. S. (U.S. 3853966 [1970/74]; C.A. **82** [1975] No. 139799).
[2] Brown, M. S. (U.S. 4159281 [1970/79]; C.A. **91** [1979] No. 91196).
[3] Cleveland, J. D. (U.S. 3897481 [1972/75]; C.A. **83** [1975] No. 163866).
[4] Chevron Research Co. (Fr. 2290424 [1974/76]; C.A. **86** [1977] No. 106204).

12.5.1.12 N-Chlorosulfanyl-N-alkyl(or aryl)-N′,N′-dialkyl-ureas

12.5.1.12.1 ClSN(CH$_3$)C(O)N(CH$_3$)$_2$

The title compound was prepared by adding (CH$_3$)$_3$SiN(CH$_3$)C(O)N(CH$_3$)$_2$ at 0 to 25 °C to SCl$_2$ and letting the mixture react at 25 °C for 30 min; 99% yield [1]; see also [2].

Yellow oil; b.p. 70 °C/0.15 Torr [1].

$ClSN(CH_3)C(O)N(CH_3)_2$ reacted with formamidines $R_2C_6H_3N{=}CHNHCH_3$ ($R_2 = 2,4\text{-}(CH_3)_2$, $2,6\text{-}(C_2H_5)_2$), and $(C_2H_5)_3N$ in THF at 0 °C to give the corresponding $R_2C_6H_3N{=}CHN(CH_3)SN\text{-}(CH_3)C(O)N(CH_3)_2$ [3]. It reacted with $CH_3N{=}C{=}O$, anhydrous HF, and $(C_2H_5)_3N$ in toluene at −50 to −20 °C to give $(CH_3)_2NC(O)N(CH_3)SN(CH_3)C(O)F$ [4].

References:

[1] Pedain, J.; Oertel, G. (Ger. 2045440 [1970/72]; C.A. **77** [1972] No. 48621).
[2] Kühle, E. (Synthesis **1970** 561/80, 573).
[3] Böger, M.; Drabek, J. (Ger. 2632692 [1972/77]; C.A. **86** [1977] No. 151527).
[4] Drabek, J.; Böger, M. (Ger. 2934729 [1977/80]; C.A. **93** [1980] No. 71056).

12.5.1.12.2 $ClSN(CH_2OCH_3)C(O)N(CH_3)_2$

The title compound was prepared by adding $(CH_3)_2NC(O)N(CH_2OCH_3)Si(CH_3)_3$ at 0 to 25 °C to SCl_2 and letting the mixture react at 25 °C for 30 min. It was not purified.

Characteristic bands in the IR spectrum are at 1680, 1585, and 1230 cm^{-1}.

Reference:

Pedain, J.; Oertel, G. (Ger. 2045440 [1970/72]; C.A. **77** [1972] No. 48621).

12.5.1.12.3 $ClSN(R)C(O)N(CH_3)_2$, $R = n\text{-}C_4H_9$, $n\text{-}C_6H_{13}$, $n\text{-}C_8H_{17}$, $c\text{-}C_3H_5$, $c\text{-}C_6H_{11}$

The preparation of the title compounds was not described.

The title compounds reacted with $2\text{-}CH_3\text{-}4\text{-}ClC_6H_3N{=}CHNHCH_3$ and $(C_2H_5)_3N$ in THF at 0 °C to produce the corresponding $2\text{-}CH_3\text{-}4\text{-}ClC_6H_3N{=}CHN(CH_3)SN(R)C(O)N(CH_3)_2$ [1]. The title compounds (exception $R = c\text{-}C_6H_{11}$) and $(C_2H_5)_3N$ reacted with $CH_3N{=}C{=}O$ and anhydrous HF in toluene at −50 to −20 °C to form the corresponding $(CH_3)_2NC(O)N(R)SN(CH_3)C(O)F$ [2].

References:

[1] Böger, M.; Drabek, J. (Ger. 2632692 [1972/77]; C.A. **86** [1977] No. 151527).
[2] Drabek, J.; Böger, M. (Ger. 2934729 [1977/80]; C.A. **93** [1980] No. 71056).

12.5.1.12.4 $ClSN(R)C(O)N(CH_3)_2$, $R = n\text{-}C_{14}H_{29}$, $4\text{-}ClC_6H_4$

The title compounds were prepared by adding a solution of $RNHC(O)N(CH_3)_2$ ($R = n\text{-}C_{14}H_{29}$, $4\text{-}ClC_6H_4$) in CH_2Cl_2 at 10 °C to SCl_2 and letting the mixture react for 30 min at 25 °C.

$R = n\text{-}C_{14}H_{29}$. Highly viscous, yellow oil. Refractive index: $n_D^{20} = 1.5017$.

$R = 4\text{-}ClC_6H_4$. Yellow-brown oil. IR: 1685 cm^{-1}, ν(CO).

Reference:

Pedain, J.; Oertel, G. (Ger. 2045440 [1970/72]; C.A. **77** [1972] No. 48621).

12.5.1.12.5 $ClSN(C_6H_3Cl_2\text{-}3,4)C(O)N(CH_3)_2$

The title compound was prepared by adding SCl_2 to an ice-cold mixture of $3,4\text{-}Cl_2C_6H_3\text{-}NHC(O)N(CH_3)_2$ and pyridine (mole ratio 1.1 : 1.0 : 1.2) in CH_2Cl_2.

Clear yellow oil. ^1H NMR (solvent ?/TMS): $\delta = 3.0$ ppm, $N(CH_3)_2$.

Reference:

Cleveland, J. D. (U.S. 3946062 [1972/76]; C.A. **85** [1976] No. 46055).

12.5.1.12.6 $ClSN(CH_3)C(O)N(C_2H_5)_2$

The title compound was prepared by adding a solution of $CH_3NHC(O)N(C_2H_5)_2$ in CH_2Cl_2 at 10 °C to SCl_2 and letting the mixture react for 30 min at 25 °C; 80% yield [1].

Liquid, b.p. 110 °C/13 Torr [1].

A mixture of $ClSN(CH_3)C(O)N(C_2H_5)_2$ and $(C_2H_5)_3N$ reacted with $CH_3N{=}C{=}O$ and anhydrous HF in toluene at −50 to −20 °C to give $(C_2H_5)_2NC(O)N(CH_3)SN(CH_3)C(O)F$ [2].

References:

[1] Pedain, J.; Oertel, G. (Ger. 2045440 [1970/72]; C.A. **77** [1972] No. 48621).
[2] Drabek, J.; Böger, M. (Ger. 2934729 [1977/80]; C.A. **93** [1980] No. 71056).

12.5.1.12.7 $ClSN(CH_3)C(O)NR_2$, $R = n\text{-}C_3H_7$, $i\text{-}C_3H_7$, $n\text{-}C_4H_9$

The preparation of the title compounds was not described.

$ClSN(CH_3)C(O)N(C_3H_7\text{-}n)_2$ reacted with $2\text{-}CH_3\text{-}4\text{-}RC_6H_3N{=}CHNHCH_3$ ($R = Cl$, CH_3) and $(C_2H_5)_3N$ in THF at 0 °C to form the corresponding $2\text{-}CH_3\text{-}4\text{-}RC_6H_3N{=}CHN(CH_3)SN(CH_3)C(O)N\text{-}(C_3H_7\text{-}n)_2$ [1]. A mixture of the title compounds and $(C_2H_5)_3N$ reacted with anhydrous HF and $CH_3N{=}C{=}O$ in toluene at −50 to −20 °C yielding the corresponding $R_2NC(O)N(CH_3)SN(CH_3)\text{-}C(O)F$ ($R = n\text{-}C_3H_7$, $i\text{-}C_3H_7$, $n\text{-}C_4H_9$) [2].

References:

[1] Böger, M.; Drabek, J. (Ger. 2632692 [1972/77]; C.A. **86** [1977] No. 151527).
[2] Drabek, J.; Böger, M. (Ger. 2934729 [1977/80]; C.A. **93** [1980] No. 71056).

12.5.1.12.8 $ClSN(CH_3)C(O)NR^1R^2$, $R^1 = R^2 = n\text{-}C_8H_{17}$; $R^1 = CH_3$, $R^2 = n\text{-}C_4H_9$, $c\text{-}C_6H_{11}$, OCH_3; $R^1 = CH_2C_3H_5\text{-}c$, $R^2 = n\text{-}C_3H_7$; $R^1R^2 = (CH_2)_5$, $(CH_2)_4O$, $CH_2CHCH_3OCHCH_3CH_2$

The preparation of the title compounds was not described.

The compounds $ClSN(CH_3)C(O)NR^1R^2$ ($R^1 = R^2 = n\text{-}C_8H_{17}$; $R^1 = CH_3$, $R^2 = n\text{-}C_4H_9$, $c\text{-}C_6H_{11}$, OCH_3; $R^1 = CH_2C_3H_5\text{-}c$, $R^2 = n\text{-}C_3H_7$; $R^1R^2 = (CH_2)_5$, $(CH_2)_4O$, $CH_2CHCH_3OCHCH_3CH_2$) reacted with formamidines $2\text{-}CH_3\text{-}4\text{-}RC_6H_3N{=}CHNHCH_3$ ($R = Cl$, CH_3) and $(C_2H_5)_3N$ in THF at 0 °C forming the corresponding $2\text{-}CH_3\text{-}4\text{-}RC_6H_3N{=}CHN(CH_3)SN(CH_3)C(O)NR^1R^2$.

Reference:

Böger, M.; Drabek, J. (Ger. 2632692 [1972/77]; C.A. **86** [1977] No. 151527).

12.5.1.12.9 $ClSN(C_4H_9\text{-}n)C(O)NR^1R^2$, $R^1 = R^2 = C_2H_5$, $n\text{-}C_4H_9$; $R^1 = CH_3$, $R^2 = n\text{-}C_4H_9$

The preparation of the title compounds was not described.

A mixture of the title compounds and $(C_2H_5)_3N$ reacted with anhydrous HF and $CH_3N{=}C{=}O$ in toluene at −50 to −20°C to give the corresponding $R^1R^2NC(O)N(C_4H_9\text{-}n)SN(CH_3)C(O)F$ ($R^1 = R^2 = C_2H_5$, $n\text{-}C_4H_9$; $R^1 = CH_3$, $R^2 = n\text{-}C_4H_9$).

Reference:

Drabek, J.; Böger, M. (Ger. 2934729 [1977/80]; C.A. **93** [1980] No. 71056).

12.5.1.12.10 $ClSN(C_2H_5)C(O)N(C_2H_5)C(Cl){=}NC_2H_5$

The title compound was prepared by adding SCl_2 to excess of the imino-1,3-diazetidin-2-one III while holding the temperature at 30°C by external water cooling.

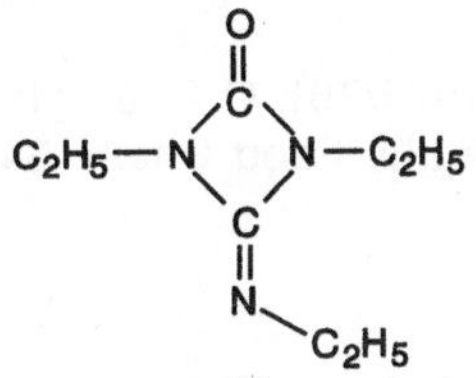

III

Reddish oil. IR bands at 5.7 and 6.0 μ.

The title compound decomposed when heated to 80 to 100°C/1 Torr.

Reference:

Farbenfabriken Bayer AG; Fischer, P. (Br. 959997 [1960/64]; C.A. **61** [1964] 6924).

12.5.1.12.11 $ClSN(CH_2CH_2Cl)C(O)N(CH_2)_4O$

The title compound was prepared by reacting SCl_2 with a solution of equimolar amounts of $ClCH_2CH_2N{=}C{=}O$ and $O(CH_2)_4NSi(CH_3)_3$ in $CHCl_3$ at 0 to 20°C.

The yellow oil can not be distilled. IR: Characteristic strong bands at 1685 and 1115 cm^{-1}. Refractive index: $n_D^{20} = 1.5634$.

Reference:

Pedain, J.; Oertel, G. (Ger. 2045440 [1970/72]; C.A. **77** [1972] No. 48621).

12.5.1.13 N-Chorosulfanyl-N-alkyl-formamidines

12.5.1.13.1 $ClSN(CH_3)CH{=}NC_6H_3(CH_3)$-2-R-4, R = Cl, CH_3

The title compounds were formed by reacting the corresponding formamidine hydrochlorides, 2-CH_3-4-$RC_6H_3N{=}CHNHCH_3 \cdot HCl$ (R = Cl, CH_3) and $(C_2H_5)_3N$ in CH_2Cl_2, with SCl_2 at −10 °C and then warming the mixture to 15 °C [1] or by reacting the corresponding formamidines, 2-CH_3-4-$RC_6H_3N{=}CHNHCH_3$ (R = Cl, CH_3), with SCl_2 in the presence of $(C_2H_5)_3N$ in THF [2]. Attempts to isolate the title compounds by distillation caused decomposition to the starting formamidines [1].

^1H NMR (CCl_4/TMS): δ (in ppm) = 2.2 (s, 3H, CH_3), 3.4 to 3.5 (s, 3H, NCH_3), 6.6 to 7.2 (m, 3H, aromatic protons), 7.9 (s, 1H, N=CH). There is no difference in the ^1H NMR of the compound with R = Cl to that with R = CH_3 [1].

The title compounds (R = Cl, CH_3) reacted with the amines R'_2NH (R′ = c-C_6H_{11} [1, 2], $CH_2C_6H_5$ [1]) in hexane or with the anilines C_6H_5NHR' (R′ = CH_3, C_2H_5) in hexane [1] or THF [2] (mole ratio 1 : 2) to give the corresponding 2-CH_3-4-$RC_6H_3N{=}CHN(CH_3)SNR'_2$ [1, 2]. The title compound with R = Cl reacted with $(C_6H_5)_2NH$ and $(C_2H_5)_3N$ (mole ratio 1 : 1 : 1) in THF below 25 °C and within 10 min at room temperature to give 32% 2-CH_3-4-$ClC_6H_3N{=}CHN(CH_3)SN$-$(C_6H_5)_2$ [1].

Both title compounds reacted with $CH_3NHC(O)OR'$ (R′ = carbofuranyl, 3-i-$C_3H_7C_6H_4$, $-N{=}C(CH_3)SCH_3$, or $-N{=}C(SCH_3)C(O)N(CH_3)_2$) to give the corresponding 2-CH_3-4-$RC_6H_3N{=}$ $CHN(CH_3)SN(CH_3)C(O)OR'$ (R = Cl, CH_3) [2].

References:

[1] Rizzo, V. L. (Ger. 2619304 [1974/76]; C.A. **86** [1977] No. 71949).
[2] Eya, B. K.; Fukuto, T. R. (J. Agric. Food Chem. **34** [1986] 947/52).

12.5.1.13.2 $ClSN(C_6H_{11}$-c$)C(Cl){=}NC_6H_{11}$-c

The title compound was obtained by adding SCl_2 to a solution of excess c-C_6H_{11}-$N{=}C{=}NC_6H_{11}$-c in CCl_4 below 35 °C; highly viscous oil.

IR: 1658 cm^{-1}, ν(C=N).

Reference:

Farbenfabriken Bayer AG; Fischer, P. (Ger. 1131661 [1960/62]; C.A. **58** [1963] 1401).

12.5.1.14 N-Chlorosulfanyl-N-trichloromethyl-cyanamide, $ClSN(CCl_3)C{\equiv}N$

$ClSN(CCl_3)CN$ was synthesized by stirring a mixture of $CCl_3C(CN){=}NCl$ and SCl_2 (mole ratio 1 : 1.25) at 60 °C for 8 h; 52.5% yield.

Liquid; b.p. 73 to 74 °C/0.1 Torr; n_D^{20} = 1.5978. In the IR spectrum the cyano group appeared in the range 1590 to 1550 cm^{-1}.

Reference:

Lazukina, L. A.; Kukhar', V. P. (Zh. Obshch. Khim. **53** [1983] 2239/43; J. Gen. Chem. USSR [Engl. Transl.] **53** [1983] 2018/21).

12.5.2 N-Chlorosulfanyl-Substituted N-Heterocycles

12.5.2.1 N¹-Chlorosulfanyl-aziridine Derivatives, $ClS\overline{N-CF_2-C}FR$, R = F, Cl, SF_5, CF_3

The title compounds were prepared by reacting $S_3N_3Cl_3$ with the respective fluorinated alkenes CF_2=CFR (R = F, Cl, SF_5, CF_3; yields are given below). The compound with R = CF_3 was also produced with 45% yield by treating monomeric ClS≡N (from FS≡N + $(CH_3)_3SiCl$) with excess CF_3CF=CF_2 at −30 to 0 °C [1]; see also [2].

R = F. Liquid (46% yield); b.p. ca. 50 °C, decomposes at room temperature. ^{19}F NMR (neat ?/ $CFCl_3$): δ = −119.5 ppm (s, CF_2) [1].

R = Cl. Liquid (41% yield); b.p. 66 °C, decomposes at room temperature. ^{19}F NMR (neat ?/ $CFCl_3$): δ (in ppm) = −106.55 (F(4), ^3J(F(2),F(4)) = 18.0 Hz, ^3J(F(3),F(4)) = 12.0 Hz), −124.45, −121.28 (F(2),F(3), ^2J(F(2),F(3)) = 72.4 Hz) [1].

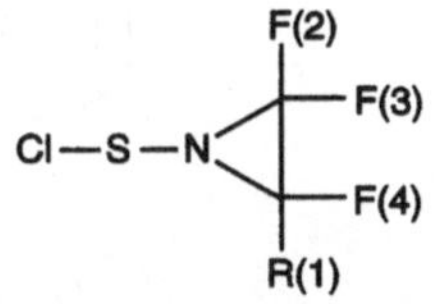

R = SF₅. Distillable, yellow liquid (65% yield); b.p. 45 to 46 °C/10 Torr. ^{19}F NMR (neat ?/ $CFCl_3$): δ (in ppm) = AB_4 system; $δ_A$ = 66.0, $δ_B$ = 50.88, J_{AB} = 146.9 Hz (F(1)), −144.13 (F(4), ^3J(F(1),F(4)) = 5.8 Hz (A), ^3J(F(1),F(4)) = 13.3 Hz (B), ^3J(F(2),F(4)) = 18.41 Hz), −118.14 (F(3), ^2J(F(2),F(3)) = 68.85 Hz, ^3J(F(3),F(4)) = 16.3 Hz), −110.95 (F(2), ^4J(F(1),F(2)) = 11.1 Hz (B)) [1].

R = CF₃. Distillable, yellow liquid (78% yield); b.p. 85 °C. ^{19}F NMR (neat ?/$CFCl_3$): δ (in ppm) = −176.1 (F(4), ^3J(F(1),F(4)) = 5.5 Hz, ^3J(F(2),F(4)) = 19.8 Hz, ^3J(F(3),F(4)) = 15.6 Hz), −123.1, −114.2 (F(3), F(2), ^4J(F(1),F(2)) = 7.5 Hz, ^4J(F(1),F(3)) = 0.6 Hz, ^2J(F(2),F(3)) = 81.0 Hz), −75.2 (F(1)) [1].

The compound reacted with mercury to give $CF_3\overline{CF-CF_2-N}$-S-S-$\overline{N-CF_2-C}FCF_3$ and $HgCl_2$; with silylamines nucleophilic exchange of the chlorine atom bonded to sulfur is possible [1].

References:

[1] Lork, A.; Gard, G.; Hare, M.; Mews, R.; Stohrer, W.-D.; Winter, R. (J. Chem. Soc. Chem. Commun. **1992** 898/9).

[2] Gard, G. L.; Hare, M.; Lork, A.; Mews, R.; Stohrer, W.-D.; Winter, R. (J. Fluorine Chem. **54** [1991] 232).

12.5.2.2 N¹-Chlorosulfanyl-1H-pyrrole Derivatives

12.5.2.2.1 $ClS\overline{N-(CH_2)_3-C}H_2$

$ClSN(CH_2)_4$ was synthesized by reacting $(CH_2)_4NSN(CH_2)_4$ with SCl_2 (no details) [1] or $(CH_2)_4NSSN(CH_2)_4$ with Cl_2 (no details given) [2].

The title compound reacted with Na^+ $((CH_3)_3Si)_2N^-$ to give $((CH_3)_3Si)_2NSN(CH_2)_4$ [1]. Adding $ClSN(CH_2)_4$ in toluene to a mixture of excess phthalimide and $(C_2H_5)_3N$ in toluene yielded 55% I after 1 h [2].

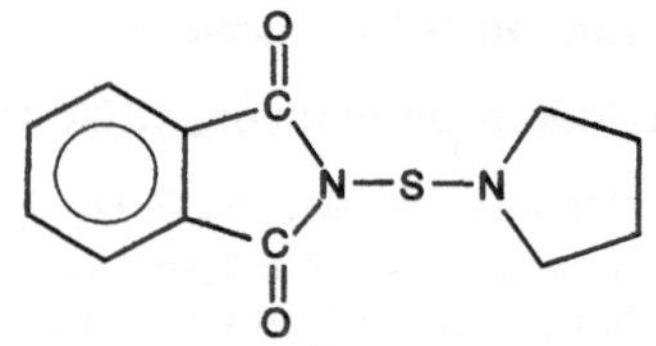

I

The title compound was reacted with carbamates II ($R^1 = R^2 = CH_3$; $R^1 = C_2H_5$, $R^2 = n\text{-}C_4H_9$; $R^1 = i\text{-}C_3H_7$, $R^2 = CH_3$) and pyridine at room temperature for 18 h to give the corresponding compound III [3].

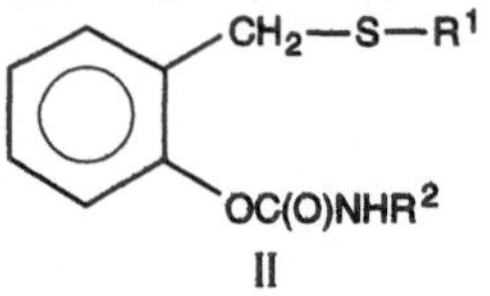

II

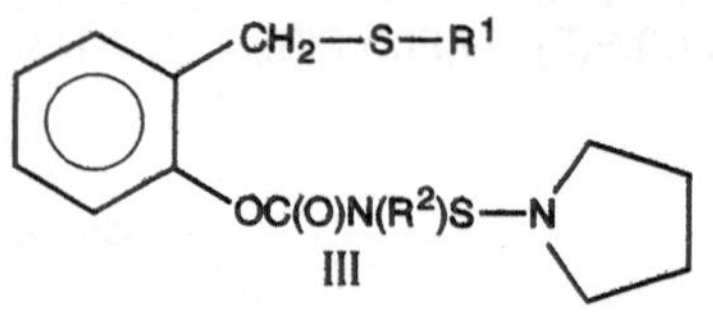

III

References:

[1] Neidlein, R.; Lenhard, T. (Chem. Ber. **116** [1983] 3133/40).

[2] Lawrence, J. P. (Rubber Chem. Technol. **49** [1976] 333/40).

[3] Hoffmann, H.; Hammann, T.; Homeyer, B.; Stendel, W. (Ger. 2737606 [1977/79]; C.A. **91** [1979] No. 56482).

12.5.2.2.2 $ClS\overline{N{-}C(O){-}(CH_2)_2{-}CH_2}$

The formation of the title compound was not described.

The title compound reacted in $(C_2H_5)_2O$ with IV and $2,6\text{-}F_2C_6H_3{-}C(O){-}N{=}C{=}O$ in the presence of $(C_2H_5)_3N$ to give V.

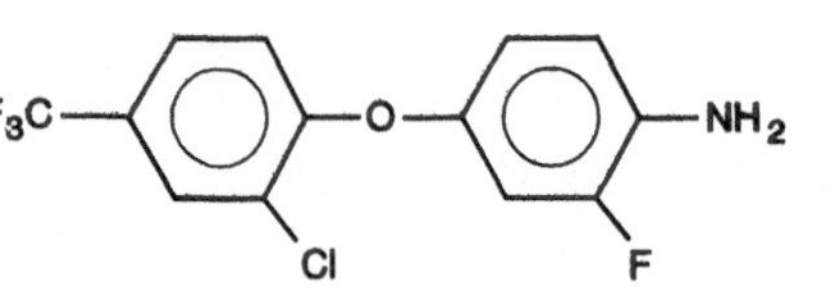

IV

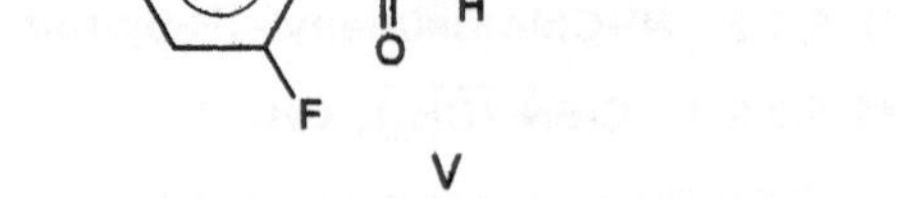

V

Reference:

Anderson, M.; Brinnand, A. G. (Eur. Appl. 216423 [1986/87]; C.A. **106** [1987] No. 213583).

12.5.2.2.3 ClSN–CH(CO$_2$CH$_3$)–(CH$_2$)$_2$–CH$_2$

The title compound was formed by adding a solution of SCl_2 in CH_2Cl_2 at room temperature to a solution of L-proline methyl ester hydrochloride, $[H_2N–CH(CO_2CH_3)–(CH_2)_2–CH_2]^+\ Cl^-$, in CH_2Cl_2. A solution of pyridine in CH_2Cl_2 was then added to the reaction mixture, and the mixture was allowed to react overnight. The product was not isolated.

The title compound reacted with IV (see p. 186) in the presence of $(C_2H_5)_3N$ in ether to give VI.

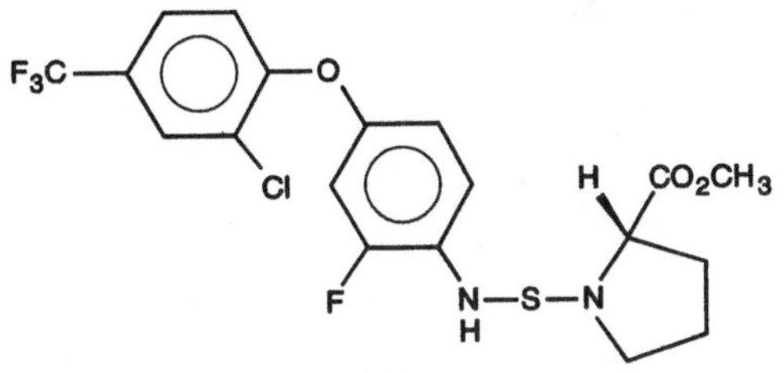

VI

Reference:

Anderson, M. (Eur. Appl. 161019 [1985]; C.A. **104** [1986] No. 148560).

12.5.2.2.4 ClSN–C(O)–(CH$_2$)$_2$–C(O)

The title compound was prepared as follows:

Method I: Cl_2 gas was passed for 8 h through a stirred solution of N,N′-disulfanyldiylbis-(succinimide), C(O)–(CH$_2$)$_2$–C(O)–NSSN–C(O)–(CH$_2$)$_2$–C(O), in $CHCl_3$ at 50 °C. The excess Cl_2 was removed by purging the mixture with inert gas (N_2) followed by evaporation in vacuum; 93.3% yield [1].

Method II: A mixture of N-chlorosuccinimide, ClN–C(O)–(CH$_2$)$_2$–C(O), elemental sulfur, and $(C_2H_5)_4N^+\ Br^-$ (or $(C_4H_9)_4N^+\ I^-$) (mole ratio 1 : 1 : 2.4) in CH_2ClCH_2Cl was heated at 70 °C until a homogeneous, pale yellow solution formed; 83% yield [2].

Method III: A mixture of ClN–C(O)–(CH$_2$)$_2$–C(O) and SCl_2 was heated in benzene at 80 °C until an orange-red solution formed (~4 h); 92% yield [2].

Yellow [1, 2] crystals [1], prisms [2]; m.p. 65 to 67 °C [1], 73 to 74 °C [2], 73 to 75 °C [2]. Thermal decomposition occurred upon recrystallization in inert solvents [2].

^{1}H NMR ($CDCl_3$/TMS): $\delta = 3.00$ ppm (s, CH_2) [1].

The reactions of the title compound are summarized in Table 40, pp. 188/91.

References:

[1] Bombala, M. U.; Ley, S. V. (J. Chem. Soc. Perkin Trans. I **1979** 3013/6).
[2] Borovikova, G. S.; Levchenko, E. S.; Kaminskaya, E. I. (Zh. Org. Khim. **22** [1986] 100/6; J. Org. Chem. USSR [Engl. Transl.] **22** [1986] 86/92).
[3] Leininger, H.; Kemmer, P.; Beck, K.; Christl, M. (Chem. Ber. **115** [1982] 3213/23).
[4] Kurita, J.; Aruga, T.; Tsuchiya, T. (Heterocycles **31** [1990] 1769/72).
[5] Kurita, J.; Iwata, K.; Sakai, H.; Tsuchiya, T. (Chem. Pharm. Bull. **33** [1985] 4572/80).
[6] Kurita, J.; Iwata, K.; Hasebe, M.; Tsuchiya, T. (J. Chem. Soc. Chem. Commun. **1983** 941/2).
[7] Saunders, D. E. (Eur. Appl. 129359 [1984]; C.A. **102** [1985] No. 185070).

Table 40
Reactions of $ClS\overline{N-C(O)-(CH_2)_2-C}(O)$.

substrate	reaction conditions	product(s)	yield (in %)	Ref.
cyclopentene	mole ratio 5.7 : 7.35; in CH_2Cl_2 at room temperature	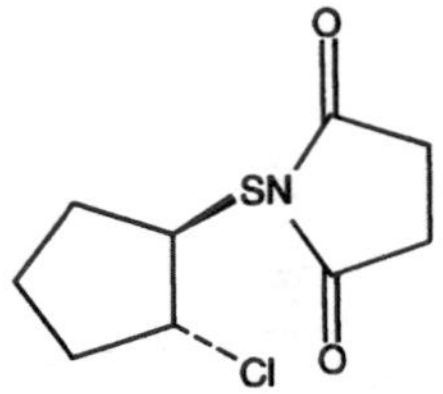	99.2	[1]
cyclohexene	mole ratio ca. 2 : 3; in CH_2Cl_2 at room temperature	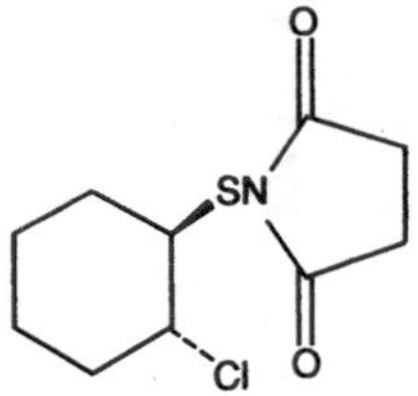	99.8	[1]
cyclooctene	mole ratio ca. 1 : 1; in CH_2Cl_2 at room temperature	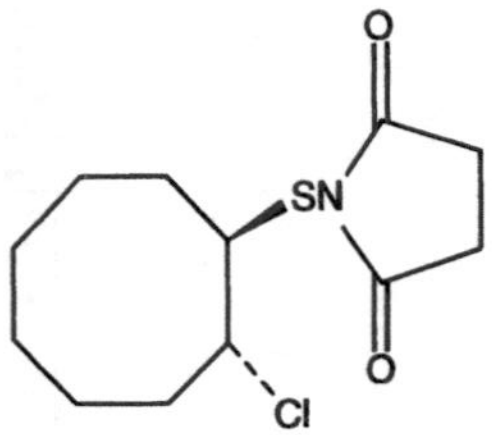	85.3	[1]
styrene	mole ratio ca. 1 : 1; in CH_2Cl_2 at room temperature	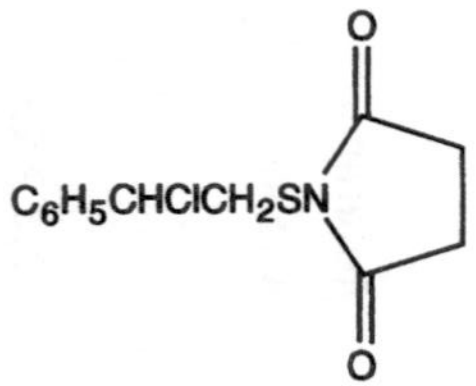	99.8	[1]

methylene cyclohexane	mole ratio 3.2 : 4.7; in CH_2Cl_2 at room temperature	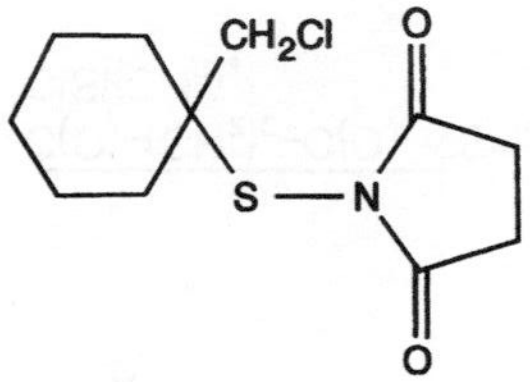	100	[1]
norbornene	mole ratio ca. 1 : 1; in CH_2Cl_2 at room temperature	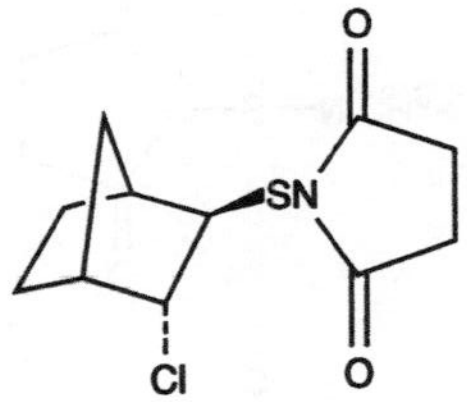	100	[1]
	mole ratio ca. 1 : 1.5; in CH_2Cl_2–ether at −78 °C, 20 h at room temperature	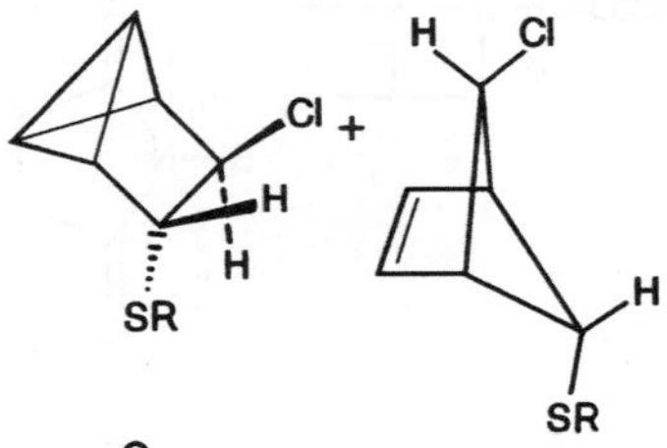 3 : 1, 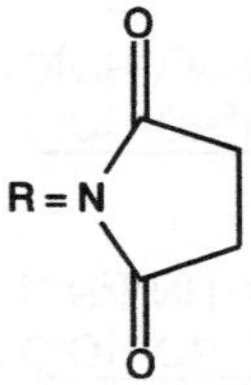		[3]
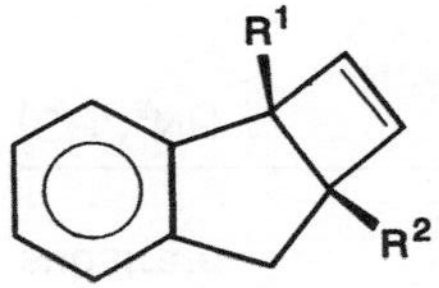 + $LiAlH_4$		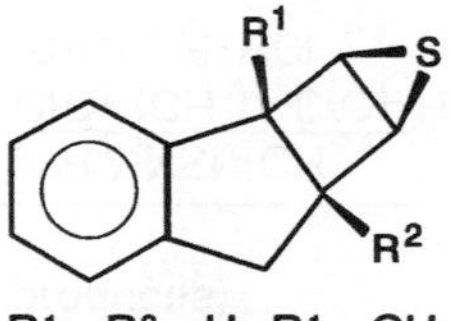 $R^1 = R^2 = H$; $R^1 = CH_3$, $R^2 = H$; $R^1 = H$, $R^2 = CH_3$		[4]

Table 40 (continued)

substrate	reaction conditions	product(s)	yield (in %)	Ref.
$(CH_2)_5NH$	mole ratio 1 : 3; in CH_2ClCH_2Cl, ~2 h at 20 to 25 °C	$(CH_2)_5NSN(CH_2)_5$ $\overline{C(O)-(CH_2)_2-C(O)-N}H$ $(CH_2)_5NH \cdot HCl$	77 92 100	[2]
$(CH_2)_5NH + (C_2H_5)_3N$	mole ratio 1 : 1 : 1; in CH_2ClCH_2Cl, ~1 h at 10 to 15 °C	$\overline{C(O)-(CH_2)_2-C(O)-N}SN(CH_2)_5$ $(CH_2)_5NH \cdot HCl$	95 100	[2]
$O(CH_2)_4NH$	mole ratio 1 : 2; in CH_2ClCH_2Cl at 10 to 15 °C, and 2 h at 20 °C	$\overline{C(O)-(CH_2)_2-C(O)-N}SN(CH_2)_4O$ $O(CH_2)_4NH \cdot HCl$	87 100	[2]
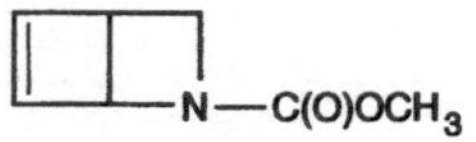	in CH_2Cl_2	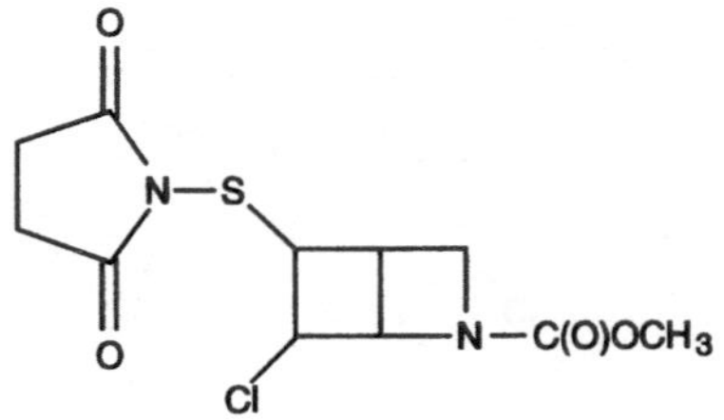		[5], see also [6]
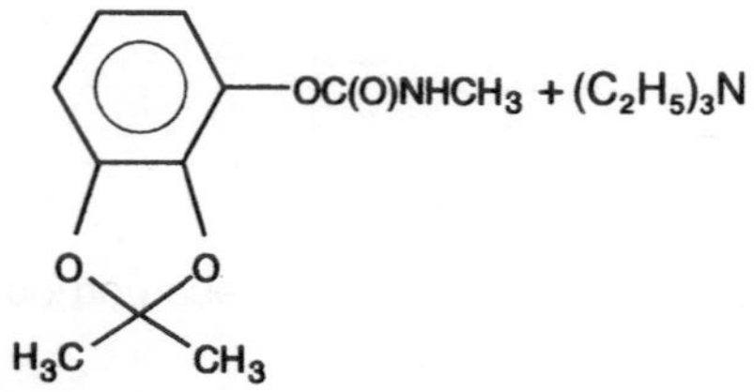	in CH_2Cl_2 at 0 to 10 °C, and 2.5 h at room temperature	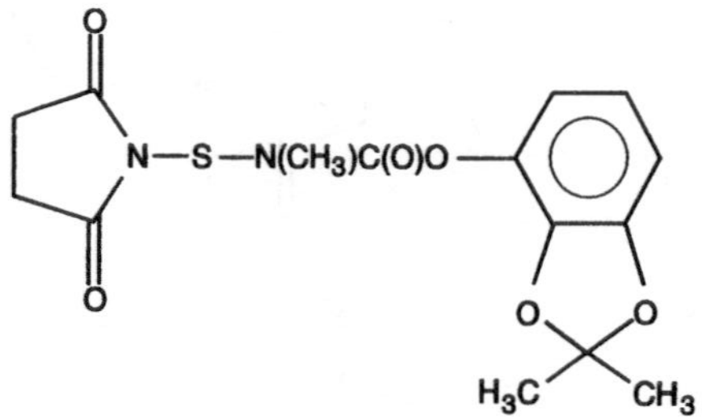		[7]
$(CH_2)_5NSi(CH_3)_3$	mole ratio 1 : 1; in CH_2ClCH_2Cl at 10 to 15 °C for 30 min	$\overline{C(O)-(CH_2)_2-C(O)-N}SN(CH_2)_5$ $ClSi(CH_3)_3$	96	[2]

$4\text{-}ClC_6H_4SH + (C_2H_5)_3N$	mole ratio 1:1:1; in CH_2ClCH_2Cl at 5 to 10 °C, and 2 h at 20 to 25 °C	$\overline{C(O){-}(CH_2)_2{-}C(O){-}N}SSC_6H_4Cl\text{-}4$ $(C_2H_5)_3N \cdot HCl$	88 90	[2]
$4\text{-}ClC_6H_4SSi(CH_3)_3$	mole ratio 1:1; in CH_2ClCH_2Cl at 10 to 15 °C, and ~1 h at 20 to 25 °C	$\overline{C(O){-}(CH_2)_2{-}C(O){-}N}SSC_6H_4Cl\text{-}4$ $ClSi(CH_3)_3$	95	[2]

12.5.2.2.5 ClSN̅–C(O)–C(CH₃)=C(CH₃)–C̅(O)

The title compound was formed with a yield of 87.7% by adding freshly distilled SCl_2 (in excess) to a precooled ($-10\,°C$) solution of dimethylmaleinic acid imide in CH_2Cl_2. Then the ca. equimolar amount of pyridine was added at 0 to $10\,°C$ and the mixture reacted for 4 h at $25\,°C$.

The solid was not characterized.

Reference:

Berger, J. (Ger. 3 314 951 [1983]; C.A. **100** [1984] No. 53 366).

12.5.2.3 N²-Chlorosulfanyl-1H-isoindole-1,3(2H)-dione, ClSN̅–C(O)–(1-C₆H₄-2)–C̅(O)

The title compound was prepared by bubbling Cl_2 gas for 8 h through a stirred solution of N,N′-disulfanediylbis(phthalimide) in $CHCl_3$ maintained at 50 to $60\,°C$. N_2 gas was passed through the mixture to remove excess Cl_2; subsequent evaporation in vacuum yielded the compound with a yield of 98% [1]. In previous papers, chlorobenzene was reported to be used as solvent with a reaction temperature of $35\,°C$ [2] (see also [3]); 85% yield [2]. Yields of 89.5 to 98% were achieved, if the latter reaction was performed in the presence of catalytic amounts of α,α′-azobisisobutyronitrile at 45 to $50\,°C$ or under UV irradiation [4, 5]. The title compound was also formed by boiling N,N′-disulfanediylbis(phthalimide) with excess SO_2Cl_2 until completely dissolved; 55% yield [6]; see also [7].

Pale yellow [6] or yellow [1, 2, 4, 5] crystals, m.p. 135 to $137\,°C$ [2, 4, 5], 138 to $139\,°C$ (from CCl_4) [6]; see also [7].

Dipole moment: $\mu = 4.56$ D, m(N–S) = 0.30 D, indicating that the N–S bond is nearly unpolar. On that basis, the molar Kerr constant was determined to be $mK = 1467 \times 10^{-12}$ esu/ mole, the molar anisotropy $\mu^2 = 332.3$ A⁶ [8].

¹H NMR ($CDCl_3$/TMS): $\delta = 7.75$ to 8.20 ppm (m, 4H) [1]; (CH_2Cl_2/HMDS): $\delta = 7.95$ ppm (center of m) [6], see also [7].

UV ($CHCl_3$): λ_{max} (in nm) (log ε) = 295 (3.35), 304 (3.31) [6], see also [7].

The title compound reacted with $(CH_3)_3SiN=S=NSi(CH_3)_3$, mole ratio 1 : 1, in CH_2Cl_2, 5 h at $20\,°C$ or in the mole ratio 2 : 1 in CH_2Cl_2, 24 h at $20\,°C$ to give I (65%) and II (80%), respectively.

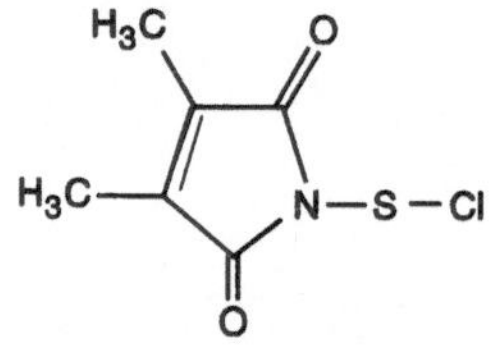

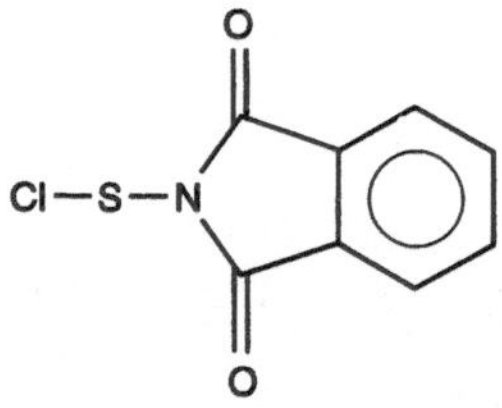

The reaction of the title compound with $(CH_3)_3SiN{=}S{=}NSN{=}S{=}NSi(CH_3)_3$ (mole ratio 1 : 1 and 2 : 1) in THF (at −50 °C for 2 h, rising to 20 °C) yielded 58% S_4N_4 along with 65% phthalimide and II (75%), respectively [6]; see also [7].

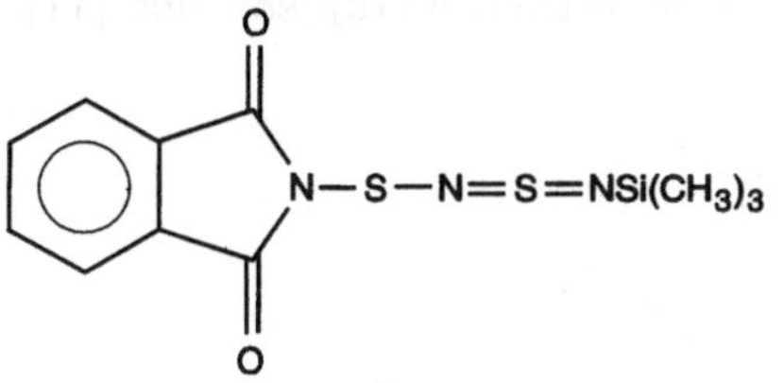

I

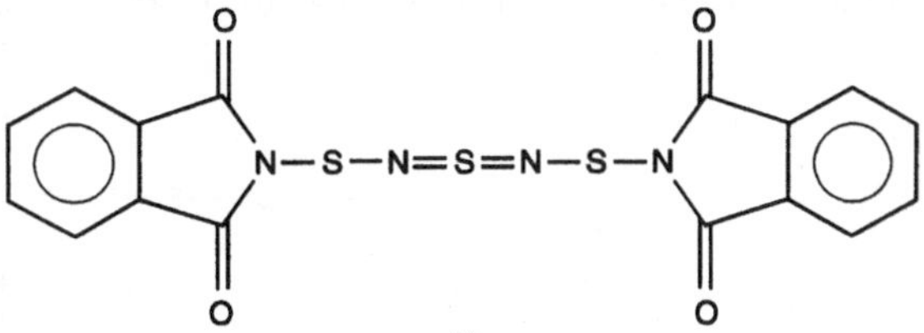

II

When a solution of the title compound in $CHCl_3$ was added to a solution of an appropriate arylamine $ArNHSi(CH_3)_3$ (Ar = 4-RC_6H_4, with R = H, NO_2, Cl, Br, OCH_3, CH_3, or Ar = 1-naphthyl, 3-pyridyl; mole ratio 1 : 1) in $CHCl_3$ within 5 min at room temperature and the mixture reacted for 12 h at room temperature, the corresponding N-(arylaminosulfanyl)-phthalimides (III) were formed with yields of 60 to >95% [9].

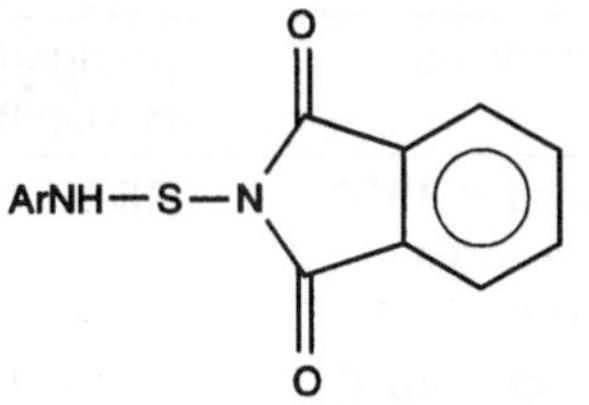

III

The reaction of the title compound with an equimolar amount of $(CH_3)_3SiN{=}C(CH_3)O$-$Si(CH_3)_3$ resulted in a ca. 1 : 1 mixture of the two tautomeric monosilylated compounds IVa and IVb [9].

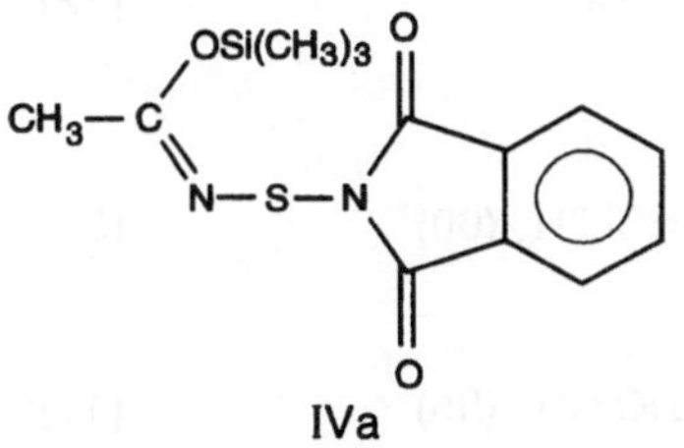

IVa

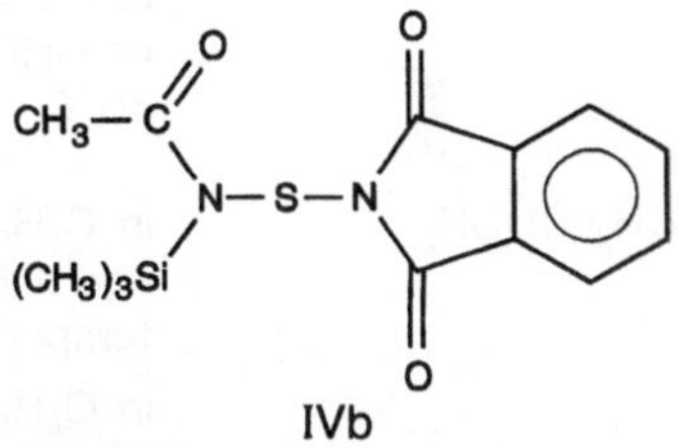

IVb

The title compound was reacted with $ArN{=}S{=}NSi(CH_3)_3$ (mole ratio 1 : 1; Ar = C_6H_5, C_6F_5, 2,6-$(CH_3)_2C_6H_3$) in CH_2Cl_2 at 20 °C for 10 h to give the corresponding V (80, 60, or 75%). Reacting the title compound with $LiN(Si(CH_3)_3)_2$ (mole ratio 1 : 1) in THF at −50 °C and raising the temperature to 20 °C within 1 h gave VI (72%) [6]; see also [7].

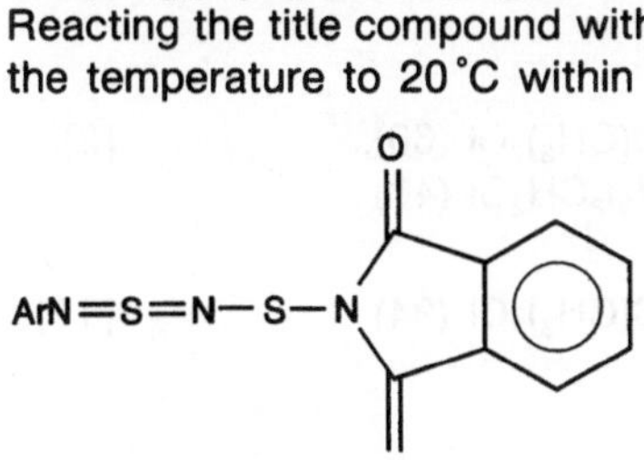

V

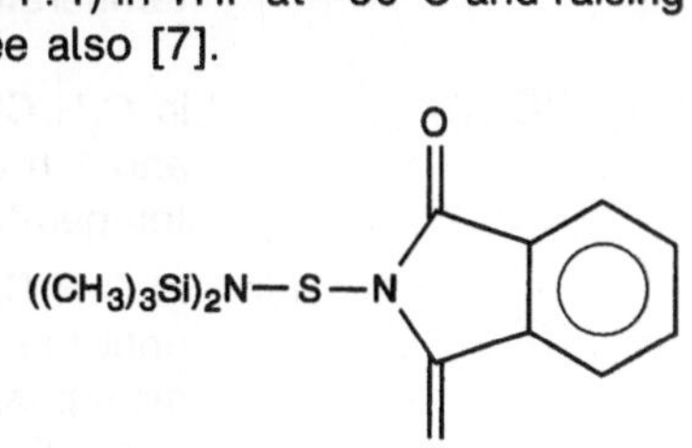

VI

Adding $(RO)_2P(S)SH$ (R = CH_3, C_2H_5, n-C_3H_7, i-C_3H_7, c-C_6H_{11}, c-C_6H_9, C_6H_5) in chlorobenzene at 20 °C or in $CHCl_3$ at 30 °C to the title compound (mole ratio 1 : 1) and letting the mixture react for additionally 15 to 20 min gave the corresponding thio-phosphoryldisulfanylphthalimides VII (99.4, 99.2, 100.0, 99.2, 97.1, 98.9, or 99.8%, respectively) [10]; see also [11].

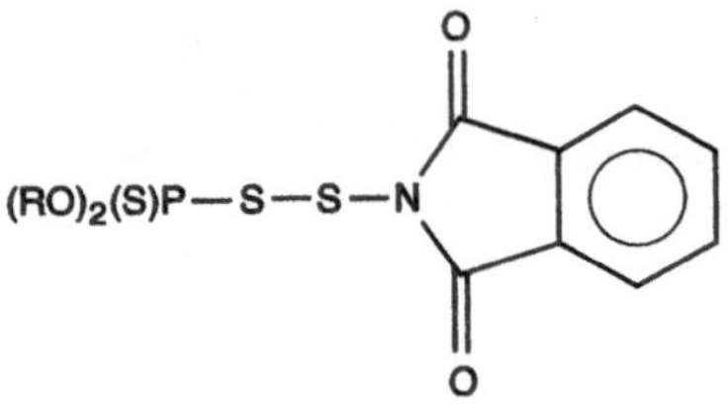

VII

The reactions of the title compound with organic compounds are summarized in Table 41.

Table 41
Reactions of $ClS\overline{N-C(O)-(1-C_6H_4-2)-C}(O)$ with Organic Compounds.

substrate	reaction conditions	product(s) (yield in %) R = $\overline{C(O)-(1-C_6H_4-2)-C(O)-N}$	Ref.
$CH_2{=}CH_2$	in C_6H_5Cl at < 40 °C and 1 h at room temperature	$RSCH_2CH_2Cl$ (90)	[3]
	in C_6H_5Cl at < 40 °C until the yellow color disappeared and 0.5 h at room temperature	$RSCH_2CH_2Cl$ (96)	[11]
	in $CCl_2{=}CCl_2$, CH_2Cl_2, or $CHCl_3$ at 25 to 40 °C	$RSCH_2CH_2Cl$	[12]
$CH_2{=}CHCH_3$	in C_6H_5Cl at < 40 °C and 1 h at room temperature	$RSCH_2CHClCH_3$ (60)	[3]
	in C_6H_5Cl at < 40 °C until the yellow color disappeared and 0.5 h at room temperature	$RSCH_2CHClCH_3$ (95)	[11]
$CH_2{=}C(CH_3)_2$	in C_6H_5Cl at < 40 °C and 1 h at room temperature	$RSCH_2C(CH_3)_2Cl$ (60), $RSC(CH_3)_2CH_2Cl$ (40)	[3]
	in C_6H_5Cl at < 40 °C until the yellow color disappeared and 0.5 h at room temperature	$RSCH_2C(CH_3)_2Cl$ (94)	[11]

Table 41 (continued)

substrate	reaction conditions	product(s) (yield in %) $R = \overline{C(O)-(1\text{-}C_6H_4\text{-}2)-C(O)-N}$	Ref.
$CH_2=CHC_6H_5$	in C_6H_5Cl at <40°C until the yellow color disappeared and 0.5 h at room temperature	$RSCH_2CH(C_6H_5)Cl$ (91)	[11]
	mole ratio 2.6 : 3.5; in CH_2Cl_2 at room temperature	$RSCH_2CH(C_6H_5)Cl$ (100)	[1]
	in chlorobenzene at <40°C and 1 h at room temperature	$RSCH_2CH(C_6H_5)Cl$ (92)	[3]
$CH_2=C(CH_3)C_6H_5$	in chlorobenzene at <40°C and 1 h at room temperature	$RSCH_2C(CH_3)ClC_6H_5$ (90)	[3]
$CH_2=CHCH_2OH$	in C_6H_5Cl at <40°C until the yellow color disappeared and 0.5 h at room temperature	$RSCH_2CH(CH_2OH)Cl$ (80)	[11]
$CH_2=CHCH_2Cl$	in chlorobenzene at 35°C and 24 h at room temperature	1 : 1 addition compound (62)	[3]
$CH_3CH=CHCH_3$	in C_6H_5Cl at <40°C until the yellow color disappeared and 0.5 h at room temperature	$RSCHCH_3CHClCH_3$ (96)	[11]
$CH_3CH=C(CH_3)_2$	in C_6H_5Cl at <40°C and 1 h at room temperature	$RSC(CH_3)_2C(CH_3)ClH$ (50), $RSCHCH_3C(CH_3)_2Cl$ (50)	[3]
$CH_3CH=CHCH_2Br$	in C_6H_5Cl at <40°C and 1 h at room temperature	$RSCHCH_3CHClCH_2Br$ (60)	[3]
$CH_2=C(CH_3)CH(CH_3)_2$	in C_6H_5Cl at <40°C and 1 h at room temperature	$RSCH_2C(CH_3)ClCH(CH_3)_2$ (75)	[3]
$CH_2=C(CH_3)CH=CH_2$	in C_6H_5Cl at 0°C and 5 h at room temperature	1 : 1 addition compound (34.5)	[3]

Table 41 (continued)

substrate	reaction conditions	product(s) (yield in %) $R = C(O)–(1-C_6H_4-2)–C(O)–N$	Ref.
trans-2-hexene (excess)	in $CCl_2{=}CCl_2$, CH_2Cl_2, or $CHCl_3$ at 25 to 40°C	mixture of $CH_3CH_2CH_2CHClC(SR)HCH_3$ and $CH_3CH_2CH_2C(SR)HCHClCH_3$	[12]
EPDM-A (an ethylene-propylene-1,4-hexadiene terpolymer)	in $CHCl_3$ containing catalytic amounts of Na_2CO_3 at 25°C and 17 h at 25 to 35°C; addition was favored by polar solvents or by $ZnCl_2$ catalysis	1,2-addition to the olefinic sites of the terpolymer	[12]
$(CH_2)_4CH{=}CH(CH_2)_4$	in C_6H_5Cl at <40°C until the yellow color disappeared and 0.5 h at room temperature	$RSCH(CH_2)_4CH(CH_2)_4Cl$ (97)	[11]
cyclopentene	in chlorobenzene at <40°C and 1 h at room temperature	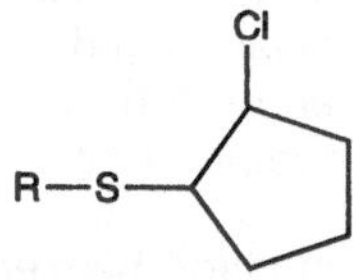 (90)	[3]
	mole ratio 1:>1; in CH_2Cl_2 at room temperature	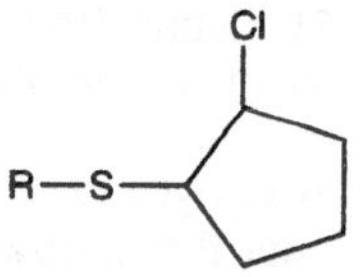 (80.4)	[1]
cyclohexene	mole ratio 5.2:7.3; in CH_2Cl_2 at room temperature	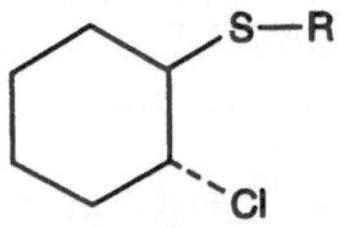 (93.3)	[1]
	in chlorobenzene at <40°C and 1 h at room temperature	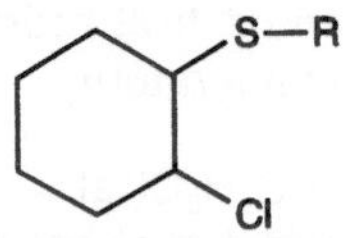	[3]
	mole ratio 1:>1; in $CCl_2{=}CCl_2$, CH_2Cl_2, or $CHCl_3$ at 25 to 40°C	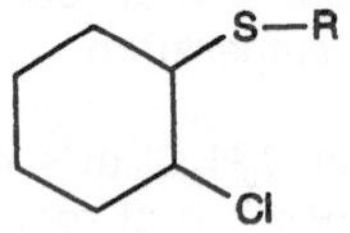	[12]
cyclooctene	mole ratio ca. 1:1; in CH_2Cl_2 at room temperature	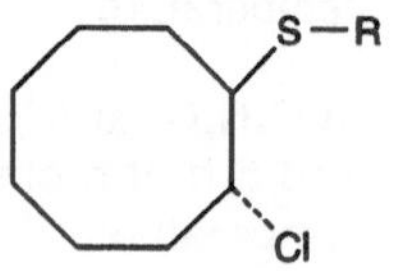 (73.7)	[1]

Table 41 (continued)

substrate	reaction conditions	product(s) (yield in %) $R = \overline{C(O)–(1\text{-}C_6H_4\text{-}2)–C(O)–N}$	Ref.
norbornene	mole ratio 1 : >1; in CH_2Cl_2 at room temperature	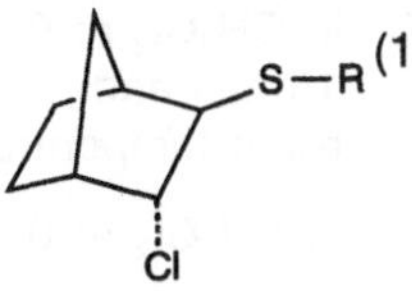 (100)	[1]
	in chlorobenzene at 0 °C and 5 h at room temperature	1 : 1 addition compound (21.5)	[3]
	in chlorobenzene at 35 °C and 24 h at room temperature	1 : 1 addition compound (61.5)	[3]
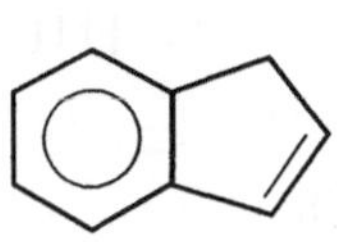	in chlorobenzene at < 40 °C and 1 h at room temperature	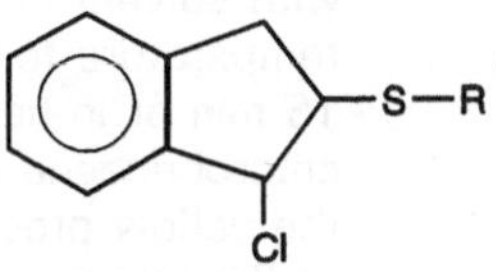 (79)	[3]
CH≡CH	in C_6H_5Cl at < 40 °C until the yellow color disappeared and 0.5 h at room temperature	RSCH=CHCl (43)	[11]
$CH≡CCH_3$	in CH_2Cl_2 at 0 °C for 15 min and 15 min at room temperature	$RSC(CH_3)$=CHCl (87)	[13]
$CH≡CC_4H_9$-t	in CH_2Cl_2 at 0 °C for 15 min and 15 min at room temperature	$RSC(C_4H_9$-t)=CHCl (92)	[13]
	in CH_2Cl_2 at 0 °C	RSCH=C(Cl)C_4H_9-t (100)	[14]
	in CH_2Cl_2 at 0 °C	$RSC(C_4H_9$-t)=CHCl, RSCH=C(C_4H_9-t)Cl (98 : 2 adduct ratio)	[15]
$CH≡CC_4H_9$-n	in CH_2Cl_2 at 0 °C	RSC(n-C_4H_9)=CHCl, RSCH=C(Cl)C_4H_9-n (11 : 89 [14], 89 : 11 [15] adduct ratio)	[14, 15]
$CH≡CC_6H_5$	in CH_2Cl_2 at 0 °C	$RSC(C_6H_5)$=CHCl, RSCH=C(Cl)C_6H_5 (15 : 85 [14], 85 : 15 [15] adduct ratio)	[14, 15]
$CH_3C≡CCH_3$	in CH_2Cl_2 at 0 °C for 15 min and 15 min at room temperature	$RSC(CH_3)$=C(Cl)CH_3 (94)	[13], see also [15]

Table 41 (continued)

substrate	reaction conditions	product(s) (yield in %) $R = \overline{C(O)-(1-C_6H_4-2)-C(O)-N}$	Ref.
$C_2H_5C{\equiv}CC_2H_5$	in CH_2Cl_2 at 0°C for 15 min and 15 min at room temperature	$RSC(C_2H_5){=}C(Cl)C_2H_5$ (86)	[13]
$CH_3C{\equiv}CC_6H_5$	in CH_2Cl_2 at 0°C	$RSC(C_6H_5){=}C(Cl)CH_3$ (100)	[14]
$C_6H_5C{\equiv}CC_6H_5$	in CH_2Cl_2 at 0°C for 15 min and 15 min at room temperature	$RSC(C_6H_5){=}C(Cl)C_6H_5$	[15]
$i\text{-}C_3H_7CHO$ + $C_6H_5CH_2N(CH_3)_2$	mole ratio ca. 1:1.1 : 1.1; in C_6H_5Cl at 30 to 35°C, cooling to ca. 20°C	$(CH_3)_2C(SR)CHO$	[2]
$R^1CH_2C(O)R^2$	with solvent at room temperature for 10 to 15 min or in hexane–chlorobenzene until the yellow precipitate of the title compound disappeared	$RSCHR^1C(O)R^2$ $R^1 = H$, $R^2 = CH_3$ (81); $R^1 = R^2 = CH_3$ (80); $R^1 = n\text{-}C_3H_7$, $R^2 = CH_3$ (76); $R^1 = i\text{-}C_3H_7$, $R^2 = CH_3$ (81); $R^1 = C_4H_9$, $R^2 = C_5H_{11}$ (48); $R^1 = C_5H_{11}$, $R^2 = CH_3$ (50); $R^1 = R^2 = (CH_2)_4$ (60); $R^1 = CH_3CO$, $R^2 = CH_3$ (88)	[11]
$N{\equiv}CCH_2CH_2NHR'$ + $(C_2H_5)_3N$	mole ratio 1:1:1; in C_6H_5Cl at 0 to 10°C	$RSN(R')CH_2CH_2CN$ $R' = CH_3$ (86.4), $R' = C_2H_5$ (84.7), $R' = n\text{-}C_4H_9$ (83.9), $R' = c\text{-}C_6H_{11}$ (85), $R' = CH_2CH_2CN$ (91.3)	[16]
R^1R^2NH + $(C_2H_5)_3N$	mole ratio 1:1:1; in C_6H_5Cl, 1 h at room temperature	$RSNR^1R^2$ $R^1 = R^2 = CH_2CH{=}CH_2$ (87); $R^1 = CH_2C_6H_5$, $R^2 = CH_2CH_2CN$ (89); $R^1 = R^2 =$ carbazolyl (84)	[11]
$O(CH_2)_4NH$	mole ratio 1:1:1; in C_6H_5Cl at 0 to 5°C and 1 h at room temperature	$RSN(CH_2)_4O$ (82.5)	[16]
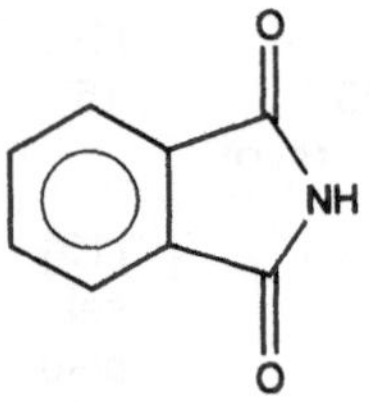 + $(C_2H_5)_3N$		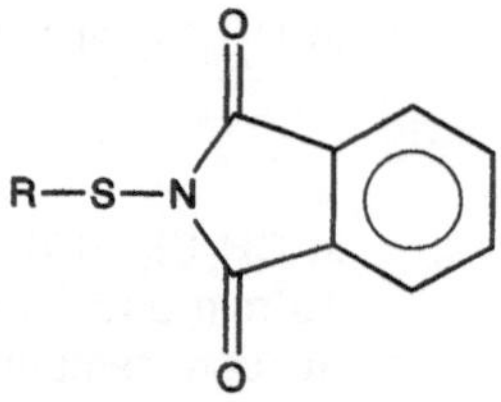	[12]

Table 41 (continued)

substrate	reaction conditions	product(s) (yield in %) $R = \overline{C(O){-}(1\text{-}C_6H_4\text{-}2){-}C(O){-}N}$	Ref.
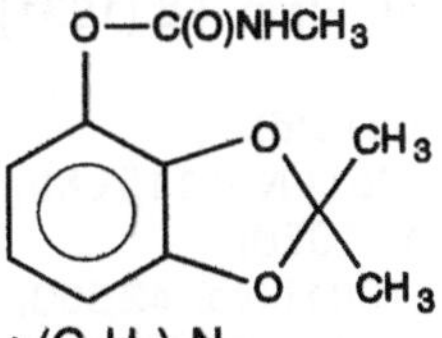 $+(C_2H_5)_3N$ $+c\text{-}C_6H_{11}NH_2$	in CH_2Cl_2 at −10 to 0 °C and 15 min at 0 °C, adding then $c\text{-}C_6H_{11}NH_2$	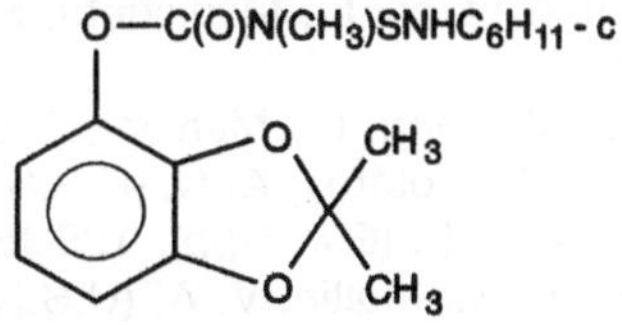	[17]
R′SH	mole ratio 1 : 1; in C_6H_5Cl at 20 °C for 30 min [11, 18] and 1 h at room temperature [18]	RSSR′ R′ = C_2H_5 (91) [11], R′ = $n\text{-}C_3H_7$ (91) [18], R′ = $i\text{-}C_3H_7$ (90) [11], R′ = $n\text{-}C_4H_9$ (90) [18], R′ = $i\text{-}C_4H_9$ (90) [18], R′ = $t\text{-}C_4H_9$ (90) [18], R′ = $c\text{-}C_6H_{11}$ (88) [18], R′ = $CH_2C_6H_5$ (94) [11], R′ = C_6H_5 (80) [18], R′ = $4\text{-}ClC_6H_4$ (86) [11], R′ = $4\text{-}BrC_6H_4$ (85) [11], R′ = $4\text{-}CH_3C_6H_4$ (80) [18], R′ = $2,6\text{-}(t\text{-}C_4H_9)_2\text{-}4\text{-}OHC_6H_2$ (80) [11]	[11, 18]
$n\text{-}C_3H_7SH$ $+(C_2H_5)_3N$	in C_6H_5Cl at 20 °C	$n\text{-}C_3H_7SSSC_3H_7$ (92.9), RSR, RH	[11]
$CH_3C(O)SH$	in C_6H_5Cl, 30 min at 45 to 50 °C	$RSSC(O)CH_3$ (57)	[19]

References:

[1] Bombala, M. U.; Ley, S. V. (J. Chem. Soc. Perkin Trans. I **1979** 3013/6).
[2] Schubart, R.; Blazejak, M.; Roos, E. (Ger. 2330241 [1973/75]; C.A. **86** [1977] No. 139844).
[3] Uhrhan, P.; Roos, E.; Abele, M.; Schubart, R.; Kempermann, T. (Ger. 2502656 [1975/76]; C.A. **85** [1976] No. 144469).
[4] Schubart, R.; Uhrhan, P.; Roos, E. (Ger. 2646349 [1976/78]; C.A. **89** [1978] No. 42839).
[5] Schubart, R.; Uhrhan, P.; Roos, E. (Ger. 2717636 [1977/78]; C.A. **90** [1979] No. 72040).
[6] Zibarev, A. V.; Miller, A. O.; Shakirov, M. M.; Furin, G. G. (Zh. Obshch. Khim. **61** [1991] 951/60; J. Gen. Chem. USSR [Engl. Transl.] **61** [1991] 864/72).
[7] Zibarev, A. V.; Miller, A. O.; Gatilov, Yu. V.; Furin, G. G. (Heteroat. Chem. **1** [1990] 443/53).
[8] Arbuzov, B. A.; Aleksandrova, L. K.; Mulin, V. A.; Zolotov, A. P.; Vereshchagin, A. N. (Izv. Akad. Nauk SSSR Ser. Khim. **1982** 1410/2; Bull. Acad. Sci. USSR [Engl. Transl.] **31** [1982] 1257/9).
[9] Bryce, M. R.; Taylor, P. C. (J. Chem. Soc. Perkin Trans. I **1990** 3225/35).
[10] Zolotov, A. N.; Mullin, V. A. (U.S.S.R. 857142 [1979/81]; C.A. **96** [1982] No. 6578).

[11] Mullin, V. A.; Zolotov, A. N. (Zh. Obshch. Khim. **56** [1986] 898/904; J. Gen. Chem. USSR [Engl. Transl.] **56** [1986] 791/6).
[12] Hopper, R. J. (Rubber Chem. Technol. **49** [1976] 341/52; C.A. **85** [1976] No. 64398).
[13] Capozzi, G.; Gori, L.; Menichetti, S. (Tetrahedron **47** [1991] 7185/96).
[14] Capozzi, G.; Gori, L.; Menichetti, S. (Phosphorus Sulfur Silicon Relat. Elem. **59** [1991] 451/4).
[15] Capozzi, G.; Gori, L.; Menichetti, S. (Tetrahedron Lett. **31** [1990] 6213/6).
[16] Mullin, V. A.; Zolotov, A. N. (U.S.S.R. 825522 [1979/81]; C.A. **95** [1981] No. 115089).
[17] Saunders, D. E. (Eur. Appl. 129359 [1984]; C.A. **102** [1985] No. 185070).
[18] Zolotov, A. N.; Mullin, V. A. (U.S.S.R. 806680 [1978/81]; C.A. **95** [1981] No. 42900).
[19] Senning, A.; Jensen, B. (Sulfur Lett. **2** [1984] 11/5; C.A. **101** [1984] No. 38301).

12.5.2.4 N[1]-Chlorosulfanyl-piperidine and Derivatives

12.5.2.4.1 $ClS\overline{N{-}(CH_2)_4{-}CH_2}$

$ClSN(CH_2)_5$ was formed directly with 63% yield by adding piperidine to a precooled (0 °C) mixture of SCl_2 and CH_3CN [1].

It was also obtained by cleaving the sulfur–sulfur bond of $(CH_2)_5NSSN(CH_2)_5$ with a chlorinating agent as follows:

a) Cl_2 was bubbled through a suspension of $(CH_2)_5NSSN(CH_2)_5$ in CCl_4 at 0 to 5 °C until there was a slight excess of Cl_2 which then was removed at 0 to 5 °C in vacuum or by purging with an inert gas. The resulting solution was cleared with animal charcoal, and after filtration the solvent was distilled at 30 to 50 °C under reduced pressure to give the crude compound with 85% yield. It was finally distilled in high vacuum [2, 3]; see also [4].

b) A solution of SO_2Cl_2 in CCl_4 was added to a solution of $(CH_2)_5NSSN(CH_2)_5$ (mole ratio 1 : 1) in CCl_4. The temperature rose to about 40 °C while SO_2 was simultaneously evolved. The mixture was then heated to 70 to 80 °C for 15 min [3, 5]; see also [4]; the same reaction was performed with an excess of SO_2Cl_2 (mole ratio ca. 2 : 3) without solvent at −78 °C [6].

c) A solution of $(CH_2)_5NSSN(CH_2)_5$ in CCl_4 was added without cooling to a suspension of powdered PCl_5 (excess) in CCl_4, and the mixture maintained at 40 to 50 °C for 20 min. After it had cooled to room temperature, the mixture was cleared with animal charcoal and filtered, the solvent evaporated, and the residue distilled in vacuum [3, 5]; see also [4].

$ClSN(CH_2)_5$ was formed with a yield as high as 84% by reacting $(CH_2)_5NSOCH_3$ with RC(O)Cl (R = CH_3, C_6H_5) at 20 to 25 °C in an inert atmosphere [7]. $ClSN(CH_2)_5$ was formed along with ClC≡N by decomposition of $(CH_2)_5NSN{=}CCl_2$ at 40 °C [8].

Orange-colored liquid [1 to 3, 5], brown liquid [8]; b.p. 37 to 39 °C/0.09 Torr [3, 5], 44 to 45 °C/0.3 Torr [2, 3] (see also [4]), 46 to 48 °C/0.4 to 0.5 Torr [3, 5], 50 to 52 °C/0.4 Torr [1], 52 to 53 °C/0.2 Torr [7], 53 °C/0.2 Torr [9].

Refractive index: $n_D^{20} = 1.5460$ [7], $n_D^{24} = 1.5470$ [9].

The title compound decomposed in the presence of moisture and when standing at 25 to 30 °C for several days [10]. Caution: An explosion occurred during the distillation after the flask temperature exceeded 80 °C [11].

Treating the title compound with S_2Cl_2 (mole ratio 1 : 1) and aqueous ammonia gave $S{=}S{=}NSN(CH_2)_5$ with low (1.2% calculated) yield; a reaction path was proposed [12].

Adding the title compound to $C_2H_5OPCl_2$ (mole ratio 1 : 2) at < 25 °C and cooling with ice overnight gave $C_2H_5OP(S)(Cl)N(CH_2)_5$ (95%), $POCl_3$ (95%), and C_2H_5Cl [13]. $ClSN(CH_2)_5$ reacted with an equimolar amount of $(C_2H_5O)_2P(O)NHNHC_6H_4R$ (R = H, 3-CF_3) and excess $(C_2H_5)_3N$ in ether at 2 to 4 °C for 2 h to give the corresponding $(C_2H_5O)_2P(O)NHN(C_6H_4R)SN(CH_2)_5$ with a yield of 84 or 29%, respectively [14, 15]. Adding a solution of $ClSN(CH_2)_5$ in CH_2Cl_2 to a precooled (0 °C) solution of $(2\text{-}CH_3OC_6H_4)_2P(O)CH_2NHCH_2C(O)OCH_3$ and $(C_2H_5)_3N$ (mole ratio ca. 3 : 1 : 1) in toluene yielded $(2\text{-}CH_3OC_6H_4)_2P(O)CH_2N(CH_2C(O)OCH_3)SN(CH_2)_5$ [16]. Adding $(C_2H_5O)_2P(S)H$ to a precooled (−5 °C) solution of $ClSN(CH_2)_5$ (mole ratio 2 : 1) or to a mixture of $ClSN(CH_2)_5$ and pyridine in CCl_4 so that the temperature did not rise above 0 °C gave $(C_2H_5O)_2P(S)SP(S)(OC_2H_5)_2$ and $(CH_2)_5NSP(S)(OC_2H_5)_2$, respectively, along with $(CH_2)_5NH \cdot HCl$ [17]. Adding a solution of $(RO)_2P(S)SH$ (R = C_2H_5, n-C_3H_7, n-C_4H_9, i-C_4H_9, i-C_5H_{11}) in CCl_4 to a to 10 °C cooled solution of $ClSN(CH_2)_5$ (mole ratio 2 : 1) in CCl_4 gave the corresponding $((RO)_2P(S))_2S_3$ along with $(CH_2)_5NH \cdot HCl$; a reaction mechanism is suggested in the paper [18]. Treating $ClSN(CH_2)_5$ with $(C_2H_5O)_2P(S)NHNHC_6H_4R$ (R = H, 3-CF_3) or $(C_6H_5O)_2P(S)NHNHC_6H_5$ (mole ratio 1 : 1) and a slight excess of $(C_2H_5)_3N$ in ether at −5 to +5 °C for 0.5 h yielded the corresponding $(C_2H_5O)_2P(S)N(SN(CH_2)_5)NHC_6H_4R$ (R = H, 3-CF_3, both ~100%) or $(C_6H_5O)_2P(S)N(SN(CH_2)_5)NHC_6H_5$ (100%) [14]; see also [15]. Adding a mixture of $ClSN(CH_2)_5$ and pyridine in petroleum ether to a precooled (10 °C) suspension of $(CH_2)_5NH_2^+$ $(C_2H_5O)_2P(S)S^-$ (mole ratio ca. 1 : 1 : 1) in petroleum ether gave 60% $(C_2H_5O)_2P(S)SSN(CH_2)_5$ along with pyridinium hydrochloride [18].

$ClSN(CH_2)_5$ reacted with $CH_2{=}CHMgBr$ (mole ratio ca. 1 : 1 in THF at −78 °C for 15 min and then at room temperature for 2.5 h) to give $(CH_2)_5NSCH{=}CH_2$ (yielding 90% crude product) [20].

The title compound reacted with alkynyl metal compounds $MC{\equiv}CR$ (R = H, n-C_4H_9, $(CH_3)_3Si$; M = Li, $LiB(OCH_3)_3$, MgBr) to $(CH_2)_5NSC{\equiv}CR$; see the following table [19].

$MC{\equiv}CR$ R	M	reaction conditions	products $(CH_2)_5NSC{\equiv}CR$	yield[a] (in %)
H	$LiB(OCH_3)_3$	−78 to +20 °C, 2.5 h	R = H	78
H	MgBr	−20 to +20 °C, 1 h	R = H	(75)
			R = $SN(CH_2)_5$	(25)
n-C_4H_9	Li	−78 to +20 °C, 2 h	R = n-C_4H_9	(98)
$(CH_3)_3Si$	Li	−78 to +20 °C, 2 h	R = $(CH_3)_3Si$	(97)

[a] Yield of crude product (given in parentheses) was estimated by ^1H NMR.

Reacting a solution of $ClSN(CH_2)_5$ in ether at −78 °C with a suspension of $NaN(Si(CH_3)_3)_2$ in ether and allowing the mixture to come to room temperature overnight yielded $(CH_2)_5NSN(Si(CH_3)_3)_2$ (69%). With a solution of $LiN(C_4H_9\text{-}t)Si(CH_3)_3$ analogously $(CH_2)_5NSN(C_4H_9\text{-}t)Si(CH_3)_3$ was formed [21]. Adding $ClSN(CH_2)_5$ to a solution of CH_3SNa in CH_3OH at −10 °C yielded $(CH_2)_5NSSCH_3$ [22]. When a solution of $ClSN(CH_2)_5$ in benzene was added to a suspension of $(C_2H_5O)_2P(S)SNa$ in benzene, the temperature rose to about 30 to 35 °C; stirring the mixture at this temperature for 1 h gave $(C_2H_5O)_2P(S)SSN(CH_2)_5$ (ca. 90%) [23]. Adding $ClSN(CH_2)_5$ to an equimolar amount of $AgN(SO_2CH_3)_2$ in CH_3CN at −50 °C and reacting the mixture for another 6 h at −50 °C gave $(CH_2)_5NSN(SO_2CH_3)_2$ [24].

Reactions of $ClSN(CH_2)_5$ with organic compounds are summarized in Table 42, pp. 202/6.

Table 42
Reactions of $ClSN(CH_2)_5$ with Organic Compounds.

organic substrate	reaction conditions	product(s) (yield in %)	Ref.
$CH_3CH{=}CH_2 + AlCl_3$ (cat.)	in CCl_4 at 0 °C	$CH_3CHClCH_2SN(CH_2)_5$ (79)	[25]
$CH{\equiv}CC_6H_5$	(nonregiospecific addition)	$C_6H_5C(Cl){=}CHSN(CH_2)_5$, $CHCl{=}C(C_6H_5)SN(CH_2)_5$	[26]
C_2H_5OH + pyridine	mole ratio 1 : 1 : 2; in petroleum ether at 0 °C and 24 h at room temperature	$(CH_2)_5NSOC_2H_5$ (65)	[27]
C_2H_5OH	mole ratio 1 : 2; in petroleum ether at 5 °C and 24 h at room temperature	$C_2H_5OSOC_2H_5$ (65)	[27]
$H_2NCH_2CH_2NH_2$	mole ratio 1 : 1; in ether at −20 °C, and stirring at −20 °C overnight	intractable tar	[28]
$H_2N(CH_2)_3NH_2$	mole ratio 3 : 1(?); in ether at −20 °C, and stirring at −20 °C overnight	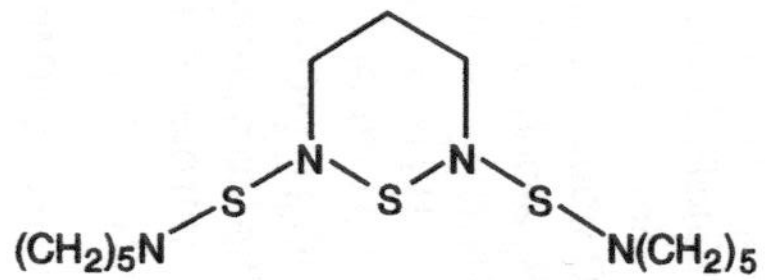	[28]
$H_2N(CH_2)_4NH_2$	mole ratio 1 : 1; in ether at −20 °C, and stirring at −20 °C overnight	intractable tar	[28]
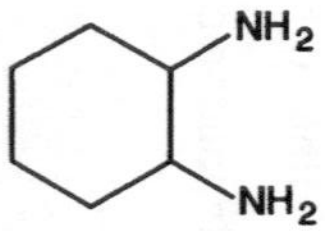	mole ratio 1 : 1; in ether at −20 °C and stirring at −20 °C overnight	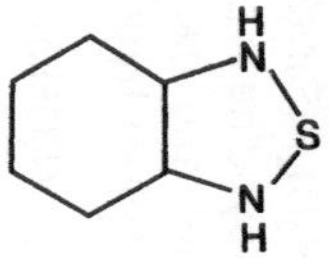 (10) (impure)	[28]
$4\text{-}XC_6H_4NH_2$ (X = H, Cl, Br, NO_2)	mole ratio 1 : 2; in anhydrous ether at −20 °C	$4\text{-}XC_6H_4NHSNHC_6H_4X\text{-}4$	[29]

4-$XC_6H_4NH_2$ ($X = H$, OCH_3, Cl, Br)	mole ratio 1:2; in ether at 10 °C and stirring for 1 h at room temperature under N_2	4-$XC_6H_4NHSNHC_6H_4X$-4 $X = H$ (75), OCH_3 (55), Cl (70), Br (67)	[11]
$XC_6H_4NH_2$ ($X = H$, 4-OCH_3, 3-NO_2, 4-Cl, 4-Br) + $(C_2H_5)_3N$	as above at −78 °C; or in mole ratio 1:1 containing $(C_2H_5)_3N$	$(CH_2)_5NSNHC_6H_4X$ $X = H$ (46), 4-OCH_3 (42), 3-NO_2 (65), 4-Cl (70), 4-Br (60)	[11]
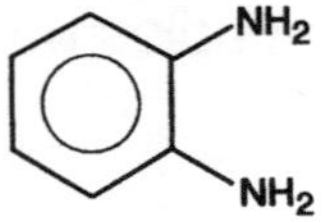	mole ratio 1:1; in ether at −20 °C and stirring at −20 °C overnight	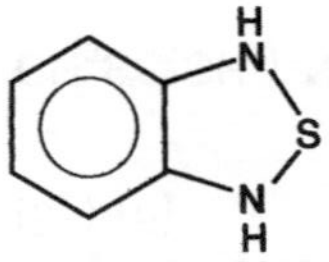 (>45), 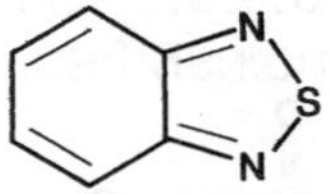 (15), $(CH_2)_5NSN(CH_2)_5$ (10)	[28]
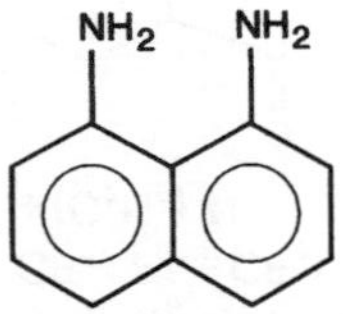	mole ratio 1:1; in ether at −20 °C and stirring at −20 °C overnight	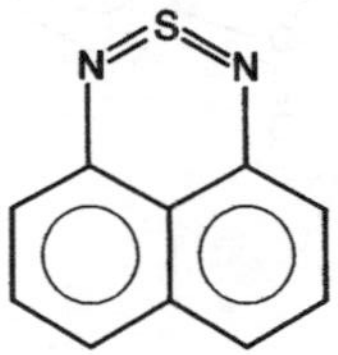 (62)	[28]
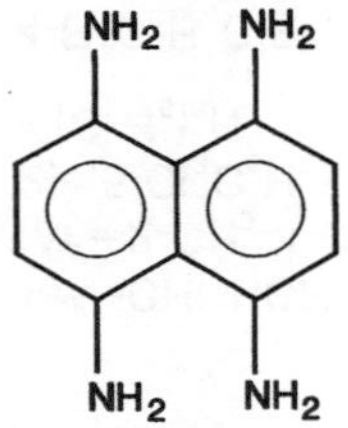	mole ratio 1:1; in ether at −20 °C and stirring at −20 °C overnight	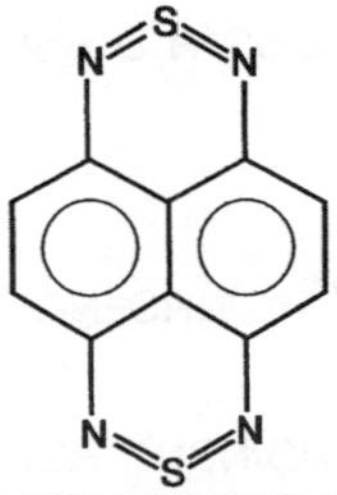 (25)	[28]
$HN(CH_2)_2 + (C_2H_5)_3N$	mole ratio ~1:1:1; in CCl_4 at 0 to 5 °C, and 3 h at room temperature	$(CH_2)_5NSN(CH_2)_2$ (55)	[30], see also [4]

Table 42 (continued)

organic substrate	reaction conditions	product(s) (yield in %)	Ref.
$HN(CH_2)_5$	in ether at room temperature	$(CH_2)_5NSN(CH_2)_5$ (58)	[8]
$ArN{=}CHNHCH_3$ ($Ar = 2\text{-}CH_3\text{-}4\text{-}ClC_6H_3$, $2,4,5\text{-}Cl_3C_6H_2$) $+ (C_2H_5)_3N$	in THF, with cooling	$ArN{=}CHN(CH_3)SN(CH_2)_5$	[31]
$4\text{-}R^1C_6H_4C(R^2){=}NH$ ($R^1 = H$, $R^2 = t\text{-}C_4H_9$, CF_3, C_6H_5; $R^1 = CH_3$, CF_3, $R^2 = CF_3$) $+ (C_2H_5)_3N$	mole ratio 1 : 1 : 1; in benzene	$4\text{-}R^1C_6H_4C(R^2){=}NSN(CH_2)_5$	[32]
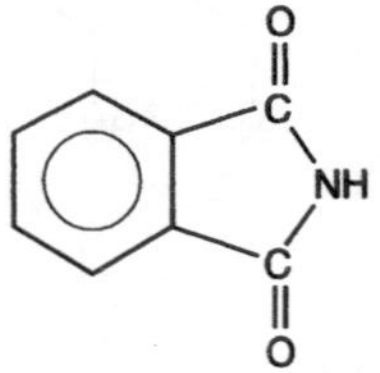 $+ (C_2H_5)_3N$	mole ratio 1 : >1 : >1.2; in CCl_4–THF at 0 to 10 °C for 1 h [33]; in toluene for 1 h [34]	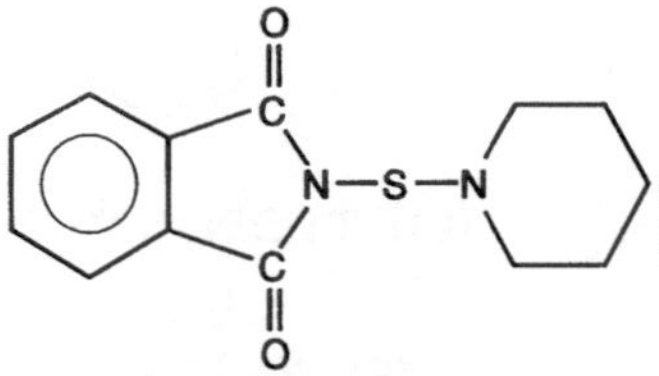 (59.6) [33], (85) [34]	[33, 34]
$CH_3N{=}C{=}O + HF + (C_2H_5)_3N$	in CH_2Cl_2 at −10 °C, then warming to room temperature [35]; see also [36]	$(CH_2)_5NSN(CH_3)C(O)F$ (46) [35], see also [36], (59) [6]	[6, 35, 36]
$CH_3SC(CH_3){=}NOC(O)NHCH_3$ $+ (C_2H_5)_3N$	in CH_3CN; at 0 to 5 °C, then warming to room temperature within 2 h	$CH_3SC(CH_3){=}NOC(O)N(CH_3)SN(CH_2)_5$	[37]
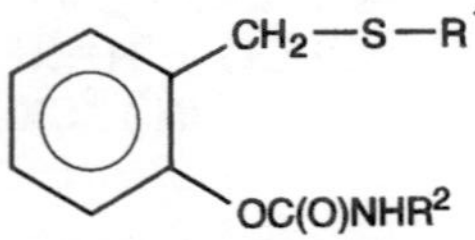 ($R^1 = C_2H_5$, $n\text{-}C_3H_7$, $R^2 = CH_3$; $R^1 = R^2 = C_2H_5$) + pyridine	18 h at room temperature	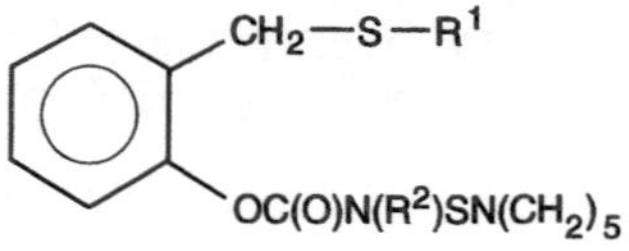	[38]

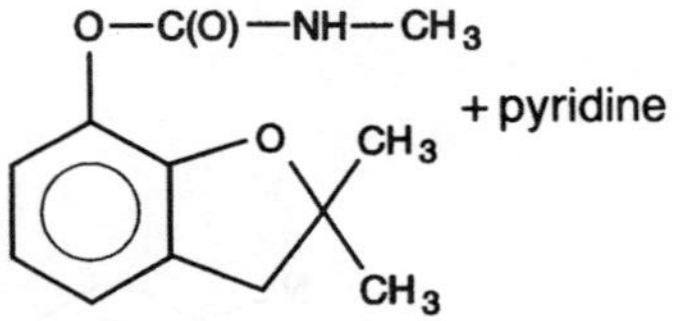	18 h at room temperature	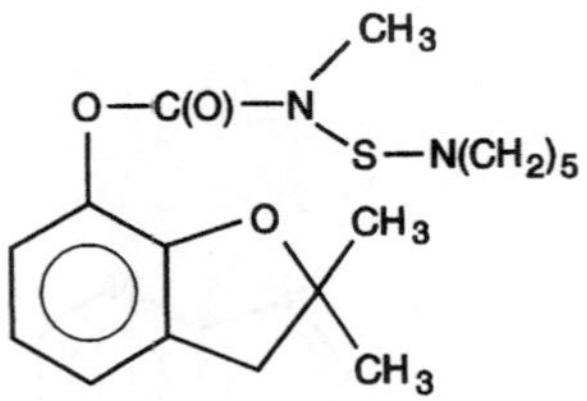	[39]
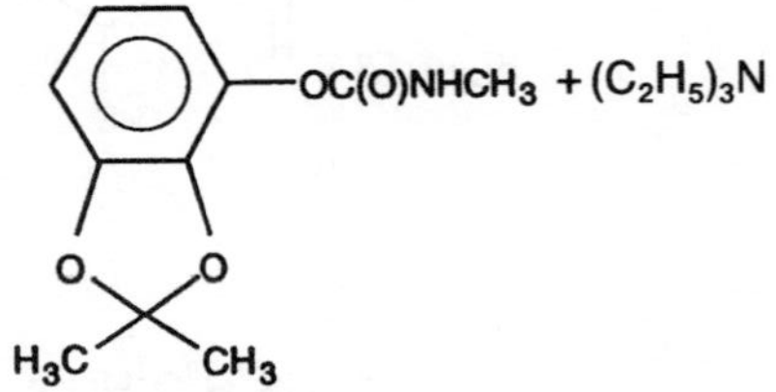	at 20 to 25 °C, and 2.5 h at 30 to 35 °C	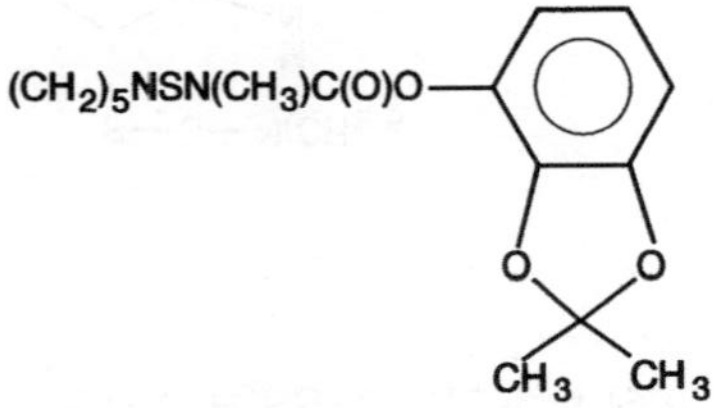	[40]
$CH_3NHCH_2CH_2NHCH_3$	mole ratio 1 : 1; in ether at −20 °C and stirring at −20 °C overnight	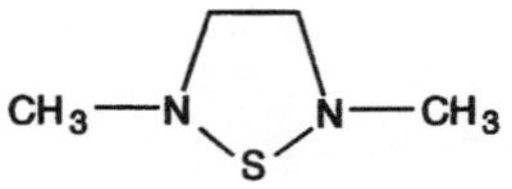 (30)	[28]
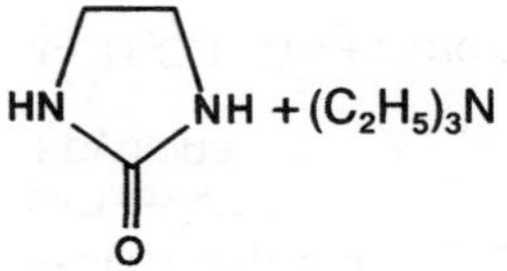	in THF	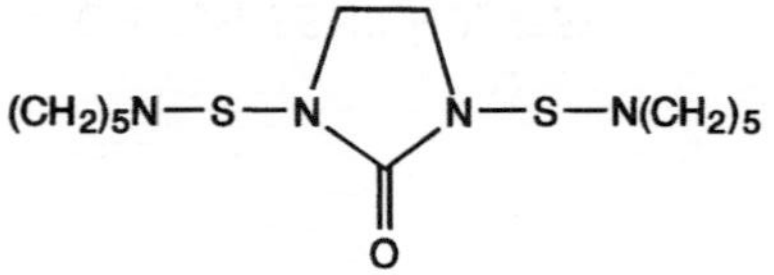 (50.5)	[41]
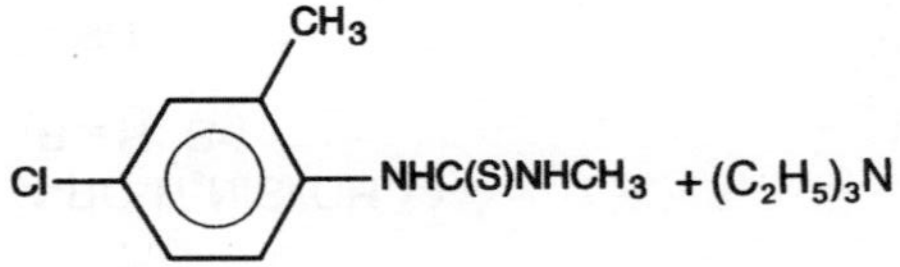	in CH_2Cl_2; at 0 to 10 °C and 2 h at room temperature	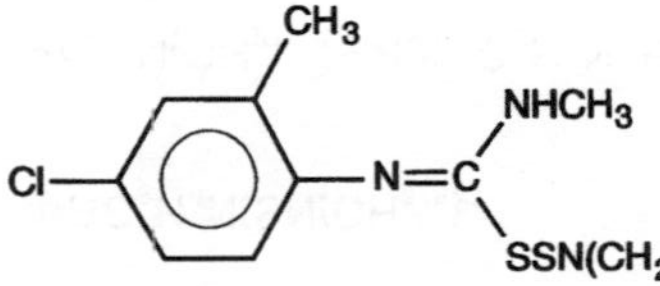	[42]
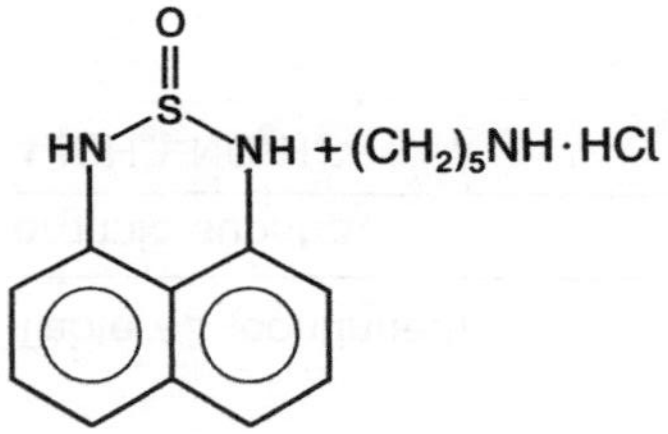	mole ratio 2 : 2(?) : 1; in ether at −20 °C and stirring at −20 °C overnight	none	[28]

Table 42 (continued)

organic substrate	reaction conditions	product(s) (yield in %)	Ref.
4-$(CH_3)_2NC_6H_5$	mole ratio 1:2; in petroleum ether at 5 °C, and 24 h at room temperature	4-$(CH_3)_2NC_6H_4SC_6H_4N(CH_3)_2$-4 (70)	[27]
4-$RC_6H_4N(Si(CH_3)_3)_2$ (R = H, Br)	mole ratio 2:1	4-$RC_6H_4N(SN(CH_2)_5)_2$	[43]
ArSH (Ar = C_6H_5, 4-BrC_6H_4, 4-$CH_3C_6H_4$, 2,5-$(CH_3)_2C_6H_3$) + pyridine	mole ratio 1:1:2; in petroleum ether at 0 °C, 1 h at room temperature (?)	$ArSSN(CH_2)_5$ (~80), $C_5H_5N \cdot HCl$	[44]
4-$CH_3C_6H_4SH$ + pyridine	mole ratio 1:1:2; in petroleum ether at 0 °C, and 24 h at room temperature	4-$CH_3C_6H_4SSN(CH_2)_5$ (75)	[27]
4-$CH_3C_6H_4SH$	mole ratio 1:2; in petroleum ether at 5 °C and 24 h at room temperature	4-$CH_3C_6H_4SSSC_6H_4CH_3$-4 (90)	[27]
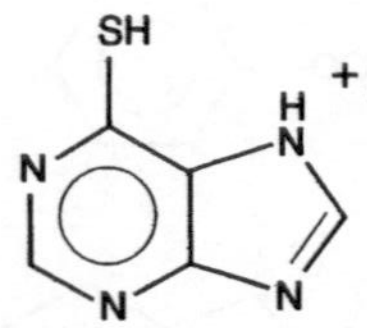 + pyridine	heating to 80 to 90 °C for 5 min	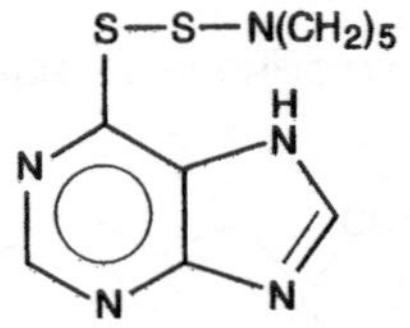	[45, 46]

References:

[1] Badische Anilin-Soda-Fabrik AG; Weiß, G.; Schulze, G. (Ger. 1 131 222 [1960/62]; C.A. **57** [1962] 13771).
[2] Freytag, H.; Lober, F. (Ger. 965968 [1954/57]; C.A. **1959** 11 417).
[3] Farbenfabriken Bayer AG (Br. 790021 [1954/58]; C.A. **1958** 13806).
[4] Dorlars, A. (Houben-Weyl Methoden Org. Chem. 4th Ed. **11** Pt. 2 [1958] 744/51).
[5] Freytag, H.; Lober, F. (Ger. 969813 [1954/58]; C.A. **1960** 5473).
[6] Hatch, C. E. (J. Org. Chem. **43** [1978] 3953/7).
[7] Musin, B. M.; Ivanov, V. B.; Ivanov, B. E. (Izv. Akad. Nauk SSSR Ser. Khim. **1988** 1693; Bull. Acad. Sci. USSR Div. Chem. Sci. [Engl. Transl.] **1988** 1509).
[8] Bacon, R. G. R.; Irwin, R. S. (J. Chem. Soc. **1960** 5079/87).
[9] Armitage, D. A.; Towle, I. D. H. (Phosphorus Sulfur **1** [1976] 37/9).
[10] Morita, E.; D'Amico, J. J. (Rubber Chem. Technol. **44** [1971] 881/8; C.A. **76** [1972] No. 15537).

[11] Davis, F. A.; Skibo, E. B. (J. Org. Chem. **41** [1976] 1333/6).
[12] Morimura, S.; Horiuchi, H.; Tamura, C.; Yoshioka, T. (Bull. Chem. Soc. Jpn. **53** [1980] 1666/9).
[13] Ivanova, Z. M.; Gusar, N. I.; Gololobov, Yu. G. (Zh. Vses. Khim. O-va. im. D.I. Mendeleeva [1973] 349/50; C.A. **79** [1973] No. 104677).
[14] Gusar', N. I.; Randina, L. V.; Shurubura, A. K. (Zh. Obshch. Khim. **59** [1989] 548/56; J. Gen. Chem. USSR [Engl. Transl.] **59** [1989] 486/92).
[15] Gusar, N. I.; Gololobov, Yu. G. (Heteroat. Chem. **3** [1992] 407/14).
[16] Hoobler, M.; Sikorski, J. A. (Belg. 894595 [1982/83]; C.A. **99** [1983] No. 6049).
[17] Almasi, L.; Hantz, A. (Chem. Ber. **97** [1964] 661/6).
[18] Almasi, L.; Hantz, A. (Omagiu' Raluca Ripan **1966** 59/64; C.A. **67** [1967] No. 116856).
[19] Baudin, J.-B.; Julia, S. A.; Lorne, R. (Bull. Soc. Chim. Fr. **1987** 181/8).
[20] Baudin, J.-B.; Julia, S. A.; Ruel, O. (Tetrahedron **43** [1987] 881/9).

[21] Neidlein, R.; Lehr, W. (Chem. Ber. **114** [1981] 80/5).
[22] Williams, H. R. (U.S. 3121661 [1961/64]; C.A. **61** [1964] 6958).
[23] Malz, H.; Bayer, O.; Freytag, H.; Lober, F. (U.S. 2891059 [1957/59]; C.A. **1960** 4387).
[24] Blaschette, A.; Näveke, M. (Chem.-Ztg. **115** [1991] 61/4).
[25] Badische Anilin-Soda-Fabrik AG; Weiß, G. (Ger. 1153744 [1961/64]; C.A. **60** [1964] 1711).
[26] Kutateladze, A. G.; Zyk, N. V.; Denisko, O. V.; Zefirov, N. S. (Izv. Akad. Nauk SSSR Ser. Khim. **1990** 1689/90; Bull. Acad. Sci. USSR Div. Chem. Sci. [Engl. Transl.] **39** [1990] 1535/6).
[27] Almasi, L.; Hantz, A. (Chem. Ber. **99** [1966] 3288/92).
[28] Bryce, M. R. (J. Chem. Soc. Perkin Trans. I **1984** 2591/3).
[29] Tavs, P. (Angew. Chem. **78** [1966] 1057/8; Angew. Chem. Int. Ed. Engl. **5** [1966] 1048/9).
[30] Farbenfabriken Bayer AG; Freytag, H.; Lober, F.; Domagk, G. (Ger. 948330 [1954/56]; C.A. **1959** 6267).

[31] Drabek, J.; Böger, M. (Ger. 2621077 [1974/76]; C.A. **86** [1977] No. 89430).
[32] Markovskii, L. N.; Shermolovich, Yu. G.; Shevchenko, V. I. (Zh. Org. Khim. **11** [1975] 2533/7; J. Org. Chem. USSR [Engl. Transl.] **11** [1975] 2603/7).
[33] Lawrence, J. P. (Ger. 2329431 [1970/74]; C.A. **81** [1974] No. 50818, Fr. 2190819 [1970/74]; C.A. **81** [1974] No. 122404, U.S. 3928340 [1972/75]; C.A. **84** [1976] No. 166003, U.S. 3944552 [1972/76]; C.A. **85** [1976] No. 22609).

[34] Lawrence, J. P. (Rubber Chem. Technol. **49** [1976] 333/40).
[35] Thurman, D. E. (Ger. 2628574 [1975/77]; C.A. **86** [1977] No. 155209, Ger. 2628575 [1975/77]; C.A. **86** [1977] No. 155665).
[36] Durden, J. A., Jr. (U.S. 4071627 [1976/78]; C.A. **89** [1978] No. 43450).
[37] Gemrich, E. G., II; Lee, B. L.; Nelson, S. J.; Rizzo, V. L. (J. Agric. Food Chem. **26** [1978] 391/5).
[38] Hoffmann, H.; Hammann, I.; Homeyer, B.; Stendel, W. (Ger. 2737606 [1977/79]; C.A. **91** [1979] No. 56482).
[39] Black, L.; Fukuto, T. R. (Ger. 2433680 [1972/75]; C.A. **82** [1975] No. 156050).
[40] Saunders, D. E. (Eur. Appl. 129359 [1984]; C.A. **102** [1985] No. 185070).

[41] Lawrence, J. P. (U.S. 4042642 [1972/77]; C.A. **87** [1977] No. 153223, U.S. 3960857 [1972/76]; C.A. **85** [1976] No. 79437).
[42] Böger, M.; Drabek, J. (Ger. 2654080 [1972/77]; C.A. **87** [1977] No. 102078).
[43] Mayer, R.; Domschke, G.; Bleisch, S.; Bartl, A. (Tetrahedron Lett. **1978** 4003/6).
[44] Almasi, L.; Hantz, A. (J. Prakt. Chem. [4] **38** [1968] 113/8).
[45] Carter Products, Inc.; Stiefel, F. J.; Ludwig, B. J.; Berger, F. M. (Br. 993302 [1964/65]; C.A. **63** [1965] 8381).
[46] Stiefel, F. J.; Ludwig, B. J.; Berger, F. M. (U.S. 3400125 [1963/68]; C.A. **70** [1969] No. 28955).

12.5.2.4.2 $ClS\overline{N-(CH_2)_2-CR^1R^2-CHR^3-C}H_2$, $R^1 = R^2 = H$, $R^3 = CH_3$; $R^1 = R^3 = H$, $R^2 = CH_3$, $CH_2C_6H_5$, C_6H_5; $R^1 = R^2 = CH_3$, $R^3 = H$

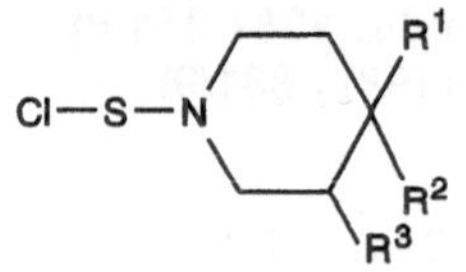

The formation of the title compounds was not described.

Reacting a mixture of the title compounds and "carbofuran" in pyridine for 18 h gave the carbamates I [1].

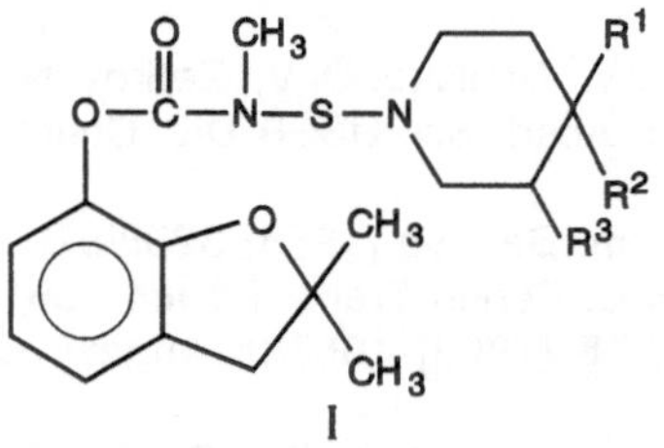

I

A solution of the title compound with $R^1 = R^3 = H$, $R^2 = CH_3$ in CCl_4 reacted with succinimide and $(C_2H_5)_3N$ in DMF yielding "N-(4-methylpiperidino-N′-succinimidyl) sulfide" [2].

References:

[1] Black, L.; Fukuto, T. R. (Ger. 2433680 [1972/75]; C.A. **82** [1975] No. 156050).
[2] Laithwaite, P.; Taylor, J. A. (Br. 1355802 [1970/74]; C.A. **82** [1975] No. 32146).

12.5.2.4.3 $ClS\overline{N-C(CH_3)_2-(CH_2)_3-C}(CH_3)_2$

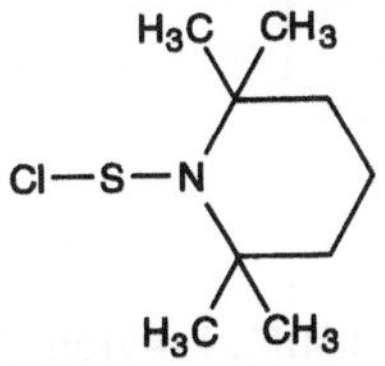

The title compound was formed by introducing Cl_2 into a solution of the disulfide I in CCl_4 with ice cooling until no more Cl_2 gas was accepted. Then the solvent was removed in vacuum, and the remaining yellow-orange solid was recrystallized.

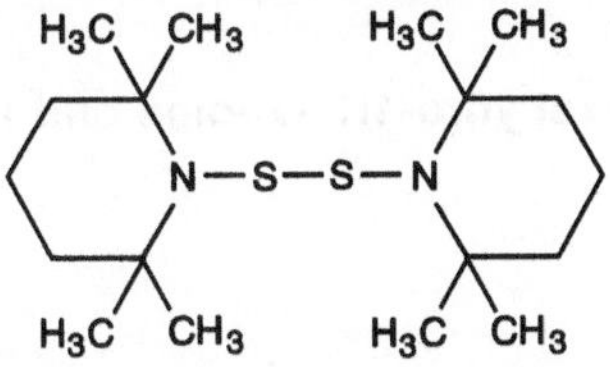

I

Solid; 65% yield; m.p. 64 °C (from methylcyclohexane).

^1H NMR (C_6H_6/TMS): δ (in ppm) = −1.55, −1.42 (CH_3, CH_2; assignment not unambiguous).

MS (70 eV): m/e (rel. int. in %), related to $^{35}Cl = M^+$ (31), $M^+ - Cl$ (84), $M^+ - Cl - CH_3$ (74), $M^+ - SCl - CH_3$ (100).

The title compound reacted with excess CsF in toluene within 2 d to give 78% II.

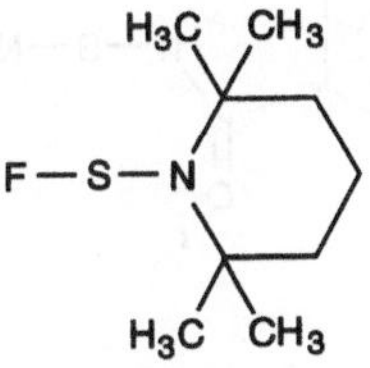

II

Reference:

Röschenthaler, G.-V.; Starke, R. (Z. Naturforsch. **32b** [1977] 721/2).

12.5.2.5 N^2-Chlorosulfanyl-1,2,3,4-tetrahydro-isoquinoline, $ClS\overline{N-CH_2(1-C_6H_4-2)CH_2-C}H_2$

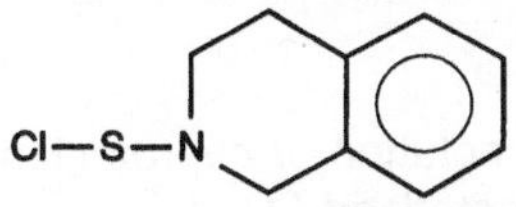

The title compound was formed by introducing dry Cl_2 at 0 to 5 °C into a solution of the disulfide I in CCl_4 until there was a slight excess (of Cl_2). Excess Cl_2 was then removed at 0 to 5 °C in vacuum or by purging with an inert gas. The resulting solution was cleared with animal charcoal, and the solvent was evaporated under reduced pressure to give the compound as a light yellow solid.

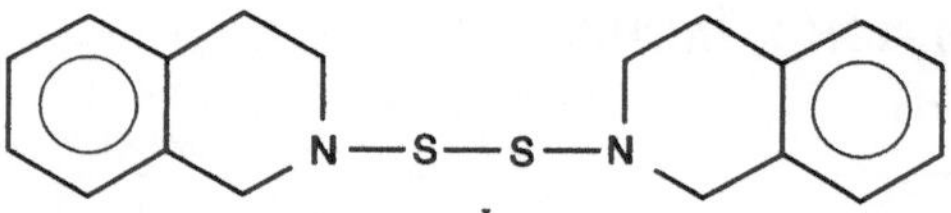

I

Reference:

Farbenfabriken Bayer AG (Br. 790021 [1954/58]; C.A. **1958** 13806), Freytag, H.; Lober, F. (Ger. 1092475 [1955/61]; C.A. **56** [1962] 462).

12.5.2.6 N¹-Chlorosulfanyl-hexahydro-1H-azepine and Derivative

12.5.2.6.1 $ClS\overline{N-(CH_2)_5-C}H_2$

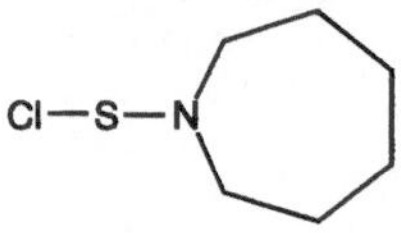

The formation of the title compound was not described.

Adding $ClSN(CH_2)_6$ in toluene to a mixture of (excess) phthalimide and $(C_2H_5)_3N$ in toluene gave compound I after 1 h (53%).

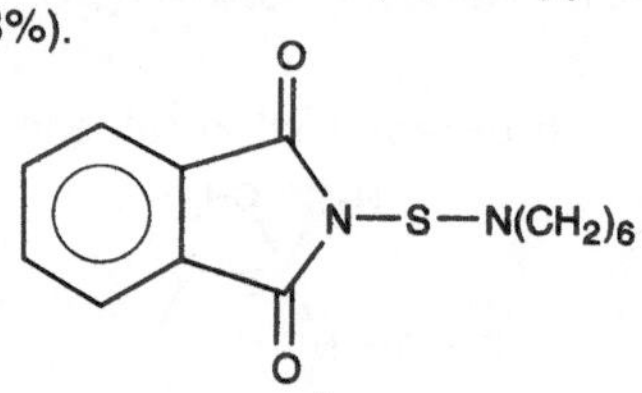

I

Reference:

Lawrence, J. P. (Rubber Chem. Technol. **49** [1976] 333/40).

12.5.2.6.2 $ClS\overline{N-C(O)-(CH_2)_4-C}H_2$

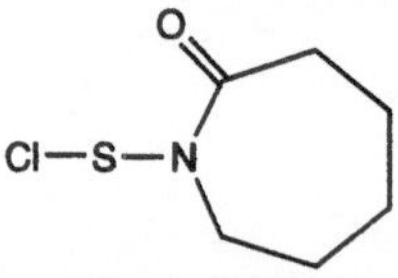

The title compound was formed by adding SO_2Cl_2 to a solution of caprolactam disulfide (mole ratio ca. 1 : 2) in CCl_4 at 20 °C. Yellow oil; not characterized [1].

Adding a solution of the title compound in CCl_4 to amino compounds, t-$C_4H_9NH_2$ [2], $(CH_2)_5NH$ [1, 2], $O(CH_2)_4NH$, or 4-$H_2NC_6H_4NH_2$ [2] gave via abstraction of HCl the corresponding diamino sulfides I (R^1 = H, R^2 = C_4H_9-t, 4-$NH_2C_6H_4$; R^1R^2 = $(CH_2)_5$, $(CH_2)_4O$) [2].

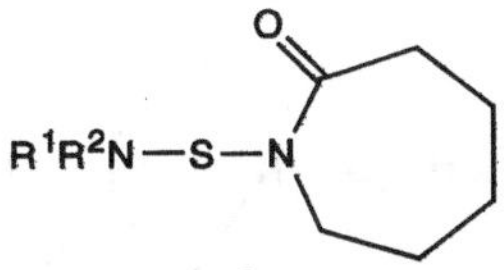

I

References:

[1] Taylor, J. A. (Br. 1355801 [1970/74]; C.A. **81** [1974] No. 105328).
[2] Laithwaite, P.; Taylor, J. A. (Br. 1355802 [1970/74]; C.A. **82** [1975] No. 32146).

12.5.2.7 N³-Chlorosulfanyl-3-azabicyclo[3.2.2]nonane, $ClS\overline{N-CH_2-CH(CH_2)_4CH-C}H_2$

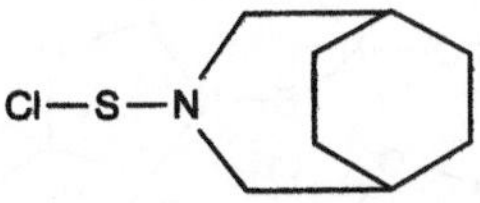

The title compound was formed by adding Cl_2 at −5 to +5 °C to a solution of the disulfide I in CCl_4 and reacting the mixture at 0 to 5 °C for 20 min; then activated carbon was added and stirring was continued at 0 to 5 °C for 15 min.

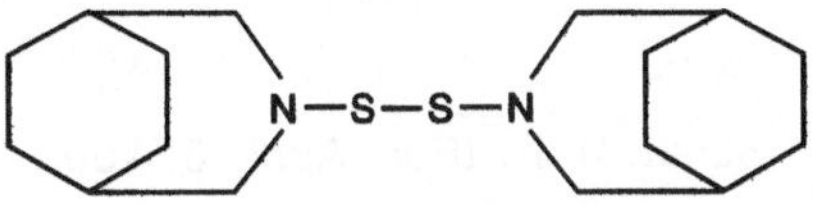

I

88.7% yield; solid; m.p. 93 to 95 °C. Recrystallization in various solvents resulted in decomposition.

Reference:

Morita, E.; D'Amico, J. J. (Rubber Chem. Technol. **44** [1971] 881/8; C.A. **76** [1972] No. 15537).

12.5.2.8 N³-Chlorosulfanyl-2-oxazolidinone and Derivatives, $ClS\overline{N-C(O)-O-CH(R^1)-C}R^2R^3$, $R^1 = R^2 = R^3 = H$; $R^1 = CH_3$, $R^2 = R^3 = H$; $R^1 = H$, $R^2 = R^3 = CH_3$

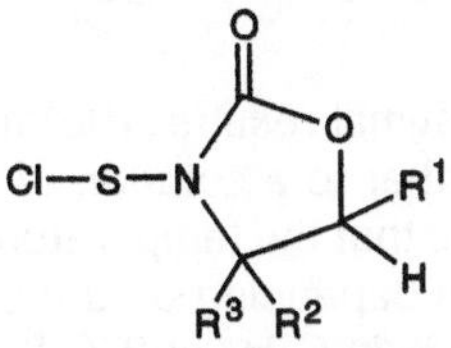

The title compounds were formed by adding trimethylsilyl-substituted oxazolidinones (I) in $(C_2H_5)_2O$ at 0 °C to a solution of SCl_2 (mole ratio ca. 1 : 1) in $(C_2H_5)_2O$ and allowing the mixture to come to 20 °C and remain at this temperature for 30 min.

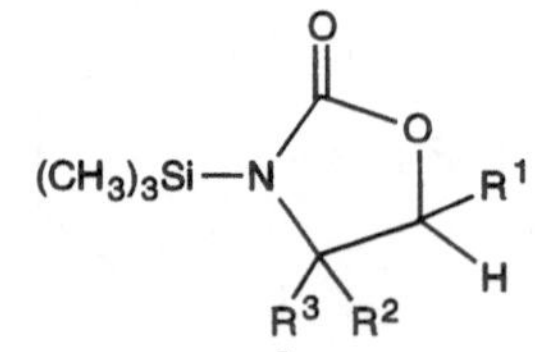

I

$R^1 = R^2 = R^3 = H$. Yellow-orange liquid, 91% yield.

$R^1 = CH_3$, $R^2 = R^3 = H$. Yellow-orange liquid, 92.9% yield.

$R^1 = H$, $R^2 = R^3 = CH_3$. Yellow liquid, 95% yield.

Solutions of the title compounds reacted in CH_2Cl_2 with II and $(CH_2)_5NH$ (mole ratio 6.5 : 5.9 : 8.8) at ambient temperature within 24 h to give the corresponding compounds III.

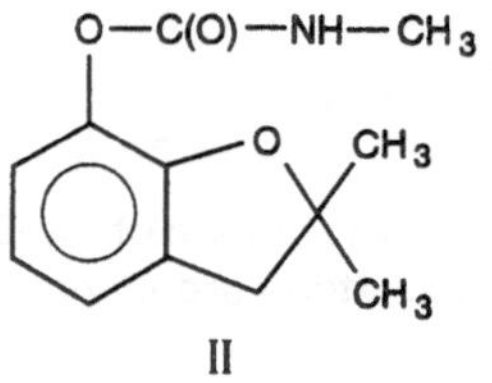

II

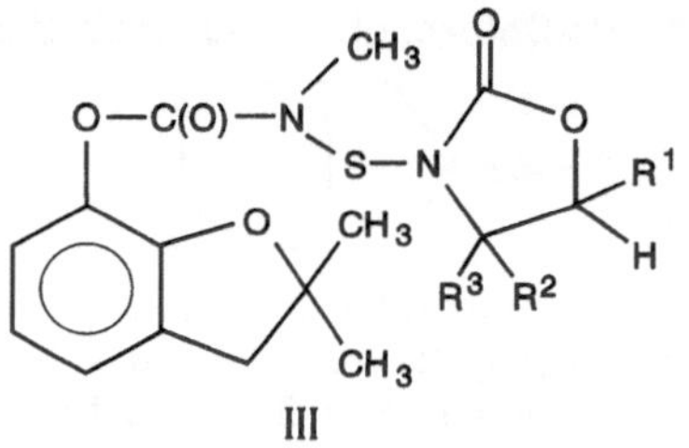

III

Reference:

Denarie, M.; Formigoni, A.; Senet, J.-P. (Eur. Appl. 374002 [1989/90]; C.A. **113** [1990] No. 211965).

12.5.2.9 N⁴-Chlorosulfanyl-morpholine and Derivative

12.5.2.9.1 $ClSN(CH_2)_4O$

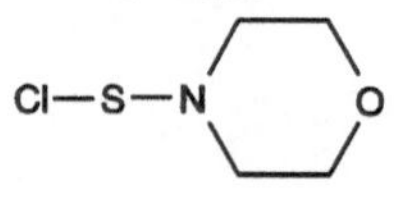

Preparation

The $ClSN(CH_2)_4O$ can be synthesized directly from $O(CH_2)_4NH$ and SCl_2 (Method I), by cleaving the sulfur–sulfur bond of $O(CH_2)_4NSSN(CH_2)_4O$ or of a derivative $RSSN(CH_2)_4O$ with a chlorinating agent (Method II), by cleaving the S–O bond of $CH_3OSN(CH_2)_4O$ with a chloro compound (Method III), or by some special procedures.

Method I: Method I was established in several variations. Optimal results (100%) have been achieved by adding morpholine and $(C_2H_5)_3N$ in ether to a solution of SCl_2 (mole ratio 1 : 1 : 1) in ether cooled to −20 °C at such a rate that the temperature remained below 0 °C [1]. Earlier attempts gave lower yields depending on the solvent, the reaction temperature, and the HCl scavenger: 81% in C_6H_6 at 0 °C [2], ~60% in ether at 0 to 5 °C [3], 35% in ether at −10 °C and then at room temperature [5], 55% in CH_2Cl_2 with pyridine as base at 5 to 10 °C and then at room temperature [6], and ~33% in ether without a base at −10 °C for 1 h and then at room temperature [4].

Instead of SCl_2, a "hidden" SCl_2 was used by adding morpholine to $[4\text{-}CH_3C_5H_4N\text{-}SCl]^+ Cl^-$ (prepared from $SCl_2 + \gamma$-picoline) in CCl_4 at 10 to 20 °C for 1 h; 91.2% yield [7].

Method II: With Cl_2 as chlorinating agent, $ClSN(CH_2)_4O$ was formed by introducing dry Cl_2 at 0 to 5 °C into a suspension of $O(CH_2)_4NSSN(CH_2)_4O$ in CCl_4 until there was a slight excess (of Cl_2). The excess Cl_2 was then removed at 0 to 5 °C in vacuum or by purging the mixture with an inert gas. The resulting solution was then cleared with animal charcoal, and after filtration the solvent was distilled off under reduced pressure at 30 to 50 °C [8 to 11] to give $ClSN(CH_2)_4O$ with a yield of 92% (crude) [8, 9]; see also [10]; the crude product was distilled at 0.1 Torr in a Kugelrohr (oven temperature 80 °C) to yield 48% of the title compound [11]; for modifications, see [12 to 16].

With SO_2Cl_2 as chlorinating agent, $ClSN(CH_2)_4O$ was synthesized by adding a solution of SO_2Cl_2 in CCl_4 to a solution of $O(CH_2)_4NSSN(CH_2)_4O$ (mole ratio 1 : 1) in CCl_4 and heating the mixture to 70 to 80 °C (bath temperature) for 15 min. Subsequently, the mixture was cooled and cleared with animal charcoal; (~72% yield) [9, 17]; see also [18 to 20]. The reaction was also performed in CCl_4 with an excess of SO_2Cl_2 (mole ratio 1.25 : 1) at 60 °C for 2 h (85% yield) [21], in benzene at 5 °C and heating the mixture to 33 °C in the course of 10 min (88% yield) [22], or without solvent with an excess of SO_2Cl_2 (mole ratio ca. 2 : 3) at −78 °C [23].

$ClSN(CH_2)_4O$ was also obtained by S–S bond cleavage of $(C_2H_5O)_2P(S)SS\text{-}N(CH_2)_4O$ with HCl (mole ratio 1 : 1 ?) in ether–petroleum ether (1 : 1) giving $(C_2H_5O)_2P(S)SSSP(S)(OC_2H_5)_2$, $(C_2H_5O)_2P(S)Cl$, and $O(CH_2)_4NH \cdot HCl$ as by-products; a reaction mechanism was suggested. When the reaction was performed with SCl_2 (mole ratio ca. 2 : 1) in heptane at room temperature the compound formed along with $(C_2H_5O)_2P(S)SSSP(S)(OC_2H_5)_2$ and $(C_2H_5O)_2P(S)Cl$ as side products [24].

Method III: $ClSN(CH_2)_4O$ was formed from $O(CH_2)_4NSOCH_3$ reacting with RC(O)Cl ($R = CH_3$, C_6H_5) at 20 to 25 °C in an inert atmosphere (70% yield) [25], or with $(CH_3)_3SiCl$ (no details) [26].

Special Procedures. The title compound was formed by boiling a mixture of equimolar amounts of N-chloromorpholine, $ClN(CH_2)_4O$, and sulfur in CCl_4 for 2 h [27] or by decomposing $O(CH_2)_4NSN{=}CCl_2$ above 60 °C whereby $ClC{\equiv}N$ was also formed [28].

Physical Properties

$ClSN(CH_2)_4O$ is a vacuum-distillable, pale yellow [11], yellow [20, 22], or orange-colored [2, 8, 9, 17, 18] liquid; it was also described as a yellow semisolid [1]; b.p. 38 °C/0.05 Torr [22], 44 °C [21] or 44 to 45 °C at 0.1 Torr [27], 45 °C/0.2 Torr [29], 55 to 56 °C [25] or 56 °C at 0.4 Torr [26], 58 to 60 °C/0.6 Torr [9, 17], see also [11, 18], 60 °C [6] or 64 to 66 °C at 0.5 Torr [8, 9], see also [10], 60 to 62 °C [4] or 64 to 65 °C at 1 Torr [7], 62 to 63 °C/0.11 Torr [3], 66 to 67 °C/1.2 Torr [2].

Refractive index: $n_D^{20} = 1.5490$ [25], 1.5572 [22], $n_D^{24} = 1.5492$ [26].

^1H NMR ($CDCl_3$/TMS): δ (in ppm) = 3.37 to 3.47 (m, 4H), 3.77 to 3.87 (m, 4H) [29].

^{13}C NMR ($CDCl_3$/TMS): δ (in ppm) = 58.7 (t), 66.4 (t) [29].

Chemical Reactions

The title compound is highly reactive. In many cases, reactions were performed with solutions of $ClSN(CH_2)_4O$ obtained in the preparation [14]. It is recommended to store the pure compound in a freezer [11]. The title compound can be stored for several weeks at −18 °C [21]. $ClSN(CH_2)_4O$ was found to be unstable above 60 °C [28]. It decomposed when left standing at 25 to 30 °C for several days [30].

It decomposes in the presence of moisture [30], especially violently with water [14].

$ClSN(CH_2)_4O$ reacted with $(n\text{-}C_4H_9)_3SnOCH_3$ (mole ratio 1 : 1.1) under cooling and then warming within 1 h to give 84% $O(CH_2)_4NSOCH_3$ along with $(n\text{-}C_4H_9)_3SnCl$ [26].

Adding $(RO)_3P$ ($R = C_2H_5$, $n\text{-}C_4H_9$) to the title compound at 6 to 13 °C gave the corresponding $O(CH_2)_4NSP(O)(OR)_2$, $O(CH_2)_4NP(O)(OR)_2$, and RCl along with a small amount of $(RO)_3PS$ [22]. Adding a solution of $ClSN(CH_2)_4O$ in CH_2Cl_2 to a precooled solution (0 °C) of $(4\text{-}CH_3\text{-}OC_6H_4)_2P(O)CH_2NHCH_2C(O)OC_2H_5$ [31] or $(C_6H_5O)_2P(O)CH_2NHCH_2CN$ [1] and $(C_2H_5)_3N$ (mole ratio ca. 3 : 1 : 1) in toluene yielded 53% $(4\text{-}CH_3OC_6H_4)_2P(O)CH_2N(CH_2C(O)OC_2H_5)SN(CH_2)_4O$ and 26% $(C_6H_5O)_2P(O)CH_2N(SN(CH_2)_4O)CH_2CN$, respectively. Adding a solution of $ClSN\text{-}(CH_2)_4O$ to $R_2P(S)H$ ($R = CH_3$, C_2H_5, $n\text{-}C_3H_7$, $n\text{-}C_4H_9$, $CH_2C_6H_5$, C_6H_5) without a base (mole ratio 1 : 2, in C_6H_6 at 5 °C) or in the presence of $(C_2H_5)_3N$ (mole ratio 1 : 1 : 1, in ether at 0 °C) gave the corresponding $R_2P(S)SP(S)R_2$ with yields of 92, 90, 95, 85, 94, or 95%, respectively, along with $O(CH_2)_4NH \cdot HCl$ or $R_2P(S)SN(CH_2)_4O$ in yields of 53, 77, 73, 63, 75, or 80%, respectively, along with $(C_2H_5)_3N \cdot HCl$ [32]. A solution of $ClSN(CH_2)_4O$ in ether reacted with a precooled (5 °C) solution of $C_2H_5(RO)P(S)H$ (mole ratio 1 : 2; $R = C_2H_5$, $n\text{-}C_3H_7$, $i\text{-}C_3H_7$) [33] or $C_6H_5(RO)P(S)H$ (mole ratio 1 : 2; $R = C_2H_5$, $n\text{-}C_3H_7$, $i\text{-}C_3H_7$, $n\text{-}C_4H_9$, $i\text{-}C_4H_9$ [34]) in ether for 5 min at room temperature to give the corresponding $C_2H_5(RO)P(S)SP(S)(C_2H_5)OR$ [33] or $C_6H_5(RO)\text{-}P(S)SP(S)(C_6H_5)OR$ with yields of 81 to 90% [34]. Reacting $ClSN(CH_2)_4O$ with $C_2H_5(RO)P(S)H$ ($R = C_2H_5$, $n\text{-}C_3H_7$, $i\text{-}C_3H_7$) and $(C_2H_5)_3N$ in the mole ratio 1 : 1 : 1 in ether at ≤0 °C gave the corresponding $RO(C_2H_5)P(S)SN(CH_2)_4O$ [33], or with $C_6H_5(RO)P(S)H$ ($R = C_2H_5$, $n\text{-}C_3H_7$, $i\text{-}C_3H_7$, $n\text{-}C_4H_9$, $i\text{-}C_4H_9$) and $(C_2H_5)_3N$ (mole ratio 1 : 2 : 1) in ether at −5 °C the corresponding $C_6H_5(RO)\text{-}P(S)SN(CH_2)_4O$ with yields of 43 to 61% [34]. Treating $ClSN(CH_2)_4O$ with $(S)_p\text{-}C_2H_5O(C_6H_5)\text{-}P(S)H$ and $(C_2H_5)_3N$ produced $(R)_p\text{-}C_2H_5O(C_6H_5)P(S)SN(CH_2)_4O$ stereospecifically [35]. The title compound reacted with $C_2H_5O(RO)P(S)H$ (R = L-menthyl, L-menthoxy) to give $C_2H_5O(RO)\text{-}P(S)SN(CH_2)_4O$ [36] and a 1 : 1 mixture of (R) and (S) diastereomers of $O(CH_2)_4NP(S)\text{-}(OC_2H_5)(R)$, respectively [37].

Treating a precooled solution (5 °C) of $R_2P(S)SH$ ($R = CH_3$, C_2H_5, $n\text{-}C_3H_7$, $n\text{-}C_4H_9$) with a solution of $ClSN(CH_2)_4O$ (mole ratio 2 : 1) in benzene at 0 to 5 °C gave the corresponding $R_2P(S)SSSP(S)R_2$ with yields of 22, 80, 75, or 70%, respectively [38]. Adding a solution of $ClSN(CH_2)_4O$ in ether at 0 °C to a mixture of $R_2P(S)SH$ ($R = C_2H_5$, $n\text{-}C_3H_7$, $n\text{-}C_4H_9$) and $(C_2H_5)_3N$ (mole ratio 1 : 1 : 1) in ether, stirring the mixture for 1 h at 0 to 5 °C, and allowing it to come to room temperature gave the corresponding $O(CH_2)_4NSSP(S)R_2$ with a yield of 67, 70, or 60%, respectively [38]. The reaction of $ClSN(CH_2)_4O$ with a precooled (−5 °C) solution of $C_2H_5O(C_2H_5)P(S)SH$ (mole ratio 1 : 2) in benzene gave $C_2H_5O(C_2H_5)P(S)SSSP(S)(C_2H_5)OC_2H_5$; reacting the same compounds and $(C_2H_5)_3N$ (mole ratio 1 : 1 : 1) yielded the corresponding $C_2H_5O(C_2H_5)P(S)SSN(CH_2)_4O$ [33]. Adding a solution of $(RO)_2P(S)SH$ ($R = C_2H_5$, $n\text{-}C_3H_7$, $n\text{-}C_4H_9$, $i\text{-}C_4H_9$, $i\text{-}C_5H_{11}$) in CCl_4 to a solution of $ClSN(CH_2)_4O$ (mole ratio 2 : 1) in CCl_4 at 10 °C yielded the corresponding $(RO)_2P(S)SSSP(S)(OR)_2$ along with $O(CH_2)_4NH \cdot HCl$; a reaction mechanism is suggested in the paper [24]. Adding a solution of $ClSN(CH_2)_4O$ in CCl_4 at 0 °C to a mixture of $(C_2H_5O)_2P(S)H$ and pyridine in CCl_4 so that the temperature did not rise above 0 °C gave $(C_2H_5O)_2P(S)SN(CH_2)_4O$ [39]. Adding a solution of $ClSN(CH_2)_4O$ in benzene to a suspension of $(C_2H_5O)_2P(S)SNa$ in benzene while maintaining the temperature below 30 to 35 °C and stirring the mixture at this temperature for 1 h yielded $(C_2H_5O)_2P(S)SSN(CH_2)_4O$ [40]. The same compound was formed with a yield of 60% along with $C_5H_5N \cdot HCl$ by adding a mixture of $ClSN(CH_2)_4O$ and pyridine in petroleum ether to a precooled (10 °C) suspension of $C_5H_5NH^+$ $(C_2H_5O)_2P(S)S^-$ (mole ratio ca. 1 : 1 : 1) in petroleum ether [24].

Adding a solution of $ClSN(CH_2)_4O$ in hexane at −20 °C to a stirred suspension of an equimolar amount of $LiN(Si(CH_3)_3)_2$ (in hexane and reacting the mixture for 15 h at 20 °C [41]) or with $NaN(Si(CH_3)_3)_2$ [45] gave 58% [41] $O(CH_2)_4NSN(Si(CH_3)_3)_2$. $ClSN(CH_2)_4O$ reacted with $NaNHC(O)C_6H_5$ to give $O(CH_2)_4NSNHC(O)C_6H_5$ [42]. A solution of $ClSN(CH_2)_4O$ in toluene reacted with a suspension of the sodium salt I in toluene to give 79% II [43].

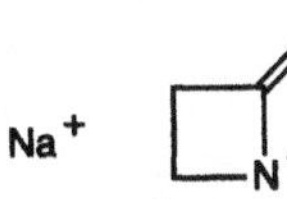

I

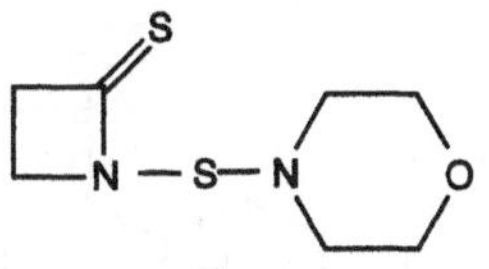

II

Adding the title compound at −10 °C to a solution of RCH_2SNa (R = H, $C(O)OC_2H_5$) in CH_3OH yielded the corresponding $O(CH_2)_4NSSCH_2R$ [44]. A solution of $ClSN(CH_2)_4O$ in CH_2Cl_2 reacted with a suspension of an equimolar amount of the sulfinate $M^+ RSO_2^-$ (M = Na, R = 4-$CH_3C_6H_4$, 4-$CH_3C(O)NHC_6H_4$; M = K, R = CH_3) in CH_2Cl_2 to yield the corresponding $O(CH_2)_4NSSO_2R$ and MCl [46]. $ClSN(CH_2)_4O$ reacted with a finely ground suspension of potassium phthalimide in benzene at reflux temperature for 1 h to give "morpholino-N′-phthalimidyl sulfide" (III) [47].

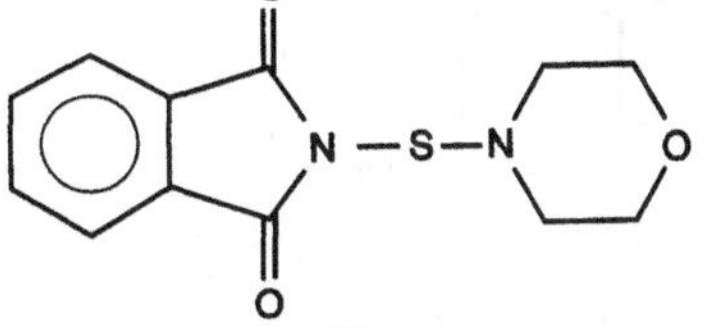

III

The title compound was reacted with $CH_2{=}CHMgBr$ (mole ratio ca. 1 : 1) in THF at −78 °C for 15 min and at room temperature for 2.5 h to give 85% (crude) $O(CH_2)_4NSCH{=}CH_2$ [29]. Adding a solution of $ClSN(CH_2)_4O$ in CH_3CN to an equimolar amount of $AgN(SO_2CH_3)_2$ in CH_3CN and allowing the mixture to react at room temperature for 24 h gave $O(CH_2)_4NSN(SO_2CH_3)_2$ [48]. The reactions of $ClSN(CH_2)_4O$ with some palladiumorganic compounds, RPdSC(S)-$N(C_3H_7\text{-}i)_2$, (mole ratio 2 : 1.5, in $CHCl_3$ at room temperature for 1 to 2 h, followed by addition of pyridine), are summarized in Table 43; the chlorides obtained were converted into the ClO_4^- salts with 70% aqueous $AgClO_4$ [11]; see also [49].

Table 43
Reactions of $ClSN(CH_2)_4O$ with Palladiumorganic Compounds $RPdSC(S)N(C_3H_7\text{-}i)_2$ [11] (see also [49]).

R	product(s) (yield in %)
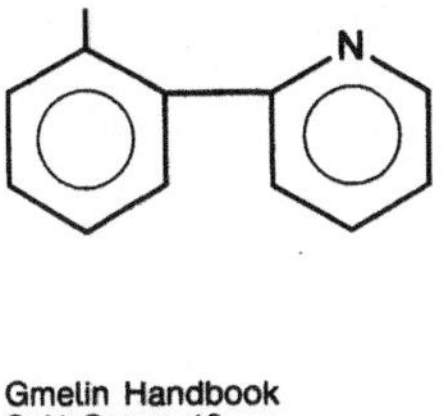	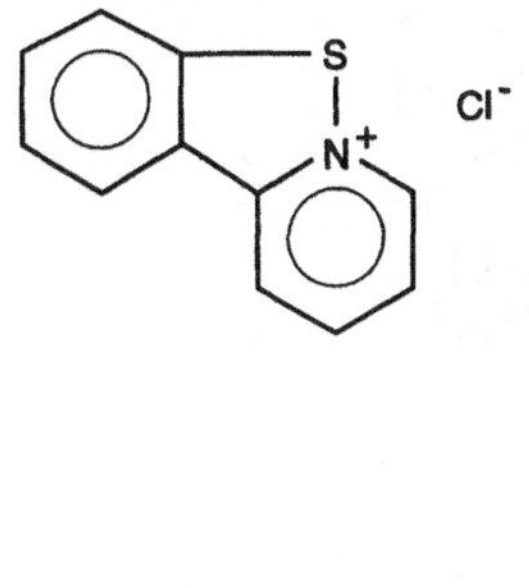

Table 43 (continued)

R	product(s) (yield in %)
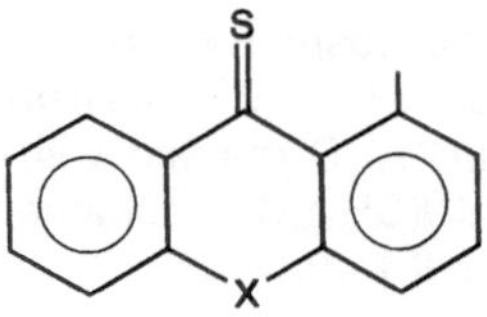	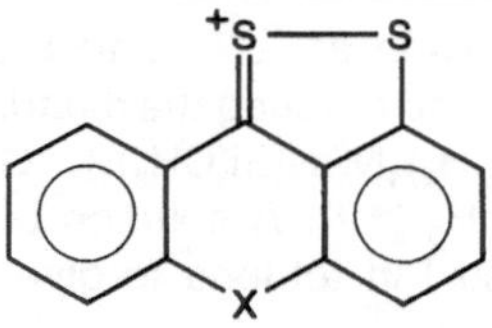 Cl^- X = O (39), X = S (61)
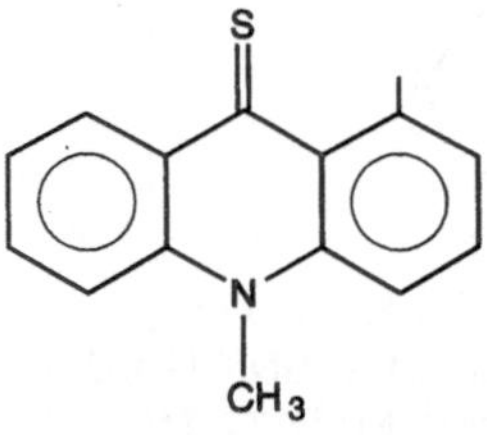	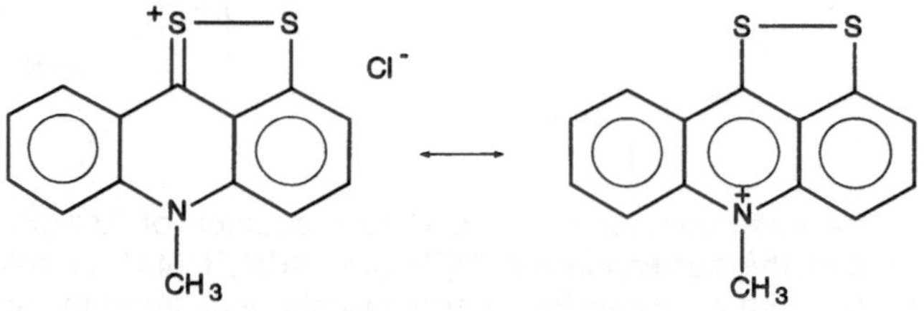 Cl^- (81)
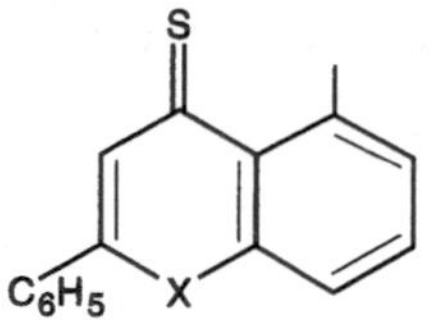	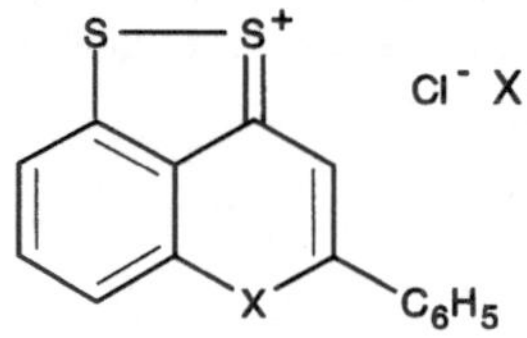 Cl^- X = O (16), X = S (66)
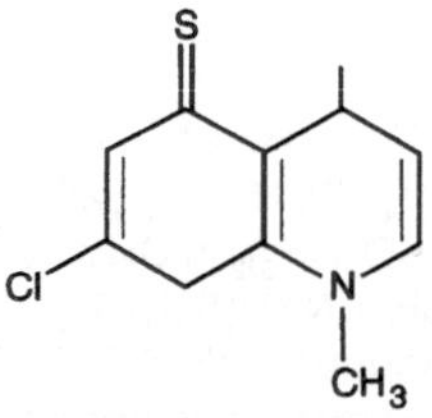	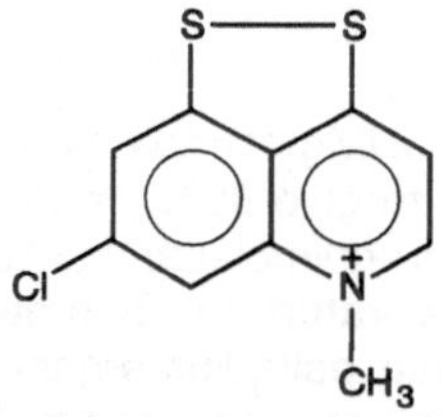 Cl^- (76)
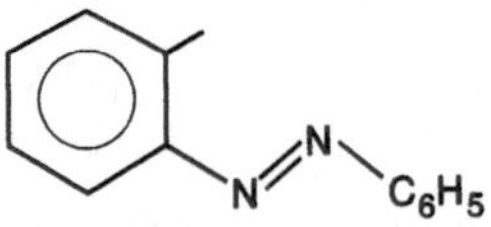	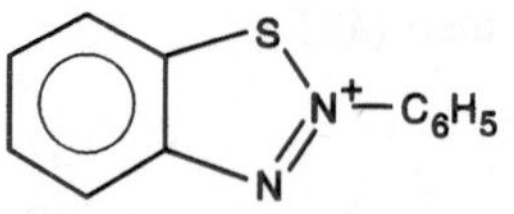 Cl^- (89)
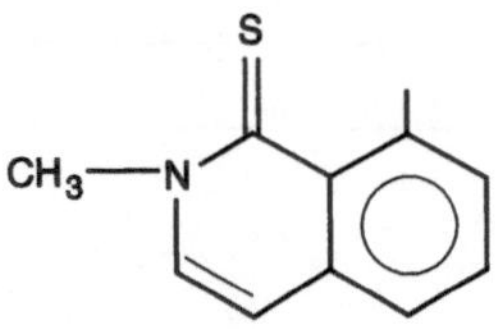 (mole ratio 1 : 1)	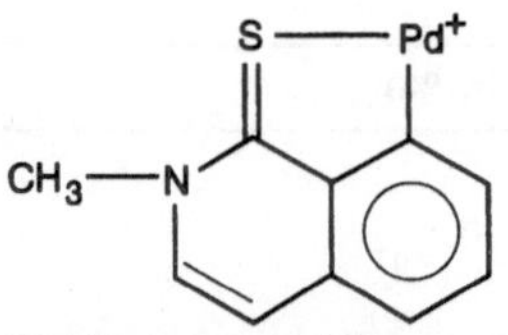 Cl^- $[Pd((i\text{-}C_3H_7)_2NCS_2)Cl]$ $[Pd((i\text{-}C_3H_7)_2NCS_2)_2]$ $O(CH_2)_4NSSN(CH_2)_4O$

Table 43 (continued)

R	product(s) (yield in %)
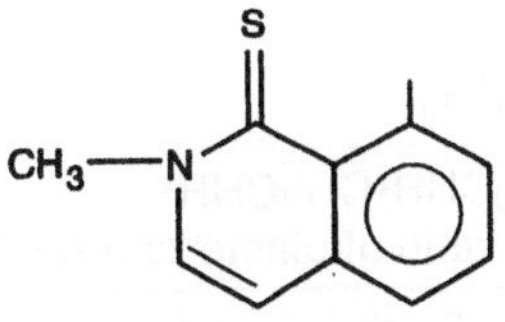	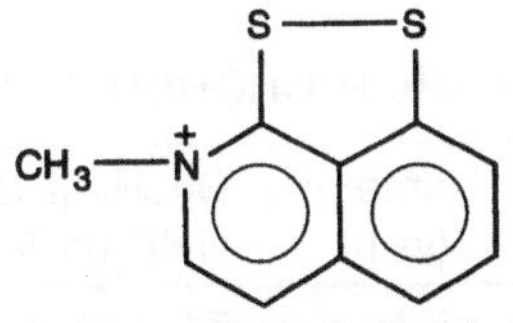 Cl^- (91)
(mole ratio 2 : 1)	$O(CH_2)_4NSSN(CH_2)_4O$ (88) $[\{Pd((i\text{-}C_3H_7)_2NCS_2)Cl\}_2]$ (78)

Adding a solution of $ClSN(CH_2)_4O$ in ether to a −78 °C-precooled solution of the allylic alcohols $CH_2{=}CR^1CR^2R^3OH$ (for R^1, R^2, R^3, see Table 44) and $(C_2H_5)_3N$ (mole ratio 1 : 1 : 2) in ether, allowing the mixture to warm to room temperature (1 h), and leaving the isolated (crude) products at room temperature under N_2 for the time indicated in Table 44 gave after purification by flash chromatography the corresponding unsaturated sulfinamides $O(CH_2)_4NS(O)CH_2\text{-}C(R^1){=}CR^2R^3$ (I) [50].

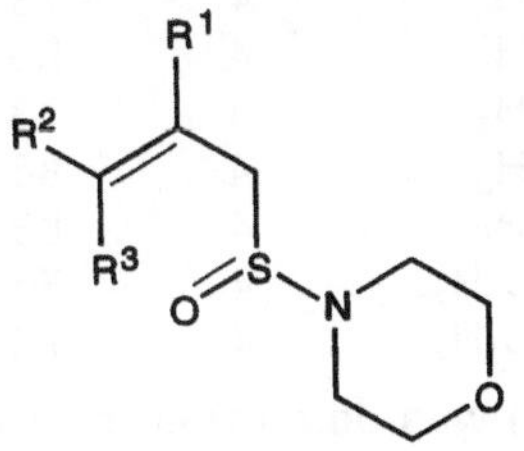

I

Table 44
Reactions of $ClSN(CH_2)_4O$ with Allylic Alcohols $CH_2{=}CR^1CR^2R^3OH$ [50].

$CH_2{=}CR^1CR^2R^3OH$ R^1	R^2	R^3	time (in h)	sulfin-amides (I) yield (in %)	E : Z ratio
H	H	H	42	83	—
H	H	CH_3	2	87	10 : 0
H	CH_3	CH_3	<1	81	—
H	CH_3	$BrCH_2$	<1	a)	
CH_3	H	$C_6H_5\text{–}CH_2\text{–}O\text{–}CH_2$	2	86	9 : 1
H	CH_3	$(CH_3)_2C{=}CH\text{–}(CH_2)_2$	<1	79	5.5 : 4.5
H	H	4-$CH_3C_6H_4$	<1	80	10 : 0
H	H	n-C_8H_{17}	2	82	10 : 0
$O{=}C\text{–}CH_3$	n-C_3H_7	H	<1	86	1 : 9
$O{=}C\text{–}OCH_3$	n-C_3H_7	H	<1	82	3.4 : 6.6
$O{=}C\text{–}CH_3$	C_6H_5	H	<1	92	0 : 10
$O{=}C\text{–}OCH_3$	C_6H_5	H	<1	91	0 : 10

a) This compound could not be purified by chromatography on silica gel (decomposition).

Adding a solution of $ClSN(CH_2)_4O$ in ether to a solution of the allylic alcohols $R^1CHOH\text{-}CH{=}CHR^2$ for R^1, R^2, see Table 45) and $(C_2H_5)_3N$ (mole ratio 1 : 1 : 2) in ether at −78 °C for 15 min and 1 h at room temperature gave the corresponding $O(CH_2)_4NS(O)CHR^2CH{=}CHR^1$ diastereomers which were left at room temperature under Ar for the time indicated in Table 45 [51].

Table 45
Reactions of $ClSN(CH_2)_4O$ with Allylic Alcohols $R^1CHOHCH{=}CHR^2$ [51].

$R^1CHOHCH{=}CHR^2$ configuration	R^1	R^2	time (in h)	$O(CH_2)_4NS(O)CHR^2CH{=}CHR^1$ yield (in %)	ratio of diastereomers[a)]
E	H	CH_3	72	71	65:35
Z	H	CH_3	72	90	10:90
E	CH_3	CH_3	48	70	55:45
Z	CH_3	CH_3	72	46	30:70
E	$i\text{-}C_3H_7$	CH_3	4	35	70:30
Z	$i\text{-}C_3H_7$	CH_3	4	33	5:95
E	$t\text{-}C_4H_9$	CH_3	4	64	65:35
Z	$t\text{-}C_4H_9$	CH_3	4	54	<5:>95
E	$n\text{-}C_7H_{15}$	CH_3	4	73	55:45
Z	$n\text{-}C_7H_{15}$	CH_3	4	60	10:90
E	$4\text{-}CH_3C_6H_4$	CH_3	4	73	77:23
Z	$4\text{-}CH_3C_6H_4$	CH_3	4	86	5:95
	$4\text{-}BrC_6H_4$	CH_3	4	83	5:95
E	$n\text{-}C_7H_{15}$	$i\text{-}C_3H_7$	4	75	65:35
Z	$n\text{-}C_7H_{15}$	$i\text{-}C_3H_7$	4	75	20:80

a) The ratio of diastereomers (E:Z) was determined by ^1H NMR.

The title compound converted the allylic alcohols $R^1R^2C(OH)CH{=}CHR^3$ ($R^1 = H$, CH_3; $R^2 = CH_3$, $n\text{-}C_7H_{15}$, $4\text{-}CH_3C_6H_4$; $R^3 = CH_3$, $i\text{-}C_3H_7$, $t\text{-}C_4H_9$, $n\text{-}C_8H_{17}$, C_6H_5) in the presence of $(C_2H_5)_3N$ in ether at −78 °C within 1 to 16 h to the corresponding sulfinamides $R^1R^2C{=}CH\text{-}CHR^3S(O)N(CH_2)_4O$ with 20 to 86% yield [52].

Adding a solution of $ClSN(CH_2)_4O$ in ether to a −78 °C-precooled solution of the propargylic alcohols $R^1CHOHC{\equiv}CR^2$ (for R^1, R^2, see Table 46) and $(C_2H_5)_3N$ (mole ratio 1:1:1) in ether, stirring at −78 °C for 15 min, warming to room temperature, and stirring for 1 h at room temperature gave the corresponding α-allenic sulfinamide diastereomers IIa and IIb [21].

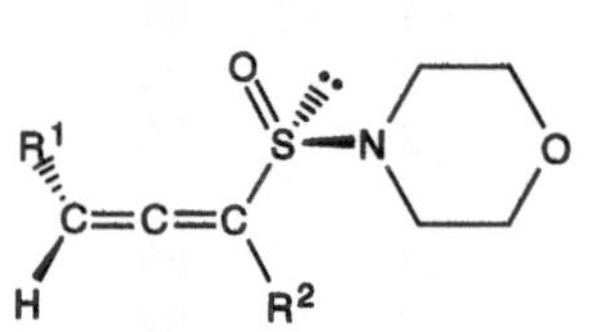

IIa

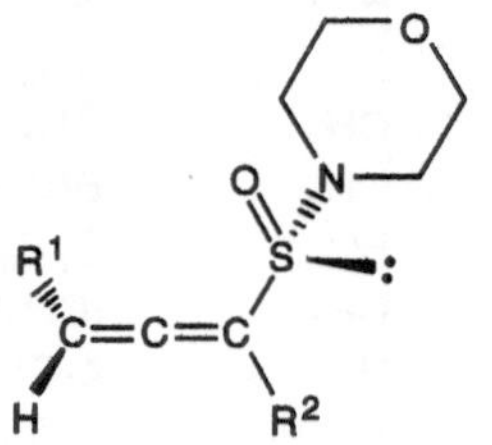

IIb

Table 46
Reaction of $ClSN(CH_2)_4O$ with the Propargylic Alcohols, $R^1CHOHC{\equiv}CR^2$ [21].

$R^1CHOHC{\equiv}CR^2$ R^1	R^2	$R^1CH{=}C{=}C(R^2)S(O)N(CH_2)_4O$ yield[a)] (in %)	ratio of diastereomers (IIa:IIb)[b)]
CH_3	CH_3	98	65:35
$i\text{-}C_3H_7$	CH_3	77	85:15

Table 46 (continued)

$R^1CHOHC{\equiv}CR^2$		$R^1CH{=}C{=}C(R^2)S(O)N(CH_2)_4O$	
R^1	R^2	yield[a] (in %)	ratio of diastereomers (IIa : IIb)[b]
$t\text{-}C_4H_9$	CH_3	76	85:15
$n\text{-}C_7H_{15}$	CH_3	75	75:25
$n\text{-}C_7H_{15}$	$i\text{-}C_3H_7$	77	75:25
$n\text{-}C_7H_{15}$	$t\text{-}C_4H_9$	85	75:25

[a] The crude products had to be kept for 1 to 2 h in Ar at room temperature. – [b] The ratio of diastereomers was determined by ^1H NMR.

Adding a solution of the compound in ether to a −78 °C-precooled solution of the propargylic alcohols $R^1C{\equiv}CCR^2R^3OH$ (for R^1, R^2, R^3, see Table 47) and $(C_2H_5)_3N$ (mole ratio 1 : 1 : 2) in ether, allowing the mixture to warm to room temperature (1 h), and leaving the isolated (crude) products at room temperature under N_2 for the time indicated in Table 47 gave after purification by flash chromatography the corresponding allenic sulfinamides $O(CH_2)_4NS(O)C(R^1){=}C{=}CR^2R^3$ (III) [53].

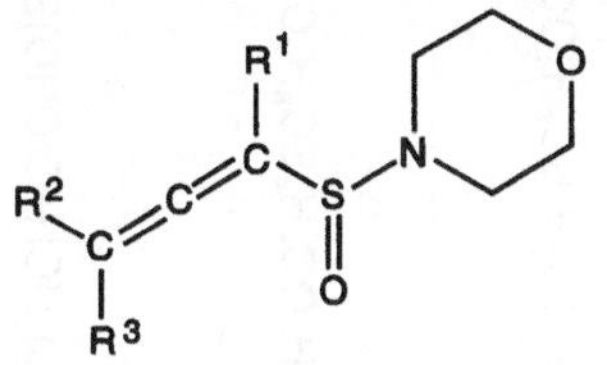

III

Table 47
Reactions of $ClSN(CH_2)_4O$ with the Propargylic Alcohols $R^1C{\equiv}CCR^2R^3OH$ [53].

$R^1C{\equiv}CCR^2R^3OH$			time	sulfinamides (III)	
R^1	R^2	R^3	(in h)	yield (in %)	ratio of diastereomers[a]
H	H	H	1	99[b]	—
H	CH_3	H	15	69[b]	60:40
H	CH_3	CH_3	1	88	—
H	$-(CH_2)_5-$		1	93	—
H	$n\text{-}C_5H_{11}$	H	1	91	[c]
H	$(CH_3)_2C{=}CH{-}(CH_2)_2-$	CH_3	1	89	75:25
H	$n\text{-}C_7H_{15}$	H	1	90	[c]
D	$n\text{-}C_7H_{15}$	H	1	94	65:35
CH_3	$4\text{-}CH_3C_6H_4$	H	1	96	55:45
CH_3	$n\text{-}C_7H_{15}$	H	1	75	75:25
$i\text{-}C_3H_7$	$n\text{-}C_7H_{15}$	H	1	77	75:25
$n\text{-}C_4H_9$	C_6H_5	H	1	79	55:45
$t\text{-}C_4H_9$	$n\text{-}C_7H_{15}$	H	1	85	75:25
$n\text{-}C_5H_{11}$	$n\text{-}C_4H_9$	H	1	96	85:15
$n\text{-}C_6H_{13}$	$i\text{-}C_3H_7$	H	1	72	90:10
$n\text{-}C_8H_{17}$	CH_3	CH_3	1	70	—

[a] The ratio of diastereomers (E : Z) was determined by ^1H NMR. – [b] This compound is very unstable, but can be stored in solution at −20 °C. – [c] The ratio could not be determined by ^1H NMR.

Reactions of $ClSN(CH_2)_4O$ with other organic compounds are summarized in Table 48, pp. 220/6.

Table 48
Reactions of $ClSN(CH_2)_4O$ with Other Organic Compounds.

substrate	reaction conditions	product(s) (yield in %)	Ref.
	in CH_2Cl_2 at −20 °C	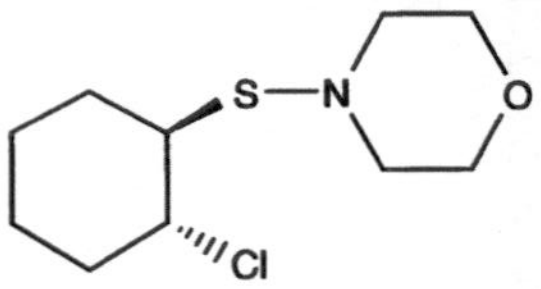	[54]
$CH{\equiv}CC_6H_5$	(nonregiospecific addition)	$C_6H_5C(Cl){=}CHSN(CH_2)_4O$, $CHCl{=}C(C_6H_5)SN(CH_2)_4O$	[55]
$C_6H_5OCH_3 + BF_3$	in CH_2Cl_2 at −78 °C, warming to room temperature	$4\text{-}CH_3OC_6H_4SC_6H_4OCH_3\text{-}4$ (66)	[56]
$C_6H_5OC_6H_5 + BF_3$	in CH_2Cl_2 at −78 °C, warming to room temperature	$4\text{-}C_6H_5OC_6H_4SC_6H_4OC_6H_5\text{-}4$ (30)	[56]
$(C_2H_5)_2NCH{=}CHCHO$	in benzene, ice cooling	$(C_2H_5)_2NCH{=}C(CHO)SC(CHO){=}CHN(C_2H_5)_2$ (89)	[57]
$CH_3C(NH_2){=}CHC(O)OCH_3$	mole ratio ca. 1:2; in benzene at 20 °C	$CH_3C(NH_2){=}C(C(O)OCH_3)\text{-}SC(C(O)OCH_3){=}C(NH_2)CH_3$ (57, crude), $O(CH_2)_4NH \cdot HCl$	[57]
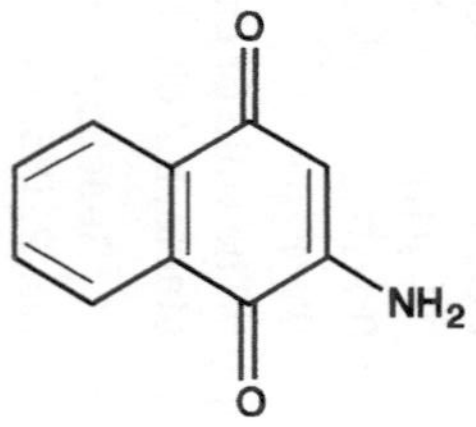		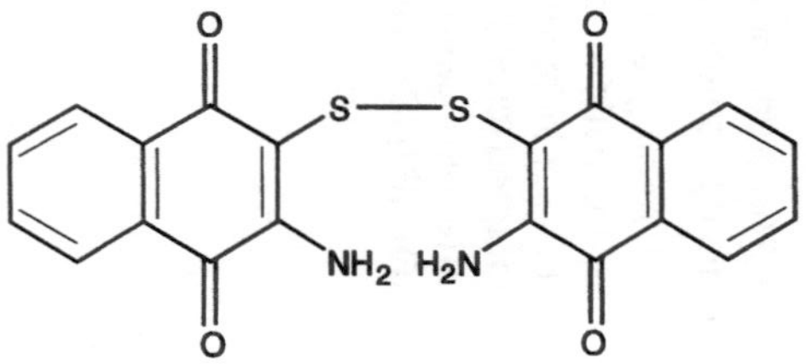	[57]

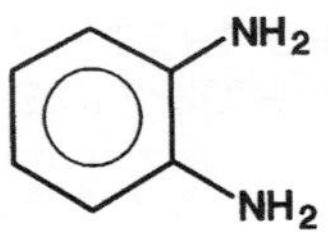	mole ratio 1 : 1; in ether at −20 °C overnight	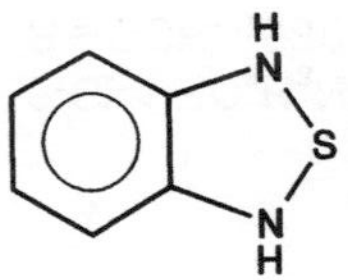 (10), 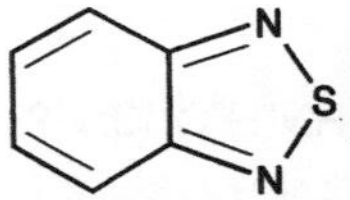 (10), $O(CH_2)_4NSN(CH_2)_4O$ (20)	[58]
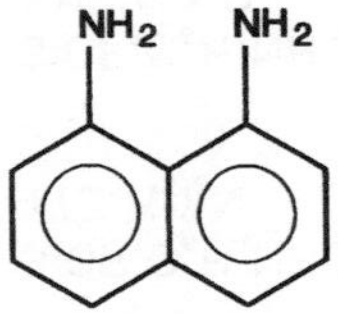	mole ratio 1 : 1; in ether at −20 °C overnight	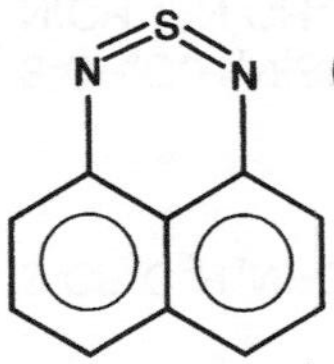 (8), $O(CH_2)_4NSN(CH_2)_4O$ (15)	[58]
$(CH_2)_2NH + (C_2H_5)_3N$	mole ratio ca. 1 : 1 : 1; in CCl_4 at 0 to 5 °C, and 3 h at room temperature	$O(CH_2)_4NSN(CH_2)_2$ (55)	[59]
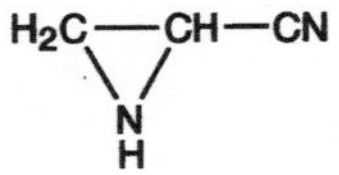	in ether	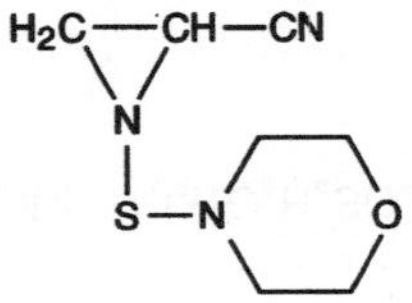	[20]
$O(CH_2)_4NH$	in ether	$O(CH_2)_4NSN(CH_2)_4O$ (25)	[27, 28]
$ArN{=}CHNHCH_3 + (C_2H_5)_3N$ (Ar = 4-$NO_2C_6H_4$, 2-CH_3-4-ClC_6H_3, 2,4-$(CH_3)_2C_6H_3$)	in THF with cooling	$ArN{=}CHN(CH_3)SN(CH_2)_4O$	[60]

Table 48 (continued)

substrate	reaction conditions	product(s) (yield in %)	Ref.
4-$R^1C_6H_4C(R^2)$=NH + $(C_2H_5)_3N$ (R^1 = H, R^2 = t-C_4H_9, CF_3, C_6H_5; R^1 = CH_3, CF_3, R^2 = CF_3)	mole ratio 1:1:1; in benzene	4-$R^1C_6H_4C(R^2)$=$NSN(CH_2)_4O$	[61]
$C_6H_5N(CH_3)_2 + BF_3$	in CH_2Cl_2 at −78°C, warming to room temperature	4-$(CH_3)_2NC_6H_4SC_6H_4N(CH_3)_2$-4 (71)	[56]
$C_6H_5N(C_2H_5)_2 + BF_3$	in CH_2Cl_2 at −78°C, warming to room temperature	4-$(C_2H_5)_2NC_6H_4SC_6H_4N(C_2H_5)_2$-4 (80)	[56]
4-$BrC_6H_4NHC(O)NHCH_3$ + $(C_2H_5)_3N$	mole ratio ca. 1:1:1:1; in THF, 16 h at 20°C	4-$BrC_6H_4NHC(O)N(CH_3)SN(CH_2)_4O$ (45)	[62], see also [63]
3-$CF_3C_6H_4NHC(O)NHCH_3$ + $(C_2H_5)_3N$	mole ratio 1:1:1:1; in THF, 16 h at 20°C	3-$CF_3C_6H_4NHC(O)N(CH_3)SN(CH_2)_4O$ (49)	[62], see also [63]
3-$CF_3C_6H_4NHC(O)N(CH_3)$-$SN(CH_2)_4O$ + NaH (excess) (50% in vaseline oil)	mole ratio ca. 3:2; in THF, 4 h at 20°C	3-$CF_3C_6H_4N(SN(CH_2)_4O)C(O)$-$N(CH_3)SN(CH_2)_4O$ (37)	[62]
3,4-$Cl_2C_6H_3NHC(O)NHCH_3$ + $(C_2H_5)_3N$	mole ratio ca. 1.2:1:1; in THF, 5 min at room temperature, and 4 h at room temperature	3,4-$Cl_2C_6H_3NHC(O)N(CH_3)SN(CH_2)_4O$ (63)	[62]
3-Cl-4-$RC_6H_3NHC(O)NHCH_3$ (R = OCH_3, CH_3) + $(C_2H_5)_3N$	mole ratio 1.1:1:1; in THF, 16 h at 20°C	3-Cl-4-$RC_6H_3NHC(O)N(CH_3)SN(CH_2)_4O$ R = OCH_3 (58), R = CH_3 (35)	[62], see also [63]

$CH_3N{=}C{=}O + HF + (C_2H_5)_3N$	in toluene or CH_2Cl_2 at −10 °C, letting it come to room temperature within 0.5 h [5, 64]; at −10 °C and 2 h at 20 °C [65]	$O(CH_2)_4NSN(CH_3)C(O)F$ (60.5) [5, 64], (70) [65], (38) [23]	[5, 23, 64, 65], see also [66, 67]
$i\text{-}C_3H_7N{=}C{=}O + HF + (C_2H_5)_3N$	in CH_2Cl_2 at −10 to 0 °C, letting it come to room temperature within 1 h	$O(CH_2)_4NSN(C_3H_7\text{-}i)C(O)F$ (27)	[5, 64]
$C_6H_5N{=}C{=}O + HF + (C_2H_5)_3N$	in CH_2Cl_2 at −10 °C, and 1 h at 0 °C	$O(CH_2)_4NSN(C_6H_5)C(O)F$ (11)	[5, 64]
$C_6H_5NHC(O)C(O)NHC_6H_5 + (C_2H_5)_3N$	mole ratio ca. 1 : 1 : 2; in $CCl_2{=}CHCl$–DMA, 1.5 h at room temperature [15]; in DMA, room temperature to 46 °C for 0.5 h [14]	$O(CH_2)_4NSN(C_6H_5)C(O)C(O)\text{-}N(C_6H_5)SN(CH_2)_4O$, $O(CH_2)_4NSN(C_6H_5)C(O)C(O)\text{-}NHC_6H_5$ (small amounts)	[14, 15]
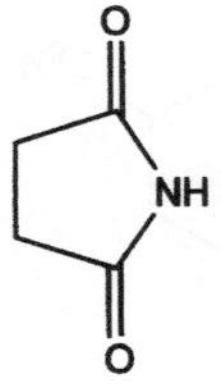 + $(C_2H_5)_3N$	with DMF in $CCl_2{=}CHCl$, room temperature to 47 °C for 2 h [14]; mole ratio 1 : >1; in toluene, 1 h at room temperature [12]	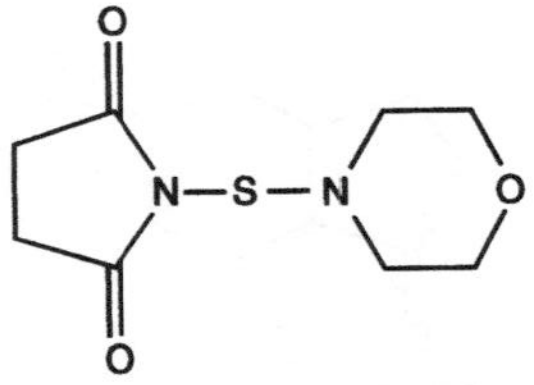 (52.4, crude) [14], (59) [12]	[12, 14]
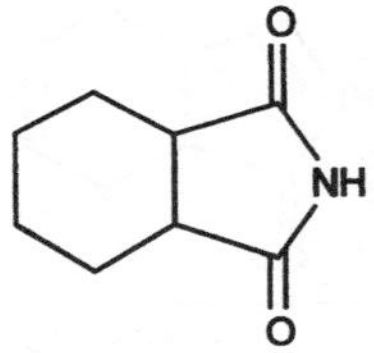 + $(C_2H_5)_3N$	excess substrate; in toluene [12], in ethylene dichloride, 1 h at room temperature [68], see also [13, 16]	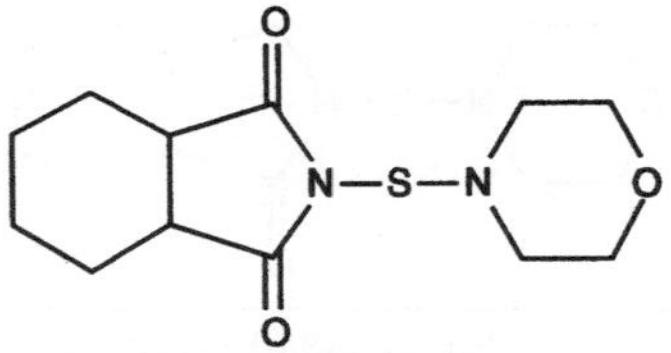 (94) [12], (50.5) [68], see also [13, 16]	[12, 68], see also [13, 16]

Table 48 (continued)

substrate	reaction conditions	product(s) (yield in %)	Ref.
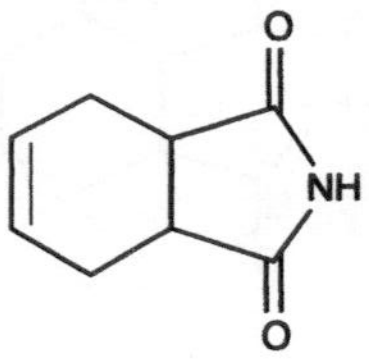 + $(C_2H_5)_3N$	excess substrate; in toluene, 1 h at room temperature	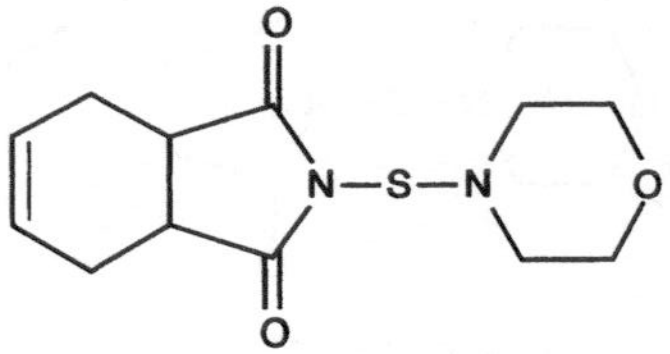 (76)	[12]
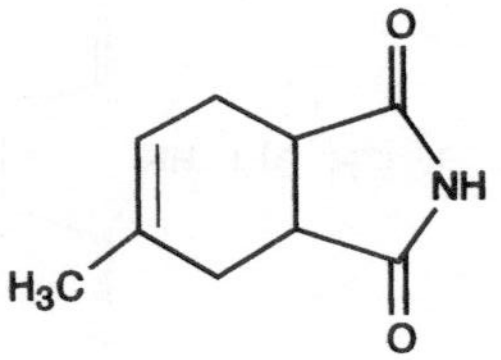 + $(C_2H_5)_3N$	excess substrate; in toluene, 1 h at room temperature	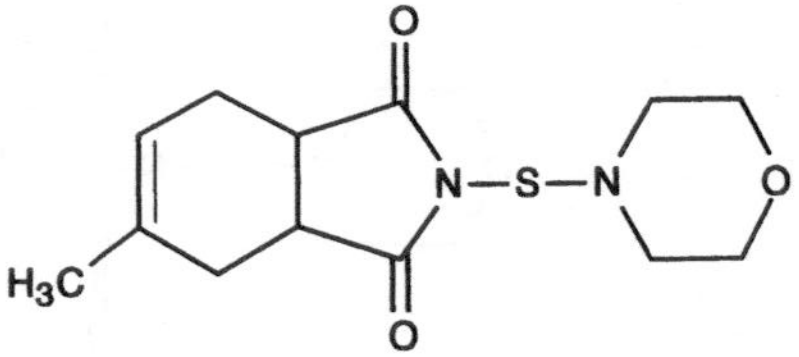 (90)	[12]
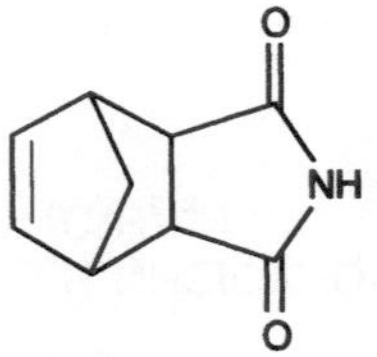 + $(C_2H_5)_3N$	excess substrate; in toluene, 1 h at room temperature	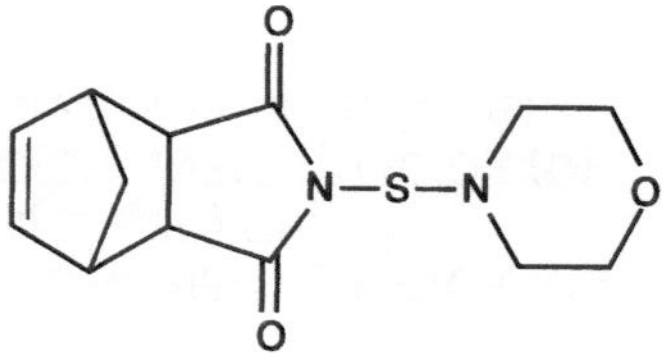 (64)	[12]
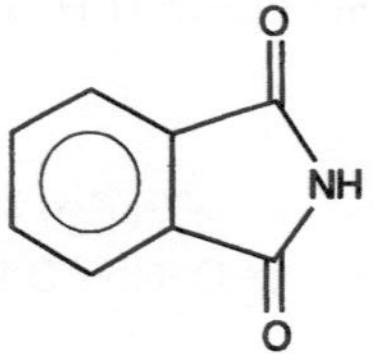 + $(C_2H_5)_3N$	excess substrate; in toluene, 1 h at room temperature [12]; excess $(C_2H_5)_3N$, in THF at 10 to 15°C for 1 h [13, 16, 68]; in CCl_4–DMS, room temperature to 36°C [14], or in CCl_4–THF [14]	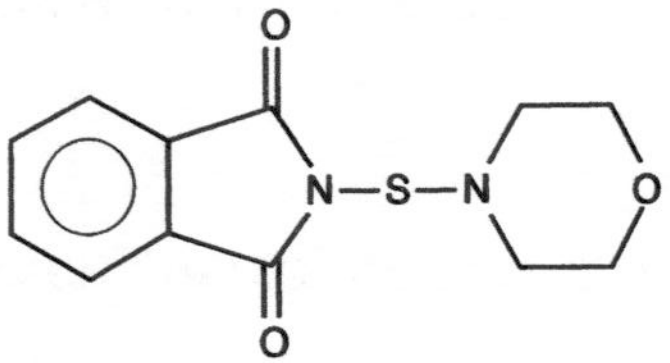 (89) [12], (61) [13, 16, 68], (87.5) [14], or (<87) [14]	[12 to 14, 16, 68]
$CH_3SC(CH_3){=}NOC(O)NHCH_3$	in CH_3CN at 0 to 5°C,	$CH_3SC(CH_3){=}NOC(O)N(CH_3)SN(CH_2)_4O$	[69], see

("methomyl") + $(C_2H_5)_3N$	letting it come to room temperature within 2 h		also [70]
$CH_3SC(CH_3)_2CH{=}NO$-$C(O)NHCH_3$ ("aldicarb")	in pyridine at ca. 5 °C, and ~20 h at room temperature	$CH_3SC(CH_3)_2CH{=}NOC(O)N(CH_3)SN(CH_2)_4O$	[3]
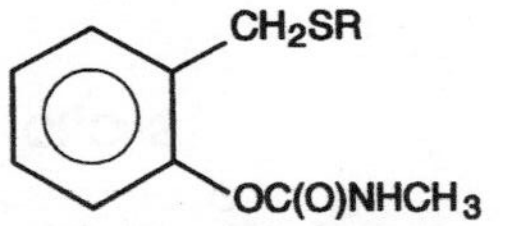 ($R = CH_3$, C_2H_5, n-C_3H_7, i-C_3H_7) + pyridine	18 h at room temperature	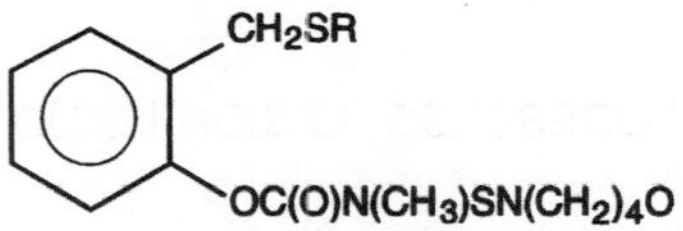	[71]
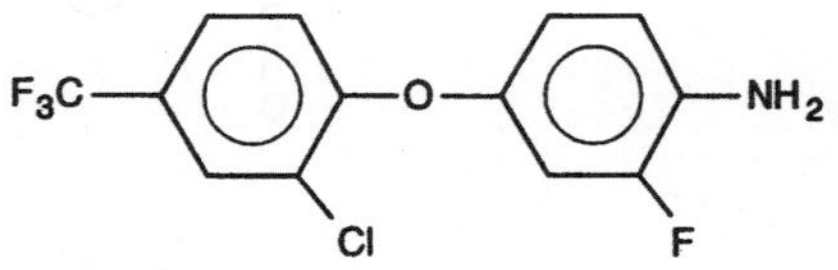 + $(C_2H_5)_3N$	in $(C_2H_5)_2O$	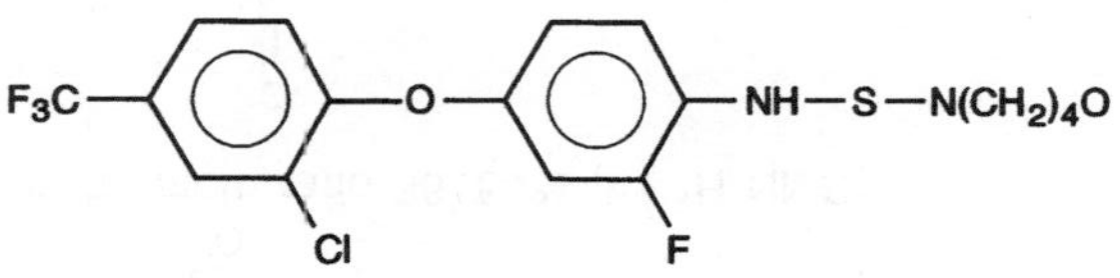	[6]
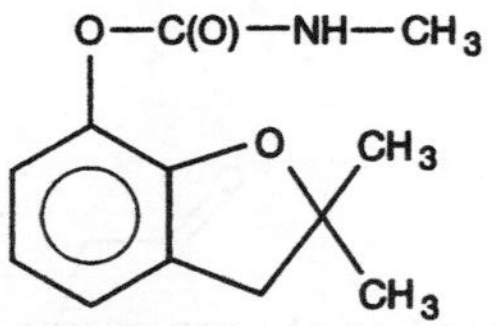 + $(C_2H_5)_3N$	in pyridine, 18 h at room temperature	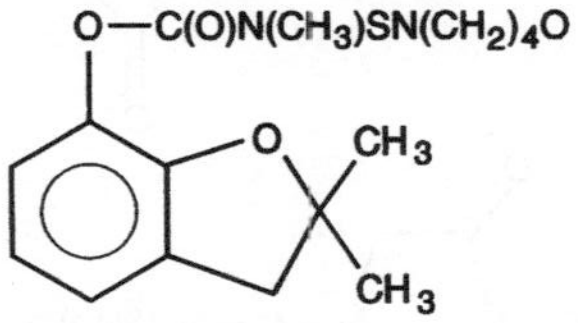	[4]
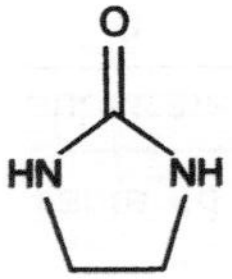 + $(C_2H_5)_3N$ + DMF	in $CCl_2{=}CHCl$, room temperature to 51 °C, 1.5 h	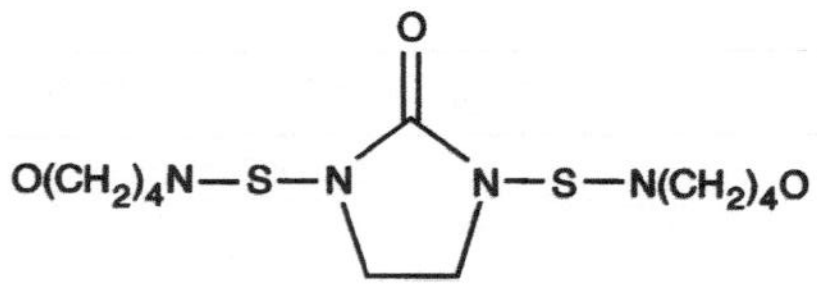	[14], see also [72]

Table 48 (continued)

substrate	reaction conditions	product(s) (yield in %)	Ref.
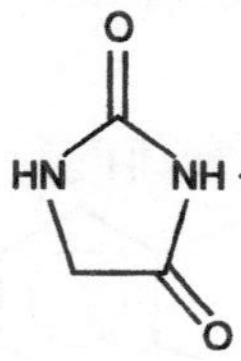 + $(C_2H_5)_3N$ + DMF	in CCl_2=CHCl, 2 h at room temperature	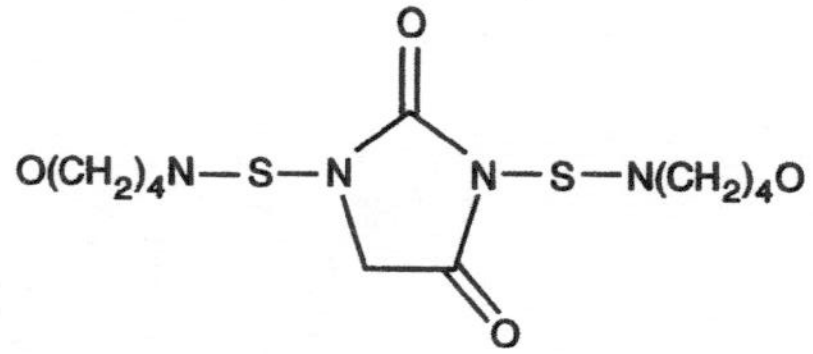 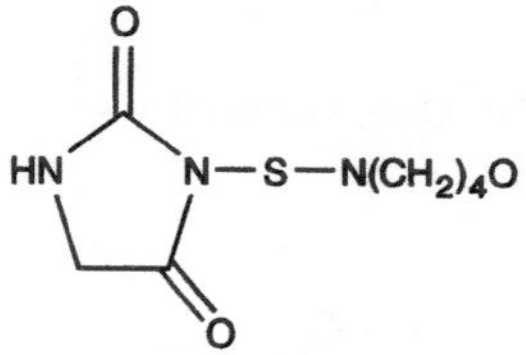 in the mole ratio 36 : 64% (by ^{1}H NMR)	[14]
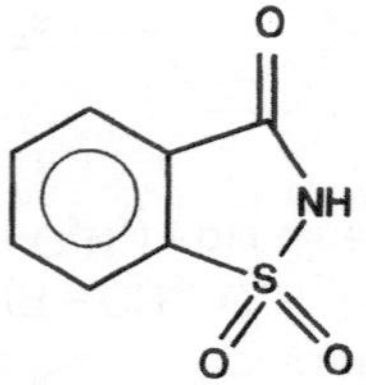 + $(C_2H_5)_3N$ + DMF	in CCl_2=CHCl, room temperature to 46 °C, 1.5 h	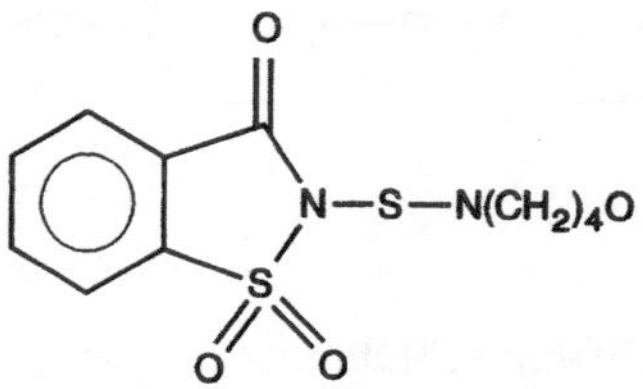	[14]
Cl_2C=S	mole ratio 1 : >1; in CH_3CN	$O(CH_2)_4NC(S)Cl$, $CCl_3SSSCCl_3$	[73]
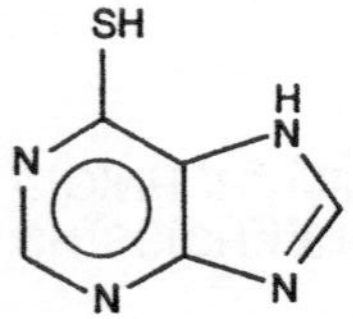	in pyridine (excess), 5 min at 80 to 90 °C	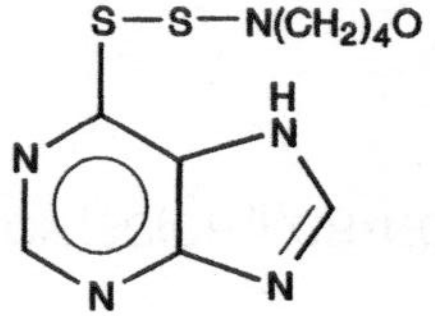	[10]
4-$CH_3C_6H_4SO_2S^-$ $^+H_2N(CH_2)_4O$	in CH_2Cl_2 at 0 to 5 °C	4-$CH_3C_6H_4SO_2SSN(CH_2)_4O$	[74]

References:

[1] Sikorski, J. A.; Hoobler, M. A. (U.S. 4433996 [1981/84]; C.A. **101** [1984] No. 23718).

[2] Badische Anilin-Soda-Fabrik AG; Weiß, G.; Schulze, G. (Ger. 1131222 [1960/62]; C.A. **57** [1962] 13771).

[3] Fukuto, T. R.; Black, A. L. (U.S. 4108991 [1976/78]; C.A. **90** [1979] No. 151581).

[4] Black, L.; Fukuto, T. R. (Ger. 2433680 [1974/75]; C.A. **82** [1975] No. 156050).

[5] Thurman, D. E. (Ger. 2628574 [1975/77]; C.A. **86** [1977] No. 155209).

[6] Anderson, M.; Brinnand, A. G. (Eur. Appl. 216423 [1986/87]; C.A. **106** [1987] No. 213583).

[7] Otsuka Chemical Co., Ltd. (Jpn. Kokai Tokkyo Koho 57-163359 [1981/82]; C.A. **98** [1983] No. 126128).

[8] Freytag, H.; Lober, F. (Ger. 965968 [1954/57]; C.A. **1959** 11417).

[9] Farbenfabriken Bayer AG (Br. 790021 [1954/58]; C.A. **1958** 13806).

[10] Carter Products, Inc.; Stiefel, F. J.; Ludwig, B. J.; Berger, F. M. (Br. 993302 [1964/65]; C.A. **63** [1965] 8381, U.S. 3400125 [1963/68]; C.A. **70** [1969] No. 28955).

[11] Davis, R. C.; Grinter, T. J.; Leaver, D.; O'Neil, R. M.; Thomson, G. A. (J. Chem. Soc. Perkin Trans. I **1990** 2881/7).

[12] Lawrence, J. P. (Rubber Chem. Technol. **49** [1976] 333/40).

[13] Lawrence, J. P. (Fr. 2190819 [1970/74]; C.A. **81** [1974] No. 122404, U.S. 3944552 [1972/76]; C.A. **85** [1976] No. 22609).

[14] Son, P. N.; Krueger, R. A. (J. Appl. Polym. Sci. **21** [1977] 1731/42), Son, P. N. (U.S. 4006140 [1973/77]; C.A. **86** [1977] No. 107832).

[15] Son, P. N. (Ger. 2425310 [1972/75]; C.A. **82** [1975] No. 141394).

[16] Lawrence, J. P. (Ger. 2329431 [1970/74]; C.A. **81** [1974] No. 50818).

[17] Freytag, H.; Lober, F. (Ger. 969813 [1954/58]; C.A. **1960** 5473).

[18] Kühle, E. (Synthesis **1970** 561/80, 566; The Chemistry of the Sulfenic Acids, Thieme, Stuttgart 1973, p. 10).

[19] Dorlars, A. (Houben-Weyl Methoden Org. Chem. 4th Ed. **11** Pt. 2, Stuttgart 1958, pp. 744/51).

[20] Berger, H.; Gall, R.; Kampe, W.; Bicker, U.; Kuhn, R. (Ger. 2948832 [1979/81]; C.A. **95** [1981] No. 115257).

[21] Baudin, J.-B.; Bkouche-Waksman, I.; Julia, S. A.; Pascard, C.; Wang, Y. (Tetrahedron **47** [1991] 3353/64).

[22] Mazitova, F. N.; Kharullin, V. K. (Zh. Obshch. Khim. **56** [1986] 788/90; J. Gen. Chem. USSR [Engl. Transl.] **56** [1986] 694/5).

[23] Hatch, C. E., III (J. Org. Chem. **43** [1978] 3953/7).

[24] Almasi, L.; Hantz, A. (Omagiu' Raluca Ripan **1966** 59/64; C.A. **67** [1967] No. 116856).

[25] Musin, B. M.; Ivanov, V. B.; Ivanov, B. E. (Izv. Akad. Nauk SSSR Ser. Khim. **1988** 1693; Bull. Acad. Sci. USSR Div. Chem. Sci. [Engl. Transl.] **1988** 1509).

[26] Armitage, D. A.; Towle, I. D. H. (Phosphorus Sulfur **1** [1976] 37/9).

[27] Levchenko, E. S.; Dorokhova, E. M. (Zh. Org. Khim. **5** [1969] 1516; J. Org. Chem. USSR [Engl. Transl.] **5** [1969] 1481/2).

[28] Bacon, R. G. R.; Irwin, R. S. (J. Chem. Soc. **1960** 5079/87).

[29] Baudin, J. B.; Julia, S. A.; Ruel, O. (Tetrahedron **43** [1987] 881/9).

[30] Morita, E.; D'Amico, J. J. (Rubber Chem. Technol. **44** [1971] 881/8; C.A. **76** [1972] No. 15537).

[31] Hoobler, M.; Sikorski, J. A. (Belg. 894595 [1982/83]; C.A. **99** [1983] No. 6049).

[32] Almasi, L.; Paskucz, L. (Chem. Ber. **102** [1969] 1489/94).

[33] Hantz, A.; Salamon, A.-M.; Raita, G.; Almasi, L. (J. Prakt. Chem. [4] **320** [1978] 183/90).
[34] Almasi, L.; Popovici, N.; Hantz, A. (Monatsh. Chem. **103** [1972] 1027/32).
[35] Lopusiński, A.; Luczak, L.; Michalski, J.; Koziol, A. E.; Gdaniec, M. (J. Chem. Soc. Chem. Commun. **1991** 889/90).
[36] Michalski, J.; Lopusiński, A.; Jezierska, B.; Luczak, L.; Potrzebowski, M. (Phosphorus Sulfur **30** [1987] 221/4).
[37] Lopusiński, A.; Luczak, L.; Michalski, J. (J. Chem. Soc. Chem. Commun. **1989** 1694/5).
[38] Almasi, L.; Paskucz, L. (Monatsh. Chem. **101** [1970] 662/7).
[39] Almasi, L.; Paskucz, L. (Chem. Ber. **98** [1965] 613/6).
[40] Malz, H.; Bayer, O.; Freytag, H.; Lober, F. (U.S. 2891059 [1957/59]; C.A. **1960** 4387).

[41] Shermolovich, Yu. G.; Solov'ev, A. V.; Borodin, A. V.; Pen'kovskii, V. V.; Trachevskii, V. V.; Markovskii, L. N. (Zh. Org. Khim. **27** [1991] 1637/41; J. Org. Chem. USSR [Engl. Transl.] **27** [1991] 1433/6).
[42] Burgess, E. M.; Penton, H. R., Jr. (J. Org. Chem. **39** [1974] 2885/92).
[43] Hopper, R. J.; Lawrence, J. P. (U.S. 3856762 [1972/74]; C.A. **82** [1975] No. 99745, U.S. 3898206 [1972/75]; C.A. **83** [1975] No. 180814, U.S. 3898205 [1972/75]; C.A. **83** [1975] No. 180815).
[44] Williams, H. R. (U.S. 3121661 [1961/64]; C.A. **61** [1964] 6958).
[45] Neidlein, R.; Lenhard, T. (Chem. Ber. **116** [1983] 3133/40).
[46] Markley, L. D.; Dunbar, J. E. (J. Org. Chem. **37** [1972] 2512/4).
[47] Laithwaite, P.; Taylor, J. A. (Br. 1355802 [1970/74]; C.A. **82** [1975] No. 32146).
[48] Blaschette, A.; Näveke, M. (Chem.-Ztg. **115** [1991] 61/4).
[49] Davis, R. C.; Grinter, T. J.; Leaver, D.; O'Neil, R. M. (Tetrahedron Lett. **1979** 3339/42).
[50] Baudin, J.-B.; Julia, S. A. (Tetrahedron Lett. **29** [1988] 3251/4).

[51] Baudin, J.-B.; Bkouche-Waksman, I.; Hareau, G.; Julia, S. A.; Lorne, R.; Pascard, C. (Tetrahedron **47** [1991] 6655/72).
[52] Baudin, J.-B.; Julia, S. A. (Tetrahedron Lett. **30** [1989] 1963/6).
[53] Baudin, J.-B.; Julia, S. A.; Wang, Y. (Tetrahedron Lett. **30** [1989] 4965/8).
[54] Kutateladze, A. G.; Zyk, N. V.; Denisko, O. V.; Zefirov, N. S. (Zh. Org. Khim. **27** [1991] 659/61; J. Org. Chem. USSR [Engl. Transl.] **27** [1991] 569/70).
[55] Kutateladze, A. G.; Zyk, N. V.; Denisko, O. V.; Zefirov, N. S. (Izv. Akad. Nauk SSSR Ser. Khim. **1990** 1689/90; Bull. Acad. Sci. USSR Div. Chem. Sci. [Engl. Transl.] **39** [1990] 1535/6).
[56] Bombala, M. U.; Ley, S. V. (Synth. Commun. **10** [1980] 291/7).
[57] Gompper, R.; Euchner, H.; Kast, H. (Justus Liebigs Ann. Chem. **675** [1964] 151/74).
[58] Bryce, M. R. (J. Chem. Soc. Perkin Trans. I **1984** 2591/3).
[59] Farbenfabriken Bayer AG; Freytag H.; Lober, F.; Domagk, G. (Ger. 948330 [1954/56]; C.A. **1959** 6267).
[60] Drabek, J.; Böger, M. (Ger. 2621077 [1974/76]; C.A. **86** [1977] No. 89430).

[61] Markovskii, L. N.; Shermolovich, Yu. G.; Shevchenko, V. I. (Zh. Org. Khim. **11** [1975] 2533/7; J. Org. Chem. USSR [Engl. Transl.] **11** [1975] 2603/7).
[62] Hainaut, D.; Demoute, J.-P.; Teche, A. (Ger. 2256275 [1972/73]; C.A. **79** [1973] No. 42215).
[63] Roussel-UCLAF (Fr. 2208604 [1971/74]; C.A. **82** [1975] No. 125134).
[64] Thurman, D. E. (Ger. 2628575 [1975/77]; C.A. **86** [1977] No. 155665).
[65] Stetter, J.; Homeyer, B.; Hammann, I. (Ger. 2824394 [1978/79]; C.A. **92** [1980] No. 181198).
[66] Durden, J. A., Jr. (U.S. 4071627 [1976/78]; C.A. **89** [1978] No. 43450).

[67] Kurtz, A. P., Jr.; D'Silva, T. D. J. (U.S. 4073930 [1976/78]; C.A. **88** [1978] No. 170204).
[68] Lawrence, J. P. (U.S. 3928340 [1972/75]; C.A. **84** [1976] No. 166003).
[69] Gemrich, E. G., II; Lee, B. L.; Nelson, S. J.; Rizzo, V. L. (J. Agric. Food Chem. **26** [1978] 391/5).
[70] Rizzo, V. L. (Ger. 2655212 [1976/77]; C.A. **88** [1978] No. 6344).

[71] Hoffmann, H.; Hammann, I.; Homeyer, B.; Stendel, W. (Ger. 2737606 [1977/79]; C.A. **91** [1979] No. 56482).
[72] Lawrence, J. P. (U.S. 4042642 [1972/77]; C.A. **87** [1977] No. 153223, U.S. 3960857 [1972/76]; C.A. **85** [1976] No. 79437).
[73] Hansen, H. C. (Sulfur Lett. **3** [1985] 181/8; C.A. **105** [1986] No. 78470).
[74] Hansen, H. C. (Sulfur Lett. **1** [1982] 15/8; C.A. **98** [1983] No. 89277).

12.5.2.9.2 ClSN–CH$_2$–CH(CH$_3$)–O–CH(CH$_3$)–CH$_2$

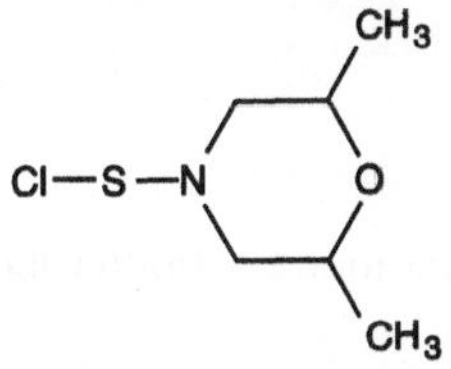

The preparation of the title compound was not reported.

The title compound reacted under cooling with a solution of $R^1R^2C_6H_3N{=}CHNHCH_3$ ($R^1 = 2$-CH_3, $R^2 = 4$-Cl, 4-CH_3; $R^1R^2 = 3,5$-$(CF_3)_2$) and $(C_2H_5)_3N$ in THF to give the corresponding $R^1R^2C_6H_3N{=}CHN(CH_3)SN$–CH$_2$–CH(CH$_3$)–O–CH(CH$_3$)–CH$_2$ [1]. Adding $(C_2H_5)_3N$ in CH_3CN to a cold (0 to 5 °C) solution of $CH_3SC(CH_3){=}NOC(O)NHCH_3$ and the compound in CH_3CN and allowing the mixture to warm to room temperature within 2 h gave $CH_3SC(CH_3){=}NOC(O)N$-(CH$_3$)SN–CH$_2$–CH(CH$_3$)–O–CH(CH$_3$)–CH$_2$ [2]. Reacting a mixture of the title compound with "carbofuran" I in pyridine at room temperature for 18 h gave the carbamate II [3].

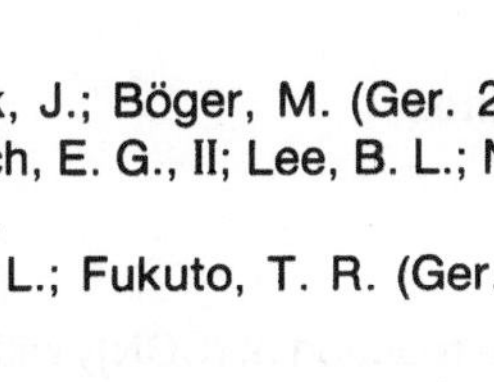

I

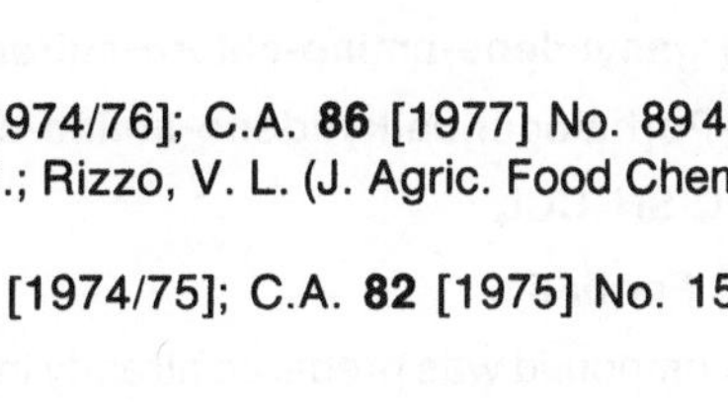

II

References:

[1] Drabek, J.; Böger, M. (Ger. 2621077 [1974/76]; C.A. **86** [1977] No. 89430).
[2] Gemrich, E. G., II; Lee, B. L.; Nelson, S. J.; Rizzo, V. L. (J. Agric. Food Chem. **26** [1978] 391/5).
[3] Black, L.; Fukuto, T. R. (Ger. 2433680 [1974/75]; C.A. **82** [1975] No. 156050).

12.5.2.10 N[1],N[3]-Bis(chlorosulfanyl)-5,5-dimethyl-2,4-imidazolidinedione, ClSN–C(O)–C(CH$_3$)$_2$–N(SCl)–C(O)

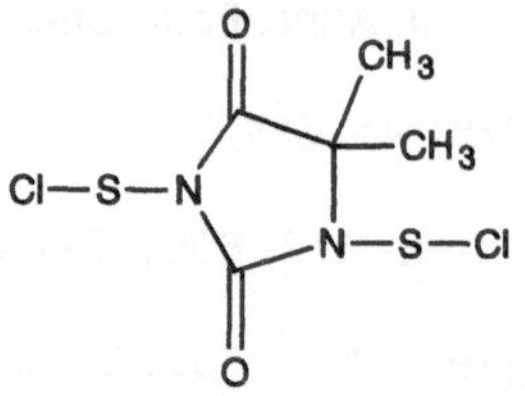

The title compound was formed by heating a mixture of 1,3-dichloro-5,5-dimethylhydantoin (I), elemental sulfur, and $(C_2H_5)_4N^+$ Br^- (mole ratio 1 : 2 : 2.4) in 1,2-dichloroethane slowly to 50 °C, keeping the mixture at this temperature for ~1 h, and then holding it at 80 °C for 30 min; 92% yield.

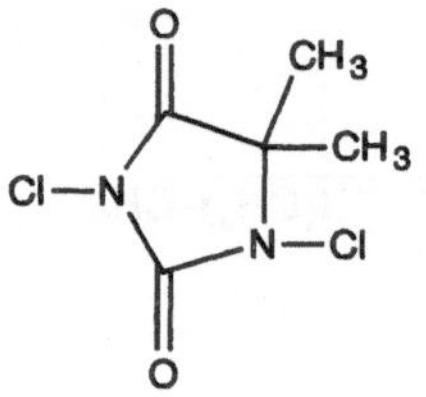

I

Yellow prisms; m.p. 103 to 104 °C. The title compound decomposes upon recrystallization from inert solvents.

Adding a solution of the title compound in 1,2-dichloroethane at 0 to 5 °C to a solution of morpholine (mole ratio 1 : 4) in 1,2-dichloroethane and letting the mixture react at 20 °C for ~2 h gave 100% morpholine hydrochloride and II (95%).

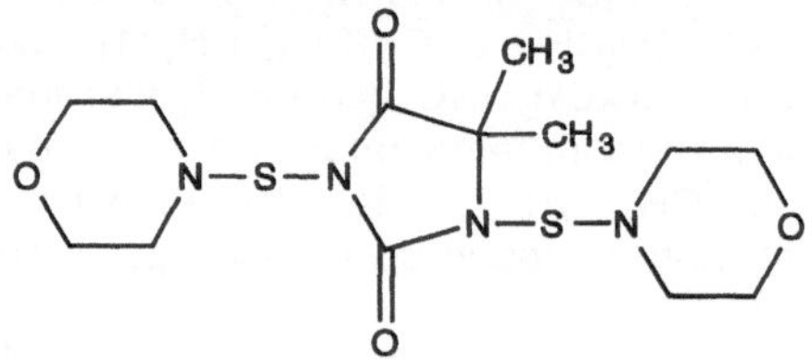

II

Reference:

Borovikova, G. S.; Levchenko, E. S.; Kaminskaya, E. I. (Zh. Org. Khim. **22** [1986] 100/6; J. Org. Chem. USSR [Engl. Transl.] **22** [1986] 86/92).

12.5.3 N-Organylidene-amino-chloro-sulfanes

12.5.3.1 N-Perhalogenoalkylidene-amino-chloro-sulfanes

12.5.3.1.1 ClSN=CCl$_2$

Preparation. Formation

The title compound was prepared already in 1924 by the reaction of $(SCN)_2$ with excess Cl_2 in C_2H_5Br while cooling [1] or at –50 °C [2]; it was erroneously thought to be "rhodane trichloride", $(SCN)Cl_3$ [1, 2]. The compound was reinvestigated: saturation of an ethyl bromide solution of $(SCN)_2$ (~ 0.4 M) with dry chlorine at room temperature or 0 °C [3], at −16 °C [4] (see also [5]), or in $CHCl_3$ or CCl_4 at room temperature or 0 °C [6] gave the compound with a yield of 70 to 80% [3] or 65% [4] (see also [5]) or 20 to 25% [6], respectively. These authors [3 to 6] proposed Cl_3SCN to be ClN=C(Cl)SCl, but [3, 6] pointed out that the isomeric $Cl_2C{=}NSCl$ can not be ruled out. $ClSN{=}CCl_2$ was also obtained with 56% yield in a single process by passing a stream of Cl_2 through a suspension of $Pb(SCN)_2$ in C_2H_5Br for 30 min at room temperature [6].

$ClSN{=}CCl_2$ was also prepared by reacting equimolar amounts of SCl_2 and ClC≡N with activated carbon as catalyst in a bomb at 60 °C for 14 h with 82% yield [18], or with a yield of 27% by condensing ClC≡N onto C_2H_5Br containing SCl_2 and keeping the mixture at 0 °C for 5 d [3].

The title compound was formed with yields of 70 and 50% along with ClC≡N, SCl_2, and $Cl_2C{=}NSN{=}CCl_2$ by mixing $Cl_2C{=}NCl$ and elemental sulfur without a solvent at 0 °C and adding catalytic amounts of DMF or $(CH_3)_4NCl$, respectively. After the sulfur dissolved, the mixture was kept overnight at room temperature and distilled [9].

$ClSN{=}CCl_2$ was formed by bubbling Cl_2 at 0 to 10 °C through $(CH_3)_3SiN{=}C{=}S$ [10].

Physical Properties. Spectra

The title compound is a yellow, penetrating, tear-irritant oil [1], or a pungent, orange liquid [2, 3]; b.p. 40 to 41 °C/9 Torr [3], 43 to 46 °C/11 Torr [10], 50 °C/23 Torr [8], 50 to 51 °C/18 Torr [3], 51 °C/18 Torr [4] (see also [5]), 54 to 56 °C/20 Torr [1], 57 to 58 °C/20 Torr [2], 152 to 153 °C [1], 154 to 155 °C [3].

Refractive index: $n_D^{19} = 1.576$; density: $D_4^{19} = 1.6208$ g/cm³; mole refraction: R = 33.55 [2].

The compound is soluble in indifferent organic solvents [1].

Evidence for the structure of N-chlorosulfanyl-chloroformimidoyl chloride, $Cl_2C{=}NSCl$, was obtained from the bands at 1600 and 900 cm^{-1} in the IR spectrum being characteristic for isocyanide dichlorides as well as from a few chemical transformations of the compound [7].

Absorption bands at 1592 to 1578 cm^{-1} and 915 to 895 cm^{-1} in the IR spectrum (displayed in the paper) have been assigned to vibrations of the C=N and C−Cl band, respectively [10]. IR: ν (in cm^{-1}) = 1603 s, 1589 s (1594 doublet); 923 vs, 905 vs (913 doublet); 834 m; 762 w; 542 ms, 524 vs (529 doublet); 473 vs; 430 w (assignments erroneous); the IR spectra for the pure liquid, vapor, and solution in CCl_4 from 4000 to 400 cm^{-1} were reproduced in a figure [3].

Raman (displayed in the paper): ν (in cm^{-1}; intensities in parentheses) = 1669 (1?), 1652 (1?), 1602 (12), 1581 (12), 834 (1), 759 (2), 609 (4), 544 (1), 522 (3), 471 (7), 385 (3), 308 (1), 277 (1); 200 (7), 126 (2), 114 (4); the intense bands at 1602 and 1581 cm^{-1} have been assigned to the C=N bond [4], and 1604 (10), 1587 (10), 832 (1dp), 757 (2), 609 (2dp), 471 (7dp), 430 (1dp), 385 (3), 311 (2), 275 (2), 204 (7) [5]; see also [3].

UV (CCl_4): ClSN=CCl absorbs in both the visible and the near-UV regions, displaying a maximum at 345 nm [3].

Chemical Reactions

$ClSN{=}CCl_2$ is preferably distilled under reduced pressure, but is reasonably stable at the normal boiling point (154 to 155 °C). Carefully redistilled $ClSN{=}CCl_2$ remained unchanged for many weeks in a sealed vessel. A reasonably pure sample of $ClSN{=}CCl_2$, which had been stored in the dark in a stoppered bottle for about one year, deposited 61% of the theoretical amount of ClC≡N and 70% SCl_2 [3].

$ClSN{=}CCl_2$ was completely hydrolyzed, exothermically, by shaking it with about ten times its weight of cold water for about 1 h, causing deposition of sulfur. The pale yellow, acidic, aqueous solution responded strongly to qualitative tests for $ClC{\equiv}N$, H_2S, and SO_4^{2-} and Cl^- ions, responded weakly to tests for CN^-, SCN^-, and SO_3^{2-} ions, and gave a weak indication for NH_3 when warmed with alkali [3]. Earlier, the hydrolysis of $ClSN{=}CCl_2$ was said to be very inactive upon heating with water. During the decomposition with 10% aqueous NaOH solution, HCl, H_2S, H_2SO_3, $H_2S_2O_3$, H_2SO_4, HSCN, H_2CO_3, HCN, and NH_3 were determined [1].

Adding a solution of $ClSN{=}CCl_2$ in ether to a large excess of a solution of ammonia in ether gave ammonium chloride and presumably $H_2NSN{=}CCl_2$, which was highly unstable and decomposed violently when the solvent was removed [6]. Reactions of $ClSN{=}CCl_2$ with H_2S or H_2S_n to get chain prolongation by splitting off HCl failed; only decomposition products were obtained [4]. $Cl_2C{=}NSN{=}CCl_2$ was formed with a yield of 11% when a solution of $ClSN{=}CCl_2$ in ethylbromide and excess $ClC{\equiv}N$ was kept in the dark at 0 °C for 10 d [6] or when $ClSN{=}CCl_2$ reacted with $ClC{\equiv}N$ in the presence of DMF [11].

The title compound was reacted with an about equimolar amount of $F_2C{=}S$ or $FClC{=}S$ in a glass bomb under UV irradiation for 30 or 42 h yielding 50% $F_2ClCSSN{=}CCl_2$ and 53% $FCl_2CSSN{=}CCl_2$, respectively [12]; see also [13].

$ClSN{=}CCl_2$ oxidized aqueous KI solutions to yield I_2 [4]. When 0.1 M solutions of $ClSN{=}CCl_2$ in CCl_4 were shaken for 3 min with an excess of 10% aqueous KI, an average of 0.975 equivalent of I_2 was found per mole of $ClSN{=}CCl_2$. When 0.1-M solutions of $ClSN{=}CCl_2$ in dry acetic acid (5% of acetic anhydride present) were treated at room temperature with an excess of 5% KI solution in the same solvent, 1.03 equivalents of I_2 per mole of $ClSN{=}CCl_2$ were formed after 3 min; the amount thereafter increased very slowly to 2 equivalents per mole. NaI was found under similar conditions to react more rapidly liberating 2.03 equivalents of I_2 per mole after 3 min reaction time. Sulfur and $ClC{\equiv}N$ were detected in this reaction. Shaking the title compound in CCl_4 for 3 min with a large excess of aqueous KI and removing the liberated I_2 immediately with aqueous thiosulfate yielded 45% $Cl_2C{=}NSSN{=}CCl_2$ [3].

$ClSN{=}CCl_2$ reacted with $[(CH_3)_3Si]_2NNa$ (mole ratio 1:1) in benzene yielding 66% $((CH_3)_3Si)_2NSN{=}CCl_2$ [14].

Adding $ClSN{=}CCl_2$ under cooling and vigorous shaking to a mixture of freshly prepared PbS or HgS in CS_2 (in ether or benzene the reaction rate is very small, in dioxane fast decomposition occurred) and C_2H_5Br gave $Cl_2C{=}NSSSN{=}CCl_2$ (nearly quantitatively). Analogous reactions with NiS, CoS, or CuS failed; with lead, polysulfides and higher sulfur homologs were obtained [4].

Stirring an excess of finely divided $Pb(SCN)_2$ for 15 h at ~20 °C with a 1.5 M solution of $ClSN{=}CCl_2$ in CCl_4 probably yielded dichlorothiocyanogen S-thiocyanate [6].

Adding the title compound while cooling and vigorously shaking to a mixture of $Hg(SCN)_2$, CS_2, and C_2H_5Br gave nearly quantitatively $Cl_2C{=}NSSC{\equiv}N$ and $HgCl_2$ [4].

Reacting the title compound in benzene solution with $AgClO_4$ in benzene solution at room temperature for 3 min showed that no more than two chlorine atoms can be split off, thereby producing AgCl. Reaction products other than AgCl were not identified [3]. Stirring an excess of finely powdered AgCN 15 h at ~20 °C with a 1.5 M solution of $ClSN{=}CCl_2$ in CCl_4 yielded 54% dichlorothiocyanogen S-cyanide [6].

Passing ethylene for 3 h into a 1.25 M solution of $ClSN{=}CCl_2$ in CCl_4 in the dark at ~20 °C and keeping the mixture in the dark for another 12 h yielded 13% $ClCH_2CH_2SN{=}CCl_2$ [6]. When a solution of $ClSN{=}CCl_2$ in CCl_4 was treated over a period of ~20 min with excess cyclohexene, the latter was added to give I (80% yield) [3].

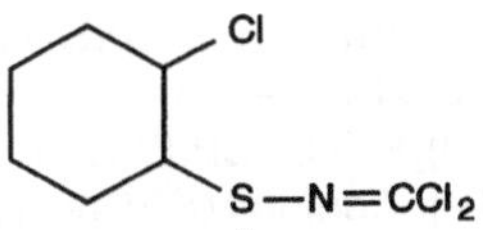

I

The reaction of the title compound with acetylene was unsuccessful under the conditions used for ethylene. Passing acetylene at ~40 °C under irradiation of an Hg vapor lamp probably resulted in the formation of $CHCl{=}CHSN{=}CCl_2$. The reactions of $ClSN{=}CCl_2$ with 3-hexyne or phenylacetylene in CCl_4 occurred rapidly and exothermally without irradiation, but in both cases the product decomposed to $ClC{\equiv}N$ and resin upon isolation [6].

When a toluene solution of $ClSN{=}CCl_2$ was maintained at 40 °C in the dark for several hours, no reaction products of toluene could be detected. But when a 0.25 M solution of $ClSN{=}CCl_2$ in toluene was irradiated in a quartz flask with an Hg vapor lamp for 8 h at ~40 °C, HCl was slowly evolved with formation of (probably) benzyl dichlorothiocyanate and benzyl thiocyanate [6].

Adding excess absolute alcohol, ROH (R = CH_3, C_2H_5, n-C_3H_7, i-C_3H_7, n-C_4H_9, i-C_4H_9, s-C_4H_9, $CH_2C_6H_5$) while cooling to $ClSN{=}CCl_2$ (mole ratio ca. 10 : 1) and allowing the mixture to react for 15 min at room temperature gave, via evolution of HCl and SO_2 and leaving sulfur chlorides and gums, the corresponding carbamates $NH_2C(O)OR$ with a yield of 66, 38, 37, 23, 27, 38, 17, or 36%, respectively; a reaction path was proposed. The analogous reaction with t-C_4H_9OH probably leads to t-$C_4H_9OSN{=}CCl_2$. Adding a solution of ethanol in CCl_4 to a solution of $ClSN{=}CCl_2$ (mole ratio 1 : 1) in CCl_4 over a period of 30 min probably yielded 16% $C_2H_5OSN{=}CCl_2$. The same compound was obtained when a solution of $ClSN{=}CCl_2$ in ether was reacted with a suspension of an equimolar amount of $NaOC_2H_5$ in ether for 3 h [6]. No reaction was observed when a mixture of $ClSN{=}CCl_2$ and triphenylmethanol was briefly held at 100 °C [6].

Adding a solution of $ClSN{=}CCl_2$ in ether to a solution of the secondary amine R_2NH (mole ratio 1 : 2, R = CH_3, C_2H_5, i-C_3H_7, s-C_4H_9, c-C_6H_{11}; R_2 = $(CH_2)_5$, $O(CH_2)_4$; R = C_6H_5; R_2 = $(CH_3)C_6H_5$) kept at ~10 °C gave high yields of the corresponding amine hydrochlorides, $R_2NH \cdot HCl$ and $R_2NSN{=}CCl_2$, the latter being unstable under the reaction conditions used to give $ClSNR_2$ and $ClC{\equiv}N$; a reaction path was proposed [6].

The title compound partly decomposed in the presence of $R_4N^+ Cl^-$ (R = CH_3, n-C_4H_9) or DMF to give equilibrium mixtures with free $ClC{\equiv}N$ and SCl_2 [9].

Adding a solution of $ClSN{=}CCl_2$ in benzene at 20 to 25 °C to a solution of phenothiazine (II) (mole ratio 1 : 2) in benzene and heating then the mixture at reflux temperature for 24 h gave 65.6% 10-cyano-10H-phenothioazine (III) with evolution of HCl [15].

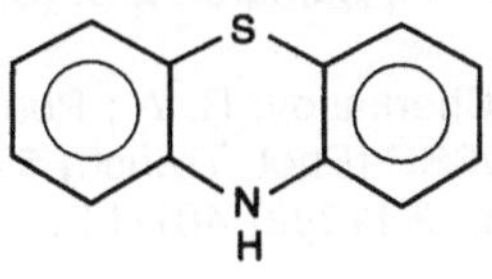

II

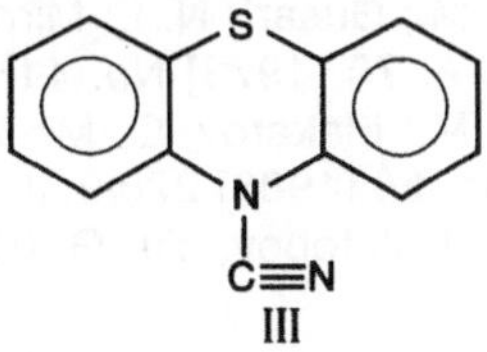

III

$ClSN{=}CCl_2$ reacted with $ROPCl_2$ (R = CH_3, C_2H_5, n-C_3H_7, i-C_3H_7, i-C_4H_9; mole ratio 1 : 2) at 15 to 20 °C and 20 min at 100 °C to give the corresponding $ROP(S)(Cl)N{=}CCl_2$ with a yield of 57 to 94% along with $POCl_3$ with 80 to 90% and RCl (a reaction mechanism was suggested) [16]; see also [17].

When $ClSN{=}CCl_2$ was added to a solution of $MSP(S)(OR)_2$ (M = Na, NH_4; R = C_2H_5, n-C_3H_7, i-C_3H_7; mole ratio 1 : 3) in anhydrous acetone and the mixture was reacted for 6 to 7 h at room temperature, the corresponding $(RO)_2P(S)SSSP(S)(OR)_2$ (73 to 81%) and $(RO)_2P(S)N{=}C{=}S$ (68 to 71% yield) were formed [18, 19]. Adding $MSP(S)(OR)_2$ (M = Na, NH_4; R = C_2H_5, i-C_3H_7) in ether to an ethereal solution of the compound in the mole ratio 1 : 1 for 4 d yielded the corresponding $(RO)_2P(S)Cl$, ClCN, and elemental sulfur [18], see also [19]. Similar results were obtained with free O,O-dialkyl dithiophosphoric acids [19].

The reaction of $ClSN{=}CCl_2$ with $(C_4H_9O)_2P(O)SNa$ in CCl_4 or benzene yielded nearly quantitative amounts of NaCl, but other products were not isolated [3].

Reacting $ClSN{=}CCl_2$ and $C_6H_5CH_2MgCl$ (mole ratio 1 : 1) in ether gave $MgCl_2$ and 27% $C_6H_5CH_2SN{=}CCl_2$ (presumably). With $C_6H_5CH_2MgCl$ in 100% excess additionally some C_6H_5-$CH_2SSCH_2C_6H_5$ was obtained [6].

The title compound reacted with Pb or Hg alkyl- or -arylmercaptides to form colorless crystalline compounds presumably according to 2 $ClSN{=}CCl_2$ + $M(SR)_2$ → 2 $Cl_2C{=}NSSR$ + MCl_2 (M = Pb, Hg; R = alkyl, aryl) [4].

References:

[1] Kaufmann, H. P.; Liepe, J. (Ber. Dtsch. Chem. Ges. **57** [1924] 923/8).
[2] Baroni, A. (Atti Accad. Naz. Lincei Cl. Sci. Fis. Mat. Nat. Rend. [6] **23** [1936] 871/3).
[3] Bacon, R. G. R.; Irwin, R. S.; Pollock, J. M.; Pullin, A. D. E. (J. Chem. Soc. **1958** 764/73).
[4] Fehér, F.; Weber, H. (Chem. Ber. **91** [1958] 2523/7).
[5] Fehér, F.; Weber, H. (Z. Naturforsch. **11b** [1956] 426).
[6] Bacon, R. G. R.; Irwin, R. S. (J. Chem. Soc. **1960** 5079/87).
[7] Kühle, E.; Anders, B.; Zumach, G. (Angew. Chem. **79** [1967] 663/80; Angew. Chem. Int. Ed. Engl. **6** [1967] 649/65).
[8] Hagemann, H. (Ger. 2126540 [1971/72]; C.A. **78** [1973] No. 57794).
[9] Geevers, J.; Trompen, W. P. (Tetrahedron Lett. **1974** 1687/90).
[10] Anders, B. (unpublished results in [7]).

[11] Kühle, E. (unpublished results in [7]).
[12] Gielow, P.; Haas, A. (Z. Anorg. Allg. Chem. **394** [1972] 53/66).
[13] Gielow, P.; Haas, A. (Chem.-Ztg. **95** [1971] 1010/1).
[14] Roesky, H. W.; Diehl, M.; Banek, M. (Chem. Ber. **111** [1978] 1503/8).
[15] Kamalov, R. M.; Makarov, G. M.; Litvinov, I. A.; Kataeva, O. N.; Pudovik, M. A.; Cherkasov, R. A. (Izv. Akad. Nauk SSSR Ser. Khim. **1991** 451/60; Bull. Acad. Sci. USSR Div. Chem. Sci. [Engl. Transl.] **40** [1991] 388/96).
[16] Ivanova, Zh. M.; Gusar', N. I.; Gololobov, Yu. G. (Z. Obshch. Khim. **44** [1974] 538/42; J. Gen. Chem. USSR [Engl. Transl.] **44** [1974] 516/9).
[17] Ivanova, Zh. M.; Gusar', N. I.; Miroshnichenko, V. V.; Samqrai, L. I. (U.S.S.R. 374321 [1971/73]; C.A. **79** [1973] No. 41905).
[18] Kamalov, R. M.; Makarov, G. M.; Zimin, M. G.; Cherkasov, R. A.; Pudovik, A. N. (Zh. Obshch. Khim. **58** [1988] 228/9; J. Gen. Chem. USSR [Engl. Transl.] **58** [1988] 202/3).
[19] Gusar', N. I.; Gololobov, Yu. G. (Heteroat. Chem. **3** [1992] 407/14).

12.5.3.1.2 ClSN=CClBr

The title compound was probably formed when a solution of equimolar amounts of BrC≡N and SCl_2 in C_2H_5Br was kept for 6 d at ~17 °C; 4% yield.

Pungent, orange-yellow liquid, b.p. 38 to 43 °C/12 Torr.

The IR spectrum showed strong doublets at 1603 and 1587 cm^{-1} and at 921 and 902 cm^{-1}.

Reference:

Bacon, R. G. R.; Irwin, R. S. (J. Chem. Soc. **1960** 5079/87).

12.5.3.1.3 $ClSN{=}C(Cl)CF_3$

The title compound was formed by shaking CF_3CN and excess SCl_2 with active charcoal as catalyst in a steel autoclave at 40 °C for 2 d until the pressure dropped to below 3 bar. The mixture was fractionated through a Vigreux column in a rough vacuum to give the compound with a yield of 61%. The compound can be obtained by dehalogenating $CF_3CCl_2N{=}SCl_2$ with metals (e.g., the steel parts of GC columns).

Pale yellow liquid, b.p. 103.9 °C (extrapolated); $\ln p/p_0$ (Torr) = −4956.6/T (K) + 13.147; $\Delta H_v = 41.21$ kJ/mol, $\Delta S_v = 109.3$ kJ·mol^{-1}·K^{-1}.

^{19}F NMR (50% in $CFCl_3/CFCl_3$): δ = −71.4 ppm (CF_3).

IR (KBr, gas): ν (in cm^{-1}) = 1671 w, 1624 s (C=N), 1323 s, 1290 vs, 1225 vs, 1181 vs, 957 vs, 864 s, 798 m, 716 s, 553 s, 476 s.

MS (70 eV): m/e (rel. int. in %) = 197 (85) M^+, 162 (100) $M^+ - Cl$, 128 (38) $M^+ - CF_3$, 127 (19), 93 (24), 69 (99) CF_3^+, 67 (71) SCl^+, 64 (35), 50 (26).

Reacting $ClSN{=}C(Cl)CF_3$ with Hg (mole ratio ca. 1 : 5) in CH_2Cl_2 for 10 d at 20 °C gave Hg_2Cl_2, $(CF_3(Cl)C{=}NS)_2$ (30%), and $(CF_3(Cl)C{=}NS)_2S$ (13%).

Reference:

Höfs, H.-U.; Mews, R.; Noltemeyer, M.; Sheldrick, G. M.; Schmidt, M.; Henkel, G.; Krebs, B. (Z. Naturforsch. **38b** [1983] 454/9).

12.5.3.1.4 $ClSN{=}C(Cl)CCl_3$

The title compound was prepared with a yield of 55% along with Cl_3CCN and SCl_2 by mixing $ClN{=}C(Cl)CCl_3$ and elemental sulfur without a solvent at 0 °C and adding 0.5% DMF as catalyst, or with a yield of 68% by reacting CCl_3CN with SCl_2 and catalytic amounts of DMF at room temperature and distilling after 24 h [1]. $ClSN{=}C(Cl)CCl_3$ was obtained with a yield of 79% by introducing Cl_2 gas into a suspension of $S_2NC_2Cl_2^+ Cl^-$ (I) (mole ratio 2 : 1) and catalytic amounts of I_2 in CH_2Cl_2 and allowing the mixture to react for another 4 h at room temperature [2].

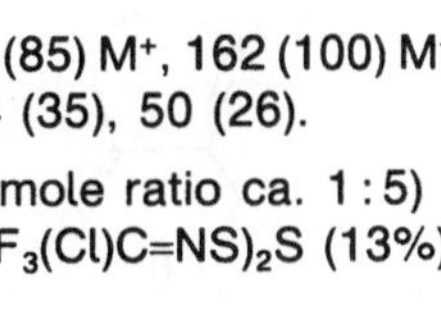

I

Liquid, b.p. 106 to 107 °C/11 Torr [1], 104 °C/10 Torr [2]; $n_D^{20} = 1.6005$ [1].

^{13}C NMR ($CDCl_3$/TMS): δ (in ppm) = 92.6 (s, C-2), 134.6 (s, C-1) [2].

IR: 1612 and 1594 cm^{-1} (doublet), ν(C=N) [1].

MS (20 eV, 25 °C): m/e (rel. int. in %) = 245 (60) M^+, 128 (100) $M^+ - CCl_3$, 93 (15) $ClCNS^+$ [2].

$ClSN{=}C(Cl)CCl_3$ partly decomposed in the presence of $R_4N^+\ Cl^-$ ($R = CH_3$, n-C_4H_9) to give equilibrium mixtures with the free nitrile, $Cl_3CC{\equiv}N$, and SCl_2. With cyclohexene an addition product formed [1].

References:

[1] Geevers, J.; Trompen, W. P. (Tetrahedron Lett. **1974** 1687/90).
[2] Appel, R.; Janssen, H.; Siray, M.; Knoch, F. (Chem. Ber. **118** [1985] 1632/43).

12.5.3.1.5 **ClSN=C(Cl)C(O)X**, X = F, Cl

ClSN=C(Cl)C(O)F was formed from the related ClSN=C(Cl)C(O)Cl by halogen exchange with KF/[18]crown-6 (no further data given) [1].

ClSN=C(Cl)C(O)Cl was formed by introducing Cl_2 into a solution of II (mole ratio 2:1) cooled to −70 °C and catalytic amounts of I_2 in CH_2Cl_2 and then heating the mixture at reflux temperature for 10 h; 90% yield [2].

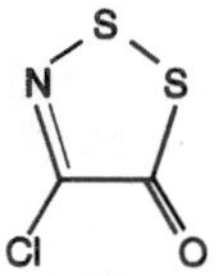

II

Yellow liquid, b.p. 70 °C/1 Torr [2].

^{13}C NMR ($CDCl_3$/TMS): δ (in ppm) = 129.8 (s, C-1), 158.8 (s, C-2) [2].

IR (film): 11 absorption bands between 3480 and 455 cm^{-1}, (unassigned) [2].

MS (70 eV, 25 °C): m/e (rel. int. in %) = 191 (32) M^+, 163 (17) $M^+ - CO$, 156 (17) $M^+ - Cl$, 128 (100) $M^+ - COCl$, 93 (34) $ClCNS^+$, 67 (67) SCl^+, 35 (24) S^+ [2].

Heating the title compounds in vacuum (Hg diffusion pump) at 800 °C gave X(O)C–CN (X = F (74%), X = Cl (88%)) along with SCl_2 [1].

References:

[1] Appel, R.; Siray, M. (Angew. Chem. **95** [1983] 807; Angew. Chem. Int. Ed. Engl. **22** [1983] 785).
[2] Appel, R.; Janssen, H.; Siray, M.; Knoch, F. (Chem. Ber. **118** [1985] 1632/43).

12.5.3.1.6 **ClSN=C(Cl)SO$_2$R**, R = CH_3, C_6H_5, 4-$NO_2C_6H_4$

The title compounds were formed by adding SCl_2 at 0 °C to a solution of the corresponding $RSO_2C{\equiv}N$ in CH_2Cl_2 in the presence of catalytic amounts of $(n\text{-}C_4H_9)_4NCl$.

R = CH$_3$. 30% yield; liquid, b.p. 101 to 103 °C/0.5 Torr, partial decomposition into SCl_2 and CH_3SO_2CN during distillation.

R = C_6H_5. >64% yield; could not be purified by distillation.

R = 4-$NO_2C_6H_4$. 63% yield; m.p. 136 to 138 °C; stable when kept at 0 °C.

The title compounds reacted with cyclohexene to give the corresponding N-(2-chlorocyclohexylthio)-sulfonyl formimidoyl chloride (III).

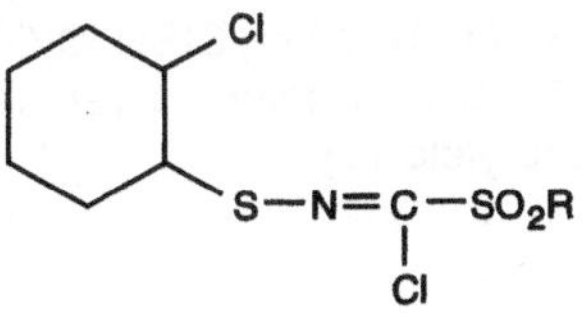

III

Reference:

Vrijland, M. S. A. (Tetrahedron Lett. **1974** 837/8).

12.5.3.1.7 $ClSN=C(CF_3)_2$

The title compound was formed by heating $(CF_3)_2C=NSSN=C(CF_3)_2$ with Cl_2 (mole ratio ca. 1 : 1.5) in a Pyrex bulb at 110 °C for 12 h; 94% yield. It was also produced when the two reactants were photolyzed at 2537 Å, but the yield was much lower and there were numerous other products [1]. $ClSN=C(CF_3)_2$ was formed with a yield of 95% by reacting $S_3N_3Cl_3$ with $(CF_3)_2CN_2$ (mole ratio 1 : 3) in a glass tube at 40 °C for 1 week; caution: explosions happened very often [2].

Yellow liquid; b.p. 95.4 °C [1], 95 °C [2]. Vapor pressure: <25 Torr at 25 °C [1]. ΔH_v = 9.0 kcal/mol; ΔS_v = 24.3 cal·mol^{-1}·K^{-1}; log p (Torr) = 8.17 − 1960/T (K) [1].

^{19}F NMR (neat ?/$CFCl_3$): δ (in ppm) = −60.7 [2], −60.8 [1] ($CF_3C(A)$); −68.4 [1], −69.2 [2] ($CF_3(B)$), $^2J(F,F)$ = 6.5 Hz [2]. The resonance band at 60.8 ppm was assigned to the CF_3 group trans to Cl [1]. The two CF_3 signals split at 0 °C into quartets [2].

IR (KBr): ν (in cm^{-1}) = 1630 m ν(C=N), 1330 s, 1270 vs, 1195 vs, 980 s, 905 m, 750 m, 715 s, 540 w, 490 m, 455 w [1].

The title compound is easily hydrolyzable. Fluorination reactions to give the corresponding $(CF_3)_2C=NSF$ failed: with F_3NO (3 h at 25 °C), AgF_2 (8 h at −20 °C), ClF (1 h at −78 °C), or CsF (1 h at −78 °C), $(CF_3)_2CFN=SF_2$ was obtained in each case along with NO and Cl_2, AgF and Cl_2, Cl_2, or $((CF_3)_2C=N)_2S_2$ and CsCl, respectively. With KF no reaction took place, even at 100 °C within 12 h [1].

Condensing $ClSN=C(CF_3)_2$ in a quartz bomb onto $(CF_3)_2CN_2$ (mole ratio 1 : 1) and UV-irradiating the mixture for 70 h gave 68% $(CF_3)_2C(Cl)SN=C(CF_3)_2$. Stirring $ClSN=C(CF_3)_2$ with a slight excess of CF_3CHN_2 in a glass bomb at room temperature for 3 d yielded 11.5% $CF_3CHClSN=C(CF_3)_2$ along with CF_3CH_2Cl and $CF_3CH=CHCF_3$ [4].

When $ClSN=C(CF_3)_2$ was condensed together with NH_3 (mole ratio 1 : 2.5) into a Pyrex vessel and the mixture reacted at −20 °C for 1.5 h, $(CF_3)_2C=NSNH_2$ was obtained with a yield of 62% [1]; see also [2, 3]. When it was allowed to stand with $(CH_3)_2NH$ (mole ratio ca. 1 : 2.5) at 25 °C for 12 h, 96% $(CF_3)_2C=NSN(CH_3)_2$ was formed [1]. $(CF_3)_2C=NSCl$ reacted with $(CF_3)_2$-$C=NSNH_2$ and $(C_2H_5)_3N$ (mole ratio 2 : 1 : 2) in ether at −70 °C for 5 h to yield 30% $[(CF_3)_2C=NS]_3N$ [3].

Stirring $ClSN{=}C(CF_3)_2$ with $(CH_3)_3SiN{=}S{=}NSi(CH_3)_3$ (mole ratio ca. 2 : 1) at room temperature for 20 h yielded 69.5% $(CF_3)_2C{=}NSN{=}S{=}NSi(CH_3)_3$. The title compound and CH_3SH (mole ratio 1 : 1), when condensed in a Pyrex vessel at −196 °C and allowed to stand at −78 °C for 6 h, gave 45% $(CF_3)_2C{=}NSSCH_3$ [1].

Heating $ClSN{=}C(CF_3)_2$ and AgNCO (mole ratio ca. 1 : 1.85) in CH_2Cl_2 at reflux temperature for 6 h yielded 81% $(CF_3)_2C{=}NSN{=}C{=}O$ [3]. When $ClSN{=}C(CF_3)_2$ was condensed onto excess AgCN, which had been dried at 75 °C in a dynamic vacuum, and reacted for 6 h, $(CF_3)_2$-$C{=}NSC{\equiv}N$ was isolated with 50.5% yield [1].

References:

[1] Metcalf, S. G.; Shreeve, J. M. (Inorg. Chem. **11** [1972] 1631/4).
[2] Varwig, J.; Steinbeißer, H.; Mews, R.; Glemser, O. (Z. Naturforsch. **29b** [1974] 813/4).
[3] Steinbeißer, H.; Mews, R.; Glemser, O. (Z. Naturforsch. **32b** [1977] 160/2).
[4] Steinbeißer, H.; Mews, R. (J. Fluorine Chem. **16** [1980] 145/51).

12.5.3.2 N-(1-Arylalkylidene)-amino-chloro-sulfanes, $ClSN{=}CR^1R^2$, $R^1 = t\text{-}C_4H_9$, $R^2 = C_6H_5$; $R^1 = CF_3$, $R^2 = C_6H_5$, $C_6H_4CH_3\text{-}4$, $C_6H_4CF_3\text{-}4$; $R^1 = CCl_3$, $R^2 = C_6H_5$; $R^1 = R^2 = C_6H_5$

The title compounds (except for the compounds with $R^1 = CCl_3$, $R^2 = C_6H_5$) were prepared by adding a solution of the respective ketimine $R^1R^2C{=}NH$ (R^1, R^2 see title) and $(C_2H_5)_3N$ in benzene at 5 to 10 °C to a solution of SCl_2 (mole ratio 1 : 1 : 1) in benzene and reacting the mixture at 20 °C for 2 h [1]. The compound with $R^1 = CCl_3$, $R^2 = C_6H_5$ was formed by heating a mixture of $ClN{=}C(CCl_3)C_6H_5$ and SCl_2 (mole ratio 1 : 1.2) in anhydrous CCl_4 at 80 °C for 20 h; separation by fractional distillation under vacuum [2]. The compound with $R^1 = R^2 = C_6H_5$ presumably is an intermediate in the reaction of $(C_6H_5)_2C{=}NSi(CH_3)_3$ with SCl_2 in CH_2Cl_2 at −78 to −60 °C (but it was not isolated) [3].

$ClSN{=}C(C_6H_5)C_4H_9\text{-}t$. 35% yield; light yellow liquid, b.p. 78 to 80 °C/0.04 Torr [1].

$ClSN{=}C(C_6H_5)CF_3$. 58% yield; light yellow liquid, b.p. 45 to 47 °C/0.08 Torr [1].

$ClSN{=}C(C_6H_4CH_3\text{-}4)CF_3$. 51% yield; light yellow liquid, b.p. 54 to 56 °C/0.05 Torr [1].

$ClSN{=}C(C_6H_4CF_3\text{-}4)CF_3$. 50% yield; light yellow liquid, b.p. 50 °C/1 Torr [1].

$ClSN{=}C(C_6H_5)CCl_3$. 60% yield; liquid, b.p. 118 to 123 °C/0.05 Torr [2].

$ClSN{=}C(C_6H_5)_2$. 38% yield; light yellow liquid, b.p. 124 to 126 °C/0.08 Torr [1].

A distinct feature in the IR spectra of the title compounds is the absence of an absorption for the C=N multiple bond. A strong absorption in the region of 450 cm^{-1} was assigned to vibrations of the S–Cl bond. In the Raman spectrum there is a doublet of bands in the regions of 1618 and 1610 cm^{-1} which undoubtedly belongs to the absorption of the benzene ring and the C=N bond [1].

The compounds (except those with $R^1 = CCl_3$, $R^2 = C_6H_5$) decompose during storage and are hydrolyzed by atmospheric moisture. They react with the initial ketimines $R^1R^2C{=}NH$ to give the corresponding sulfanediylbis(ketimines) $R^1R^2C{=}NSN{=}CR^1R^2$ [1].

Solutions of the title compounds in benzene were reacted with a solution of $(CH_3)_3SiNR_2$, $R_2 = (CH_2)_5$, $(CH_2)_4O$ (mole ratio 1 : 1), at 5 to 10 °C overnight to give the corresponding $R^1R^2C{=}NSNR_2$ [1]. In situ reactions of the compound with $R^1 = CCl_3$, $R^2 = C_6H_5$ with $O{=}S{=}N$-$Si(CH_3)_3$ or $(CH_3)_3SiN{=}S{=}NSi(CH_3)_3$ led to $(C_6H_5)_2C{=}NSN{=}S{=}O$ or $((C_6H_5)_2C{=}NSN{=})_2S$, respectively [3].

The compounds with $R^1 = CF_3$; $R^2 = C_6H_5$, $C_6H_4CH_3$-4, $C_6H_4CF_3$-4 reacted with $(C_2H_5O)_3PO$ (mole ratio 1 : 1) at 20 °C and 20 min at 100 °C to give the corresponding $R^2(CF_3)C{=}NSP(O)(OC_2H_5)_2$. By reacting with Cl_2 (mole ratio 1 : 3) in CCl_4 at 20 °C for 2 d, the corresponding compounds $Cl_2S{=}NCCl(CF_3)R^2$ ($R^2 = C_6H_5$, $C_6H_4CH_3$-4, $C_6H_4CF_3$-4) were obtained. $ClSN{=}C(C_6H_5)_2$ does not react with Cl_2 under analogous conditions [1].

References:

[1] Markovskii, L. N.; Shermolovich, Yu. G.; Shevchenko, V. I. (Zh. Org. Khim. **11** [1975] 2533/7; J. Org. Chem. USSR [Engl. Transl.] **11** [1975] 2603/7).

[2] Shermolovich, Yu. G.; Talanov, V. S.; Pirozhenko, V. V.; Markovskii, L. N. (Zh. Org. Khim. **18** [1982] 2539/47; J. Org. Chem. USSR [Engl. Transl.] **18** [1982] 2240/7).

[3] Chivers, T.; Oakley, R. T.; Pieters, R.; Richardson, J. F. (Can. J. Chem. **63** [1985] 1063/7).

12.5.3.3 N-(Hexachloro-3-cyclopentenyl-1-idene)-amino-chloro-sulfane, $ClSN{=}C_5Cl_6$

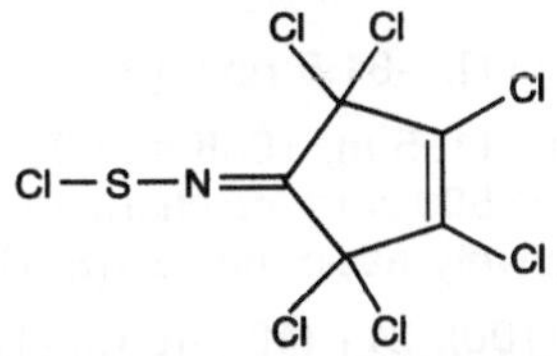

The title compound was prepared by reacting $S_3N_3Cl_3$ with hexachlorocyclopentadiene, C_5Cl_6, (mole ratio ca. 1.4 : 1) in toluene for 3 h at 110 °C. Then the solvent was removed in vacuum to get a residue which was sublimed at 70 °C/10^{-3} Torr and recrystallized to give the compound with a yield of 38% based on C_5Cl_6 [1]; see also [2]. Reacting $S_3N_3Cl_3$ with C_5Cl_6 in the mole ratio ca. 1 : 3 without a solvent at 65 °C for 48 h yielded 43% product [3]. The title compound formed with 20% yield by stirring a mixture of C_5Cl_6 and FS≡N (mole ratio 1 : 2) in SO_2 in a sealed glass pressure tube at 20 °C for 2 d [1].

Single crystals suitable for an X-ray structure analysis were obtained by subsequent sublimation at 30 °C/10^{-3} Torr. The crystals were found to be triclinic, space group $P\bar{1} - C_i^1$ (No. 2) with a = 9.566(12), b = 9.691(13), c = 13.67(19) Å, α = 79.51(15)°, β = 89.50(13)°, and γ = 75.89(12)°; Z = 4; V = 1208.1 Å³; R = 0.056, R_w = 0.069 from 8710 observed reflections. The molecular structure with selected bond lengths and angles is shown in **Fig. 8**, p. 240. For additional bond lengths and angles and atomic coordinates, see the paper [1]. The S–N distance (1.56 Å) is considerably short, possibly due to strong p_π–d_π interactions. The angles ∢(CNS) (147.3°) and ∢(NSCl) (114.9°) are remarkably widened, consistent with the higher electron density of the S–N bond and some delocalization from lone-pair electron density on sulfur to chlorine.

Pale yellow crystals, m.p. 68 °C (from CH_2Cl_2) [1, 2], m.p. 69 °C (from CH_2Cl_2) [3].

^{13}C NMR ($(CD_3)_2CO$/TMS): δ (in ppm) = 80.0 and 82.0 (C(2) and C(5)), 136.0 (C(3), C(4)), 143.2 (C(1)) [1]; ($CDCl_3$/TMS, 25 °C): δ (in ppm) = 78.4 (C(2), C(5)), 135.4 (C(3), C(4)), 142.3 (C(1)); ($CDCl_3$/TMS, −50 °C): δ (in ppm) = 76.3 (C(5)), 79.6 (C(2)), 134.6 (C(4)), 135.8 (C(3)), 142.3 (C(1)) [3]. Upon warming the solution, the resonances for C(3) and C(4) broaden and collapse symmetrically to give a singlet at 135.2 ppm, while the chemical shift of C(1) is independent of temperature. This change is reversible and can be explained in terms of a fluxional process either by rotation about the S–N bond or inversion at the nitrogen center. The coalescence temperature is ca. 30 °C which corresponds to an interconversion barrier of 68.1 ± 1.2 kJ/mol [3].

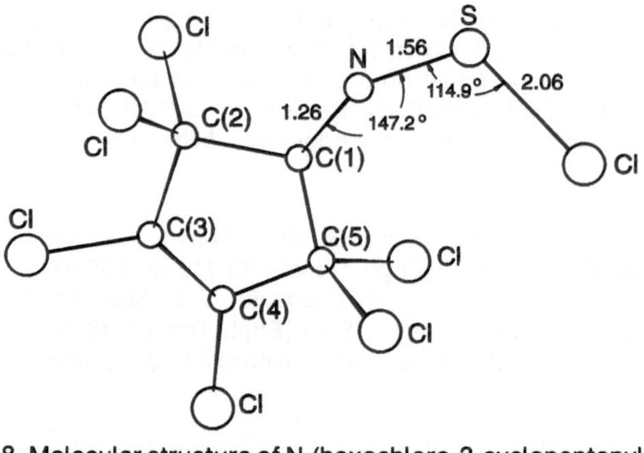

Fig. 8. Molecular structure of N-(hexachloro-3-cyclopentenyl-1-idene)-amino-chloro-sulfane with selected bond lengths (in Å) and angles (in °) [1].

^{15}N NMR ($CDCl_3$/external CH_3NO_2): $\delta = -81.0$ ppm (s) [1], -81.5 ppm [3].

IR (KBr): ν (in cm^{-1}) = 1646 s, 1608 s, 1205 s, 1175 m, 1155 m, 1038 m, 800 vs, 693 s, 636 s, 590 m, 582 m, 451 s [1]; bands at 1205, 800, 636, and 590 cm^{-1} are characteristic of the C_5Cl_6NS moiety, an intense band at 452 cm^{-1} was tentatively assigned to ν(S–Cl) [3].

EIMS (70 eV): m/e (rel. int. in %) = 355 (12) M^+, 318 (100), 211 (40), 46 (8) [1]. The MS showed a weak parent ion and the cleavage of the S–Cl bond as the major fragmentation process [3].

The substance can be handled in air, but should be stored under an inert atmosphere to avoid decomposition [2].

The reactivity of the compound is comparable to that of other aminochlorosulfanes. It slowly decomposed in $(CD_3)_2CO$ within 12 h [1].

Reactions of $ClSN{=}C_5Cl_6$ are given in Table 49:

Table 49
Reactions of $ClSN{=}C_5Cl_6$.

reactant	conditions	product(s) (yield in %)	Ref.
$(i\text{-}C_3H_7)_2NH$	mole ratio 1:2; in CH_2Cl_2, 16 h at 23 °C	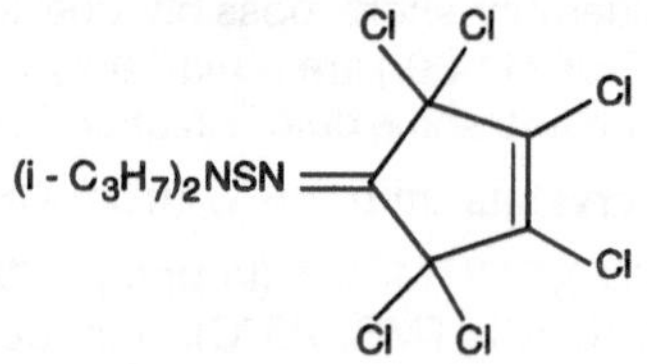 (61)	[3]
$(CH_3)_3SiBr$	mole ratio 1:1.1; in n-hexane, 4 h at 23 °C	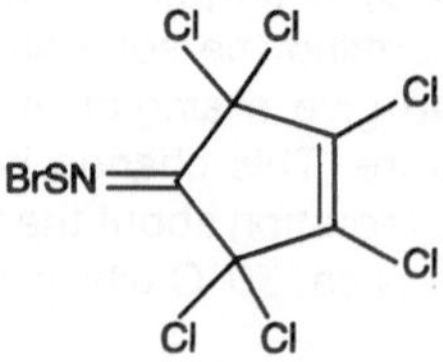 (96)	[3]

Table 49 (continued)

reactant	conditions	product(s) (yield in %)	Ref.
$(CH_3)_3SiI$	mole ratio 1 : 1.04; in n-hexane at 78 °C, 1 h at −78 °C	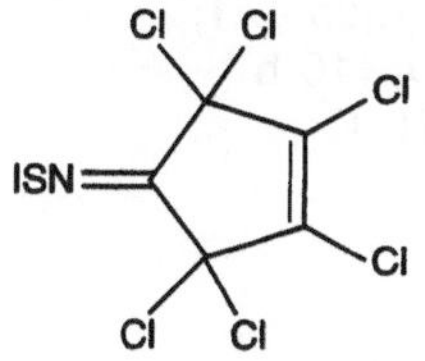 (39)	[3]
$(CH_3)_3SiN{=}S{=}O$	no reaction		[1]
$SbCl_5$	mole ratio 1 : >1; in SO_2 at 20 °C	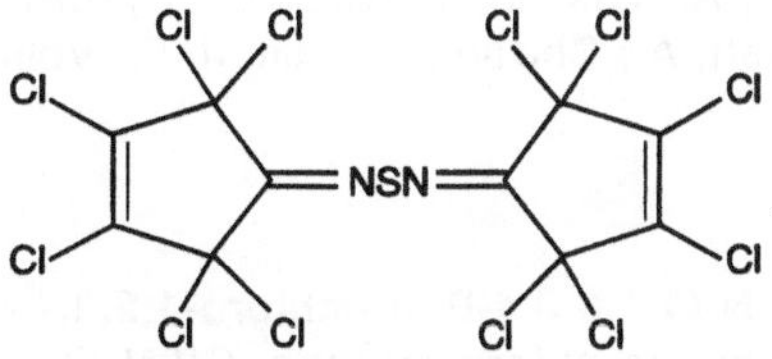 (62)	[1]
$(C_6H_5)_3Sb$	mole ratio 2 : 1; in CH_2Cl_2, 4 d at 23 °C	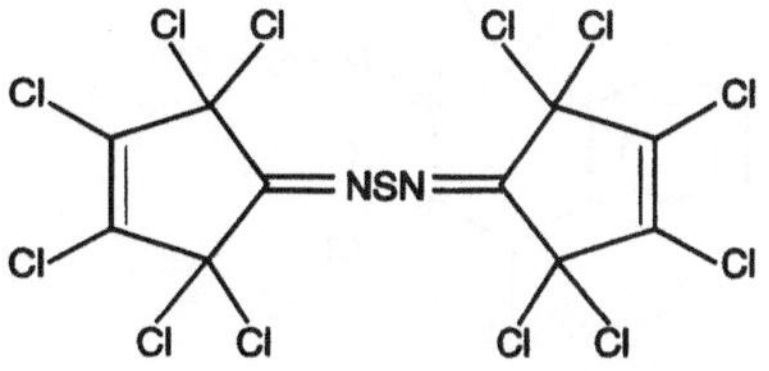	[3]
$Hg(NSO)_2$	mole ratio 1 : 1; in CH_3CN, 10 h at 20 °C	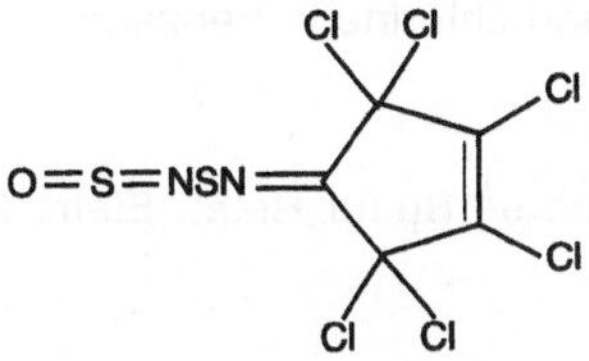 (82)	[1]
$Hg(SCF_3)_2$	mole ratio 1 : 1; in CH_2Cl_2, 48 h at 20 °C	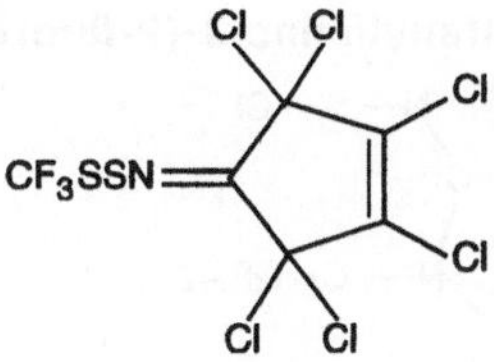 (86)	[1]
$(C_6H_5)_2Hg$	mole ratio 1 : 1; in 1,2-dichloroethane, 2 d heating at reflux	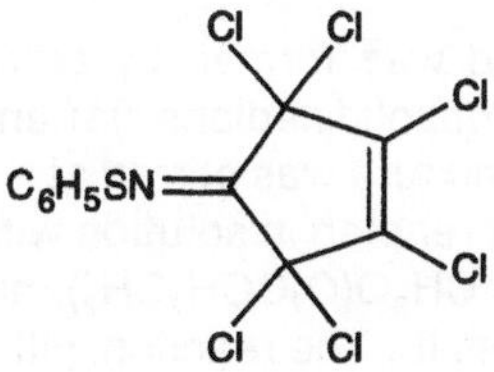 (54), C_6H_5HgCl	[3]

Table 49 (continued)

reactant	conditions	product(s) (yield in %)	Ref.
$Ag^+ AsF_6^-$	mole ratio 1 : 1; in SO_2, 16 h at −78 °C	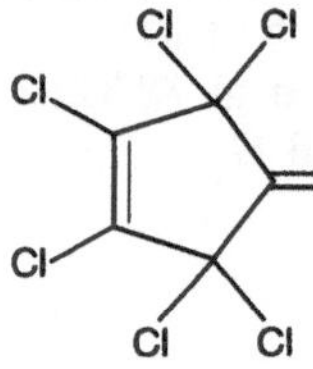 $NS^+AsF_6^-$ (98), AgCl	[3]

References:

[1] Haas, A.; Mischo, T. (Can. J. Chem. **67** [1989] 1802/8).
[2] Haas, A.; Mischo, T. (Chimia **41** [1987] 344/5).
[3] Apblett, A.; Chivers, T.; Fait, J. F.; Vollmerhaus, R. (Can. J. Chem. **69** [1991] 1022/7).

12.5.3.4 N-(1,1,3,3,4-Pentachloro-1,2,3,4-tetrahydro-naphthalen-2-ylidene)-amino-chloro-sulfane, $ClSN{=}C_{10}H_5Cl_5$

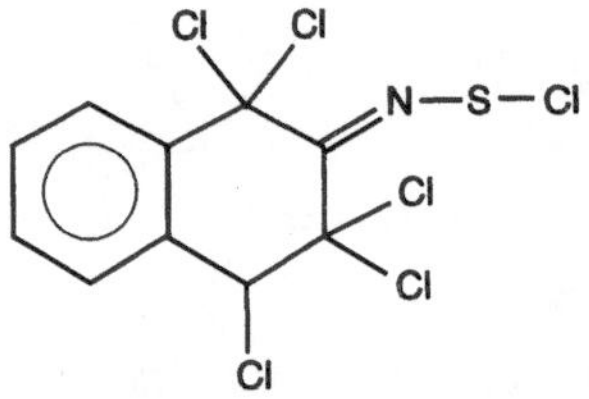

The title compound was formed with a high yield by reacting 2-naphthylamine with elemental sulfur and chlorine in benzene.

Reference:

Mayer, R. (Phosphorus Sulfur Relat. Elem. **23** [1985] 277/96).

12.5.3.5 5-Chlorosulfanylimino-1-(2-fluorophenyl)-3-methyl-2,4-imidazoledione

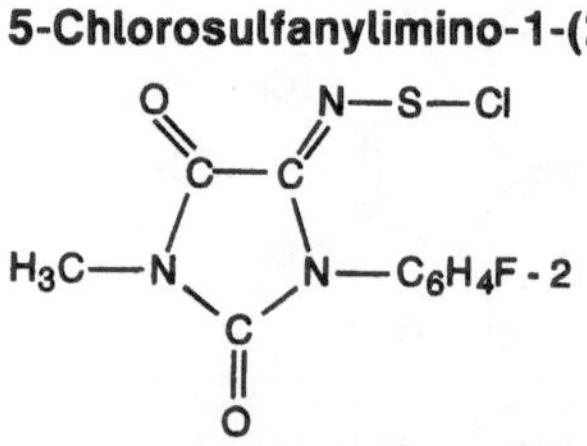

The title compound was formed by adding pyridine to a slurry of 1-(2-fluorophenyl)-3-methyl-5-imino-2,4-imidazolidinedione (Ia) and SCl_2 (mole ratio ca. 1 : 1 : 1) in CH_2Cl_2. After addition, the reaction mixture was stirred at ~25 °C for 30 min and filtered. The title compound was not isolated, but it reacted in solution with mercaptanes, R′SH (R′ = CH_3, n-C_3H_7, n-C_4H_9, C_6H_5, $C_2H_5O(O)CCH_2$, $CH_3O(O)CCH_2CH_2$), and pyridine in CH_2Cl_2 to give the corresponding 5-dithioimino derivatives Ib. The reaction with methanol and pyridine gave Ic [1]; see also [2].

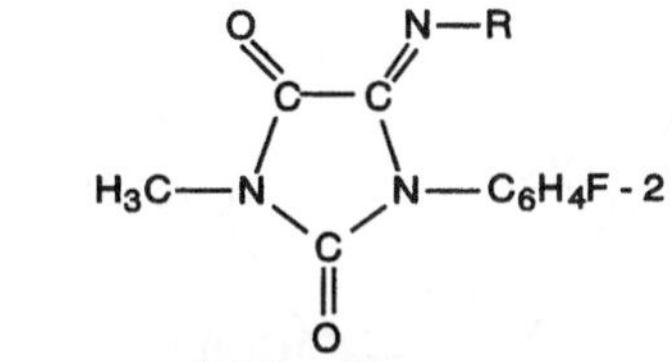

Ia: R = H
Ib: R = SSR′
Ic: R = $SOCH_3$

References:

[1] Cleveland, J. D. (U.S. 3843677 [1972/74]; C.A. **82** [1975] No. 43424).
[2] Cleveland, J. D. (U.S. 3956308 [1972/76]; C.A. **85** [1976] No. 63070).

12.5.3.6 2,5-Bis(chlorosulfanylimino)-2,5-dihydro-3,4-thiophenedicarbonitrile, $(ClSN=)_2SC_4(CN)_2$

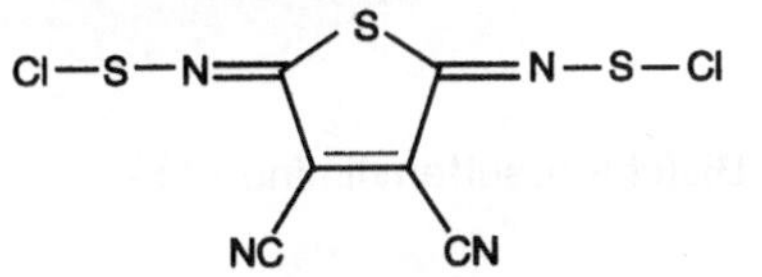

The title compound was prepared with 50% yield by adding $\overline{C(CN)=C(NH_2)-S-C(NH_2)=C}(CN)$ to an excess of SCl_2 and stirring the mixture gently and leaving it stand overnight [1].

Recrystallization of the title compound from ethylene chloride with and without chlorine gas yielded completely different morphologies, yet the chemical analyses were identical [1].

An X-ray diffraction study showed the crystals to be orthorhombic, C-centered, with the space group Cmcm – D_{2h}^{17} (No. 63) and the parameters a = 6.327(3), b = 9.678(8), c = 17.337(17) Å, and α = β = γ = 90°; Z = 4; V = 1061.7 Å³; D_x = 1.847 g/cm³; R_1 = 0.029, R_2 = 0.042 from 1354 independent reflections [2].

The molecular structure of the title compound is shown in **Fig. 9**, p. 244, and the intramolecular bond lengths and angles are listed in Table 50, p. 244.

The most striking features of the molecular structure are the planarity and the inward folding of the S–Cl bonds, such that the S(1)–Cl(1) distance of 3.29 Å is 0.26 Å shorter than the sum of the van der Waals radii [2].

The solid state structure (given in the paper) reveals uniform stacks along the a axis and sheets along the b-c plane. The closest intermolecular contact (3.11 Å) is between Cl(1) and N(2) of two molecules within a sheet in the b direction [2].

Shiny red crystals [1]; see also [2].

IR (KBr): ν (in cm^{-1}) = 2230 w, 1530 s, 1500 s, 1345 s, 1240 w, 892 s, 841 s, 820 m [1, 2].

UV-VIS (CH_2Cl_2): λ (in nm) (ε in $L \cdot mol^{-1} \cdot cm^{-1}$) = 282 sh (5880), 297 (7720), 400 (4680), 507 (22700) [1].

MS: m/e = 296 M^+ + 2, 294 M^+, 259 M^+ – Cl, 224 M^+ – 2 Cl, 192 M^+ – SCl_2, 146 M^+ – $NSCl_2$ [1, 2].

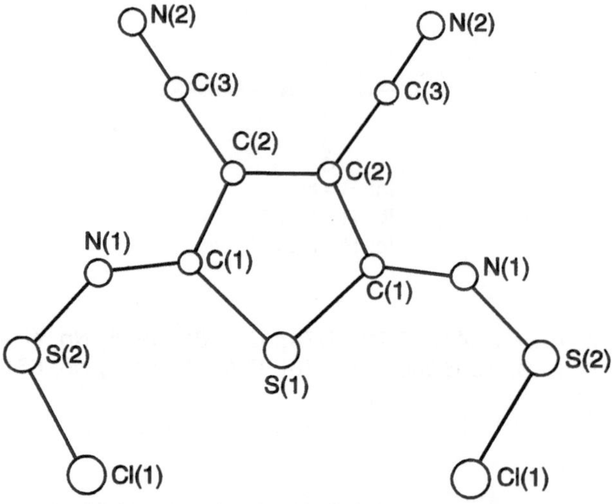

Fig. 9. Molecular structure of 2,5-bis(chlorosulfanylimino)-2,5-dihydro-3,4-thiophenedicarbonitrile [2].

Table 50
Intramolecular Bond Lengths and Angles of 2,5-Bis(chlorosulfanylimino)-2,5-dihydro-3,4-thiophenedicarbonitrile [2].

bond length	in Å	bond angle	in °
S(1)–Cl(1)	3.290(1)	C(1)–S(1)–C(1)	91.3(1)
S(1)–C(1)	1.775(1)	Cl(1)–S(2)–N(1)	111.29
S(2)–Cl(1)	2.049(1)	S(2)–N(1)–C(1)	138.5(1)
S(2)–N(1)	1.583(1)	S(1)–C(1)–N(1)	129.0(1)
N(1)–C(1)	1.283(2)	S(1)–C(1)–C(2)	110.2(1)
N(2)–C(3)	1.136(2)	N(1)–C(1)–C(2)	120.8(1)
C(1)–C(2)	1.439(2)	C(1)–C(2)–C(2)	114.13(8)
C(2)–C(3)	1.431(2)	C(1)–C(2)–C(3)	122.5(1)
C(2)–C(2)	1.363(3)	C(2)–C(2)–C(3)	123.39(9)
		N(2)–C(3)–C(2)	178.5(2)

Half-wave potentials: $E^{P}_{1/2} = -0.32, -0.77, -1.1$ V vs. SCE, in CH_2Cl_2 with $(n\text{-}C_4H_9)_4N^+ BF_4^-$ as supporting electrolyte. The first two waves are reversible, the last is not. The solutions deeply color during these experiments [1]. For an electron density distribution of the compound, see [4].

The title compound is stable in ambient atmosphere in the solid state and can be recrystallized from chlorinated solvents but is unstable in solution, e.g. in polar CH_3CN, or Lewis bases, e.g. THF solvents. Generally, the compound has low solubility in most appropriate solvents. Nucleophiles and reducing agents readily attack the molecule [2].

Adding $(n\text{-}C_4H_9)_4N^+ I^-$ to a warm solution of the title compound (mole ratio ca. 3:1) in CH_2Cl_2 and allowing the mixture to stand for 2 d completely dechlorinated and reduced it to the oligomeric radical anion $[C_{22}N_{17}S_9]^{-\bullet}$ with $(n\text{-}C_4H_9)_4N^+$ as counterion [1]; see also [2].

A solution of the title compound in CH_2Cl_2 reacted in N_2 atmosphere with a solution of $(CH_3)_3SiN{=}S{=}NSi(CH_3)_3$ (mole ratio 1:1) in CH_2Cl_2 at room temperature [5] or at 30 °C [6] to give 59% [5] or 66% [6] of the polymer I, $n > 10$ [5], $n \approx 3$ [6], and $(CH_3)_3SiCl$.

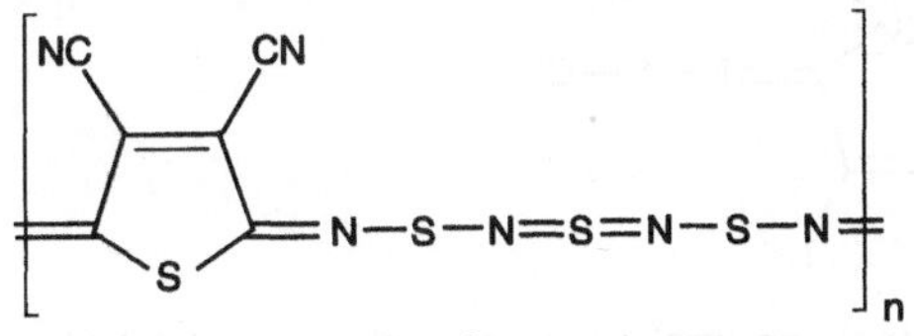

I

A solution of the title compound in CH_2Cl_2 reacted with a solution of 1,4-$((CH_3)_3SiN{=})_2C_6H_4$, $(CH_3)_3SiN{=}S{=}NSN{=}S{=}NSi(CH_3)_3$, or 1,4-$((CH_3)_3SiN{=}S{=}NS)_2C_6H_4$ (mole ratio 1:1 each) in CH_2Cl_2 at 30 °C to give the oligomers II (89% yield), III (84% yield), and IV (85% yield), respectively, with $n \approx 3$, and $(CH_3)_3SiCl$ [6].

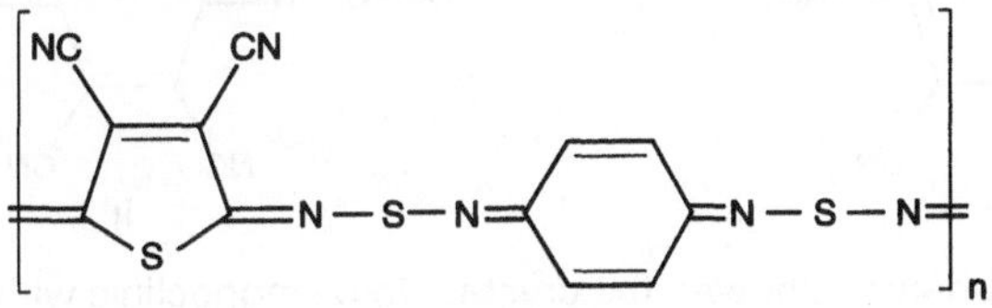

II

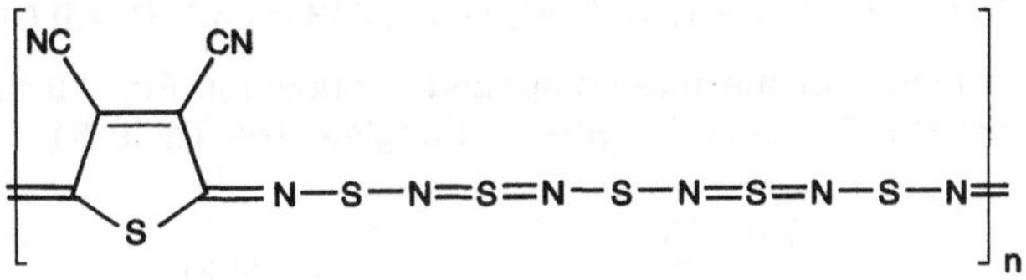

III

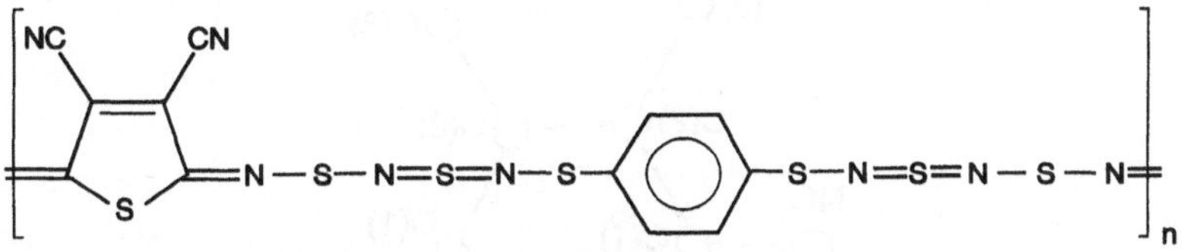

IV

Immersing clean Cu foils into a saturated solution of the title compound in CH_3CN produced a unique metalloorganic thin-film structure which demonstrates bistable memory switching behavior. The structure exhibits a rapid (50 ns) transition from a low-conductance to a high-conductance state with a resistivity ratio of approximately 10^3 to 10^4 [3].

References:

[1] Wudl, F.; Zellers, E. T.; Nalewajek, D. (J. Org. Chem. **45** [1980] 3211/5).
[2] Wudl, F.; Zellers, E. T. (J. Am. Chem. Soc. **102** [1980] 4283/4).
[3] Zellers, E. T.; Roedel, R. J.; Wudl, F. (J. Non-Cryst. Solids **46** [1981] 361/9).
[4] Troup, J. M.; Extine, M. W.; Ziolo, R. F. (Electron Distrib. Chem. Bond Proc. Symp., Atlanta 1981 [1982], pp. 285/96; C.A. **97** [1982] No. 203382).
[5] Wolmershäuser, G.; Jotter, R.; Wilhelm, T. (J. Phys. Colloq. [Paris] **44** [1983] C3-729/C3-732).
[6] Chien, J. C. W.; Zhou, M.-Y. (J. Polym. Sci. A **24** [1986] 2947/57).

12.5.3.7 2,5-Bis(chlorosulfanylimino)-1,1-dichloro-1,5-dihydro-2H-1λ⁴-selenophene-3,4-dicarbonitrile, $(ClSN{=})_2SeCl_2C_4(CN)_2$

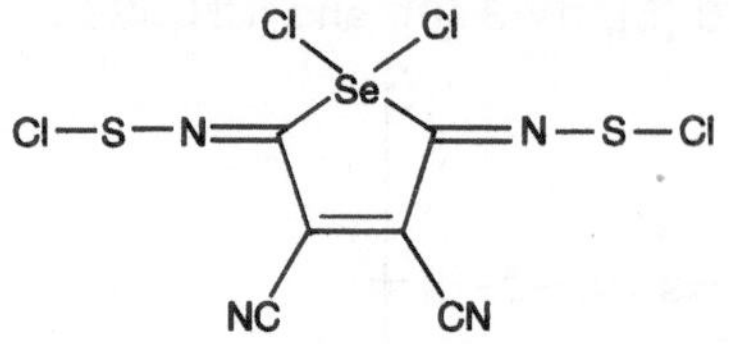

The title compound (black prisms) was formed when the heterocycle I was exposed to excess SCl_2 (mole ratio 1 : 3) under mild conditions [1]. Initially, the title compound was thought to have structure II in analogy to the related 2,5-bis(chlorosulfanylimino)-2,5-dihydro-3,4-thiophenedicarbonitrile [2].

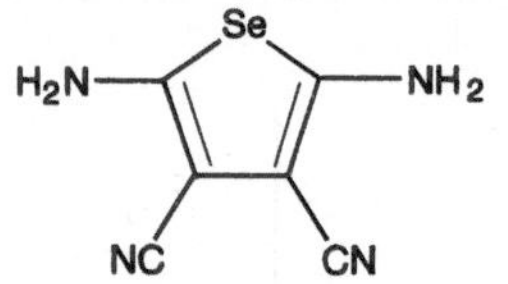

I

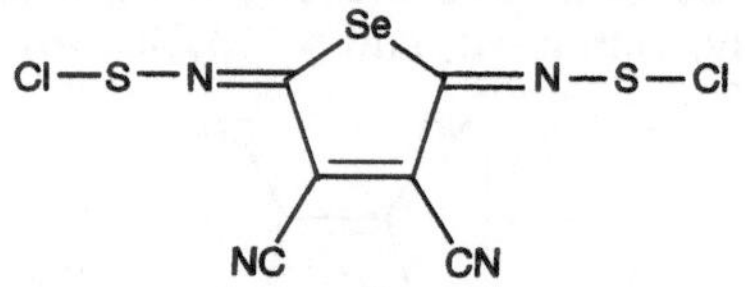

II

An X-ray structure analysis showed the crystals to be monoclinic with the space group I2/a (standard setting C2/c) – C^6_{2h} (No. 15) and the parameters a = 14.206(6), b = 13.737(3), c = 6.882(4) Å, β = 91.60(5)°; Z = 4; V = 1342.5 Å³; D_x = 2.043 g/cm³; R_1 = 0.041, R_2 = 0.047 [1].

The molecular structure of the title compound is shown in **Fig. 10** and the solid state structure in the paper [1]. For bond lengths and angles, see Table 51.

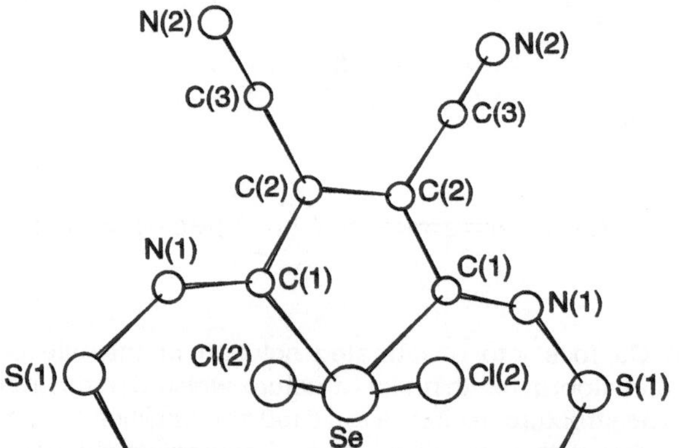

Fig. 10. Molecular structure of 2,5-bis(chlorosulfanylimino)-1,1-dichloro-1,5-dihydro-2H-1λ⁴-selenophene-3,4-dicarbonitrile [1].

The Se–Cl(2) and the Se–C distances are unusual, but the Se–Cl(1) distance is remarkably shorter (by 0.54 Å) than the sum of the van der Waals radii. The S–Cl bonds are directed inward towards the chalcogen atom of the heterocycle. The C–Se–C angle is normal, but the pseudooctahedral coordination around Se and the Cl(2)–Se–Cl(2) angle are unusual. It is possible that these chlorine atoms are bent towards the ring because of electron-electron repulsion between the remaining lone pair on selenium and the nonbonded electrons on the chlorines or because of electrostatic attraction between these chlorine atoms and the electron-deficient carbon atoms (C(2)) [1].

Table 51
Intramolecular Bond Distances and Angles of 2,5-Bis(chlorosulfanylimino)-1,1-dichloro-1,5-dihydro-2H-1λ⁴-selenophene-3,4-dicarbonitrile [1].

bond length	in Å	bond angle	in °
Se–Cl(2)	2.369(1)	Cl(2)–Se–Cl(2)	173.46(5)
Se–C(1)	1.970(3)	Cl(2)–Se–C(1)	87.16(9)
C(1)–C(2)	1.445(4)	Cl(2)–Se–C(1)	88.07(9)
C(2)–C(2)	1.362(6)	Cl(2)–Se–C(1)	88.07(9)
C(1)–N(1)	1.266(4)	Cl(2)–Se–C(1)	87.16(9)
N(1)–S(1)	1.585(3)	C(1)–Se–C(1)	86.4(2)
S(1)–Cl(1)	2.041(1)	Cl(1)–S(1)–N(1)	108.3(1)
C(2)–C(3)	1.427(4)	S(1)–N(1)–C(1)	137.4(3)
C(3)–N(2)	1.138(4)	Se–C(1)–N(1)	127.4(2)
		Se–C(1)–C(2)	109.3(2)
		N(1)–C(1)–C(2)	123.3(3)
		C(1)–C(2)–C(2)	117.5(2)
		C(1)–C(2)–C(3)	120.3(3)
		C(2)–C(2)–C(3)	122.2(2)
		N(2)–C(3)–C(2)	179.2(4)

There are several unusual features in the crystal packing. The molecules form stacks along the c axis and sheets within the a-b plane. The title compound forms chlorine "clusters" by head-to-head ordering; very short intermolecular Cl–Cl contacts are thus generated [1].

IR (KBr): ν (in cm^{-1}) = 2225 vs, 1520 m, 1490 s, 1330 s, 1227 w, 849 m, 812 s [1]; see also [2].

UV-VIS (CH_2Cl_2): λ (in nm) (ε in $L \cdot mol^{-1} \cdot cm^{-1}$) = 296 (6270), 400 (6840), 427 (10200), 454 (15200), 511 (10900) [2].

MS: m/e = 344 $M^+ + 2$, 342 M^+, 307 $M^+ - Cl$, 272 $M^+ - 2\ Cl$ [1]; see also [2].

Apart from the loss of chlorine, the title compound is stable in the solid state in the ambient atmosphere and can be kept in a plastic-capped vial at ambient temperature for months (the plastic cap turned yellow after ca. 1 d at room temperature (Cl_2 evaluation), but a microscopic examination showed that the crystals do not change in appearance [1].

Adding excess $(n\text{-}C_4H_9)_4N^+ I^-$ to a warm solution of the title compound in CH_2Cl_2 and leaving the mixture stand for 2 d under Ar completely dechlorinated it with formation of black crystals with the composition $C_{25}H_{25}N_{10}S_9Se_9$ [2].

References:

[1] Wudl, F.; Zellers, E. T. (J. Am. Chem. Soc. **102** [1980] 5430/1).
[2] Wudl, F.; Zellers, E. T.; Nalewajek, D. (J. Org. Chem. **45** [1980] 3211/5).

12.5.4 N-Organylsulfonyl-N-alkyl(or aryl)-amino-chloro-sulfanes

12.5.4.1 $ClSN(CH_3)SO_2C_6H_5$

The title compound was formed by adding a solution of SCl_2 in ether at 0 °C to a solution of $C_6H_5SO_2NHCH_3$ and $(C_2H_5)_3N$ (mole ratio ca. 1 : 1.75 : 1); 93% yield; brown, oily liquid.

Reference:

Badische Anilin-Soda-Fabrik AG; Schulze, G.; Weiß, G. (Ger. 1 156 403 [1961/64]; C.A. **60** [1964] 4061).

12.5.4.2 $ClSN(CH_3)SO_2C_6H_4CH_3$-4

The title compound was prepared with a yield of 98.7% (crude) by bubbling Cl_2 through a suspension of 4-$CH_3C_6H_4SO_2N(CH_3)SSN(CH_3)SO_2C_6H_4CH_3$-4 in CCl_4 until the disulfide dissolved and the color turned yellow. The mixture was left standing for 30 min at room temperature [1]. Instead of Cl_2, SO_2Cl_2 in CH_2Cl_2, $CCl_2{=}CCl_2$, or $CHCl_3$ could also be used [2].

Solid, m.p. 68 to 69 °C (from petroleum ether) [1].

The ^{1}H NMR ($CCl_2{=}CCl_2$/TMS) showed a singlet at 3.36 ppm due the N–CH_3 group [2].

Reactions of the title compound are compiled in Table 52.

Table 52
Reactions of $ClSN(CH_3)SO_2C_6H_4CH_3$-4.

substrate	reaction conditions	product(s) (tos = 4-$CH_3C_6H_4SO_2$) (yield in %)	Ref.
H_2O	12 h at room temperature	(tos)$N(CH_3)S(O)SN(CH_3)$(tos) (84.7 (crude))	[1]
NH_3	in $CHCl_3$ at −15 °C and warm-up to room temperature	(tos)$N(CH_3)SNH_2$ (81.8)	[1]
AgSCN	mole ratio 1 : 1; in ether–benzene; 20 min at room temperature	(tos)$N(CH_3)SSC{\equiv}N$ (100 (crude))	[1]
$CH_2{=}CH_2$	mole ratio 1 : >1; in CH_2Cl_2, $CHCl_3$, or $CCl_2{=}CCl_2$ at 25 to 40 °C	(tos)$N(CH_3)SCH_2CH_2Cl$	[2]
trans-2-C_6H_{12}	mole ratio 1 : >1; in $CCl_2{=}CCl_2$ at 25 °C overnight or at room temperature in the presence of powdered anhydrous $ZnCl_2$ for 45 min	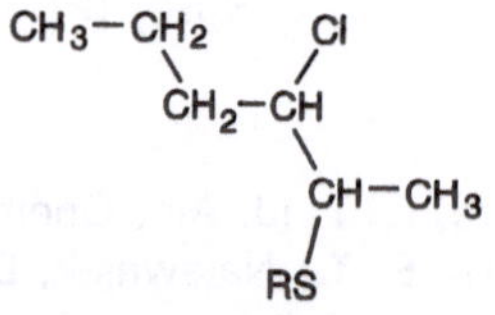 (R = $N(CH_3)$(tos))	[2]

Table 52 (continued)

substrate	reaction conditions	product(s) (tos = $4\text{-}CH_3C_6H_4SO_2$) (yield in %)	Ref.
trans-2-C_6H_{12} (continued)		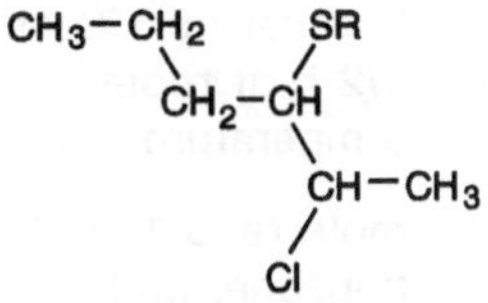 ($R = N(CH_3)(tos)$)	[2]
c-C_6H_{10}	mole ratio 1 : >1; in CH_2Cl_2, $CHCl_3$, or $CCl_2{=}CCl_2$ at 25 to 40 °C	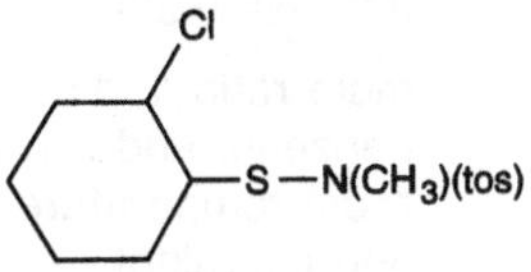	[2]
n-$C_{12}H_{25}NH_2$	mole ratio 1 : 2; in ether with cooling and 1/2 h at room temperature	$(tos)N(CH_3)SNHC_{12}H_{25}$-n (97.5)	[1]
4-$NO_2C_6H_4NH_2$ + $(C_2H_5)_3N$	mole ratio 1 : 1 : 1; in dioxane–ether under ice cooling and 1/2 h at room temperature	$(tos)N(CH_3)SNHC_6H_4NO_2$-4 (53.8)	[1]
(c-$C_6H_{11})_2NH$	mole ratio 1 : 2; in ether, 1/2 h at room temperature	$(tos)N(CH_3)SN(C_6H_{11}$-c$)_2$ (47.9)	[1]
$(CH_2)_5NH$	mole ratio 1 : 2; in ether, 1/2 h at room temperature	$(tos)N(CH_3)SN(CH_2)_5$ (86.4)	[1]
$C_6H_5NHCH_3$ + $(C_2H_5)_3N$	mole ratio 1 : 1 : 1; in ether under ice cooling and 1 h at room temperature	$(tos)N(CH_3)SN(CH_3)C_6H_5$ (99.2)	[1]
$((CH_2)_5N)_2CH_2$	mole ratio 1 : 1; in ether, 1/2 h at room temperature	$(tos)N(CH_3)SN(CH_2)_5$ (100)	[1]
$O(CH_2)_4NH$	mole ratio 1 : 2; in ether	$(tos)N(CH_3)SN(CH_2)_4O$ (92.6)	[1]
$(O(CH_2)_4N)_2CH_2$	mole ratio 1 : 1; in ether, 1/2 h at room temperature	$(tos)N(CH_3)SN(CH_2)_4O$ (92.6)	[1]

Table 52 (continued)

substrate	reaction conditions	product(s) (tos = 4-$CH_3C_6H_4SO_2$) (yield in %)	Ref.
CH_2CN_2	in ether at 0 °C and 1/2 h at room temperature	(tos)$N(CH_3)SCH_2Cl$ (86.8)	[1]
$CH_3C(O)NHC_6H_5$ + $(C_2H_5)_3N$	mole ratio 1 : 1 : 1; in benzene, and 5 h at room temperature and overnight	(tos)$N(CH_3)SN(C_6H_5)C(O)CH_3$ (45.7)	[1]
$C_6H_5C(O)NHCH_3$ + $(C_2H_5)_3N$	mole ratio 1 : 1 : 1; in benzene, and 5 h at room temperature and overnight	(tos)$N(CH_3)SN(CH_3)C(O)C_6H_5$ (37.1)	[1]
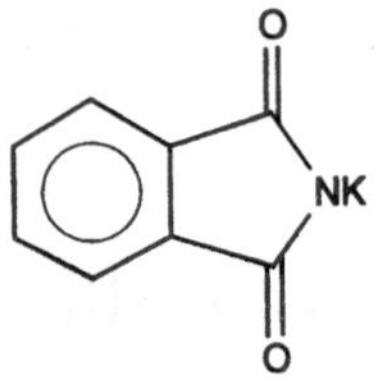	mole ratio 1 : 1; in toluene, 3 h at 80 °C and overnight at room temperature	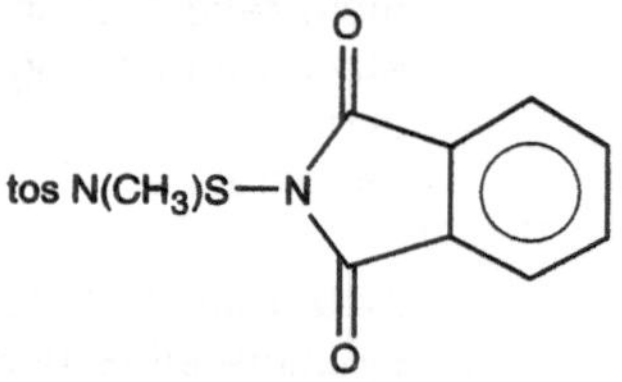 (49.7)	[1]
4-$CH_3C_6H_4SO_2NH_2$ + K_2CO_3	mole ratio ca. 1 : 1 : 1; in benzene and 2 h at room temperature	(tos)$N(CH_3)SNH$(tos) (54.4)	[1]
$(CH_3)_2NSO_2N(R)Na$ (R = C_6H_5, $CH_2C_6H_5$)	mole ratio 1 : 1; in toluene and 2 h at 80 °C	$(CH_3)_2NSO_2NRSN(CH_3)$(tos) R = C_6H_5 (55.2), R = $CH_2C_6H_5$ (48.9)	[1]
4-$CH_3C_6H_4SO_2$-$N(CH_3)Na$	mole ratio 1 : 1; in benzene and 2 h at room temperature	(tos)$N(CH_3)SN(CH_3)$(tos) (65)	[1]
$C_6H_5SO_2N(C_6H_5)Na$	mole ratio 1 : 1; in toluene and 2 h at room temperature	(tos)$N(CH_3)SN(C_6H_5)SO_2C_6H_5$ (52.4)	[1]
$(CH_2)_5NSN(CH_2)_5$	mole ratio 1 : 1; in benzene–ether, 3 h at room temperature	(tos)$N(CH_3)SN(CH_2)_5$ (100 (crude))	[1]
$O(CH_2)_4NSN(CH_2)_4O$	mole ratio 1 : 1; in benzene–ether, 3 h at room temperature	(tos)$N(CH_3)SN(CH_2)_4O$ (92.6)	[1]
$C_6H_5SO_2Na \cdot 2\ H_2O$	mole ratio 1 : 1; in benzene, 1/2 h at room temperature	(tos)$N(CH_3)SSO_2C_6H_5$ (98)	[1]

Table 52 (continued)

substrate	reaction conditions	product(s) (tos = $4\text{-}CH_3C_6H_4SO_2$) (yield in %)	Ref.
$4\text{-}CH_3C_6H_4SO_2Na \cdot 4\,H_2O$	mole ratio 1 : 1; in benzene, 1/2 h at room temperature	$(tos)N(CH_3)SSO_2C_6H_4CH_3\text{-}4$ (89)	[1]
$c\text{-}C_6H_{11}NHC(O)SR$ ($R = CH_3$, C_2H_5)	mole ratio 1 : 1; in benzene ($R = CH_3$, C_2H_5) and 3 h at room temperature ($R = C_2H_5$)	$(tos)N(CH_3)SSC(O)NHC_6H_{11}\text{-}c$ $R = CH_3$ (82.9), $R = C_2H_5$ (66.8)	[1]
$C_6H_5CH_2NHC(O)\text{-}SCH_3$	mole ratio 1 : 1; in benzene, 12 h at room temperature	$(tos)N(CH_3)SSC(O)NHCH_2C_6H_5$ (86.4)	[1]
$O(CH_2)_4NC(O)SR$ ($R = CH_3$, C_2H_5)	mole ratio 1 : 1; in benzene; briefly warmed and left to stand overnight	$(tos)N(CH_3)SSC(O)N(CH_2)_4O$ $R = CH_3$ (66.3), $R = C_2H_5$ (63.5)	[1]
$CH_3C(O)CHN_2$	mole ratio 1 : 1; in ether, 1/2 h at room temperature	$(tos)N(CH_3)SCHClC(O)CH_3$ (52)	[1]
$CH_3SC(O)SR$ ($R = CH_3$, C_2H_5)	mole ratio 1 : 1; warming to 80 °C	$(tos)N(CH_3)SSC(O)SCH_3$ $R = CH_3$ (74.3), $R = C_2H_5$ (54.3)	[1]
$C_2H_5SC(O)SR$ ($R = CH_3$, C_2H_5)	mole ratio 1 : 1; without solvent at room temperature	$(tos)N(CH_3)SSC(O)SC_2H_5$ $R = CH_3$ (53.4), $R = C_2H_5$ (62.3)	[1]
$C_6H_5CH_2SC(O)SR$ ($R = CH_3$, C_2H_5)	mole ratio 1 : 1; warming to 80 °C and 2 h at room temperature	$(tos)N(CH_3)SSC(O)SCH_2C_6H_5$ $R = CH_3$ (35.1), $R = C_2H_5$ (32.6)	[1]
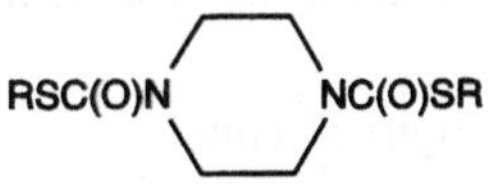 ($R = CH_3$, C_2H_5)	mole ratio 2 : 1; in benzene, with warming	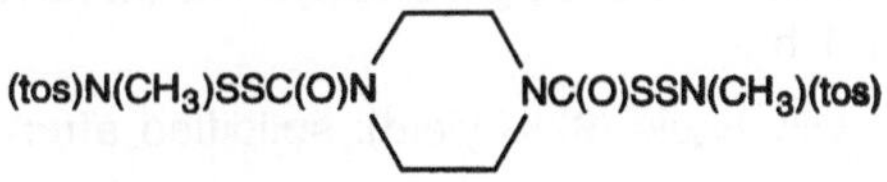 $R = CH_3$ (88), $R = C_2H_5$ (66)	[1]

References:

[1] Brandt, U. (Diss. Univ. Hamburg 1969).
[2] Hopper, R. J. (Rubber Chem. Technol. **49** [1976] 341/52; C.A. **85** [1976] No. 64398).

12.5.4.3 **$ClSN(CH_3)SO_2NR_2$**, $R_2=(CH_3)_2$, $(CH_2)_5$

The title compounds were prepared by adding a solution of SCl_2 in CH_2Cl_2 to a precooled (0 °C) solution of $R_2NSO_2NHCH_3$ ($R_2=(CH_3)_2$ [1] (see also [2]), $(CH_2)_5$ [1]) and $(C_2H_5)_3N$ (mole ratio ca. 1 : 1.4 : 1) in CH_2Cl_2 and ether, respectively. The compound with $R_2=(CH_3)_2$ was also formed by passing Cl_2 at 30 °C through a solution of $(CH_3)_2NSO_2N(CH_3)SSN(CH_3)SO_2N(CH_3)_2$ in CCl_4, while the temperature rose to 36 °C [3].

$R_2=(CH_3)_2$. Yellow oil [3], brownish liquid (93% yield); b.p. 105 °C/0.005 Torr [1]; see also [2].

$R_2=(CH_2)_5$. Brownish liquid (97% yield) [1].

References:

[1] Badische Anilin-Soda-Fabrik AG; Schulze, G.; Weiß, G. (Ger. 1156403 [1961/64]; C.A. **60** [1964] 4061).
[2] Kühle, E. (Synthesis **1970** 561/80, 573).
[3] Farbenfabriken Bayer AG (Ger. 1101407 [1957/61]; C.A. **56** [1962] 12805).

12.5.4.4 **$ClSN(CH_2C_6H_5)SO_2C_6H_4CH_3$-4**

The title compound was formed when a suspension of 4-$CH_3C_6H_4SO_2N(CH_2C_6H_5)SSN$-$(CH_2C_6H_5)SO_2C_6H_4CH_3$-4 in CCl_4 was saturated with Cl_2 under ice cooling and left standing for 1 h.

Solid (32% yield); m.p. 51 to 53 °C (from ether–petroleum ether 1 : 1).

Reactions of the title compound are summarized in Table 53, p. 253.

Reference:

Brandt, U. (Diss. Univ. Hamburg 1969).

12.5.4.5 **$ClSN(CH_2C(O)OC_2H_5)SO_2C_6H_4CH_3$-4**

The title compound was prepared by adding SCl_2 at ~20 °C to a solution of 4-$CH_3C_6H_4$-$SO_2NHCH_2C(O)OC_2H_5$ and pyridine (mole ratio 1.2 : 3 : 0.9) in THF and letting the mixture react for 1 h.

Oily liquid (97% yield); solidified after standing for an extended length of time.

Reference:

Badische Anilin-Soda-Fabrik AG; Schulze, G.; Weiß, G. (Ger. 1156403 [1961/64]; C.A. **60** [1964] 4061).

Table 53
Reactions of $ClSN(CH_2C_6H_5)SO_2C_6H_4CH_3$-4.

substrate	reaction conditions	product(s) (tos = 4-$CH_3C_6H_4SO_2$) (yield in %)
NH_3	in ether, 1/2 h at room temperature	$C_6H_5CH_2N(tos)SNH_2$ (42.4)
CH_2N_2	in ether at 0 °C and 1/2 h at room temperature	$C_6H_5CH_2N(tos)SCH_2Cl$ (55.6)
$CH_3C(O)CHN_2$	mole ratio 1 : 1; in ether, 1/2 h at room temperature	$C_6H_5CH_2N(tos)SCHClC(O)CH_3$ (54.7)
$CH_3SC(O)SH$	mole ratio 1 : 1; in benzene, 1 h at 80 °C	$C_6H_5CH_2N(tos)SSC(O)SCH_3$ (90.2)
$C_6H_5CH_2SC(O)SH$	mole ratio 1 : 1; 1 h at 80 °C	$C_6H_5CH_2N(tos)SSC(O)SCH_2C_6H_5$ (50.5)
$C_6H_5CH_2NHC(O)SH$	mole ratio 1 : 1; in ether at room temperature	$C_6H_5CH_2N(tos)SSC(O)NHCH_2C_6H_5$ (39.4)
c-$C_6H_{11}NHC(O)SH$	mole ratio 1 : 1; in ether at room temperature	$C_6H_5CH_2N(tos)SSC(O)NHC_6H_{11}$-c (86.7)
$(CH_2)_5NC(O)SH$	mole ratio 1 : 1; in benzene, 48 h at room temperature	$C_6H_5CH_2N(tos)SSC(O)N(CH_2)_5$ (61.9)
	mole ratio ca. 2 : 1; in ether, briefly warming	 (74.6)

12.5.4.6 ClSN(C_6H_{11}-c)SO_2Ar, Ar = $C_{10}H_7$-1, $C_6H_4NO_2$-2, $C_6H_4CH_3$-4

The title compounds were prepared by adding SCl_2 at ~30 °C to solutions of $ArSO_2NH$-C_6H_{11}-c (Ar = $C_{10}H_7$-1, $C_6H_4NO_2$-2 (92% yield, crude)) and pyridine (mole ratio ca. 1 : 3 : 0.8) and then letting the mixtures react for 1 and 2 h, respectively [1]. The compound with Ar = $C_6H_4CH_3$-4 was prepared with 43.8% yield (crude) by adding a solution of SO_2Cl_2 in CCl_4 to a suspension of 4-$CH_3C_6H_4SO_2N(C_6H_{11}$-c)SSN($C_6H_{11}$-c)$SO_2C_6H_4CH_3$-4 in CCl_4 and holding the mixture at reflux temperature for 2 h [2].

Ar = $C_{10}H_7$-1. Oily, undistillable liquid [1].

Ar = $C_6H_4CH_3$-4. Solid, m.p. 80 to 82 °C (from ligroine) [2].

The compound with Ar = $C_6H_4CH_3$-4 was reacted with HSC(O)N$(CH_2)_4$NC(O)SH (mole ratio 2 : 1) in benzene (briefly warming the mixture and leaving it stand overnight) to yield 82.9% 4-$CH_3C_6H_4SO_2N(C_6H_{11}$-c)SSC(O)N$(CH_2)_4$NC(O)SSN($C_6H_{11}$-c)$SO_2C_6H_4CH_3$-4 [2].

References:

[1] Badische Anilin-Soda-Fabrik AG; Schulze, G.; Weiß, G. (Ger. 1 156 403 [1961/64]; C.A. **60** [1964] 4061).
[2] Brandt, U. (Diss. Univ. Hamburg 1969).

12.5.4.7 ClSN(C_6H_5)SO_2R, R = N$(CH_3)_2$, CH_3

The title compounds were prepared by adding SO_2Cl_2 to suspensions of $RSO_2N(C_6H_5)$-SSN$(C_6H_5)SO_2$R (R = N$(CH_3)_2$, CH_3) in CCl_4 and refluxing the mixtures for 1/2 to 2 h until the evolution of SO_2 was completed [1].

R = N$(CH_3)_2$. Solid, m.p. 53 to 55 °C [1]; see also [2].

R = CH_3. Oil [1].

References:

[1] Farbenfabriken Bayer AG; Kühle, E.; Wegler, E. (Ger. 1 101 407 [1957/61]; C.A. **56** [1962] 12 805).
[2] Kühle, E. (Synthesis **1970** 561/80, 566).

12.5.4.8 (ClSN($SO_2C_6H_4CH_3$-4)CH_2)$_2$

The title compound was formed by bubbling Cl_2 for 30 min through a solution of 4-CH_3-$C_6H_4SO_2N(CH_2)_2S_2NSO_2C_6H_4CH_3$-4 (I) in CCl_4.

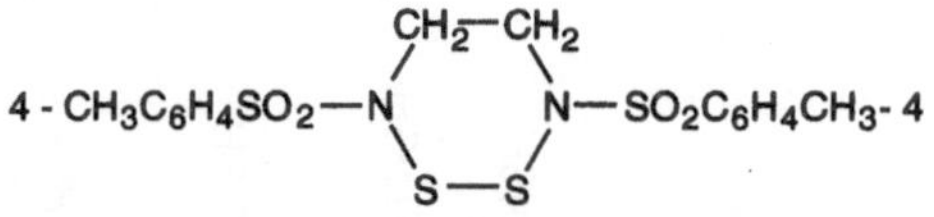

I

White solid (100% yield); m.p. 166 to 168 °C (from benzene–ligroine).

Reference:

Brandt, U. (Diss. Univ. Hamburg 1969).

12.5.5 Sulfur Chloride Hydrazide Derivatives

12.5.5.1 N,N'-Bis(benzoyl)-N-*tert*-butyl-chloro-hydrazino-sulfane, $ClSN(C(O)C_6H_5)N(C_4H_9\text{-t})C(O)C_6H_5$

The preparation of the title compound was not described.

Adding the title compound to the reaction product of $C_6H_5C(O)NHN(C_4H_9\text{-t})C(O)C_6H_5$ in THF with a suspension of NaH in mineral oil and letting the mixture react at room temperature for 5 h gave $C_6H_5C(O)N(C_4H_9\text{-t})N(C(O)C_6H_5)SN(C(O)C_6H_5)N(C_4H_9\text{-t})C(O)C_6H_5$.

Reference:

Drabek, J. (Eur. Appl. 395581 [1990]; C.A. **114** [1991] No. 185026).

12.5.6 Phosphorus-Substituted Amino-chloro-sulfanes

12.5.6.1 N-Dialkoxyphosphoryl-N-alkyl-amino-chloro-sulfanes, $ClSN(R)P(O)(OR^1)_2$

R and R^1 are given in Table 54.

Preparation. Properties

The title compounds were prepared as follows:

Method I: Adding SCl_2 (Method Ia) or S_2Cl_2 (Method Ib) to mixtures of $(R^1O)_2P(O)NHR$ and $(C_2H_5)_3N$ in C_6H_6 at 0 to 5°C [1, 2].

Method II: Adding SO_2Cl_2 (in excess) to $(R^1O)_2P(O)NRSSNRP(O)(OR^1)_2$ at 20 to 25°C. The SO_2 was removed leaving the crude product which was distilled in vacuum [3].

Table 54
Preparation and Properties of $ClSN(R)P(O)(OR^1)_2$.

$ClSN(R)P(O)(OR^1)_2$ R	R^1	method of preparation (yield in %)	remarks
CH_3	CH_3	—	— [4, 5]
CH_3	C_2H_5	Ia (82) [1] Ib [2]	yellow oil, b.p. 71 to 74°C/0.03 Torr [1], 72 to 82°C/0.11 Torr [2] $n_D^{20}=1.4825$, $D_D^{20}=1.2592$ g/cm³ [1]
CH_3	$n\text{-}C_3H_7$	Ib [2]	—
CH_3	$i\text{-}C_3H_7$	Ib [2]	—
$CH_2C_3H_5\text{-c}$	CH_3	—	— [5]
C_2H_5	CH_3	Ib [2]	—
C_2H_5	C_2H_5	Ib [2] II (62.2) [3]	mobile yellow liquid [2, 3], with an odor of S_2Cl_2 [3]; b.p. 78 to 80°C/0.14 Torr [2], 105 to 106°C/2 Torr [3] $n_D^{20}=1.4718$, $D_D^{20}=1.2107$ g/cm³ [3]
$(CH_2)_2C_6H_5$	C_2H_5	Ib [2]	—

Table 54 (continued)

$ClSN(R)P(O)(OR^1)_2$ R	R^1	method of preparation (yield in %)	remarks
n-C_3H_7	C_2H_5	Ib [2]	—
	n-C_3H_7	Ib [2]	—
	C_6H_5	Ib [2]	—
n-C_4H_9	CH_3	—	— [5]
i-C_4H_9	C_2H_5	—	— [4, 5]
n-C_8H_{17}	C_2H_5	—	— [4, 5]
c-C_3H_5	C_2H_5	—	— [4, 5]
c-C_6H_{11}	C_2H_5	—	— [4, 5]

Chemical Reactions

$ClSN(C_2H_5)P(O)(OC_2H_5)_2$ reacted with ROH (R = C_2H_5, i-C_3H_7, i-C_4H_9, in the presence of $(C_2H_5)_3N$ [3]; R = $ClCH_2CH_2$, $BrCH_2CH_2$, CH_2=$CHCH_2$, or $(C_2H_5)_2NCH_2CH_2$ and pyridine [7]) in ether at −5 to 0 °C gave after warm-up to room temperature [7] the corresponding $(C_2H_5O)_2P$-(O)$N(C_2H_5)SOR$. It reacted with R_2NH (R = CH_3, C_2H_5 in the presence of $(C_2H_5)_3N$ [3]; R_2 = CH_2CH_2, $(CH_2$=$CHCH_2)_2$ in the presence of pyridine [7]) in ether at −5 to 0 C and warm-up to room temperature to give the corresponding $(C_2H_5O)_2P(O)N(C_2H_5)SNR_2$; with aniline $(C_2H_5O)_2$-$P(O)N(C_2H_5)SNHC_6H_5$ was obtained analogously [3].

Adding $ClSN(C_2H_5)P(O)(OC_2H_5)_2$ to $(C_2H_5O)_2P(O)C_2H_5$ gradually at 8 to 10 °C and heating the mixture to 40 °C gave $(C_2H_5O)_2P(O)N(C_2H_5)SP(O)(OC_2H_5)_2$ (31%) with evolution of C_2H_5Cl [7].

$ClSN(CH_3)P(O)(OC_2H_5)_2$ was reacted with $C_2H_5OPF_2$ (mole ratio 1 : 2) at 0 to 2 °C and 4 h at 20 to 25 °C to give $(C_2H_5O)_2P(O)N(CH_3)P(O)F_2$ (79.3%) and $C_2H_5OP(S)F_2$ (91.5%). It was reacted with $(C_2H_5O)_2PF$ (mole ratio 1 : 2) at 10 to 15 °C for 2 d to give $(C_2H_5O)_2P(O)N(CH_3)P(O)$-$(OC_2H_5)F$ (92%) and $(C_2H_5O)_2P(S)F$ (65.5%) [6].

Adding the same compound at 10 to 15 °C to $C_2H_5OPCl_2$ (mole ratio 1 : 2) gave after 3 d $(C_2H_5O)_2P(O)N(CH_3)P(S)(Cl)OC_2H_5$ (80%), $POCl_3$, and C_2H_5Cl; the reaction with $(C_2H_5O)_2PCl$ (mole ratio 1 : 2) at 5 to 10 °C over 2 d gave $(C_2H_5O)_2P(O)N(CH_3)P(S)(OC_2H_5)_2$ (70%) along with $C_2H_5OP(O)Cl_2$ and C_2H_5Cl [1].

The compounds $ClSN(R)P(O)(OR^1)_2$ with R = R^1 = CH_3 [3]; R = CH_3 [4], i-C_4H_9, n-C_8H_{17}, c-C_3H_5, c-C_6H_{11} [2], R^1 = C_2H_5 [2, 4] reacted with ArN=$CHNHCH_3$ (R = R^1 = CH_3: Ar = 2,4-$(CH_3)_2C_6H_3$ [3]; R = CH_3, R^1 = C_2H_5: Ar = 2-CH_3-4-Cl(or Br)C_6H_3 [4]; R = CH_3, R^1 = i-C_4H_9, n-C_8H_{17}, c-C_3H_5, c-C_6H_{11}: Ar = 2-CH_3-4-Cl(Br or CH_3)C_6H_3, 2,3-$(CH_3)_2C_6H_3$) and $(C_2H_5)_3N$ in CH_2Cl_2 at 0 °C to give the corresponding ArN=$CHN(CH_3)SN(R)P(O)(OR^1)_2$ [4].

The compounds $ClSN(R)P(O)(OR^1)_2$ were reacted with R′SC(R″)=NOC(O)$NHCH_3$ and pyridine in CH_2Cl_2 at 0 °C and subsequently for 12 h at 0 to 5 °C to give the corresponding $(R^1O)_2P$-(O)$NRSN(CH_3)C(O)ON$=C(R″)SR′ [5]. R, R^1, R′, R″ are given in the table below [5]:

R	R^1	R′	R″	R	R^1	R′	R″
CH_3	CH_3	CH_3	C_2H_5	n-C_4H_9	CH_3	C_2H_5	CH_3
CH_3	CH_3	C_2H_5	CH_3	i-C_4H_9	C_2H_5	CH_3	CH_3
CH_3	C_2H_5	CH_3	CH_3	n-C_8H_{17}	C_2H_5	CH_3	CH_3
CH_3	C_2H_5	C_2H_5	CH_3	c-C_3H_5	C_2H_5	CH_3	CH_3
$CH_2C_3H_5$-c	CH_3	CH_3	CH_3	c-C_6H_{11}	C_2H_5	CH_3	CH_3
C_2H_5	CH_3	CH_3	CH_3				

The compounds ($R = CH_3$, $R^1 = C_2H_5$, n-C_3H_7, i-C_3H_7; $R = C_2H_5$, $R^1 = CH_3$; $R = R^1 = C_2H_5$; $R = (CH_2)_2C_6H_5$, $R^1 = C_2H_5$; R = n-C_3H_7, $R^1 = C_2H_5$, n-C_3H_7, C_6H_5) reacted with "carbofuran" in pyridine at 0 °C to give compound I [2].

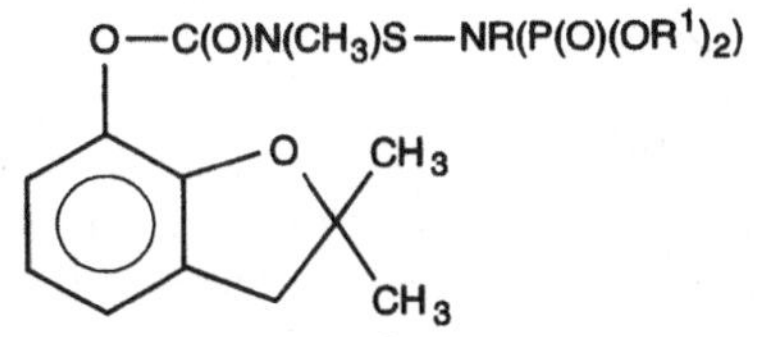

I

References:

[1] Gusar, N. I.; Ivanova, Z. M.; Gololobov, Yu. G. (Zh. Obshch. Khim. **44** [1974] 1456/9; J. Gen. Chem. USSR [Engl. Transl.] **44** [1974] 1430/2).
[2] Engel, J. F. (U.S. 4024277 [1975/77]; C.A. **87** [1977] No. 184359).
[3] Alimov, P. I.; Antokhina, L. A. (Izv. Akad. Nauk SSSR Ser. Khim. **1963** 1132/4; Bull. Acad. Sci. USSR Div. Chem. Sci. [Engl. Transl.] **1963** 1034/6).
[4] Böger, M.; Drabek, J. (Ger. 2753065 [1974/78]; C.A. **89** [1978] No. 129284).
[5] Drabek, J.; Böger, M. (Ger. 2751028 [1975/78]; C.A. **89** [1978] No. 108198).
[6] Gusar, N. I.; Ivanova, Z. M.; Chaus, M. P.; Gololobov, Yu. G. (Zh. Obshch. Khim. **46** [1976] 1981/6; J. Gen. Chem. USSR [Engl. Transl.] **46** [1976] 1910/4).
[7] Alimov, P. I.; Antokhina, L. A. (Izv. Akad. Nauk SSSR Ser. Khim. **1964** 1316/7; Bull. Acad. Sci. USSR Div. Chem. Sci. [Engl. Transl.] **1964** 1220/1).

12.5.6.2 N-Dialkoxythiophosphoryl-N-alkyl(or phenyl)-amino-chloro-sulfanes

12.5.6.2.1 $ClSN(R)P(S)(OC_2H_5)_2$, R = n-C_3H_7, i-C_3H_7, C_6H_5

The title compounds were prepared by adding a solution of the respective $RNHP(S)(OC_2H_5)_2$ (R = n-C_3H_7, i-C_3H_7, C_6H_5) and $(C_2H_5)_3N$ in hexane (for R = n-C_3H_7 or i-C_3H_7) or ether ($R = C_6H_5$) to chilled (0 °C for n-C_3H_7 or i-C_3H_7 and −10 °C for $R = C_6H_5$) solutions of SCl_2 (mole ratio ca. 1 : 1 : 1) in hexane or ether and stirring the mixture at 0 to 5 °C for 30 min. Workup gave the crude compounds as pale yellow (R = n-C_3H_7), orange (R = i-C_3H_7), or yellow oils ($R = C_6H_5$), respectively, which were not purified [1]; see also [2 to 6].

The title compounds reacted with aryl methylcarbamates $CH_3NHC(O)OR'$ (mole ratio ca. 1 : 1; R' = aryl, especially 2-i-$C_3H_7C_6H_4$ (for R = n-C_3H_7 or C_6H_5 [2, 3]), 3-i-$C_3H_7C_6H_4$ (for R = i-C_3H_7 [2, 3, 5]), 3,5-$(CH_3)_2$-4-$(CH_3)_2NC_6H_2$ (for R = n-C_3H_7 [2, 3, 6]) in cold (0 °C) DMF solutions and sometimes with $(C_2H_5)_3N$ at 25 °C to give the corresponding $(C_2H_5O)_2P(S)N(R)SN(CH_3)C(O)OR'$ [1 to 6].

A mixture of $(C_2H_5O)_2P(S)N(C_3H_7$-i$)SCl$ and $CH_3NHC(O)F$ in DMF was reacted with a solution of $(C_2H_5)_3N$ in DMF at 0 °C for 2 h to give $(C_2H_5O)_2P(S)N(C_3H_7$-i$)SN(CH_3)C(O)F$ [5, 7].

References:

[1] Nelson, S. J. (U.S. 4201733 [1977/80]; C.A. **93** [1980] No. 166891).
[2] Nelson, S. J. (U.S. 4208409 [1977/80]; C.A. **94** [1981] No. 102864).
[3] Nelson, S. J. (U.S. 4081536 [1977/78]; C.A. **89** [1978] No. 43787).
[4] Nelson, S. J. (Ger. 2902647 [1978/79]; C.A. **92** [1980] No. 76561).
[5] Nelson, S. J. (Ger. 3019634 [1979/80]; C.A. **95** [1981] No. 132314).
[6] Nelson, S. J. (Braz. Pedido 7800665 [1976/78]; C.A. **90** [1979] No. 138023).
[7] Nelson, S. J. (Ger. 3019590 [1978/80]; C.A. **94** [1981] No. 191729).

12.5.6.2.2 $ClSN(R)\overline{P(S)\text{-}O\text{-}CH_2\text{-}C(CH_3)_2\text{-}CH_2\text{-}O}$, $R = i\text{-}C_3H_7$, $n\text{-}C_4H_9$, $t\text{-}C_4H_9$

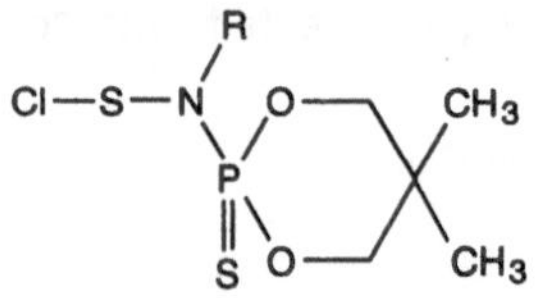

R = i-C_3H_7. The title compound was formed (but not isolated) by introducing Cl_2 into a suspension of the disulfide I ($R = C_3H_7$-i) in CH_2Cl_2 cooled to 0 °C until a yellow homogeneous solution resulted [1].

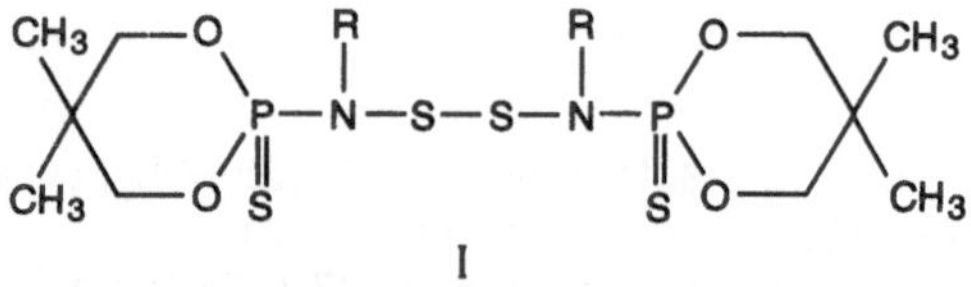

I

Adding this solution at 0 °C to a solution of $CH_3SC(CH_3)$=NOC(O)$NHCH_3$, $(C_2H_5)_3N$, and CuCl in THF yielded II (R = i-C_3H_7) (62%) [1].

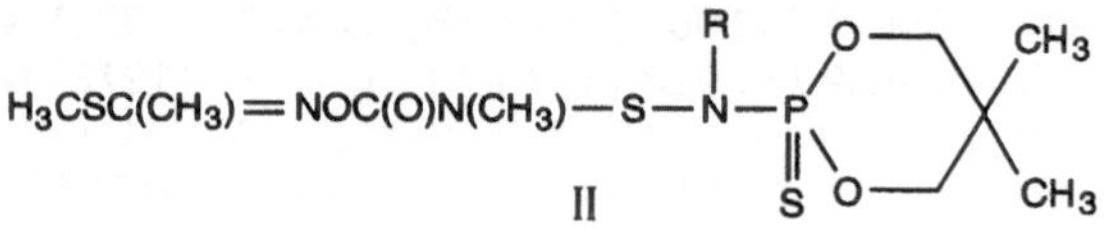

II

R = n-C_4H_9. The title compound was formed by stirring a mixture of n-$C_4H_9NHP(S)O_2(CH_2)_2$-$C(CH_3)_2$, S_2Cl_2, and $(C_2H_5)_3N$ in THF at 0 to 5 °C for 0.5 h; 78% yield [2].

R = t-C_4H_9. The title compound was formed by adding Cl_2 in CCl_4 to a suspension of the disulfide I (R = t-C_4H_9) in CCl_4 cooled to −10 °C until the mixture became homogeneous [1].

^{1}H NMR (CCl_4 ?/TMS): δ (in ppm) = 1.12 (s, 3H, CH_3), 1.16 (s, 3H, CH_3), 1.66 (s, 9H, t-C_4H_9), 4.10 (dd, 4H, CH_2O) [1].

Adding a solution of the title compound in THF at 3 °C to a solution of $CH_3SC(CH_3)$=NOC-(O)$NHCH_3$, $(C_2H_5)_3N$, and CuCl in THF yielded II (R = t-C_4H_9) (76%) [1].

References:

[1] Nelson, S. J.; Sacks, C. E. (U.S. 4308217 [1980/81]; C.A. **96** [1982] No. 162937).
[2] Mitsubishi Chemical Industries Co., Ltd. (Jpn. Kokai Tokkyo Koho 57-130990 [1981/82]; C.A. **98** [1983] No. 126377).

12.5.7 Silicon-Substituted Amino-chloro-sulfanes

12.5.7.1 N-Silyl-N-alkyl-amino-chloro-sulfanes, $ClSN(R)SiH_3$, $R = CH_3$, $t\text{-}C_4H_9$, CH_2CF_3

The title compounds were not synthesized.

The geometrical and electronic structures were investigated quantum-chemically by the MNDO method with complete geometry optimization. The geometrical parameters of ClSN-$(CH_3)SiH_3$ are: d(Cl–S) = 200 pm, d(S–N) = 159 pm, d(N–Si) = 177 pm; ∢(ClSN) = 107°, ∢(SNC) = 119°, ∢(CNSi) = 121°. The geometrical parameters of the molecules $ClSN(C_4H_9\text{-}t)SiH_3$ and $ClSN(CH_2CF_3)SiH_3$ are not significantly different. The nitrogen hybridization is of the sp^2 type in $ClSN(CH_3)SiH_3$ and $ClSN(C_4H_9\text{-}t)SiH_3$ and close to the sp^2 type in $ClSN(CH_2CF_3)SiH_3$. The S–N and C–N bonds are simply covalent, the S–Cl bond is practically orthogonal to the CNS plane. Negatively charged centers are the nitrogen atom and the chlorine atom.

The decomposition of the title compounds may be schematically described as an S–Cl bond weakening and subsequent silyl group transfer to the chlorine atom. The activation barrier of this process is estimated to be 274, 260, and 294 kJ/mol for the compounds with $R = CH_3$, C_4H_9-t, and CH_2CF_3, respectively.

Reference:

Penkovsky, V. V.; Shermolovitch, Yu. G. (Phosphorus Sulfur Silicon Relat. Elem. **80** [1993] 255/8).

12.5.7.2 $ClSN(R)Si(CH_3)_3$, $R = CH_3$, n-C_3H_7, i-C_3H_7, t-C_4H_9, n-C_5H_{11}, n-C_7H_{15}, $CH_2(CF_2)_nH$ (n = 2,6)

The compounds with $R = i\text{-}C_3H_7$, t-C_4H_9 [2,3] were formed by adding mixtures of $RNHSi(CH_3)_3$ and $(C_2H_5)_3N$ in ether to solutions of freshly distilled SCl_2 (mole ratio 1 : 1 : 1) in ether at −10 °C [1], −15 °C [6], or −78 °C [3] and letting the mixtures react at −10 °C for 0.5 h and at 20 °C for 1 h [1], at 20 °C for 0.5 h [6], or reach room temperature overnight [3]. Then the hydrochloride was filtered off, the filtrate evaporated under vacuum (10 to 20 Torr) below 20 °C, and the residue distilled. The compounds with $R = CH_2(CF_2)_nH$ (n = 2,6) are red liquids which were not isolated [6]. The compound with $R = CH_3$ was not obtained under the same conditions as used for the production of the isopropyl or tert-butyl derivative (from $CH_3NHSi(CH_3)_3$ with SCl_2 and $(C_2H_5)_3N$); it evidently decomposed to thionitrosomethane forming polymeric products [1]. The formation of the compounds with $R = n\text{-}C_3H_7$, n-C_5H_{11}, and n-C_7H_{15} were not described.

$R = i\text{-}C_3H_7$. Dark red, mobile liquid (70% yield) [1].

^1H NMR ($CDCl_3$/TMS): δ (in ppm) = 0.41 (s, $Si(CH_3)_3$), 1.22 (d, CH_3, $^3J(H,H) = 6$ Hz), 3.60 (quint, CH, $^3J(H,H) = 6$ Hz) [1].

$R = t\text{-}C_4H_9$. Red liquid [1] (yield: 73% [1], 77%; much lower yield when purified by column distillation [3]); b.p. 30 to 31 °C/0.01 Torr [3].

^1H NMR ($CDCl_3$/TMS): δ (in ppm) = 0.42 (s, $Si(CH_3)_3$), 1.39 (s, $C(CH_3)_3$) [1]; (10% CCl_4/TMS): 0.45 (s, $Si(CH_3)_3$), 1.42 (s, $C(CH_3)_3$) [3].

Thermolysis of the compounds with $R = n\text{-}C_3H_7$, n-C_5H_{11}, and n-C_7H_{15} led to unstable S=NR [2]. The compound with $R = i\text{-}C_3H_7$ slowly decomposed at 20 °C and rapidly when heated to 100 °C with elimination of $(CH_3)_3SiCl$ to form a polymeric material [1]. The compounds with $R = CH_2(CF_2)_nH$ (n = 2, 6) decomposed at room temperature with the formation of polymeric products [6]; see also [7].

On heating the compound with $R = t\text{-}C_4H_9$ decomposed with formation of t-$C_4H_9N{=}S{=}N$-C_4H_9-t, $ClSi(CH_3)_3$, and sulfur [1].

Heating a solution of the compounds with $R = i\text{-}C_3H_7$ and t-C_4H_9 with 2,3-dimethyl-1,3-butadiene (mole ratio 1 : 2.5) in toluene at 100 °C for 1 h [1] or boiling for 2 h and 15 h at 20 °C [6] gave I with 53% ($R = C_3H_7$-i), 55% ($R = C_4H_9$-t), 38% ($R = CH_2(CF_2)_2H$), or 36% ($R = CH_2(CF_2)_6H$) yield, respectively [1, 6].

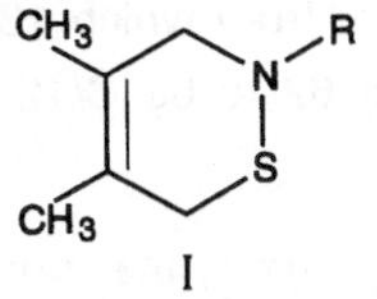

I

When isoprene was added to a solution of $ClSN(R)Si(CH_3)_3$ in toluene, where R = t-C_4H_9, $CH_2(CF_2)_6H$, and the mixture reacted for 1 h at 20 °C or boiled for 2 h, a mixture of the isomeric dihydrothiazines IIa and IIb was obtained with a total yield of 79% or 41% [6].

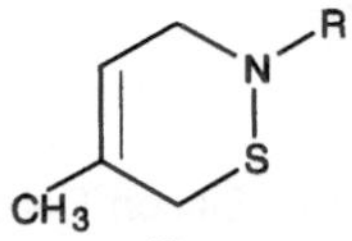

IIa

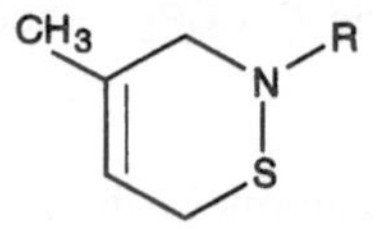

IIb

Adding a solution of $ClSN(C_4H_9\text{-t})Si(CH_3)_3$ in ether at −78 °C to a solution of t-$C_4H_9NH_2$ and $(C_2H_5)_3N$ (mole ratio 1 : 1 : 1) in ether gave 75% $(CH_3)_3SiN(C_4H_9\text{-t})SNHC_4H_9$-t [3]. Adding a solution of the same (crude) compound (excess) in ether at −78 °C to a solution of c-$C_6H_{11}NH_2$ and $(C_2H_5)_3N$ (mole ratio 1 : 1) and letting the mixture come to room temperature overnight yielded c-$C_6H_{11}NHSN(C_4H_9\text{-t})Si(CH_3)_3$ (55%) [4].

A solution of $ClSN(C_4H_9\text{-t})Si(CH_3)_3$ in ether reacted at −78°C with $(CH_3)_3SiN(M)R$ (M = Li, R = i-C_3H_7, $C(CH_3)_2C_2H_5$, $C(CH_3)_2CH_2C(CH_3)_3$, c-C_6H_{11}; M = Na, R = $Si(CH_3)_3$; reacting $(CH_3)_3Si$-NHR and LiC_4H_9-n or NaH (?)) in ether and letting the reaction mixture cool overnight to room temperature yielded the corresponding $(CH_3)_3SiN(t\text{-}C_4H_9)SN(R)Si(CH_3)_3$ [5].

References:

[1] Markovskii, L. N.; Solov'ev, A. V.; Kaminskaya, E. I.; Borodin, A. V.; Shermolovich, Yu. G. (Zh. Org. Khim. **26** [1990] 2083/6; J. Org. Chem. USSR [Engl. Transl.] **26** [1990] 1799/801).

[2] Markovskii, L. N.; Borodin, A. V.; Solov'ev, A. V.; Kolesnik, N. P.; Shermolovich, Ju. G. (J. Fluorine Chem. **54** [1991] 403).

[3] Scherer, O. J.; Wolmershäuser, G. (Z. Naturforsch. **29b** [1974] 277/8).

[4] Neidlein, R.; Lehr, W. (Chem. Ber. **114** [1981] 80/5).

[5] Neidlein, R.; Lehr, W. (Heterocycles **16** [1981] 1179/85).

[6] Markovskii, L. N.; Solov'ev, A. V.; Pen'kovskii, V. V.; Kolesnik, N. P.; Borodin, A. V.; Iksanov, S. V.; Shermolovich, Yu. G. (Zh. Org. Khim. **28** [1992] 1388/95; J. Org. Chem. USSR [Engl. Transl.] **28** [1992] 1092/8).

[7] Penkovsky, V. V.; Shermolovitch, Yu. G. (Phosphorus Sulfur Silicon Relat. Elem. **80** [1993] 255/8).

12.6 Sulfur Amide Bromides, Amino-bromo-sulfane Derivatives

12.6.1 N,N-Diorganyl-amino-bromo-sulfanes

12.6.1.1 N,N-Dialkyl-amino-bromo-sulfanes

12.6.1.1.1 $BrSN(CH_3)_2$

$BrSN(CH_3)_2$ was formed by thermal decomposition of $[((CH_3)_2N)_2SBr]^+ Br^-$ (obtained from $(CH_3)_2NSN(CH_3)_2$ and Br_2 in ether below −10 °C) at about 80 °C.

When the reaction of $(CH_3)_2NSN(CH_3)_2$ with Br_2 was performed in benzene at room temperature, a greasy product was obtained, from which 13% of $BrSN(CH_3)_2$ could be distilled.

Liquid; b.p. 58 to 60 °C/10 Torr or 62 to 63 °C/12 Torr.

Reference:

Nöth, H.; Mikulaschek, G. (Chem. Ber. **97** [1964] 202/6).

12.6.1.1.2 $BrSN(C_2H_5)_2$

$BrSN(C_2H_5)_2$ (erroneously quoted in the paper) was prepared by adding a solution of $(C_2H_5)_2NSSN(C_2H_5)_2$ in CCl_4 at 0 to 5 °C to a solution of Br_2 in CCl_4 and stirring the mixture for another 5 min at 0 to 5 °C. The solution was cleared with animal charcoal, the solvent evaporated under reduced pressure at about 30 to 40 °C, and the residue distilled in vacuum to give $BrSN(C_2H_5)_2$ with 81.5% yield [1]; see also [2, 3].

Dark red liquid, b.p. 32 to 33 °C/0.1 Torr [1].

The ^1H NMR spectrum at 35 °C in $CD_3C_6D_5$ (0.68 M) showed a sharp triplet and quartet for the ethyl signal [4].

A mixed solution of $BrSN(C_2H_5)_2$ and $ClSN(C_2H_5)_2$ of similar concentrations in $CD_3C_6D_5$ showed sharp signals at τ values intermediate between those of the pure compounds at temperatures above 50 °C, indicating rapid halogen exchange. Below −40 °C distinct signals were observed, attributable to each of the separate components [4].

Mixtures of $BrSN(C_2H_5)_2$ and $ClSN(C_2H_5)_2$ in CH_2Cl_2 (0.58 or 1.15 M) showed evidence for much faster halogen exchange than in $CD_3C_6D_5$. Individual spectra of the two compounds were only obtained at temperatures of ca. −80 °C [4].

A solution of the compound in CCl_4 reacted with I to give as much as 72% II ($R = N(C_2H_5)_2$) [5].

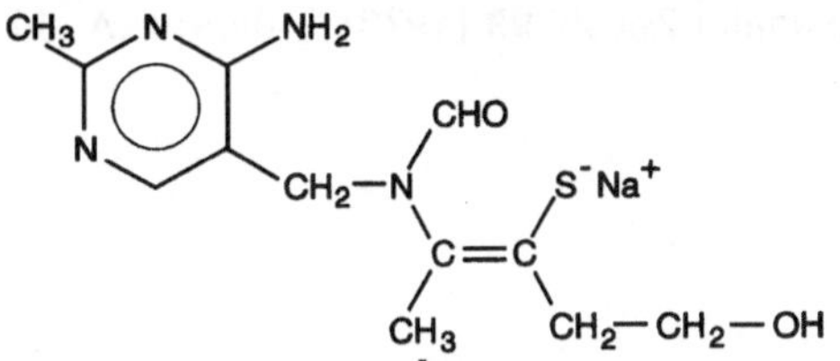

I

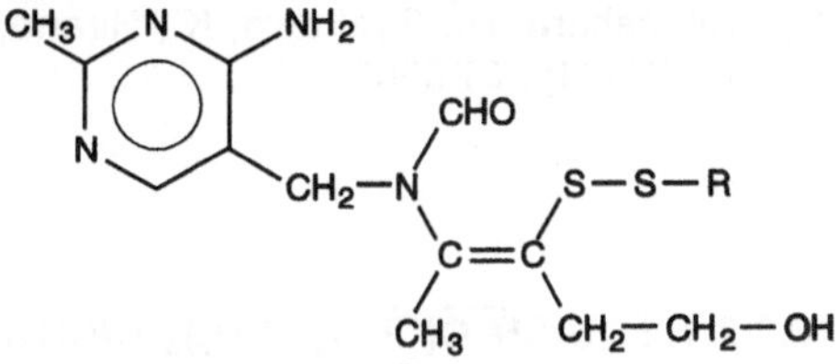

II

References:

[1] Farbenfabriken Bayer AG (Br. 790021 [1954/58]; C.A. **1958** 13806).
[2] Kühle, E. (Synthesis **1970** 561/80, 566).
[3] Dorlars, A. (Houben-Weyl Methoden Org. Chem. 4th Ed. **11** Pt. 2 [1958] 744/51).
[4] Jackson, W. R.; Kee, T. G.; Spratt, R.; Jennings, W. B. (Tetrahedron Lett. **1973** 3581/4).
[5] Shimahara, N.; Sirakawa, K.; Hirano, H. (Yakugaku Zasshi **93** [1973] 1484/9; C.A. **80** [1974] No. 37060).

12.6.1.2 N-(2,5-Dichlorophenyl)-N-trifluoromethyl-amino-bromo-sulfane, $BrSN(CF_3)C_6H_3Cl_2$-2,5

The title compound was formed by adding a solution of $(C_2H_5)_3N$ in CCl_4 under cooling to a mixture of $CF_3NHC_6H_3Cl_2$-2,5 and S_2Cl_2 in CCl_4 or benzene while holding the temperature below 25 °C, removing the hydrochloride by filtration, adding a solution of Br_2 in CCl_4, and stirring the mixture for 3 h at 40 to 50 °C [1]; see also [2].

Liquid, b.p. 125 to 130 °C/14 Torr [1].

References:

[1] Farbenfabriken Bayer AG; Kühle, E.; Klauke, E. (Ger. 1187627 [1962/65]; C.A. **62** [1965] 16118).
[2] Kühle, E. (Synthesis **1970** 561/80, 567).

12.6.2 N-Bromosulfanyl-Substituted N-Heterocycles

12.6.2.1 N[1]-Bromosulfanyl-piperidine and Derivative

12.6.2.1.1 $BrSN(CH_2)_5$

$BrSN(CH_2)_5$ was prepared by adding $(CH_2)_5NSSN(CH_2)_5$ in CCl_4 at 0 to 5 °C to a solution of Br_2 (mole ratio 1 : 1) in CCl_4. After 5 min animal charcoal was added to the solution yielding after a standard workup the crude compound with 90% yield; it was distilled in vacuum to give a dark red liquid; b.p. 65 to 68 °C/0.4 Torr [1].

Adding a solution of $BrSN(CH_2)_5$ in CCl_4 to a precooled (0 to 5 °C) solution of $(CH_2)_2NH$ and $(C_2H_5)_3N$ (mole ratio ca. 1 : 1 : 1) in CCl_4 and stirring the mixture at room temperature for 3 h yielded $(CH_2)_2NSN(CH_2)_5$ [2].

A solution of $BrSN(CH_2)_5$ in CCl_4 reacted with I to give 50% II (R = $N(CH_2)_5$) [3]; see p. 261.

References:

[1] Farbenfabriken Bayer AG; Freytag, H.; Lober, F. (Ger. 965968 [1954/57]; C.A. **1959** 11417), Farbenfabriken Bayer AG (Br. 790021 [1954/58]; C.A. **1958** 13806).

[2] Farbenfabriken Bayer AG; Freytag, H.; Lober, F.; Domagk, G. (Ger. 948330 [1954/56]; C.A. **1959** 6267).

[3] Shimahara, N.; Sirakawa, K.; Hirano, H. (Yakugaku Zasshi **93** [1973] 1484/9; C.A. **80** [1974] No. 37060).

12.6.2.1.2 $BrS\overline{N{-}C(CH_3)_2{-}(CH_2)_3{-}C}(CH_3)_2$

The title compound was prepared with 65% yield by adding a solution of Br_2 in CCl_4 to a solution of $(CH_3)_2\overline{C{-}(CH_2)_3{-}C(CH_3)_2{-}N}SS\overline{N{-}C(CH_3)_2{-}(CH_2)_3{-}C}(CH_3)_2$ (mole ratio 2 : 1) in CCl_4 while stirring and cooling to 0 to 5 °C. Stirring was continued for 2 h at room temperature, the solvent was removed in vacuum, and the remaining orange-red solid was recrystallized.

Solid; m.p. 48 °C (from methylcyclohexane).

^1H NMR (C_6H_6/TMS): δ (in ppm) = 1.45, 1.35 (CH_3, CH_2; assignment not unambiguous).

MS (70 eV): m/e (rel. int. in %), related to ^{79}Br: M^+ (33), $M^+ - Br$ (100), $M^+ - Br - CH_3$ (17), $M^+ - SBr - CH_3$ (74).

Reference:

Röschenthaler, G.-V.; Starke, R. (Z. Naturforsch. **32b** [1977] 721/2).

12.6.2.2 N[4]-Bromosulfanyl-morpholine, $BrSN(CH_2)_4O$

$BrSN(CH_2)_4O$ was prepared by reacting $O(CH_2)_4NSSN(CH_2)_4O$ with Br_2 (no details) [1].

It reacted with cyclohexene in CH_2Cl_2 at −20 °C to give the trans adduct III [2].

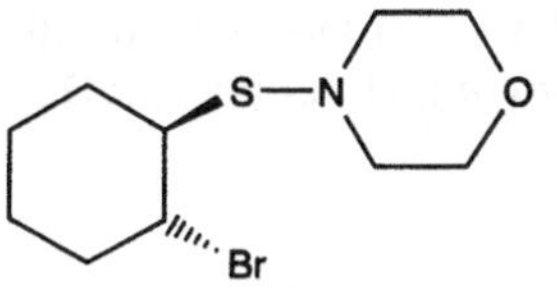

III

A solution of $BrSN(CH_2)_4O$ in CCl_4 reacted with I to give (up to) 69% II (R = $N(CH_2)_4O$); see p. 261 [3].

A solution of $BrSN(CH_2)_4O$ reacted with a solution of (8-chloro-1-mercaptomethyl)-6-(2-chlorophenyl)-4H-s-triazolo[4,3-a][1,4]benzodiazepine IV in CH_2Cl_2 at −20 to −13°C for 1.5 h to give V [1].

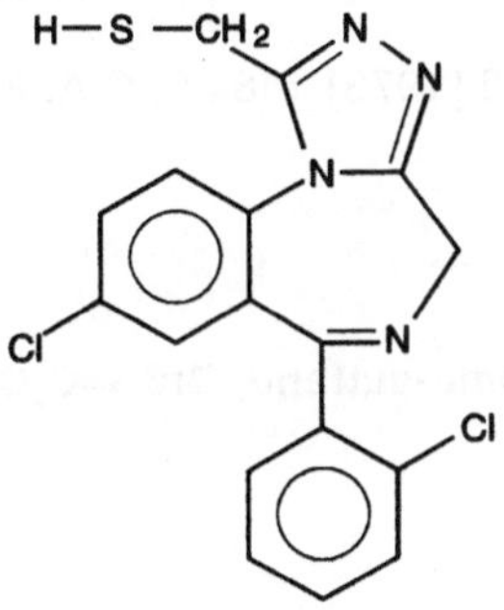

IV

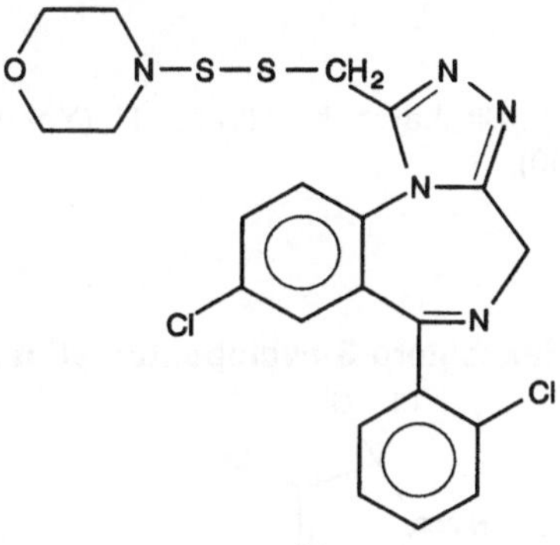

V

References:

[1] Hirai, K.; Fujishita, T.; Ishiba, T. (Eur. Appl. 4320 [1978/79]; C.A. **92** [1980] No. 94447).
[2] Kutateladze, A. G.; Zyk, N. V.; Denisko, O. V.; Zefirov, N. S. (Zh. Org. Khim. **27** [1991] 659/61; J. Org. Chem. USSR [Engl. Transl.] **27** [1991] 569/70).
[3] Shimahara, N.; Sirakawa, K.; Hirano, H. (Yakugaku Zasshi **93** [1973] 1484/9; C.A. **80** [1974] No. 37060).

12.6.2.3 N[4]-Bromosulfanyl-thiomorpholine, $BrSN(CH_2)_4S$

The title compound was prepared by adding a solution of Br_2 in CCl_4 at 0 to 5°C to a suspension of $S(CH_2)_4NSSN(CH_2)_4S$ (mole ratio 1 : 1) in CCl_4. After 5 min charcoal was added to the solution, and the solvent was removed by distillation at 30 to 50°C under reduced pressure to give the crude compound (86.5% yield) which is distillable in vacuum [1].

Deep red liquid; b.p. 103 to 105°C/0.45 Torr [1]; see also [2, 3].

Adding a solution of (crude) $BrSN(CH_2)_4S$ in CCl_4 to a precooled (0 to 5°C) solution of $(CH_2)_2NH$ and $(C_2H_5)_3N$ (mole ratio ca. 1 : 1 : 1) in CCl_4 and stirring the mixture at room temperature for 3 h yielded $(CH_2)_2NSN(CH_2)_4S$ [4].

References:

[1] Farbenfabriken Bayer AG; Freytag, H.; Lober, F. (Ger. 965968 [1954/57]; C.A. **1959** 11417, Br. 790021 [1954/58]; C.A. **1958** 13806).
[2] Kühle, E. (Synthesis **1970** 561/80, 566).

[3] Dorlars, A. (Houben-Weyl Methoden Org. Chem. 4th Ed. **11** Pt. 2 [1958] 744/51).
[4] Farbenfabriken Bayer AG; Freytag, H.; Lober, F.; Domagk, G. (Ger. 948330 [1954/56]; C.A. **1959** 6267).

12.6.2.4 N[1]-Bromosulfanyl-N[4]-methyl-piperazine, $BrSN(CH_2)_4NCH_3$

The preparation of the title compound was not described.

A solution of $BrSN(CH_2)_4NCH_3$ in CCl_4 reacted with I to give 27% II ($R = N(CH_2)_4NCH_3$); see p. 261.

Reference:

Shimahara, N.; Sirakawa, K.; Hirano, H. (Yakugaku Zasshi **93** [1973] 1484/9; C.A. **80** [1974] No. 37060).

12.6.3 N-(Hexachloro-3-cyclopenten-ylidene)-amino-bromo-sulfane, $BrSN{=}C_5Cl_6$

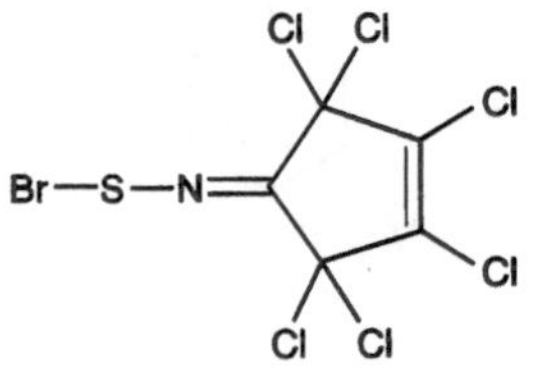

The title compound was prepared with 96% yield by stirring a mixture of $ClSN{=}C_5Cl_6$ and $(CH_3)_3SiBr$ (mole ratio 1 : 1.1) in hexane at 23 °C for 4 h. Removal of the solvent under vacuum produced $BrSN{=}C_5Cl_6$ as a yellow solid.

Transparent yellow plates; m.p. 74 °C (from n-hexane).

^{13}C NMR ($CDCl_3$/TMS): δ (in ppm) = 78.3 (C(2), C(5)), 135.8 (C(3), C(4)), 142.7 (C(1)).

^{14}N NMR ($CDCl_3$/external CH_3NO_2): $\delta = -75.0$ ppm.

IR: ν (in cm^{-1}) = 1646 s, 1608 s, 1206 s, 1034 s, 810 vs, 692 s, 637 s, 583 s, 419 s, 355 s; the bands at 1206, 810, 637, and 583 are characteristic of the C_5Cl_6NS moiety, that at 355 cm^{-1} for ν(S–Br). The mass spectrum showed a weak parent ion and the cleavage of the S–Br bond to be the major fragmentation process.

An X-ray structure analysis (25 °C) showed the crystals to be triclinic, space group $P\bar{1}-C_i^1$ (No. 2), with a = 9.626(2), b = 9.810(2), c = 13.736(2) Å, α = 79.68(1)°, β = 88.62(1)°, and γ = 74.73(1)°; Z = 4; V = 1230.69 Å^3; D_x = 2.152 g/cm^3; R = 0.075, R_w = 0.063 from 4850 observed reflections. For atom coordinates, see the paper.

The structure of $BrSN{=}C_5Cl_6$ is shown in **Fig. 11** and consists of two independent molecules in the asymmetric unit. Selected bond lengths and bond angles for the two $BrSN{=}C_5Cl_6$ molecules are listed in Table 55. The structure exhibits a normal C=N bond length, a short S–N bond length, anomalously large bond angles at N and S, and a syn geometry for the BrSN=C unit which was explained as a result of negative hyperconjugation ($n_N \rightarrow \sigma^*_{SBr}$). This interaction involves donation from the in-plane nonbonding orbital on N to the antibonding σ orbital of the S–Br bond. It is bonding with respect to the S–N bond and is enhanced by wider angles at N and/or S.

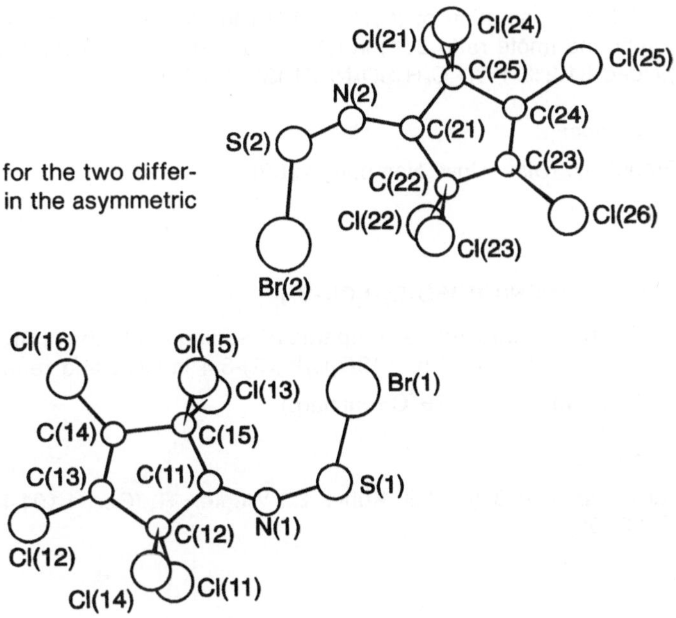

Fig. 11. Molecular structures for the two different molecules of $BrSN{=}C_5Cl_6$ in the asymmetric unit.

Table 55
Selected Bond Lengths and Bond Angles for the Two Different Molecules of $BrSN{=}C_5Cl_6$ in the Asymmetric Unit.

molecule 1		molecule 2	
bond length in pm			
Br(1)–S(1)	2.229(2)	Br(2)–S(2)	2.235(2)
S(1)–N(1)	1.560(6)	S(2)–N(2)	1.569(6)
N(1)–C(11)	1.278(8)	N(2)–C(21)	1.253(8)
bond angle in °			
N(1)–S(1)–Br(1)	116.8(2)	N(2)–S(2)–Br(2)	116.5(2)
C(11)–N(1)–S(1)	146.7(4)	C(21)–N(2)–S(2)	148.6(4)
N(1)–C(11)–C(15)	130.5(5)	N(2)–C(21)–C(22)	130.8(6)
N(1)–C(11)–C(12)	120.2(5)	N(2)–C(21)–C(25)	120.2(5)

Reference:

Apblett, A.; Chivers, T.; Fait, J. F.; Vollmerhaus, R. (Can. J. Chem. **69** [1991] 1022/7).

12.6.4 N-Organylsulfonyl-N-alkyl(or phenyl)-amino-bromo-sulfanes

12.6.4.1 $BrSN(CH_3)SO_2C_6H_4CH_3$-4

The title compound was prepared with 100% yield by treating a suspension of 4-$CH_3C_6H_4$-$SO_2N(CH_3)SSN(CH_3)SO_2C_6H_4CH_3$-4 in CCl_4 with a solution of Br_2 in CCl_4 at 0 °C and stirring the mixture for another 10 min at 0 °C.

Crystalline solid; m.p. 76 to 77 °C (from ligroin).

Adding a solution of $(C_6H_5CH_2)_2NH$ in CCl_4 under ice cooling to a solution of the title compound (mole ratio 2:1) in CCl_4 and stirring the mixture for 1/2 h at room temperature yielded 81.2% $4\text{-}CH_3C_6H_4SO_2N(CH_3)SN(CH_2C_6H_5)_2$.

Reference:

Brandt, U. (Diss. Univ. Hamburg 1969).

12.6.4.2 $BrSN(CH_3)SO_2C_6H_3Cl_2\text{-}3,4$

The title compound was prepared by adding a solution of Br_2 in CCl_4 to a suspension of $3,4\text{-}Cl_2C_6H_3SO_2N(CH_3)SSN(CH_3)SO_2C_6H_3Cl_2\text{-}3,4$ in CCl_4 and refluxing the mixture for 1 h.

Solid, m.p. 70 to 78 °C (unsharp).

Reference:

Farbenfabriken Bayer AG; Kühle, E.; Wegler, R. (Ger. 1101407 [1957/61]; C.A. **56** [1962] 12805).

12.6.4.3 $BrSN(C_6H_5)SO_2N(CH_3)_2$

The title compound was prepared by adding a solution of Br_2 in CCl_4 at room temperature to a suspension of $(CH_3)_2NSO_2N(C_6H_5)SSN(C_6H_5)SO_2N(CH_3)_2$ in CCl_4 while the temperature rose from 20 to 25 °C.

Solid, m.p. 78 to 84 °C.

Reference:

Farbenfabriken Bayer AG (Ger. 1101407 [1957/61]; C.A. **56** [1962] 12805).

12.7 Sulfur Amide Jodide, Amino-iodo-sulfane Derivative

12.7.1 N-(Hexachloro-3-cyclopenten-ylidene)-amino-iodo-sulfane, $ISN{=}C_5Cl_6$

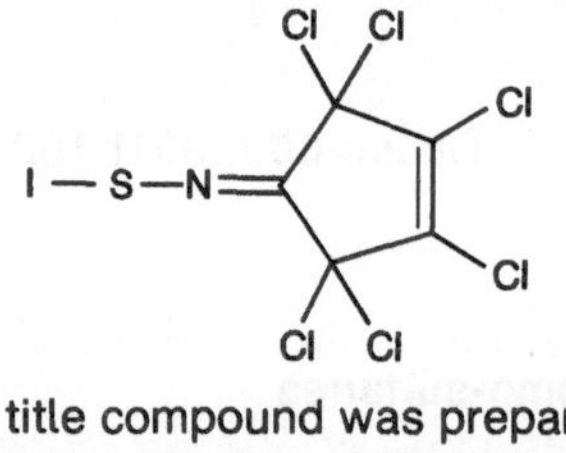

The title compound was prepared with 39% yield by adding a cold solution of $(CH_3)_3SiI$ in n-hexane to a solution of $ClSN{=}C_5Cl_6$ (mole ratio ca. 1:1) in n-hexane at −78 °C and letting the mixture react at −78 °C for 1 h. The solvent was removed under vacuum, and the brown solid was recrystallized from n-hexane at −20 °C to give red crystals.

^{13}C NMR ($CDCl_3$/TMS, 25 °C): δ (in ppm) = 79.8 (C(2), C(5)), 138.2 (C(3), C(4)), 174.0 (C(1)).

^{14}N NMR ($CDCl_3/CH_3NO_2$): δ = −78.9 ppm.

IR: ν (in cm^{-1}) = 1617 m, 1208 vs, 1154 m, 982 m, 824 s, 809 vs, 653 m, 636 vs, 595 s, 581 m; the bands at 1208, 809, 636, and 595 cm^{-1} are characteristic of the C_5Cl_6NS moiety.

The mass spectrum showed a weak parent ion and the cleavage of the S–I bond to be the major fragmentation process.

Reference:

Apblett, A.; Chivers, T.; Fait, J. F.; Vollmerhaus, R. (Can. J. Chem. **69** [1991] 1022/7).

12.8 Salts of N-(Halogenosulfanylidene)-amido-halogeno-sulfur Ions (1+), $[X{-}S{=}N{-}S{-}X]^+$, X = Cl, Br

Only the salts of $(ClS)_2N^+$ and $(BrS)_2N^+$ are known. Attempts to prepare $(FS)_2N^+$ were unsuccessful. The reaction of $S_2N^+\ AsF_6^-$ with F_2 in SO_2F_2 at low temperatures yielded SF_6, $SF_3^+\ AsF_6^-$, and $(F_2S)_2N^+\ AsF_6^-$. Fluorination of $S_2N^+\ AsF_6^-$ with XeF_2 in SO_2F_2 at 0 °C led to the formation of $(F_2S)_2N^+\ AsF_6^-$. $(ClS)_2N^+\ AsF_6^-$ reacted with ClF in SO_2ClF at room temperature to give SF_4, SF_5Cl, OSF_2, and $SF_3^+\ AsF_6^-$. These reactions are inhomogeneous, since $(FS)_2N^+\ AsF_6^-$ is initially formed on various surfaces and preferentially reacts further to give the observed products.

Attempts to prepare $(IS)_2N^+\ AsF_6^-$ from $S_2N^+\ AsF_6^-$ and I_2 in SO_2 at room temperature were unsuccessful. $(IS)_2N^+\ AsF_6^-$ is not stable towards dissociation into $S_2N^+\ AsF_6^-$ and I_2 under the prevailing reaction conditions.

Reference:

Brooks, W. V. F.; MacLean, G. K.; Passmore, J.; White, P. S.; Wong, Chi-Ming (J. Chem. Soc. Dalton Trans. **1983** 1961/8).

12.8.1 Salts of N-(Chlorosulfanylidene)-amido-chloro-sulfur Ion (1+), $[Cl{-}S{=}N{-}S{-}Cl]^+$

12.8.1.1 $(ClS)_2N^+\ CF_3SO_3^-$

$(ClS)_2N^+\ CF_3SO_3^-$ was prepared with 75% yield by adding SCl_2 (a slight molar excess) to a stirred mixture of $S_3N_3Cl_3$ and $Ag^+\ CF_3SO_3^-$ (mole ratio 3 : 1 : 3) in liquid SO_2. After stirring the mixture for 18 h, the AgCl precipitate was filtered off and washed three times with back-condensed SO_2. Removal of the solvent left the title compound as a yellow solid [1].

IR (Nujol mull): ν (in cm^{-1}) = 1400 w, 1250 s (br), 1175 s, 1022 vs, 940 m, 750 m, 720 m, 635 vs, 615 m, 572 m, 515 m, 500 w, 470 w, 422 m [1].

When $(ClS)_2N^+\ CF_3SO_3^-$ and $SnCl_2$ (molar equivalents) were stirred in liquid SO_2 for 18 h, a cream yellow precipitate of $S_2N^+\ CF_3SO_3^-$ formed with 70% yield [1].

12.8.1.2 $((ClS)_2N^+)_2\ SeCl_6^{2-}$

$((ClS)_2N^+)_2\ SeCl_6^{2-}$ was obtained with 84% yield when SCl_2 (6.6 mmol) was added to a stirred mixture of $S_3N_3Cl_3$ (6.6 mmol) and $SeCl_4$ (10 mmol) in $SOCl_2$. The mixture was heated at reflux temperature for 6 h before allowing it to cool. The title compound was isolated as a lemon yellow powder which was filtered off and washed with small quantities of ice-cold $SOCl_2$ [1].

IR (Nujol mull): ν (in cm^{-1}) = 1136 m, 730 s, 720 m, 660 s, 650 m, 524 s, 510 s, 495 s [1].

Reduction of $((ClS)_2N^+)_2$ $SeCl_6^{2-}$ with $SnCl_2$ (mole ratio 1 : 2) in CH_2Cl_2 at room temperature over a period of 24 h gave $[S_3N_2Cl]_2^{2+}$ $SeCl_6^{2-}$, $SnCl_4$, and minor by-products (mostly $[S_3N_2]_2^{2+}$ $SeCl_6^{2-}$) [1].

12.8.1.3 $(ClS)_2N^+$ AsF_6^-

Preparation

$(ClS)_2N^+$ AsF_6^- was prepared with 96% yield by adding SCl_2 to a stirred mixture of $S_3N_3Cl_3$ and Ag^+ AsF_6^- (mole ratio 3 : 1 : 3) in liquid SO_2. The mixture was stirred for 18 h to furnish an orange-yellow solution over a white precipitate of AgCl. After filtration, the white precipitate was washed three times with back-condensed SO_2. Removal of the solvent left the title compound as a yellow solid which was dried in vacuum [1].

$(ClS)_2N^+$ AsF_6^- was obtained with 100% yield by stirring SN^+ AsF_6^- (8.51 mmol) and excess SCl_2 (3 mL) in SO_2 (10 mL) at room temperature for 48 h [2, 3] or 10 h [4]. A crystalline solid was left after removing the solvent SO_2 and excess SCl_2 in vacuum [2 to 4].

$(ClS)_2N^+$ AsF_6^- could be prepared with 99% yield by reacting S_2N^+ AsF_6^- with Cl_2 in SO_2. Typically, SO_2 (63.3 mmol) and an excess of Cl_2 (3.96 mmol) were consecutively condensed onto S_2N^+ AsF_6^- (2.16 mmol). The mixture was warmed to room temperature while shaking, and after 1 h a yellow solution was filtered through a sintered glass frit. SO_2 was condensed back from the solution which contained the product. Other volatiles were removed by evacuation, leaving yellow, needle-shaped crystals of the title compound [5].

Fundamental Vibrations. Force Constants. Vibrational Spectra

For the $(ClS)_2N^+$ ion (point group symmetry C_{2v}) nine IR- and Raman-active fundamental vibrations ($4\ A_1 + A_2 + 3\ B_1 + B_2$) are expected. Their frequencies are listed in Table 56.

Table 56
Frequencies (in cm^{-1}) of the Fundamental Vibrations of $(ClS)_2N^+$ in $(ClS)_2N^+$ AsF_6^- [2, 3].

IR (Nujol)	Raman (SO_2 solution)	assignment[a)] I	II
1127 m	1128 dp w	$\nu_{as}(NS_2)$	$\nu_{as}(NS_2)$
720 ? sh	727 p m	$\{\nu_s(NS_2)\}$	$\{\nu_s(NS_2)\}$
665 m	661 p s	$\{\delta(SNS)\}$	$\{\delta(SNS)\}$
527 s	529 m	$\nu_{as}(SCl)$	$\nu_{as}(SCl)$ +
509 m	512 sh	$\nu_s(SCl)$	$2\ \delta_s(ClSN)$
450 w	459 p m	$\delta_s(ClSN)$	$\nu_s(SCl)$
358 m	358 vw	$\delta_{as}(ClSN)$	$\delta_{as}(ClSN)$
199	202 m	τ_{as}	t_{as}
	117 m	τ_s	τ_s

[a)] Assignment II assuming Fermi resonance [3]. $\{\nu_s(NS_2)\}$, $\{\delta(SNS)\}$ indicates strong coupling of both vibrations.

The frequencies (in cm^{-1}) of in-plane vibrations of $(ClS)_2N^+$ given below were observed in the IR and Raman spectra of $(ClS)_2N^+ AsF_6^-$: Symmetry class A_1: $\nu_1 = 721$ (N–S), $\nu_2 = 659$ (S–N–S), $\nu_3 = 452$ (S–Cl), $\nu_4 = 218$ (Cl···Cl); symmetry class B_1: $\nu_5 = 1128$ (N–S), $\nu_6 = 527$ (S–Cl), $\nu_7 = 202$ (N–S–Cl) [5]. The assignment differs from the one proposed by [2, 3], most notably for the two lowest frequencies [5]. There are insufficient data to make the assignments definitive [5].

A set of force constants (in N/m) for $(ClS)_2N^+$ was calculated with the method of Fadini on the basis of the author's frequencies with two possible assignments (assignment II assuming Fermi resonance; see Table 56). Assignment I: $f_{NS} = 617$, $f_{SCl} = 293$, $f_{SNS} = 90$, $f_{ClSN} = 141$, $f_{NS/NS} = 161$, $f_{NS/SCl} = 50$, $f_{NS/SNS} = 57$, $f_{ClSN/ClSN} = 140$, $f_{\tau} = 21$, $f_{\tau\tau} = 19$; assignment II: $f_{NS} = 625$, $f_{SCl} = 232$, $f_{SNS} = 99$, $f_{ClSN} = 56$, $f_{NS/NS} = 169$, $f_{NS/SCl} = 11$, $f_{NS/SNS} = 68$, $f_{ClSN/ClSN} = 29$, $f_{\tau} = 21$, $f_{\tau\tau} = 19$ [3].

Another set of force constants for the $(ClS)_2N^+$ cation was obtained from the authors' frequencies (see Table 57) by vibrational analysis using the Wilson F and G matrix methods and a least squares fitting procedure. The force constants (in N/m) include a Cl–Cl stretching constant. Diagonal constants: N–S 657, S–Cl 300, Cl···Cl 44, S–N–S 44, N–S–Cl 50; interaction constants: N–S/N–S 155, N–S/S–Cl 189, N–S/S–N–S 60, N–S/N–S–Cl 69 [5].

Table 57
Vibrational Spectra of $(ClS)_2N^+ AsF_6^-$ (ν in cm^{-1}) [5].

IR Nujol	Raman solid 6328 Å	 solid 5145 Å	 SO_2 solution 5145 Å	assignment
1128 m		~1133 ? vvw		$\nu_{as}(NS_2)$
1091 w		~1077 ? vvw		$\nu_{as}(NS_2)$
817 w				AsF_6^- ?
718 ? sh	728 (60)	728 (100)	721 (100) p	$\nu_s(NS_2)$[+ δ_s(SNS)]
698 vs				$\nu_3(AsF_6^-)$
	681 (25)	684 (9)	679 (9) p	$\nu_1(AsF_6^-)$
660 mw	661 (65)	664 (17)	659 (57) p	δ_s(SNS)(+ SCl or SN)
574 w	576 (6)	578 (1)		$\nu_2(AsF_6^-)$
529 ms	530 (30)	525 (7)	527 (18) p	ν_{as}(SCl)(+ SN)
512 m	518 sh			ν_{as}(SCl)(+ SN)
455 w	457 (17)	458 (2)	452 (10) p	ν_s(SCl)(+ SN)
	405 (3)	407 (1)	405 (1)	2 δ_{as}(NSCl)
390 s				$\nu_4(AsF_6^-)$
356 w	370 (8)	372 (2)	371 (1)	$\nu_5(AsF_6^-)$
	227 (14)	226 (8)	218 (7) p	ν_s(Cl···Cl)
203 mw	204 (100)	204 (4)	202 (13) dp?	δ_{as}(NSCl)
	129 (18)			τ_s, or lattice mode

Vibrational frequencies (in cm^{-1}) of AsF_6^- in $(ClS)_2N^+ AsF_6^-$ were also determined by [3]. Raman: $\nu_1(A_{1g}) = 682$ p m, $\nu_2(E_g) = 577$ dp w, $\nu_5(F_{2g}) = 378$ dp w; IR: $\nu_3(F_{1u}) = 694$ vs, $\nu_4(F_{1u}) = 393$ vs.

Crystal Structure

$(ClS)_2N^+ AsF_6^-$ crystallizes in a twinned, rhombohedral, or pseudo-rhombohedral structure with the parameters $a_{hex} = 9.702(3)$, $c_{hex} = 8.346(3)$ Å, $V = 680.3$ Å^3, $Z = 3$. The $(ClS)_2N^+$ ions are disordered [2].

Melting Point

The yellow, crystalline solid melts at 189 °C [2, 3].

Mass Spectrum

The mass spectrum was obtained at 130 °C using the direct inlet method: m/e (rel. int. in %) = 151 (90) AsF_4^+, 101 (8) S_2Cl^+, 99 (20) S_2Cl^+, 46 (54) SN^+ [5].

Chemical Reactions

Thermolysis. $(ClS)_2N^+\ AsF_6^-$ decomposed or dissociated at 130 °C under dynamic vacuum conditions into products that included sulfur chlorides but no Cl_2 [5].

$SnCl_2$. When $(ClS)_2N^+\ AsF_6^-$ and the molar equivalent of $SnCl_2$ were stirred in liquid SO_2 for 18 h, a cream yellow precipitate of $S_2N^+\ AsF_6^-$ (94% yield) and $SnCl_4$ were produced [1].

Tetrahydrofuran. The title compound is an efficient initiator for the polymerization of THF. The ESR spectrum at 20 °C showed the $S_3N_2^{+\bullet}$ radical cation and another not identified radical to be present in the solution prior to consolidation [6].

Alkenes. $(ClS)_2N^+\ AsF_6^-$ was reacted with $H_2C{=}CH_2$, $CH_3CH{=}CH_2$, E-$CH_3CH{=}CHCH_3$, Z-$CH_3CH{=}CHCH_3$, and $F_2C{=}CF_2$ in SO_2 at room temperature. Cycloaddition was observed to some extent in solution for all reactants except the last one. ^1H and ^{13}C NMR were used to characterize the products of these reactions [16].

$(ClS)_2N^+\ AsF_6^-$ reacted quantitatively with $H_2C{=}CH_2$ (mole ratio 1 : 1) in SO_2 at room temperature to give $[H_2C{-}S(Cl){\doteq}N{\doteq}S(Cl){-}CH_2]^+\ AsF_6^-$ (I). This reaction was also performed on a preparative scale [16].

A solution of $(ClS)_2N^+\ AsF_6^-$ (0.10 mmol) and $H_2C{=}CH_2$ (0.21 mmol) in SO_2 gave $[H_2C{-}S(Cl){\doteq}N{\doteq}S(Cl){-}CH_2]^+\ AsF_6^-$ (I), $[H_2C{-}S{\doteq}N{\doteq}S{-}CH_2]^+$ (IIa), $[S_2N(C_2H_4)_2]^+$ (III), and $ClCH_2CH_2Cl$ [16].

A mixture of $(ClS)_2N^+\ AsF_6^-$ (0.392 mmol) and $CH_3CH{=}CH_2$ (0.340 mmol) in SO_2 gave $[CH_3CH{-}S(Cl){\doteq}N{\doteq}S(Cl){-}CH_2]^+\ AsF_6^-$ in the diastereomeric forms IVa and IVb and $[CH_3CH{-}S{\doteq}N{\doteq}S{-}CH_2]^+$ (IIb) [16].

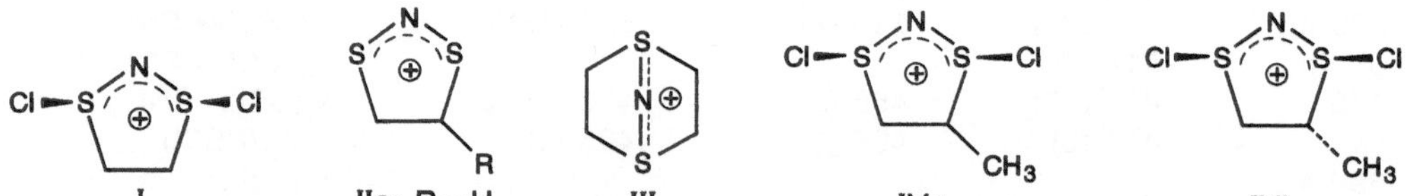

I — IIa: R = H; IIb: R = CH_3 — III — IVa — IVb

$(ClS)_2N^+\ AsF_6^-$ (0.377 mmol) and E-$CH_3CH{=}CHCH_3$ (0.320 mmol) reacted in SO_2 to give *anti*-$[CH_3CH{-}S(Cl){\doteq}N{\doteq}S(Cl){-}CHCH_3]^+\ AsF_6^-$, $[CH_3CH{-}S{\doteq}N{\doteq}S{-}CHCH_3]^+$, and $CH_3CHClCH$-$ClCH_3$ [16].

$(ClS)_2N^+\ AsF_6^-$ (0.369 mmol) and Z-$CH_3CH{=}CHCH_3$ (0.326 mmol) reacted in SO_2 to give *syn-endo*-$[CH_3CH{-}S(Cl){\doteq}N{\doteq}S(Cl){-}CHCH_3]^+$, *syn-exo*-$[CH_3CH{-}S(Cl){\doteq}N{\doteq}S(Cl){-}CHCH_3]^+$, $[CH_3CH{-}S{\doteq}N{\doteq}S{-}CHCH_3]^+$, and $CH_3CHClCHClCH_3$ [16].

$F_2C{=}CF_2$, a very poor electron donor, did not react with $(ClS)_2N^+\ AsF_6^-$ in SO_2 [16].

Alkynes. $(ClS)_2N^+\ AsF_6^-$ was reacted with $HC{\equiv}CH$, $CH_3C{\equiv}CH$, and $CF_3C{\equiv}CCF_3$ in SO_2 at room temperature. Cycloaddition occurred with $HC{\equiv}CH$ and $CH_3C{\equiv}CH$, but no reaction was observed with $CF_3C{\equiv}CCF_3$, even in the presence of an excess of alkyne and heating to 70 °C for 14 h. The reactions were monitored by ^1H or ^{19}F NMR. The NMR results were confirmed on a preparative scale in the reaction of $(ClS)_2N^+$ AsF_6^- with HC≡CH [16].

The reaction of $(ClS)_2N^+$ AsF_6^- with HC≡CH (mole ratio 1:1) in SO_2 furnished $[HC{\cdots}S{\cdots}N{\cdots}S{\cdots}CH]^+$ AsF_6^- (Va) and Cl_2. When this reaction was performed on a preparative scale in the mole ratio 1.0:3.3, $[HC{\cdots}S{\cdots}N{\cdots}S{\cdots}CH]^+$ AsF_6^- (Va) was obtained with 97% yield. The volatile products included SO_2, SO_2Cl_2, HC≡CH, and ClCH=CHCl [16].

$(ClS)_2N^+$ AsF_6^- reacted with $CH_3C≡CH$ (mole ratio 1:1) in SO_2 to give $[CH_3C{\cdots}S{\cdots}N{\cdots}S{\cdots}CH]^+$ AsF_6^- (Vb) and Cl_2 [16].

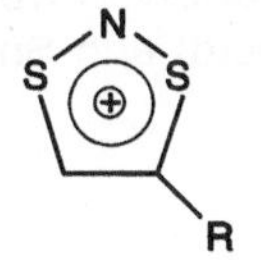

Va: R = H
Vb: R = CH_3

$CH_3C≡N$. $(ClS)_2N^+$ AsF_6^- did not react with $CH_3C≡N$, consistent with the known reluctance of nitriles to act as dienophiles [16].

12.8.1.4 $(ClS)_2N^+$ SbF_6^-

Preparation

$(ClS)_2N^+$ SbF_6^- was prepared with 80% yield by adding a slight molar excess of SCl_2 to a stirred mixture of $S_3N_3Cl_3$ and $AgSbF_6$ (3 equivalents) in liquid SO_2. The mixture was stirred 18 h yielding an orange-yellow solution and a white precipitate of AgCl. After filtering, the white precipitate was washed three times with back-condensed SO_2. Removal of the solvent left the title compound as a yellow solid which was dried in vacuum [1].

$(ClS)_2N^+$ SbF_6^- was obtained quantitatively by stirring a mixture of SN^+ SbF_6^- (9.01 mmol) and excess SCl_2 (3 mL) in liquid SO_2 at room temperature for 48 h. A crystalline solid was left after the solvent SO_2 and excess SCl_2 were removed in vacuum [2].

Vibrational Spectra

The frequencies of the fundamental vibrations of $(ClS)_2N^+$ are listed in Table 58.

Table 58
Frequencies (in cm^{-1}) of the Fundamental Vibrations of $(ClS)_2N^+$ in $(ClS)_2N^+$ SbF_6^- [2].

IR Nujol	Raman solid	assignment[a)]
1128 m		$\nu_{as}(NS_2)$
721 w	728 s	$\{\nu_s(NS_2)\}$
(covered)		$\{\delta(SNS)\}$
527 s	530 m	$\nu_{as}(SCl)$
508 m	515 sh	
454 w	457 m	$\nu_s(SCl)$
355 w		$\delta_{as}(ClSN)$
278 vw	$\delta_s(ClSN)$	

a) $\{\nu_s(NS_2)\}$, $\{\delta(SNS)\}$ indicates strong coupling of both vibrations.

Similar frequencies were measured in the IR spectrum of $(ClS)_2N^+$ SbF_6^- by [1]: (Nujol): ν (in cm^{-1}) = 1122 s, 750 m (sh), 715 m (sh), 692 s (sh), 640 vs (br), 505 ms, 360 m, 280 vs.

Melting Point

The yellow, crystalline solid melts at 177 °C [2].

Chemical Reactions

When $(ClS)_2N^+$ SbF_6^- and molar equivalents of $SnCl_2$ were stirred in liquid SO_2 for 18 h, a cream yellow precipitate of S_2N^+ SbF_6^- (75% yield) and $SnCl_4$ were produced [1].

12.8.1.5 $(ClS)_2N^+$ $SbCl_6^-$

Preparation

$(ClS)_2N^+$ $SbCl_6^-$ was prepared with ca. 80% by adding SCl_2 to a stirred mixture of $S_3N_3Cl_3$ and $SbCl_5$ (mole ratio 3 : 1 : 3) in $SOCl_2$. Stirring for ca. 30 min precipitated $(ClS)_2N^+$ $SbCl_6^-$. The bright yellow powder is sparingly soluble in $SOCl_2$ and is purified by solvent extraction with this solvent [7].

$(ClS)_2N^+$ $SbCl_6^-$ was obtained quantitatively by reacting $H_2N_3^+$ $SbCl_6^-$ with SCl_2 (mole ratio 1 : 2). SCl_2 was condensed at 77 K onto a frozen solution of $H_2N_3^+$ $SbCl_6^-$ in SO_2. The mixture was then warmed to 233 K, and after 72 h yellow crystals of the compound precipitated. The reaction proceeds via $H_2N_3^+$ $SbCl_6^-$ + 2 SCl_2 → $(ClS)_2N^+$ $SbCl_6^-$ + 2 HCl + N_2. The analogous reaction of $H_2N_3^+$ $SbCl_6^-$ with S_2Cl_2 (mole ratio 1 : 2) in SO_2 at 243 K after 1 h reaction gave a deep red solid. After two weeks the solid brightened in color and reacted to $(ClS)_2N^+$ $SbCl_6^-$ and S_8 [8].

$(ClS)_2N^+$ $SbCl_6^-$ could be obtained with 51% yield by slowly adding SbF_3 (0.1 mmol) to a stirred and ice-cooled solution of $(ClS)_2N^+$ BCl_4^- (0.05 mmol) in $SbCl_5$ (ca. 100 mL). The mixture was stirred at room temperature until BF_3 ceased to evolve. The resulting yellow precipitate was filtered off and dried in vacuum. The reaction proceeds via $(ClS)_2N^+$ BCl_4^- + $SbCl_5$ + SbF_3 → $(ClS)_2N^+$ $SbCl_6^-$ + BF_3 + $SbCl_3$ [9].

$(ClS)_2N^+$ $SbCl_6^-$ was formed together with $S_3N_2Cl^+$ $SbCl_6^-$ and S_2N^+ $SbCl_6^-$ by reacting $S_3N_3Cl_3$, $SbCl_5$, and sulfur (mole ratio 1 : 3 : 3). $SbCl_5$ (15 mmol) was added to a solution of $S_3N_3Cl_3$ (5 mmol) in $SOCl_2$ (30 mL) at room temperature to give a thick, pale orange precipitate which was stirred for additional 30 min. The major constituent of this precipitate was shown by Raman spectroscopy to be $S_4N_4^{2+}$ $SbCl_6^{2-}$. Addition of sulfur (15 mmol) and stirring for 24 h gave a pale yellow-green, insoluble solid and a dark green solution. Evaporation of the filtrate gave $S_3N_2Cl^+$ $SbCl_6^-$. The title compound was isolated from the insoluble solid by extraction with SO_2. The remaining solid was S_2N^+ AsF_6^-. A similar procedure in CH_2Cl_2 solution gave additionally the adduct $S_4N_4 \cdot SbCl_5$ (the reaction conditions were varied to maximize the yield of S_2N^+ $SbCl_6^-$) [10].

Crystal and Molecular Structure

The crystal structure was determined by X-ray diffraction. The yellow salt crystallizes at 233 K in the monoclinic space group $P\,2_1/c - C_{2h}^5$ (No. 14) with the parameters a = 7.297(7), b = 11.221(8), c = 15.756(11) Å, β = 99.43(6)°; V = 1273(2) Å^3; Z = 4; D_x = 2.5234 g/cm^3; $R_{int.}$ = 0.1113 from 2009 reflections with F ≥ 5 σ(F) [8].

The sulfur atoms in $(ClS)_2N^+$ $SbCl_6^-$ are linked to three Cl atoms in addition to the N–S bonds; see **Fig 12**. These atomic distances are as much as 5.94% shorter than the sum of the van der Waals radii of S and Cl of 3.55 Å [8].

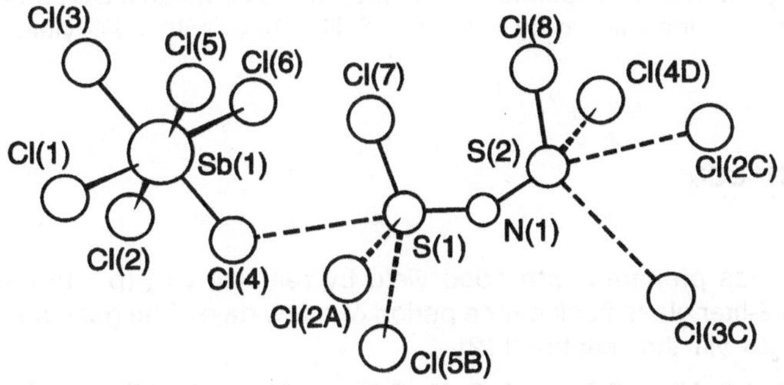

Fig. 12. Asymmetric unit of $(ClS)_2N^+$ $SbCl_6^-$ with interionic contacts; a = 1 + x, y, z; b = 1 − x, −1/2 + y, 1/2 − z; c = 1 + x, 1/2 − y, −1/2 + z; d = x, 1/2 − y, −1/2 + z [8].

The $(ClS)_2N^+$ ion possesses nearly C_{2v} symmetry. The following geometrical parameters were determined: d(S–Cl) = 1.911(7) Å, d(S–N) = 1.533(4) Å, ∢(Cl–S–N) = 110.4(2)°, ∢(S–N–S) = 149.9(3)°, τ = 2.3(7)° [8].

The $SbCl_6^-$ ion forms an almost ideal octahedron. The octahedral angles are maximally distorted by 2.96(8)°. The average Sb–Cl distance is 2.367(2) Å [8].

Vibrational Spectra

Frequencies of the fundamental vibrations of $(ClS)_2N^+$ are compiled in Table 59.

Table 59
Frequencies (in cm^{-1}) of the Fundamental Vibrations of $(ClS)_2N^+$ in $(ClS)_2N^+$ $SbCl_6^-$.

IR [7] solid	IR [8] solid	Raman [8] solid	assignment [8]
1130 m	1117 m	—	$\nu_{as}(NS_2)$
735 sh	—	718 vs	$\nu_s(NS_2)$
721 m			
654 w	670 m	658 m	δ(SNS)
521 w			ν_{as}(SCl) and
500 w	519 s	510 m	2 δ_s(ClSN)
494 w	462 m	446 w	ν_s(SCl)
—	350 sh	355 sh	δ_{as}(ClSN)
—	228 s	222 s	τ_{as}
—	—	130 m	τ_s

Mass Spectrum

The mass spectrum shows the following fragments: m/e (rel. int. in %) = 148 (14) $(ClS)_2N^+$, 46 (77) NS^+, 67 (77) SCl^+, 78 (32) NS_2^+, 81 (10) $NSCl^+$, 102 (57) SCl_2^+, 121 (66) Sb^+, 156 (57) $SbCl^+$, 193 (100) $SbCl_2^+$, 228 (66) $SbCl_3^+$, 264 (69) $SbCl_4^+$ [9].

Chemical Reactions

The salt decomposed at 98 °C in a sealed tube [9].

$(ClS)_2N^+ SbCl_6^-$ reacted with equimolar amounts of $SnCl_2$ in liquid SO_2 under stirring over a period of 36 h to produce the highly insoluble $S_2N^+ SbCl_6^-$ with 70% yield [11].

12.8.1.6 $(ClS)_2N^+ BCl_4^-$

Preparation

$(ClS)_2N^+ BCl_4^-$ was prepared with good yield by reacting BF_3 (p = 450 Torr) with $F_3S{\equiv}N$ (p = 300 Torr) in a 2-liter glass flask over a period of a few days. The gaseous products are N_2, Cl_2, BF_3, and BF_2Cl via the reaction [12]:

$$2\ F_3S{\equiv}N + 3\ BCl_3 \rightarrow (ClS)_2N^+ BCl_4^- + 1/2\ N_2 + 3/2\ Cl_2 + 2\ BF_3$$

The salt was also formed by reacting $S_3N_3Cl_3$ with Cl_2 and BCl_3 or SCl_2 and BCl_3 [12].

Crystal and Molecular Structure

The structural parameters of the salt were determined by X-ray diffraction. The yellow solid crystallizes in the monoclinic space group $P2_1/c - C_{2h}^5$ (No. 14) with a = 6.441(5), b = 16.008(10), c = 9.864(7) Å, β = 103.30(5)°; Z = 4; D_x = 2.024 g/cm³; D_m = 1.98 g/cm³; R = 0.059 from about 1800 independent reflections. The structures of the cation and the anion are shown in **Fig. 13**. Two doubly degenerate resonance structures are possible for the $(ClS)_2N^+$ ion. The $(ClS)_2N^+$ ion, within the limits of error, is planar and has cis configuration with approximate C_{2v} symmetry. The BCl_4^- forms a tetrahedron with virtually no distortion, the bond angles varying between 109.0° and 109.9° [12].

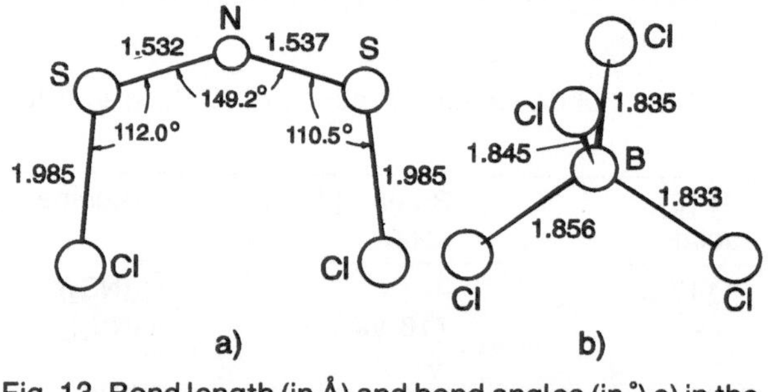

Fig. 13. Bond length (in Å) and bond angles (in °) a) in the $(ClS)_2N^+$ ion and b) in the BCl_4^- ion [12].

Vibrational Spectra

IR: ν (in cm⁻¹) = 1380 s, 1340 s, 1325 s, 1265 m, 705 w, 525 s, 423 s, 408 s [12].

Raman: ν (in cm⁻¹) = 720 s, 706 w, 652 w, 516 s, 499 m, 452 s, 408 m, 200 s, 132 s, 83 s, 38 m, 21 m [12].

Mass Spectrum

MS (rel. int. in %): SN^+ (100), SCl^+ (79), BCl_2^+ (33), $NSCl^+$ (33), Cl^+ (33), S^+ (25), SCl_2^+ (10), BCl_3^+ (3) [12].

Chemical Reactions

$(ClS)_2N^+ BCl_4^-$ is very hygroscopic and rapidly decomposes at room temperature with incomplete evolution of BCl_3. Heating the salt in a bomb at 80 °C released stoichiometric amounts of N_2, S_2Cl_2, SCl_2, and BCl_3 [12].

A solution of $(ClS)_2N^+ BCl_4^-$ (0.05 mol) in ca. 100 mL $SbCl_5$ reacted with SbF_3 (0.1 mol) at room temperature precipitating $(ClS)_2N^+ SbCl_6^-$ with 51% yield. The reaction proceeds according to the equation [9]:

$$(ClS)_2N^+ BCl_4^- + SbCl_5 + SbF_3 \rightarrow (ClS)_2N^+ SbCl_6^- + BF_3 + SbCl_3$$

12.8.1.7 $(ClS)_2N^+ AlCl_4^-$

Preparation

$(ClS)_2N^+ AlCl_4^-$ was prepared with 72% yield by reacting $AlCl_3$ (0.246 mol) with a solution of $S_3N_3Cl_3$ (0.082 mol) in SCl_2 (100 mL) while stirring the mixture for 1 h at 20 °C. The solid was filtered off and washed with CCl_4. The title compound was purified by several melt and crystallization cycles. The salt forms according to the equation [9]:

$$S_3N_3Cl_3 + 3\ SCl_2 + 3\ AlCl_3 \rightarrow 3\ (ClS)_2N^+ AlCl_4^-$$

$(ClS)_2N^+ AlCl_4^-$ was prepared with ca. 80% yield by the above method except using the solvent $SOCl_2$. SCl_2 was added to a stirred mixture of $S_3N_3Cl_3$ and $AlCl_3$ (mole ratio 3 : 1 : 3) in $SOCl_2$. $(ClS)_2N^+ AlCl_4^-$ precipitated after stirring for ca. 30 min. The yellow product was recrystallized by heating the solution to reflux temperature (78 °C), adding extra $SOCl_2$ if necessary, and cooling slowly to −10 °C. The salt was obtained in the form of yellow needles [7].

Crystal and Molecular Structure

The salt crystallizes in the monoclinic space group $P2_1/m - C_{2h}^2$ (No. 11) with a = 12.725(3), b = 13.568(3), c = 6.200(2) Å, β = 97.33(3)°; V = 1061.7 Å³; Z = 4; D_x = 1.99 g/cm³; R = 0.029 from 2431 independent reflections [2].

The $(ClS)_2N^+$ ion has a planar cis configuration (maximal deviation from the optimal molecular plane is 0.022 Å) and C_{2v} symmetry. The tetrahedral $AlCl_4^-$ ions are slightly but noticeably distorted. Interatomic distances and bond angles are listed in Table 60 [2].

Table 60
Interatomic Distances and Bond Angles of $(ClS)_2N^+ AlCl_4^-$ (bond lengths corrected to account for the effect of liberation vibration of rigid ions) [2].

atomic distance	in Å	corr. in Å	bond angle A–B–C	in °	d(A···C) in Å
N–S(1)	1.530(2)	1.531	S(1)–N–S(2)	151.0(2)	2.959(1)
N–S(2)	1.526(2)	1.527	N–S(1)–Cl(1)	110.8(1)	2.915(2)
S(1)–Cl(1)	1.997(1)	1.998	N–S(2)–Cl(2)	110.5(1)	2.898(2)
S(2)–Cl(2)	1.987(1)	1.988	Cl(3)–Al(1)–Cl(4)	108.1(1)	
Al(1)–Cl(3)	2.128(1)	2.143	Cl(3)–Al(1)–Cl(5)	109.8(1)	
Al(1)–Cl(4)	2.141(1)	2.160	Cl(4)–Al(1)–Cl(5)	108.6(1)	

Table 60 (continued)

atomic distance	in Å	corr. in Å	bond angle A–B–C	in °	d(A···C) in Å
Al(1)–Cl(5)	2.124(1)	2.140	Cl(5)–Al(1)–Cl(5^{I})	111.9(1)	
Al(2)–Cl(6)	2.142(1)	2.153	Cl(6)–Al(2)–Cl(7)	108.7(1)	
Al(2)–Cl(7)	2.122(1)	2.131	Cl(6)–Al(2)–Cl(8)	108.3(1)	
Al(2)–Cl(8)	2.140(1)	2.152	Cl(7)–Al(2)–Cl(8)	109.0(1)	
			Cl(8)–Al(2)–Cl(8^{I})	113.4(1)	

Symmetry code: I: x, 1/2 – y, z.

Vibrational Spectra

Fundamental vibrations could be observed in the IR and Raman spectra; see Table 61.

Table 61
Frequencies of the Fundamental Vibrations of $(ClS)_2N^+$ in $(ClS)_2N^+\ AlCl_4^-$ [2, 3].

IR [7] Nujol	IR [2, 3] Nujol	Raman [2, 3] SO_2 solution	assignment$^{a)}$ [2, 3] I	II
1136 m	1124 m	1123 dp w	$\nu_{as}(NS_2)$	$\nu_{as}(NS_2)$
721 m	720 w	723 p m	$\{\nu_s(NS_2)\}$	$\{\nu_s(NS_2)\}$
658 s	659 m	663 p s	$\{\delta(SNS)\}$	$\{\delta(SNS)\}$
524 s	524 s	517 m (524 and 510)	$\nu_{as}(SCl)$	$\nu_{as}(SCl)$ + 2 $\delta_s(ClSN)$
510 s	510 s		$\nu_s(SCl)$	
	445 sh	448 p m	$\delta_s(ClSN)$	$\nu_s(SCl)$
207			τ_{as}	t_{as}

$^{a)}$ Assignment II assuming a Fermi Resonance [3]. $\{\nu_s(NS_2)\}$, $\{\delta(SNS)\}$ indicates strong coupling of both vibrations.

The vibrational frequencies (in cm^{-1}) of $AlCl_4^-$ in $(ClS)_2N^+\ AlCl_4^-$ were also measured by [3]. Raman: $\nu_1(A_1) = 350$ p, $\nu_2(E) = 140$ dp, $\nu_3(F_2) = 496$ dp, $\nu_4(F_2) = 178$ dp. IR: $\nu_3(F_2) = 490$ s.

Melting Point

The yellow solid melts at ~80 °C [9].

Mass Spectrum

The mass spectrum: m/e (rel. int. in %) = 27 (100) Al^+, 35 (77) Cl^+, 62 (60) $AlCl^+$, 81 (4) $NSCl^+$, 97 (100) $AlCl_2^+$, 102 (17), SCl_2^+, 132 (60) $AlCl_3^+$ and $S_2Cl_2^+$, 194 (2) $Al_2Cl_4^+$, 227 (50) $Al_2Cl_5^+$ [9].

Chemical Reactions

The title compound is sensitive to moisture. It decomposes when reacted with several organic compounds [9].

$(ClS)_2N^+\ AlCl_4^-$ reacted with equimolar amounts of $SnCl_2$ in CH_2Cl_2 while stirring the reaction mixture at room temperature for 24 h to give $S_2N^+\ AlCl_4^-$ with 86% yield and $SnCl_4$ [11].

12.8.1.8 $((ClS)_2N^+)_3\ [V_2Cl_9]^{3-}$

The title compound was obtained with 100% yield by reacting VCl_4 (20.8 mmol) with a solution of $S_3N_3Cl_3$ (13.8 mmol) and SCl_2 (68 mmol) in CCl_4 at reflux temperature for 2 h. The precipitate was filtered, washed with CCl_4, and dried in vacuum [13].

$((ClS)_2N^+)_3\ [V_2Cl_9]^{3-}$ was also the main product in the thermolysis of $[VCl_3(NSCl)_2]_2$, suspended in CCl_4 in a sealed tube at 105°C [13].

$((ClS)_2N^+)_3\ [V_2Cl_9]^{3-}$ is a violet, hygroscopic, crystalline powder which is almost insoluble in CCl_4 and CH_2Cl_2. The salt was identified by its IR spectrum [13].

12.8.1.9 $(ClS)_2N^+\ MCl_6^-$, M = Nb, Ta

$(ClS)_2N^+\ MCl_6^-$, where M = Nb, Ta, was prepared quantitatively by adding a solution of SCl_2 (ca. 10 mL) and $S_3N_3Cl_3$ (5.32 mmol or 4.70 mmol) in CCl_4 (30 mL) to a suspension of MCl_5 (14.81 mmol or 11.72 mmol) in 25 mL CCl_4. After stirring the mixture for 1 h, the deep yellow solid was filtered, washed with CCl_4, and dried in vacuum [14].

The title compounds were also obtained by reacting $MCl_5(N{\equiv}SCl)$ and $S_2N_2\cdot(MCl_5)_2$ with SCl_2 via the equations $MCl_5(N{\equiv}SCl) + SCl_2 \rightarrow (ClS)_2N^+\ MCl_6^-$ and $S_2N_2\cdot(MCl_5)_2 + 4\ SCl_2 \rightarrow 2\ (ClS)_2N^+\ MCl_6^- + S_2Cl_2$, M = Nb, Ta [14].

The deep yellow compounds are very sensitive to moisture. The crystal powders are slightly soluble in CCl_4 [14].

$(ClS)_2N^+\ NbCl_6^-$. IR (Nujol): ν (in cm^{-1}) = 1125 s, $\nu_{as}(NS_2)$; 720 w, $\nu_s(NS_2)$; 648 m, δ(SNS); 514 vs, ν(SCl); 492 m to s, ν(SCl); 450 vw, ν(SCl); 368 m, ν(NbCl); 330 vs, $\nu_{as}(NbCl_6)$; 278 m, γ(NSCl) [14].

$(ClS)_2N^+\ TaCl_6^-$. IR (Nujol): ν (in cm^{-1}) = 1132 m, $\nu_{as}(NS_2)$; 725 vw, $\nu_s(NS_2)$; 652 m, δ(SNS); 528 vs, ν(SCl); 503 s, ν(SCl); 385 w, ν(TaCl); 330 vs, $\nu_{as}(TaCl_6)$; 296 m, γ(NSCl) [14].

12.8.1.10 $(ClS)_2N^+\ [MCl_5(NSCl)]^-$, M = Mo, W

$(ClS)_2N^+\ [MoCl_5(NSCl)]^-$ and $(ClS)_2N^+\ [WCl_5(NSCl)]^-$ are described in "Sulfur-Nitrogen Compounds" Part 5, 1990, pp. 258/60 and p. 263, respectively.

12.8.1.11 $(ClS)_2N^+\ [Re_2Cl_9]^-$

$(ClS)_2N^+\ [Re_2Cl_9]^-$ was prepared with 90% yield by stirring a mixture of $S_3N_2Cl^+\ Cl^-$ and Re_2Cl_{10} (mole ratio 1 : 1) in CH_2Cl_2 with some drops of $POCl_3$ as solving agent for five days. The deep brown, crystalline powder was filtered, extracted with CH_2Cl_2, and dried in vacuum (black needles from CH_2Cl_2) [15].

The substance is sensitive to moisture. It is slightly soluble in CH_2Cl_2 [15].

$(ClS)_2N^+\ [Re_2Cl_9]^-$ crystallizes in the monoclinic space group C2/c – C^6_{2h} (No. 15) with the parameters a = 1197(1), b = 1288(1), c = 1144(1) pm, β = 107.83(4)°; Z = 4; R = 0.031 from 1021 independent reflections [15].

The $(ClS)_2N^+$ cations have exactly C_2 and approximately C_{2v} symmetry; the N–S bond length of 162 pm and the bond angles SNS (133.6°) and NSCl (117.6°) considerably deviate from the values of the known $(ClS)_2N^+$ structures (e.g., longer S–N bonds). The $[Re_2Cl_9]^-$ anion forms two face-sharing octahedra and has an Re–Re distance of 270 pm [15].

The molecular structures of the $(ClS)_2N^+$ and $[Re_2Cl_9]^-$ ions are displayed in **Fig. 14**. The bond lengths and bond angles are compiled in Table 62.

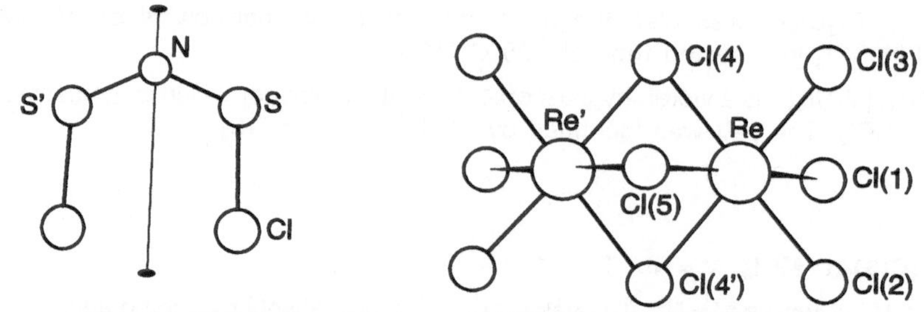

Fig. 14. Molecular structures of $(ClS)_2N^+$ and $[Re_2Cl_9]^-$ [15].

Table 62
Bond Lengths and Bond Angles of $(ClS)_2N^+$ $[Re_2Cl_9]^-$ [15].

bond length	in pm	bond angle	in °	bond angle	in °
N–S	161.5(11)	S–N–S′	133.6(17)	dihedral angle	
S–Cl	198.4(6)	S–N–Cl	117.6(9)	Cl–S–N/N–S′–Cl′	2.5
Re–Re′	270.1(1)	Cl(1)–Re–Cl(2)	91.2(1)	Cl(3)–Re–Cl(4)	87.6(1)
Re–Cl(1)	230.6(4)	Cl(1)–Re–Cl(3)	92.1(1)	Cl(3)–Re–Cl(4′)	177.9(1)
Re–Cl(2)	229.5(3)	Cl(1)–Re–Cl(4)	88.3(1)	Cl(3)–Re–Cl(5)	88.4(1)
Re–Cl(3)	228.9(3)	Cl(1)–Re–Cl(4′)	88.5(1)	Cl(4)–Re–Cl(4′)	93.3(1)
Re–Cl(4)	242.1(3)	Cl(1)–Re–Cl(5)	179.0(1)	Cl(4)–Re–Cl(5)	90.9(1)
Re–Cl(4′)	241.3(3)	Cl(2)–Re–Cl(3)	91.2(1)	Cl(4′)–Re–Cl(5)	91.1(1)
Re–Cl(5)	240.7(3)	Cl(2)–Re–Cl(4)	178.7(1)	Re–Cl(4)–Re′	68.0(1)
		Cl(2)–Re–Cl(4′)	88.0(1)	Re–Cl(5)–Re′	68.3(1)
		Cl(2)–Re–Cl(5)	89.7(1)		

IR (Nujol): ν (in cm^{-1}) = 1108 m, $\nu_{as}(NS_2)$; 720 w (covered by Nujol), $\nu_s(NS_2)$; 650 m, $\delta(SNS)$; 512 s and 494 m, $\nu_{as}(SCl)$; 448 w, $\nu_s(SCl)$; 358 sh, $\delta_{as}(NSCl)$; 343 vs, $\nu_{as}(ReCl_3)$; 325 sh, $\nu_s(ReCl_3)$; 250 s, $\nu(ReCl_3Re)$ [15].

12.8.1.12 $(ClS)_2N^+$ $FeCl_4^-$

$(ClS)_2N^+$ $FeCl_4^-$ was prepared with ca. 80% yield by adding SCl_2 to a stirred mixture of $S_3N_3Cl_3$ and $FeCl_3$ (mole ratio 3 : 1 : 3) in $SOCl_2$. After stirring for ca. 30 min, $(ClS)_2N^+$ $FeCl_4^-$ precipitated. The yellow product was recrystallized by heating the solution to reflux temperature (78 °C), adding extra $SOCl_2$ if necessary, and cooling slowly to –10 °C. The salt was obtained in the form of yellow needles [7].

The salt was also obtained by stirring a 1 : 1 mixture of S_4N_4 (1.05 g) and $FeCl_3$ (0.93 g) with 80 mL SCl_2 at room temperature for 65 h. The solid $FeCl_3$ was slowly replaced by a yellow powder which after filtration was recrystallized from $SOCl_2$ to give yellow needles; m.p. 99 to 101 °C [7].

IR: ν (in cm^{-1}) = 1130 m, $\nu_{as}(NS_2)$; 735 sh; 721 m, $\nu_s(NS_2)$; 704 sh; 645 sh, 654 s, $\delta(SNS)$; 517 s and 505 s, $\nu_{as}(SCl)$; 494 s [2, 7].

When $(ClS)_2N^+$ $FeCl_4^-$ and molar equivalents of $SnCl_2$ were stirred in liquid SO_2 for 18 h, a cream yellow precipitate of S_2N^+ $FeCl_4^-$ (72% yield) and $SnCl_4$ was produced [1].

References:

[1] Ayres, B.; Banister, A. J.; Coates, P. D.; Hansford, M. I.; Rawson, J. M.; Rickard, C. E. F.; Hursthouse, M. B.; Malik, K. M. A; Motevalli, M. (J. Chem. Soc. Dalton Trans. **1992** 3097/103).

[2] Glemser, O.; Kindler, E.; Krebs, B.; Mews, R.; Schnepel, F.-M.; Wegener, J. (Z. Naturforsch. **35b** [1980] 657/60).

[3] Schnepel, F.-M. (Spectrochim. Acta A **36** [1980] 895/8).

[4] Mews, R. (Angew. Chem. **88** [1976] 757/8; Angew. Chem. Int. Ed. Engl. **15** [1976] 691/2).

[5] Brooks, W. V. F.; MacLean, G. K.; Passmore, J.; White, P. S.; Wong, Chi-Ming (J. Chem. Soc. Dalton Trans. **1983** 1961/8).

[6] Fairhurst, S. A.; Hume-Lowe, A.; Johnson, K. M.; Sutcliffe, L. H.; Passmore, J.; Schriver, M. J. (Magn. Reson. Chem. **23** [1985] 828/31).

[7] Alange, G. G.; Banister, A. J.; Dainty, P. J. (Inorg. Nucl. Chem. Lett. **15** [1979] 175/8).

[8] Minkwitz, R.; Budde, U.; Bäck, B.; Preut, H. (Z. Naturforsch. **48b** [1993] 243/6).

[9] Glemser, O.; Wegener, J. (Inorg. Nucl. Chem. Lett. **7** [1971] 623/5).

[10] Banister, A. J.; Kendrick, A. G. (J. Chem. Soc. Dalton Trans. **1987** 1565/7).

[11] Banister, A. J.; Rawson, J. M. (J. Chem. Soc. Dalton Trans. **1990** 1517).

[12] Glemser, O.; Krebs, B.; Wegener, J.; Kindler, E. (Angew. Chem. **81** [1969] 568; Angew. Chem. Int. Ed. Engl. **8** [1969] 598).

[13] Beber, G.; Hanich, J.; Dehnicke, K. (Z. Naturforsch. **40b** [1985] 9/12).

[14] Hanich, J.; Dehnicke, K. (Z. Naturforsch. **39b** [1984] 1467/71).

[15] Hauck, H. G.; Klingelhöfer, P.; Müller, U.; Dehnicke, K. (Z. Anorg. Allg. Chem. **510** [1984] 180/8).

[16] Parsons, S.; Passmore, J.; Schriver, M.; White, P. S. (Can. J. Chem. **68** [1990] 852/62).

12.8.2 Salt of N-(Bromosulfanylidene)-amido-bromo-sulfur Ion (1+), $[Br{-}S{=}N{-}S{-}Br]^+$

12.8.2.1 $(BrS)_2N^+$ AsF_6^-

Preparation

$(BrS)_2N^+$ AsF_6^- was prepared with 99% yield by reacting of S_2N^+ AsF_6^- with excess Br_2 in SO_2 via $S_2N^+ AsF_6^- + Br_2 \rightarrow (BrS)_2N^+ AsF_6^-$. Typically, SO_2 (150 mmol) and excess Br_2 (5.34 mmol) were consecutively condensed onto frozen S_2N^+ AsF_6^- (4.88 mmol). The mixture was warmed to room temperature while shaking it, and after 1 h the solution was filtered through a sintered glass frit. SO_2 was recondensed from the solution back into the first bulb held at 0 °C. The volatiles were removed by evacuation yielding orange-red, needle-shaped crystals [1].

Molecular Properties

Based on the similarity of the vibrational spectra of $(ClS)_2N^+$ AsF_6^- (see pp. 268/9) and $(BrS)_2N^+$ AsF_6^-, both cations were concluded to have the same general cis planar conformation. A provisional $(BrS)_2N^+$ structure was constructed using the geometrical data of $(ClS)_2N^+$: r(N–S) = 1.528 Å, ∢(S–N–S) = 151°, r(S–Br) = 2.142 Å, r(Br···Br) = 3.578 Å, ∢(S–N–Br) = 112.81 Å [1].

A set of force constants for the $(BrS)_2N^+$ cation was obtained from the authors' frequencies (see Table 63, p. 280) by vibrational analysis using the Wilson F and G matrix methods and a least squares fitting procedure. The force constants (in N/m) include a Br–Br stretching constant. Diagonal constants: N–S 676, S–Br 276, Br···Br 69, S–N–S 68, N–S–Br 38; interaction constants: N–S/N–S 169, N–S/S–Br 133, N–S/S–N–S 99, N–S/N–S–Br 57 [1].

The frequencies (in cm^{-1}) of in-plane vibrations of $(BrS)_2N$ given below were observed in the IR and Raman spectra of $(BrS)_2N^+ AsF_6^-$: symmetry class A_1: $\nu_1 = 714$ (N–S), $\nu_2 = 645$ (S–N–S), $\nu_3 = 414$ (S–Br), $\nu_4 = 165$? (Br···Br); symmetry class B_1: $\nu_5 = 1126$ (N–S), $\nu_6 = 465$ (S–Br), $\nu_7 = 155$ (N–S–Br). There are insufficient data available to make the assignments definitive [1].

The frequencies of the vibrational spectra are listed in Table 63.

Table 63
Vibrational Spectra of $(BrS)_2N^+ AsF_6^-$ (frequencies in cm^{-1}) [1].

IR Nujol	Raman solid, 6328 Å	Raman SO_2 solution, 5145 Å	assignment
1126 m			$\nu_{as}(NS_2)$
1089 w			$\nu_{as}(NS_2)$
811 w			AsF_6^- ?
	716 (100)	714 (100) p	$\nu_s(NS_2)$[+ δ_s(SNS)]
698 vs			$\nu_3(AsF_6^-)$
	680 (19)		$\nu_1(AsF_6^-)$
664 mw	647 (70)	645 (18) p	δ_s(SNS)(+ SBr or SN)
570 vw	575 (4)		$\nu_2(AsF_6^-)$
461 ms	461 (2)	465 (1)	ν_{as}(SBr)(+ SN)
413 ? sh	414 (20)	414 (2)	ν_s(SBr)(+ SN)
346 vw		341 (3) dp ?	2 δ_{as}(NSBr)
393 s			$\nu_4(AsF_6^-)$
362 vw	369 (35)	365 (4) p	$\nu_5(AsF_6^-)$
		165 (6) dp ?	ν_s(Br···Br)
155 ? mw	155 (100)		δ_{as}(NSBr)

Mass Spectrum

The mass spectrum was obtained at 80 °C with the direct inlet method and showed peaks at m/e (rel. int. in %) = 162 (50), 160 (100), 158 (53), Br_2^+; 78 (1.3) S_2N^+; 151 (1.0) AsF_6^+ [1].

Chemical Reactions

Powdered $(BrS)_2N^+ AsF_6^-$ slowly decomposed to Br_2 and $S_2N^+ AsF_6^-$ when subjected to a dynamic vacuum at room temperature [1].

$(BrS)_2N^+ AsF_6^-$ (1.56 mmol) reacted with CsF (2.77 mmol) in SO_2 (57.6 mmol) while stirring the mixture for 2 days at room temperature to give a small quantity of $(O{=}S{=}N)_2S$ (other volatile products identified: SO_2, Br_2, and OSF_2, and $CsAsF_6$) [2].

References:

[1] Brooks, W. V. F.; MacLean, G. K.; Passmore, J.; White, P. S.; Wong, Chi-Ming (J. Chem. Soc. Dalton Trans. **1983** 1961/8).

[2] MacLean, G.; Passmore, J.; White, P. S.; Banister, A.; Durrant, J. A. (Can. J. Chem. **59** [1981] 187/90).

13 Sulfur Amide Hydroxide Derivatives, N,N-Substituted Amino-organyloxy-sulfanes

This class of compounds was reviewed by Fengler, G. (Houben-Weyl Methoden Org. Chem. 4th Ed. E **11** [1985] 5/31, 13/4).

13.1 N,N-Dialkyl-amino-alkoxy-sulfanes, $R^1OSNR^2R^3$, R^1, R^2, R^3 = alkyl

Preparation

The compounds compiled in Table 64, p. 282, were prepared by the following methods:

Method I: Reaction of $ClSNR^2R^3$ with R^1OM (M = H, Na, and $Sn(C_4H_9\text{-}n)_3$)

a. For $C_2H_5OSN(CH_2)_5$:

A solution of $ClSN(CH_2)_5$ and pyridine (mole ratio 1 : 1) in petroleum ether is added dropwise to a solution of equimolar amounts of C_2H_5OH in petroleum ether cooled to 0 °C and stirred vigorously. The mixture is stored for 24 h at room temperature. Then, pyridine hydrochloride is filtered off, and the filtrate is washed with water and dried over $CaCl_2$. The solvent is removed at 25 Torr, and the residue distilled [1].

b. For $i\text{-}C_3H_7OSN(CH_3)CH_2C_6H_5$:

A solution of $ClSN(CH_3)CH_2C_6H_5$ in dry THF is added dropwise to a solution of $NaOC_3H_7\text{-i}$ in THF (mole ratio 1 : 1) and allowed to react at room temperature for 3 h. The solvent is removed in vacuum and replaced with ether. After washing the solution with water, 10% aqueous $NaHCO_3$, and saturated aqueous NaCl, it is dried over anhydrous $MgSO_4$. The solvent is then removed in vacuum, and the residue distilled under reduced pressure [2].

c. For $CH_3OSN(CH_2)_4O$. Other compounds prepared analogously:

$ClSN(CH_2)_4O$ (1 mol) is added dropwise to $(n\text{-}C_4H_9)_3SnOCH_3$ (1.1 mol) under cooling. The mixture is stirred and warmed for 1 h, and subsequent vacuum distillation gives $CH_3OSN(CH_2)_4O$ with good yield [3].

Method II: Reaction of $R^2R^3NSNR^2R^3$ with R^1OH and $CuCl_2$ (or HCl):

a. A solution of $R^2R^3NSNR^2R^3$ in an inert solvent (CCl_4 for compounds with $R^2R^3 = (CH_2)_4O$ and $(C_2H_5)_2$, $(C_2H_5)_2O$ for compounds with $R^2R^3 = (CH_2)_5$) is added to a suspension of $CuCl_2$ and R^1OH (mole ratio 1 : 1 : 1) in the same solvent. The mixture is then stirred for 24 h at room temperature. After filtering the $CuCl_4$-amine complex, the solution is transferred to a column of basic alumina which is then eluted with CCl_4. The solvent is removed under vacuum, and the crude products fractionated at reduced pressure [4, 5].

b. The equivalent quantity of dry HCl is added to a solution of equimolar quantities of $R^2R^3NSNR^2R^3$ and R^1OH in n-pentane. The solution is stirred for about 15 min. After filtering the amine hydrochloride, the solvent is removed under vacuum. The products can be purified by fractional distillation under reduced pressure [4].

Table 64
Methods of Preparation, Yields, Boiling Points, and Refractive Indices of N,N-Dialkyl-amino-alkoxy-sulfanes, $R^1OSNR^2R^3$.

No.	R^1	R^2R^3	method of preparation (yield in %)	b.p. in °C/Torr	n_D (t in °C)	Ref.
1	CH_3	$(CH_3)_2$	Ic (85)	35/32	1.4390 (24)	[3]
			IIb	64/20	1.442 (25)	[6]
2	C_2H_5	$(CH_3)_2$	Ic (80)	32/20	1.4409 (18)	[3]
3	n-C_3H_7	$(CH_3)_2$	Ic (86)	32/3.5	1.4440 (17)	[3]
4	n-C_4H_9	$(CH_3)_2$	Ic (88)	46/4	1.4445 (23)	[3]
5	CH_3	$(C_2H_5)_2$	Ic (65)	30/4.5	1.4479 (21)	[3]
			IIa (10), IIb (67)	21/0.05	1.451 (25)	[4]
6	C_2H_5	$(C_2H_5)_2$	IIa (38), IIb (69)	27/0.05	1.449 (25)	[4]
7	i-C_3H_7	$(C_2H_5)_2$	IIa (25), IIb (66)	25/0.05	1.449 (25)	[4]
8	CH_3	$(C_3H_7\text{-n})_2$	Ic (92)	36/1	1.4490 (24)	[3]
9	CH_3	$(CH_2)_5$	IIa (32), IIb (66)	35/0.1	1.489 (25)	[4]
10	C_2H_5	$(CH_2)_5$	IIa (40), IIb (74)	40/0.08	1.480 (25)	[4]
			Ia (65)	43/0.4	1.4811 (20)	[1]
11	i-C_3H_7	$(CH_2)_5$	IIa (30), IIb (40)	45/0.09	1.473 (25)	[4]
12	CH_3	$(CH_2)_4O$	Ic (84)	42/0.8	1.4861 (24)	[3]
			IIa (27), IIb (50)	25/0.2	1.486 (25)	[4]
13	C_2H_5	$(CH_2)_4O$	IIa (23), IIb (56)	42/0.2	1.481 (25)	[4]
			IIa (25)	42/0.2		[5]
14	i-C_3H_7	$(CH_2)_4O$	IIa (40), IIb (60)	90/0.2	1.475 (25)	[4]
15	i-C_3H_7	$C_6H_5CH_2(CH_3)$	Ib (51)	60 to 61/0.02		[2]

^{1}H NMR Spectra

$CH_3OSN(CH_3)_2$. ^{1}H NMR (CCl_4/TMS): δ (in ppm) = 3.07 [7], 3.06 [6] (s, $N(CH_3)_2$); 3.68 [7], 3.71 [6] (s, OCH_3).

n-$C_3H_7OSN(CH_2)_5$. ^{1}H NMR (CCl_4/TMS): δ (in ppm) = 1.55 (broad, $(CH_2)_3$), 3.36 (broad, CH_2NCH_2), 0.85 (t, J = 7 Hz, $OCH_2CH_2\underline{CH_3}$), 1.56 (m, J = 7 Hz, $OCH_2\underline{CH_2}CH_3$), 3.85 (t, J = 7 Hz, $O\underline{CH_2}CH_2CH_3$) [7].

n-$C_3H_7OSN(CH_2)_4O$. ^{1}H NMR (CCl_4/TMS): δ (in ppm) = 3.31 (broad, CH_2NCH_2), 3.56 (broad, CH_2OCH_2), 0.36 (t, J = 7 Hz, $OCH_2CH_2\underline{CH_3}$), 1.55 (m, J = 7 Hz, $OCH_2\underline{CH_2}CH_3$), 3.75 (t, J = 7 Hz, $O\underline{CH_2}CH_2CH_3$) [7].

$R^1OSNR^2R^3$, $R^1 = CH_3$, C_2H_5, i-C_3H_7; $R^2R^3 = (C_2H_5)_2$, $(CH_2)_5$, $(CH_2)_4O$. The ^{1}H NMR spectra are given in Table 65.

i-$C_3H_7OSN(CH_3)CH_2C_6H_5$. ^{1}H NMR ($CDCl_3$/TMS): δ (in ppm) = 2.05 (d, J = 6.2 Hz, $CH(\underline{CH_3})_2$), 2.97 (s, CH_3), 4.15 (sept, J = 6.2 Hz, $\underline{CH}(CH_3)_2$), 4.29 (s, $\underline{CH_2}C_6H_5$), 7.27 (s, $CH_2C_6\underline{H_5}$) [2].

The low-temperature NMR spectra exhibit chemical shift nonequivalence (AB quartet) of the diastereotopic benzyl methylene protons in $CDCl_3$ or $C_6D_5CD_3$. The two isopropyl methyl groups are also diastereotopic and appear as two overlapping doublets. The chemical shift nonequivalence is the result of the molecular chirality which results from slow rotation about the S–N bond. The barrier to degenerate racemization, $\Delta G^* = 14.3$ kcal/mol, was determined by observing the coalescence of the benzyl methylene protons (at 12 °C in $CDCl_3$ and at 15 °C in $C_6D_5CD_3$) which is associated with degenerate racemization. The barrier for degenerate racemization of i-$C_3H_7OSN(CH_3)CH_2C_6H_5$ is insensitive to the nature of the solvent [2].

Table 65
^1H NMR Spectra of Other N,N-Dialkyl-amino-alkoxy-sulfanes, $R^1OSNR^2R^3$. Chemical shifts[a] in ppm (CCl_4/TMS) [4].

NR^2R^3	alkoxy-group (R^1O) protons				
	CH_3O	C_2H_5O		i-C_3H_7O	
	CH_3, s	CH_3, t	CH_2, q	CH_3, d	CH, sept
$N(C_2H_5)_2$	3.66	1.23	3.83	1.20	3.87
$N(CH_2)_5$	3.70	1.18	3.85	1.18	3.80
$N(CH_2)_4O$	3.75	1.21	3.90	1.20	3.70

R^1O	dialkylamino-group (NR^2R^3) protons					
	$N(C_2H_5)_2$		$N(CH_2)_5$		$N(CH_2)_4O$	
	CH_3, t	CH_2, q	$(CH_2)_3$, m	$N(CH_2)_2$, m	$N(CH_2)_2$, m	$(CH_2)_2O$, m
CH_3O	1.16	3.33	1.58	3.41	3.46	3.58
C_2H_5O	1.23	3.30	1.58	3.38	3.36	3.60
i-C_3H_7O	1.20	3.30	1.58	3.36	3.33	3.56

[a] Almost identical values of chemical shifts for $CH_3OSN(CH_2)_5$ and $CH_3OSN(CH_2)_4O$ are given by [7, 8].

IR Spectra

$CH_3OSN(CH_3)_2$. IR (thin film): ν (in cm^{-1}) = 2990 w, 2930 s, ν_{as}(C–H); 2885 m, 2820 w, overtone and combination bands; 2790 w, ν_s(C–H); 1475 w, sh, 1460 m, sh, 1447 s, 1428 w, sh, 1420 w, sh, $\delta(CH_3)$; 1253, w, 1192 m, 1135, $\rho(CH_3)$; 1032 m, ν_{as}(C–N); 997 s, ν_s(C–N); 970 vs, ν(S–N); 940 w, ν(S–O); 689 s, δ(N–S–O); 640 s, δ(S–O–C) [6].

Force constant: f(SN) = 5.45 mdyn/Å; S–N bond order = 1.73 [6].

Table 66
IR Spectra of N,N-Dialkyl-amino-alkoxy-sulfanes (thin films, ν in cm^{-1}).

compound	ν(S–N)	ν(S–O)	ν(C–O)	Ref.
$CH_3OSN(C_2H_5)_2$	930 st	920 m, sh	990 vst	[4]
$C_2H_5OSN(C_2H_5)_2$	930 m	827 st	1015 st	[4]
i-$C_3H_7OSN(C_2H_5)_2$	925 m, sh	820 st	910 st	[4]
$CH_3OSN(CH_2)_5$	948 st	920 w	992 st	[4]
$C_2H_5OSN(CH_2)_5$	943 st	870 st	1012 st	[4]
	860	885	1025	[1]
i-$C_3H_7OSN(CH_2)_5$	951 st	826 st	915 st	[4]
$CH_3O(CH_2)_4O$	952 st	920 w	990 st	[4]
$C_2H_5O(CH_2)_4O$[a]	952 st	875 st	1080 st	[4]
i-$C_3H_7O(CH_2)_4O$	956 st	825 st	910 st	[4]

[a] The IR spectrum was given without assignments in the range 2970 to 400 cm^{-1} [5].

Chemical Reactions

HCl. The reaction of $C_2H_5OSN(CH_2)_5$ in petroleum ether solution with dry HCl gas at 0 °C over a period of 30 min gave $(CH_2)_5NSCl$, $(C_2H_5O)_2S$, and $(CH_2)_5NH_2^+\ Cl^-$. The reaction supposedly proceeds according to the following equations [1]:

$$(CH_2)_5N{-}S{-}OC_2H_5 + 2\ HCl \rightarrow [C_2H_5OSCl] + (CH_2)_5NH_2^+\ Cl^-$$
$$[C_2H_5OSCl] + (CH_2)_5N{-}S{-}OC_2H_5 \rightarrow C_2H_5O{-}S{-}OC_2H_5 + (CH_2)_5NSCl$$

C_2H_5SH. n-$C_3H_7OSN(CH_3)_2$ reacted with excess C_2H_5SH upon warming over a period of 3 h to give $(C_2H_5)_2S_2$ as the major product and $(C_2H_5)_2S_3$ in small yield. $(C_2H_5)_2S_2$ formed from $(C_2H_5)_2S_3$ in the presence of amine [3].

$(C_2H_5)_2NH$. A mixture of $CH_3OSN(C_2H_5)_2$ (1 mol) and $(C_2H_5)_2NH$ (<1 mol) heated for 2 h left a residue and CH_3OH which was pumped off. The residue was subsequently heated with excess $(C_2H_5)_2NH$ and vacuum-distilled to give $((C_2H_5)_2N)_2S$ [3].

$CF_3SO_2OCH_3$. $CH_3OSN(CH_2)_5$ or $CH_3OSN(CH_2)_4O$ in CD_3NO_2 reacted with $CF_3SO_2OCH_3$ (mole ratio 1:1) at −20 °C to give the intermediary sulfonium salts $[CH_3O(CH_3)SN(CH_2)_5]^+$ $CF_3SO_3^-$ and $[CH_3O(CH_3)SN(CH_2)_4O]^+$ $CF_3SO_3^-$. After adding a solution of equimolar amounts of NaI in $(CH_3)_2CO$ and allowing the resulting mixtures to stand at room temperature for 2 h, $(CH_2)_5NS(O)CH_3$ and $O(CH_2)_4NS(O)CH_3$ could be isolated after working up of the respective reaction mixtures. When no NaI was added, the methylation products decomposed within 30 min at room temperature [8].

Attempts to methylate $CH_3OSN(CH_3)_2$ with an equimolar amount of $CF_3SO_2OCH_3$ in CD_3NO_2 and CD_3CN at −40 °C were unsuccessful. The reagents immediately decomposed even at low temperature [8].

$C_6H_5SO_2NClNa$. CH_3OSNR_2 ($R_2 = (C_2H_5)_2$, $(CH_2)_5$, $(CH_2)_4O$) reacted with equimolar amounts of $C_6H_5SO_2NClNa$ in acetone at reflux temperature while stirring over a period of 2 h to give $C_6H_5SO_2N{=}S(OCH_3)NR_2$. The products, $C_6H_5SO_2N{=}S(OCH_3)NR_2$ ($R_2 = (C_2H_5)_2$, $(CH_2)_4O$), were purified by column chromatography. $C_6H_5SO_2N{=}S(OCH_3)N(CH_2)_5$ was recrystallized from ether [9].

$(C_2H_5O)_2P(S)SH$. $C_2H_5OSN(CH_2)_5$ reacted with $(C_2H_5O)_2P(S)SH$ (mole ratio 1:3) in petroleum ether–diethyl ether (1:10) as follows:

$$C_2H_5OSN(CH_2)_5 + 3\ (C_2H_5O)_2P(S)SH \rightarrow (C_2H_5O)_2P(S){-}S{-}S{-}S{-}P(S)(OC_2H_5)_2 + C_2H_5OH + (CH_2)_5NH_2^+\ (C_2H_5O)_2P(S)S^-$$

When the analogous reaction was performed in the mole ratio 1:2, a portion of C_2H_5-$OSN(CH_2)_5$ remained unreacted [1].

$(CH_3)_3SiCl$. $CH_3OSN(CH_2)_4O$ reacted with $(CH_3)_3SiCl$ (mole ratio 1:1) upon dropwise addition while cooling the mixture to give $(CH_3)_3SiOCH_3$ and $ClSN(CH_2)_4O$ [3].

CH_3I. $CH_3OSN(CH_2)_5$ and n-$C_3H_7OSN(CH_2)_5$ as well as $CH_3OSN(CH_2)_4O$ and n-C_3H_7OSN-$(CH_2)_4O$ were reacted with equimolar amounts of CH_3I in the dark at 75 to 80 °C over a period of 15 to 18 h to give $CH_3S(O)N(CH_2)_5$ and $CH_3S(O)N(CH_2)_4O$, respectively [7].

The analogous reaction of $CH_3OSN(CH_3)_2$ with CH_3I gave $(CH_3)_4N^+\ I^-$, S_8, and sulfur-containing polymers [7].

C_2H_5I. The reactions of $CH_3OSN(CH_2)_5$ or $CH_3OSN(CH_2)_4O$ with C_2H_5I (mole ratio 1:1) led to the formation of $CH_3S(O)N(CH_2)_5$ and $CH_3S(O)N(CH_2)_4O$ rather than to the ethyl derivative, even after varying the mole ratios. CH_3I, generated in the reactions, is more reactive than C_2H_5I [7].

The analogous reaction of $CH_3OSN(CH_3)_2$ with C_2H_5I gave $(CH_3)_2(C_2H_5)_2N^+ I^-$, S_8, and sulfur-containing polymers [7].

CH_2=$CHCH_2Br$. $C_6H_5CH_2Br$. The reactions of $CH_3OSN(CH_3)_2$ with CH_2=$CHCH_2Br$ and $C_6H_5CH_2Br$ did not lead to the expected N,N-dimethyl sulfinic amides but to $(CH_3)_2R_2N^+ Br^-$ (R = CH_2=$CHCH_2$ or $C_6H_5CH_2$), S_8, and sulfur-containing polymers [7].

$CH_3C(O)Cl$. $C_6H_5C(O)Cl$. The reactions of $CH_3OSN(C_2H_5)_2$, $CH_3OSN(CH_2)_5$, and CH_3OSN-$(CH_2)_4O$ with $CH_3C(O)Cl$ or $C_6H_5C(O)Cl$ (mole ratio 1 : 1) at 20 to 25 °C gave good yields of $ClSNR_2$ ($R_2 = (C_2H_5)_2$, $(CH_2)_5$, $(CH_2)_4O$) and $R'C(O)OCH_3$ (R' = CH_3, C_6H_5) [10].

C_6H_5NCS. $CH_3OSN(C_2H_5)_2$ or $CH_3OSN(CH_2)_5$ was reacted with C_6H_5NCS (mole ratio 1 : 1) at 90 °C over a period of 200 h to give C_6H_5N=$C(OCH_3)SSN(C_2H_5)_2$ with 49% yield and C_6H_5-N=$C(OCH_3)SSN(CH_2)_5$ with 42% yield [11].

$[Cu(H_2O)_6]^{2+} (ClO_4^-)_2$. $C_2H_5OSN(CH_2)_4O$ reacted with $[Cu(H_2O)_6]^{2+} (ClO_4^-)_2$ (mole ratio 1 : 2) in CH_3CN solution at ambient temperature to give $[Cu(CH_3CN)_4]^+ ClO_4^-$ with 60% yield and SO_2 [12].

References:

[1] Almasi, L.; Hantz, A. (Chem. Ber. **99** [1966] 3288/92).
[2] Raban, M.; Noyd, D. A.; Bermann, L. (J. Org. Chem. **40** [1975] 752/5).
[3] Armitage, D. A.; Towle, I. D. H. (Phosphorus Sulfur Relat. Elem. **1** [1976] 37/9).
[4] Diaz, C.; Copaja, S.; González, G. (Phosphorus Sulfur Relat. Elem. **22** [1985] 317/21).
[5] González, G.; Díaz, C.; Copaja, S. (Monatsh. Chem. **114** [1983] 177/83).
[6] Diaz, C. M. (Acta Sud Am. Quim. **3** [1983] 15/20).
[7] Wenschuh, E.; Kühne, U.; Mikolajczyk, M.; Bujnicki, B. (Z. Chem. **21** [1981] 217/8).
[8] Mikolajczyk, M.; Kielbasinski, P.; Bujnicki, B.; Wenschuh, E.; Kühne, U. (Phosphorus Sulfur Relat. Elem. **25** [1985] 85/9).
[9] Musin, B. M.; Yudina, P. V.; Ivanov, V. B. (Izv. Akad. Nauk SSSR Ser. Khim. **1990** 2150/1; Bull. Acad. Sci. USSR Div. Chem. Sci. [Engl. Transl.] **1990** 1959/60).
[10] Musin, B. M.; Ivanov, V. B.; Ivanov, B. E. (Izv. Akad. Nauk SSSR Ser. Khim. **1988** 1693; Bull. Acad. Sci. USSR Div. Chem. Sci. [Engl. Transl.] **1988** 1509).
[11] Musin, B. M.; Yudina, P. V.; Ivanov, V. B.; Nekhoroshkov, V. M.; Efremov, Y. Y. (Izv. Akad. Nauk SSSR Ser. Khim. **1990** 1450; Bull. Acad. Sci. USSR Div. Chem. Sci. [Engl. Transl.] **1990** 1313).
[12] Diaz, C. (Polyhedron **4** [1985] 1269/70).

13.2 N-(2,3-Dichlorophenyl)-N-trifluoromethyl-amino-organyloxy-sulfanes, $R^1OSN(CF_3)C_6H_3Cl_2$-2,3, $R^1 = CH_3$, 4-$NO_2C_6H_4$

The title compounds were prepared by reacting $ClSN(CF_3)C_6H_3Cl_2$-2,3 with equimolar amounts of $NaOCH_3$ [1, 2] and $NaOC_6H_4NO_2$-4 [3], respectively, in inert solvents.

4-$NO_2C_6H_4OSN(CF_3)C_6H_3Cl_2$-2,3. 58% yield, m.p. 108 °C [3].

$CH_3OSN(CF_3)C_6H_3Cl_2$-2,3 showed insecticidal activity [1, 2].

References:

[1] Kühle, E.; Klauke, E.; Farbenfabriken Bayer AG (Belg. 640471 [1962/64]; C.A. **63** [1965] 6920).

[2] Kühle, E.; Klauke, E.; Farbenfabriken Bayer AG (Fr. 1382390 [1964] 4 pp.; C.A. **62** [1965] 16118).

[3] Kühle, E. (The Chemistry of the Sulfenic Acids, 1st Ed., Thieme Verlag, Stuttgart 1973, 163 pp., p. 79).

13.3 N-(4-Chloro-6-ethylamino-1,3,5-triazin-2-yl)-N-ethyl-amino-methoxy-sulfane

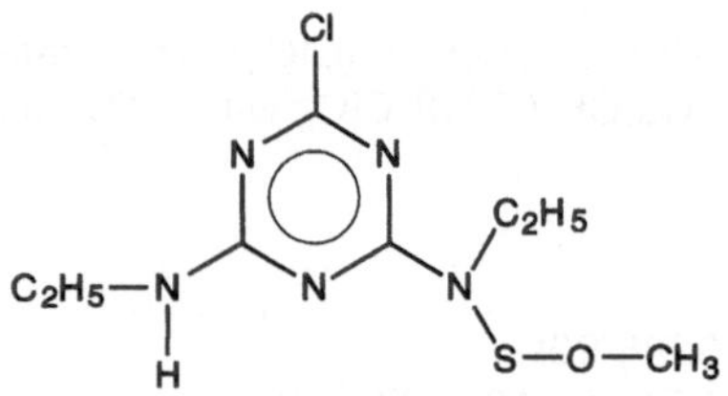

The title compound was obtained by dropwise adding a solution of CH_3OH and pyridine in CH_2Cl_2 to a solution of an equimolar amount of 2-(N-chlorothio-N-ethylamino)-4-ethylamino-6-chloro-1,3,5-triazine in CH_2Cl_2. Ten minutes later the reaction mixture was washed with water and aqueous $NaHCO_3$, dried over $MgSO_4$, and evaporated under reduced pressure giving a brown oil. Chromatography over silica gel with C_6H_6 and C_6H_6–$(C_2H_5)_2O$ (95 : 5) as eluents yielded the title compound as a white solid; m.p. 95 to 97.5°C [1 to 3].

References:

[1] Cleveland, J. D. (U.S. 3796712 [1972/74] 3 pp.; C.A. **80** [1974] No. 133486).

[2] Cleveland, J. D. (U.S. 3864342 [1972/75] 5 pp.; C.A. **82** [1975] No. 171090).

[3] Cleveland, J. D. (U.S. 3909237 [1972/75] 6 pp.; C.A. **83** [1975] No. 206335).

13.4 N-Benzyloxysulfanyl-2-acetyloxy-4-oxo-azetidine, $C_6H_5CH_2OS\overline{N{-}C(O){-}CH_2{-}C}HO(O)CCH_3$

The title compound was prepared with 64% yield by reacting Phth–S$\overline{N{-}C(O){-}CH_2{-}C}$-HO(O)$CCH_3$ (Phth = phthalimide) with $C_6H_5CH_2OH$ (mole ratio 1 : 5) in C_6H_6 in the presence of catalytic amounts of $(C_2H_5)_3N$ at room temperature over a period of 24 h. An oil was separated by thin-layer chromatography on silica gel with ethyl acetate–hexane (1 : 1) as eluent (R_f = 0.76).

^{1}H NMR ($CDCl_3$/TMS): δ (in ppm) = 2.10 (s, 3H), 3.10 (dd, 1H), 3.45 (dd, 1H), 4.97 (s, 2H), 6.17 (dd, 1H), 7.38 (s, 5H).

IR (neat): ν = 1805, 1760 cm^{-1}.

Reference:

Iwagami, H.; Woulfe, S. R.; Miller, M. J. (Tetrahedron Lett. **27** [1986] 3095/8).

13.5 N-Organyloxycarbonyl-N-methyl-amino-alkoxy-sulfanes

13.5.1 $R^1OSN(CH_3)C(O)OC_6H_3$–(2-O–$C(CH_3)_2$–CH_2-3), R^1 = alkyl (I). $R^1OSN(CH_3)C(O)OC_6H_4(OC_3H_7$-i)-2, R^1 = alkyl. $R^1OSN(CH_3)C(O)OC_6H_4(C_3H_7$-i)-3, R^1 = alkyl

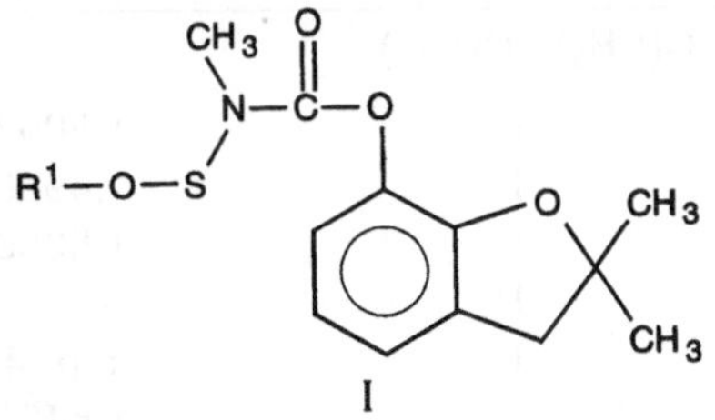

I

The compounds compiled in Table 67 were prepared by two methods:

Method I: A solution of equimolar amounts of R^1OH and a slight excess of $(C_2H_5)_3N$ in CH_2Cl_2 is added dropwise to a solution of $ClSN(CH_3)C(O)OC_6H_3$–(2-O–$C(CH_3)_2$–CH_2-3), $ClSN(CH_3)C(O)OC_6H_4(OC_3H_7$-i)-2, or $ClSN(CH_3)C(O)OC_6H_4(C_3H_7$-i)-3 in CH_2Cl_2 while stirring at 0 °C. After stirring for another 3 h, the mixture is allowed to warm to room temperature, washed three times with water, dried over Na_2SO_4, concentrated, and filtered. Evaporation of the filtrate yields a viscous oily substance. The crude substance containing the polysulfide analogs is purified by preparative thin-layer chromatography on silica gel with CH_3CN as eluent yielding 35 to 50%. The starting compounds are obtained by reacting $CH_3NHC(O)OC_6H_3$(2-O–$C(CH_3)_2$–CH_2-3), $CH_3NHC(O)OC_6H_4(OC_3H_7$-i)-2, or $CH_3NHC(O)OC_6H_4(C_3H_7$-i)-3 with equimolar amounts of SCl_2 in the presence of a slight excess of $(C_2H_5)_3N$ in CH_2Cl_2 solution at −5 to 5 °C and subsequently at room temperature. The resulting mixture is used for the following reactions with alcohols [1].

Method II: Solvolysis of $(n\text{-}C_4H_9)_2NSN(CH_3)C(O)OC_6H_3$(2-O–$C(CH_3)_2$–$CH_2$-3) in R^1OH–CH_3CO_2H (9:1) mixtures at 23 °C for 24 h gives $R^1OSN(CH_3)C(O)OC_6H_3$(2-O–C-$(CH_3)_2$–CH_2-3) and polysulfanes. Purification is performed by preparative thin-layer chromatography on silica gel and hexane–ether as eluent [1, 2].

Table 67
Methods of Preparation and Refractive Indices of $R^1OSN(CH_3)C(O)OC_6H_3$(2-O–$C(CH_3)_2$–CH_2-3), $R^1OSN(CH_3)C(O)OC_6H_4(OC_3H_7$-i)-2, and $R^1OSN(CH_3)C(O)OC_6H_4(C_3H_7$-i)-3.
Further information on compounds marked with an asterisk is given at the end of the table.

No.	R^1	method of preparation	n_D^{23} or m.p.	Ref.
$R^1OSN(CH_3)C(O)OC_6H_3$(2-O–$C(CH_3)_2$–CH_2-3)				
*1	CH_3	I, II	1.5342	[1, 2]
	CD_3	II		[2]
*2	C_2H_5	I, II	1.5250	[1, 2]
*3	i-C_3H_7	I, II	1.5402	[1]
4	n-C_4H_9	I	—	[1]
5	t-C_4H_9	I	1.5190	[1]
6	$(C_2H_5)_2CH$	I	1.5152	[1]
7	c-C_6H_{11}	I, II	1.5112	[1]
*8	n-C_6H_{13}	I, II	1.5164	[1]

Table 67 (continued)

No.	R^1	method of preparation	n_D^{23} or m.p.	Ref.
$R^1OSN(CH_3)C(O)OC_6H_3(2\text{-}O\text{–}C(CH_3)_2\text{–}CH_2\text{-}3)$				
9	n-C_8H_{17}	I	1.5093	[1]
10	n-$C_{10}H_{21}$	I	1.4988	[1]
11	n-$C_{12}H_{25}$	I	1.5030	[1]
12	n-$C_{18}H_{37}$	I	—	[1]
13	$C_6H_5CH_2$	I	m.p. 42 to 45°C	[1]
14	4-(t-C_4H_9)$C_6H_4CH_2$	I	1.5491	[1]
15	4-$CH_3OC_6H_4CH_2$	I	1.5648	[1]
$R^1OSN(CH_3)C(O)OC_6H_4(OC_3H_7\text{-i})\text{-}2$				
16	n-C_4H_9	I	1.5103	[1]
17	$C_6H_5CH_2$	I	1.5552	[1]
$R^1OSN(CH_3)C(O)OC_6H_4(C_3H_7\text{-i})\text{-}3$				
18	n-C_4H_9	I	1.5128	[1]
19	$C_6H_5CH_2$	I	1.5572	[1]

*Further information:

$CH_3OSN(CH_3)C(O)OC_6H_3(2\text{-}O\text{–}C(CH_3)_2\text{–}CH_2\text{-}3)$ (Table **67**, No. **1**). The compound was also formed when HCl was used as acid instead of CH_3COOH; however, a number of polysulfide analogs were also formed [2].

^{1}H NMR ($CDCl_3$/TMS): δ (in ppm) = 1.50 (s, gem-di-CH_3), 3.02 (s, CH_2), 3.55 (s, NCH_3), 3.95 (s, OCH_3), 6.70 to 7.10 (m, 3H arom.) [2].

EIMS (70 eV): m/e (rel. int. in %) = 284 (2.2), 283 (9.8) M^+, 252 (1.5), 226 (5.7), 208 (1.5), 195 (2.6), 167 (5.9), 164 (13.2), 163 (100), 145 (7.5), 135 (39.2), 120 (14.2), 117 (10.3), 107 (26.3), 91 (21.6), 77 (13.2), 63 (33.1) [2].

CIMS in CH_4: m/e = 324 $[M+41]^+$, 312 $[M+29]^+$, 284 $[M+1]^+$ [2].

$C_2H_5OSN(CH_3)C(O)OC_6H_3(2\text{-}O\text{–}C(CH_3)_2\text{–}CH_2\text{-}3)$ (Table **67**, No. **2**). The compound was also obtained from $ClSN(CH_3)C(O)OC_6H_3(2\text{-}O\text{–}C(CH_3)_2\text{–}CH_2\text{-}3)$ (synthesized from carbofuran and SCl_2) and sodium ethoxide [1]. ^{1}H NMR ($CDCl_3$/TMS): δ (in ppm) = 1.33 (t, $OCH_2\underline{CH_3}$), 1.50 (s, gem-di-CH_3), 3.02 (s, CH_2), 3.58 (s, NCH_3), 4.20 (q, $O\underline{CH_2}CH_3$), 6.70 to 7.10 (m, 3H arom.) [2].

i-$C_3H_7OSN(CH_3)C(O)OC_6H_3(2\text{-}O\text{–}C(CH_3)_2\text{–}CH_2\text{-}3)$ (Table **67**, No. **3**). ^{1}H NMR ($CDCl_3$/TMS): δ (in ppm) = 1.25 (d, J = 6.1 Hz, $OCH(\underline{CH_3})_2$), 1.42 (s, gem-di-CH_3), 3.02 (s, br, CH_2), 3.53 (s, NCH_3), 4.40 (sept, J = 6.1 Hz, $OC\underline{H}(CH_3)_2$), 6.7 to 7.1 (m, 3H arom.) [1].

EIMS (70 eV): m/e (rel. int. in %) = 311 (4.1), 269 (16.3), 195 (8.0), 191 (2.3), 164 (100) [1].

n-$C_6H_{13}OSN(CH_3)C(O)OC_6H_3(2\text{-}O\text{–}C(CH_3)_2\text{–}CH_2\text{-}3)$ (Table **67**, No. **8**). ^{1}H NMR ($CDCl_3$/TMS): δ (in ppm) = 0.90 (t, br, $O(CH_2)_5\underline{CH_3}$), 1.0 to 1.8 (m, $OCH_2(\underline{CH_2})_4CH_3$), 1.48 (s, gem-di-$CH_3$), 3.03 (s, CH_2), 3.57 (s, NCH_3), 4.10 (t, br, $O\underline{CH_2}(CH_2)_4$), 6.7 to 7.1 (m, 3H arom.) [1].

EIMS (70 eV): m/e (rel. int. in %) = 353 (5.6) M^+, 269 (11.3), 195 (6.9), 191 (3.1), 164 (100) [1].

The title compounds showed insecticidal activity [1, 2]. Results of biological tests are also presented in a patent [3].

References:

[1] Kawata, M.; Umetsu, N.; Goto, T.; Fukuto, T. R. (J. Pesticide Sci. **13** [1988] 595/603).
[2] Umetsu, N.; Nishioka, T.; Fukuto, T. R. (J. Agric. Food Chem. **32** [1984] 765/8).
[3] University of California, Berkeley (Jpn. 58-67665 [1982/83] 18 pp.; C.A. **99** [1983] No. 70552).

13.5.2 $C_2H_5OSN(CH_3)C(O)ON{=}C(CH_3)SCH_3$

The title compound was prepared by reacting methomyl, $CH_3SC(CH_3){=}NOC(O)NHCH_3$, with SCl_2 and adding ethanol in the presence of triethylamine (see Chapter 13.5.1, Method I, p. 287). The yield was 5 to 10%. The compound was not stable; it completely decomposed within 3 days at 23 °C.

^1H NMR ($CDCl_3$/TMS): δ (in ppm) = 1.23 (t, $OCH_2C\underline{H}_3$, J = 7 Hz), 2.30 (s, =C–$C\underline{H}_3$), 2.40 (s, =C–S–$C\underline{H}_3$), 3.51(s, $NC\underline{H}_3$), 4.09 (q, J = 7 Hz, $OC\underline{H}_2CH_3$).

Reference:

Kawata, M.; Umetsu, N.; Goto, T.; Fukuto, T. R. (J. Pesticide Sci. **13** [1988] 595/603).

13.6 N-Alkoxysulfanyl-N′-aryl-N-methyl-ureas, $R^1OSN(CH_3)C(O)NHR^2$, R^1 = alkyl, R^2 = aryl

The preparation of the title compounds (see Table 68) is identical with that of n-C_3H_7-$OSN(CH_3)C(O)NHC_6H_3Cl_2$-3,4. The other compounds were prepared by a similar procedure.

A solution of 0.1 mol n-C_3H_7OH and 0.12 mol of pyridine in 20 mL of CH_2Cl_2 was added dropwise to a solution of about 0.1 mol of $ClSN(CH_3)C(O)NHC_6H_3Cl_2$-3,4 in CH_2Cl_2 at 0 °C. After completing the addition, the resulting reaction mixture was stirred in an ice bath for about 10 min, washed with water and $NaHCO_3$ solution, dried over $MgSO_4$, and evaporated under reduced pressure giving a dark oil residue. The oil residue was purified by column chromatography on silica gel with hexane–chloroform as eluent to give the product as a white, low-melting solid with good yield [1 to 4].

The melting points of the title compounds are given in Table 68.

Table 68
Melting Points of N-Alkoxysulfanyl-N′-aryl-N-methyl-ureas, $R^1O–S–N(CH_3)–C(O)–NHR^2$ [1 to 4].

R^1	R^2	m.p. in °C	R^1	R^2	m.p. in °C
CH_3	C_6H_4F-2	oil	CH_3	$C_6H_3Cl_2$-3,4	72 to 77
C_2H_5	C_6H_4F-2	oil	C_2H_5	$C_6H_3Cl_2$-3,4	60 to 61
n-C_3H_7	C_6H_4F-2	oil	n-C_3H_7	$C_6H_3Cl_2$-3,4	<25
i-C_3H_7	C_6H_4F-2	oil	i-C_3H_7	$C_6H_3Cl_2$-3,4	oil
c-C_6H_{11}	C_6H_4F-2	69 to 72	n-C_8H_{17}	$C_6H_3Cl_2$-3,4	oil
n-C_8H_{17}	C_6H_4F-2	oil	norbornyl	$C_6H_3Cl_2$-3,4	oil
norbornyl	C_6H_4F-2	oil			

References:

[1] Brown, M. S. (U.S. 3853966 [1970/74] 11 pp.; C.A. **82** [1975] No. 139799).
[2] Cleveland, J. D. (U.S. 3857883 [1972/74] 11 pp.; C.A. **82** [1975] No. 155429).
[3] Brown, M. S.; Kohn, G. K. (U.S. 3928407 [1970/75] 7 pp.; C.A. **84** [1976] No. 135344).
[4] Brown, M. S.; Kohn, G. K. (U.S. 3997324 [1970/76] 7 pp.; C.A. **86** [1977] No. 139655).

13.7 N-(2-Chloroethoxysulfanyl)-N′-chloro-N-cyano-chloroformamidine, $ClCH_2CH_2OSN(CN)C(Cl)=NCl$

The title compound was prepared by treating a solution of $ClSN(CN)C(Cl)=NCl$ in CCl_4 with an excess of $\overline{CH_2CH_2O}$ at temperatures below 20 °C. The filtrate was evaporated under reduced pressure at 25 to 30 °C. The resulting crude product was crystallized from CCl_4 with 77% yield.

Solid, m.p. 54 to 55.5 °C.

^1H NMR (CCl_4/TMS): δ (in ppm) = 3.55 to 3.77 (m, CH_2), 4.02 to 4.23 (m, CH_2).

MS: m/e = 247 M^+, 198 [$M^+ - CH_2Cl$], 168 [$M^+ - OCH_2CH_2Cl$], 151 [$M^+ - C(Cl)=NCl$], 136 [$M^+ - SOCH_2CH_2Cl$], 125 [151 − CN]. In addition, the expected ^{37}Cl-isotope peaks were observed.

The title compound reacted in refluxing methanol to the O-methylisobiuret hydrochloride.

Reference:

Geevers, J.; Hackmann, J. T.; Trompen, W. P. (J. Chem. Soc. C **1970** 875/8).

13.8 N-(Diethoxyphosphoryl)-N-ethyl-amino-alkoxy(or alkenyloxy)-sulfanes, $R^1OSN(C_2H_5)P(O)(OC_2H_5)_2$, $R^1 = C_2H_5$, i-C_3H_7, i-C_4H_9, $ClCH_2CH_2$, $BrCH_2CH_2$, $(C_2H_5)_2NCH_2CH_2$, $CH_2=CHCH_2$

The title compounds were prepared by adding an equimolar amount of $ClSN(C_2H_5)P(O)(OC_2H_5)_2$ to an ethereal solution of ROH and $(C_2H_5)_3N$ at −5 to 0 °C and stirring the mixture at room temperature for 2 h. The products were separated by distillation [1, 2]. Yields, boiling points, densities, and refractive indices are listed in Table 69.

Table 69
Yields, Boiling Points, Densities, and Refractive Indices of $R^1OSN(C_2H_5)P(O)(OC_2H_5)_2$.

R^1	yield in %	b.p. in °C/Torr	D_4^{20} in g/cm³	n_D^{20}	Ref.
C_2H_5	24.2	85 to 86/1	1.1072	1.4494	[1]
i-C_3H_7	39.1	90 to 91/1	1.079	1.4469	[1]
i-C_4H_9	30.0	98 to 99/1	1.0701	1.4490	[1]
$ClCH_2CH_2$	50.0	105/1	1.1914	1.4670	[2]
$BrCH_2CH_2$	35.0	126 to 128/2	1.3413	1.4789	[2]
$(C_2H_5)_2NCH_2CH_2$	29.0	115 to 116/1.5	1.0678	1.4600	[2]
$CH_2=CHCH_2$	33.0	116 to 117/2	1.1524	1.7640	[2]

The title compounds are almost colorless liquids with a weak ethereal odor [1]. They are stable during distillation and do not change when stored; they dissolve readily in organic solvents, but do not dissolve in water [2].

References:

[1] Alimov, P. I.; Antokhina, L. A. (Izv. Akad. Nauk SSR Ser. Khim. **1963** 1132/4; Bull. Acad. Sci. USSR Div. Chem. Sci. **1963** 1034/6).

[2] Alimov, P. I.; Antokhina, L. A. (Izv. Akad. Nauk SSR Ser. Khim. **1964** 1316/7; Bull. Acad. Sci. USSR Div. Chem. Sci. **1964** 1220/1).

13.9 N-Methoxysulfanyl-organylideneamines

13.9.1 $CH_3OSN{=}C(C_6H_5)CO_2CH_3$

The title compound was obtained with 29% yield by reacting $CH_3O_2CCH(C_6H_5)NHSO_2NH$-$CH(C_6H_5)CO_2CH_3$ (I) first with $SOCl_2$–$(C_2H_5)_3N$ and then with CH_3OH–$(C_2H_5)_3N$ as follows: To a stirred solution of the sulfamide I (400 mg) and $SOCl_2$ (300 mg) in CH_2Cl_2 (8 mL) was added dropwise a solution of $(C_2H_5)_3N$ (600 mg) in CH_2Cl_2 (2 mL) under ice cooling. The reaction mixture was kept at 0 to 5 °C for 1 h and cooled to −78 °C. Then a solution of $(C_2H_5)_3N$ (600 mg) in CH_3OH (3 mL) was added, and the mixture was stirred at −78 °C for 15 min and gradually warmed to room temperature. The solution was poured into a mixture of $CH_3CO_2C_2H_5$ and saturated aqueous $NaHCO_3$. The organic layer was washed with saturated aqueous NaCl and dried over $MgSO_4$. After evaporating the solvents, the resulting residue was separated by thin-layer chromatography on silica gel with C_6H_6–$CH_3CO_2C_2H_5$ (20 : 1) to yield the title compound (133 mg, 29%) in addition to $CH_3O_2C(C_6H_5){=}NSN{=}C(C_6H_5)CO_2CH_3$ (65 mg, 18%), CH_3O_2CH-$(C_6H_5)NHS(O)_2OCH_3$ (116 mg, 22%), and $C_6H_5C(O)C(O)OCH_3$ (14 mg, 4%).

Pale yellow crystals from n-hexane, m.p. 52 to 53 °C.

^1H NMR ($CDCl_3$/TMS): δ (in ppm) = 3.80 (3H, s), 3.88 (3H, s), 7.1 to 7.8 (5H, m).

IR (Nujol): ν (in cm^{-1}) = 1680, 1595, 1580.

UV (CH_3CN): λ_{max} (in nm) (ε in $L \cdot mol^{-1} \cdot cm^{-1}$) = 236 (10040), 339 (12320).

MS: m/e = 225 M^+.

Reference:

Saito, T.; Hiraoka, T. (Chem. Pharm. Bull. **25** [1977] 792/9).

13.9.2 $CH_3OSN{=}\overline{C{-}C(O){-}N(CH_3){-}C(O){-}N}(C_6H_4F\text{-}2)$

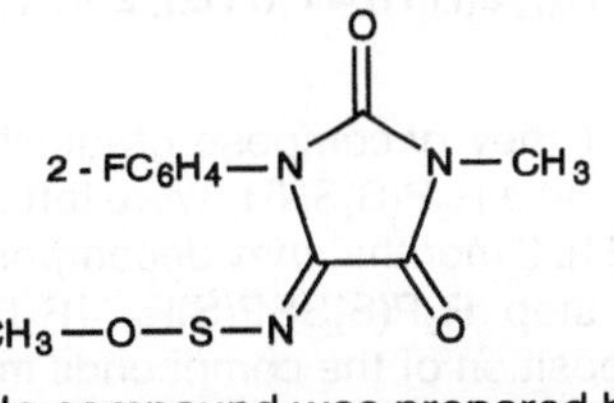

The title compound was prepared by dropwise adding a solution of CH_3OH and pyridine in CH_2Cl_2 to a solution of 1-(2-fluorophenyl)-3-methyl-5-chlorothioimino-2,4-imidazolidinedione in CH_2Cl_2 at 0 °C (mole ratio ~1 : 1 : 1). After stirring for about 30 min at 0 °C, the mixture was washed with H_2O and $NaHCO_3$, dried over $MgSO_4$, and evaporated under reduced pressure to give a yellow solid which was purified by washing with isopropyl alcohol; m.p. 124 to 127 °C [1, 2].

References:

[1] Cleveland, J. D. (U.S. 3843677 [1972/74] 5 pp.; C.A. **82** [1975] No. 43424).
[2] Cleveland, J. D. (U.S. 3596308 [1972/76] 5 pp.; C.A. **85** [1976] No. 63070).

14 S-Phosphorus-Substituted Thiohydroxylamines and Derivatives

14.1 S-Phosphorus-Substituted Thiohydroxylamines

14.1.1 S-Phosphoryl-, Thiophosphinoyl-, and Thiophosphoryl-thiohydroxylamines, $R_2P(X)SNH_2$, X = O, S; R = R′, R′O; R′ = alkyl, aryl

The compounds listed in Table 70 were prepared by passing dry NH_3 for 45 min (No. 3) or 1 h (Nos. 1, 2, 4 to 6) through a solution of the corresponding phosphorus-containing disulfides, $R_2P(X)SSP(X)R_2$, in dry C_6H_6 (Nos. 1 to 5) or in dry ether (No. 6). Liquid products (Nos. 1 to 3) were separated and purified by vacuum distillation and solid products (Nos. 5, 6) by recrystallization from hexane (compound No. 4 decomposed during distillation) [1]. The compound with R = 4-ClC_6H_4, X = S formed analogously, but was not isolated [2].

Table 70
Yields and Properties of $R_2P(X)SNH_2$ [1].

No.	R	X	yield (in %)	b.p. (in °C/Torr) m.p. (in °C)	n_D^{20}	D_4^{20} (in g/cm³)	MR_D exp.	calc.
1	CH_3O	S	95	70 to 71/0.1	1.5600	1.3463	41.60	41.77
2	C_2H_5O	S	97	76 to 77/0.1	1.5340	1.2264	51.00	51.01
3	i-C_3H_7O	S	92	85/0.3	1.5120	1.1420	60.26	60.24
4	i-C_3H_7O	O	60	dec.	1.4580	1.1285	51.50	51.81
5	C_6H_5O	S	93	74 to 75	—	—	—	—
6	C_6H_5	S	98	59 to 60	—	—	—	—

The compounds show an IR absorption band in the region around 3400 cm^{-1}, which was assigned to $\nu(NH_2)$ [1].

^{1}H NMR of $(C_2H_5O)_2P(S)SNH_2$: δ (in ppm) = 1.03 (t, CH_3, $^3J(H,H)$ = 7.0 Hz), 2.47 (s, br, NH_2), 3.87 (m, OCH_2, $^3J(P,H)$ = 7.0 Hz, $^3J(H,H)$ = 7.0 Hz) [1].

Under cooling the compounds are fairly stable, but they decompose gradually at room temperature. When the compounds with the general formula $R_2P(S)SNH_2$ were left in a closed vessel in an atmosphere of Ar at room temperature for 2 to 3 months, they decomposed to give $R_2P(S)NH_2$ and S (desulfurization) and in a second step $R_2P(S)SSP(S)R_2$, $[R_2PS_2]^- NH_4^+$, N_2, and H_2. Based on the reaction products the decomposition of the compounds may involve the formation and reaction of $R_2P(S)S^\bullet$- and $H_2N^\bullet$-type radicals [1].

$(C_2H_5O)_2P(S)SNH_2$ reacted with excess C_2H_5ONa in ether to give $(C_2H_5O)_3P(S)$, C_2H_5OH, Na_2S, NH_3, and N_2 [3], with C_6H_5NCS in C_6H_6 to give $(C_2H_5O)_2P(S)SNHC(S)NHC_6H_5$ [5].

$(R'O)_2P(S)SNH_2$ (R′ = C_2H_5, i-C_3H_7, C_4H_9, C_6H_5) reacted with $(CH_3C(O))_2O$, or (R′ = C_2H_5, i-C_3H_7) with $(C_6H_5C(O))_2O$ or maleic anhydride to give the corresponding N-acyl-substituted compounds $(R'O)_2P(S)SNHC(O)R''$ (R″ = CH_3, C_6H_5, CH=CHCOOH) [4].

$R_2P(S)SNH_2$ (R = C_2H_5O, i-C_3H_7O, C_4H_9O, C_6H_5 [5], 4-ClC_6H_4O [2]) reacted with isocyanates, R′NCO (R′ = C_2H_5 [5], C_6H_4Cl-4, $C_6H_4CH_3$-4 [2, 5], $C_6H_3Cl_2$-3,4 [5]) to give the corresponding N-ethyl- and N-arylureas, $R_2P(S)SNHC(O)NHR'$.

References:

[1] Torgasheva, N. A.; Khaskin, B. A.; Mel'nikov, N. N.; Kosminskaya, G. A. (Zh. Obshch. Khim. **46** [1976] 1467/72; J. Gen. Chem. USSR [Engl. Transl.] **46** [1976] 1440/5).
[2] Khaskin, B. A.; Torgasheva, N. A.; Sheluchenko, O. D. (Zh. Obshch. Khim. **58** [1988] 1778/84; J. Gen. Chem. USSR [Engl. Transl.] **58** [1988] 1586/91).
[3] Khaskin, B. A.; Torgasheva, N. A.; Mel'nikov, N. N. (Zh. Obshch. Khim. **46** [1976] 1472/5; J. Gen. Chem. USSR [Engl. Transl.] **46** [1976] 1445/6).
[4] Khaskin, B. A.; Torgasheva, N. A.; Mel'nikov, N. N. (Zh. Obshch. Khim. **49** [1979] 1250/2; J. Gen. Chem. USSR [Engl. Transl.] **49** [1979] 1096/8).
[5] Khaskin, B. A.; Torgasheva, N. A.; Mel'nikov, N. N. (Zh. Obshch. Khim. **47** [1977] 2176/8; J. Gen. Chem. USSR [Engl. Transl.] **47** [1977] 1987/9).

14.2 S-Phosphorus-N-Substituted Thiohydroxylamines

14.2.1 S-Phosphoryl-, Thiophosphinoyl-, and Thiophosphoryl-N-organyl-thiohydroxylamines

14.2.1.1 S-Phosphoryl-, Thiophosphinoyl-, and Thiophosphoryl-N-alkyl(or aryl)-thiohydroxylamines, $R'_2P(X)SNHR$, X = O, S; R = alkyl, aryl; R' = R", R"O

Thiophosphinoyl-N-alkyl-thiohydroxylamines of the general formula $(R'O)_2P(S)SNHR$, where R, R' = alkyl, were claimed in patents to improve the vulcanization of rubber; see, for example [1 to 3].

References:

[1] Campbell, R. H.; Wise, R. W. (U.S. 4065443 [1976/77] 6 pp.; C.A. **88** [1978] No. 106550).
[2] Morita, E. (U.S. 4077924 [1976/78] 5 pp.; C.A. **89** [1978] No. 216627).
[3] Foster, W. R.; Williams, D. J. (Br. 1505147 [1975/78] 13 pp.; C.A. **90** [1979] No. 40017).

14.2.1.1.1 $(CH_3O)_2P(S)SNHR$, R = C_2H_5, i-C_3H_7

The title compounds were prepared with yields of 84% (R = C_2H_5) and 86% (R = i-C_3H_7) by treating a solution of $(CH_3O)_2P(S)SSP(S)(OCH_3)_2$ in ether with ethereal solutions of $C_2H_5NH_2$ or i-$C_3H_7NH_2$ (mole ratio 1 : 2) at room temperature. The ammonium salts were filtered off, the filtrates evaporated in vacuum, and the residues distilled [1].

$(CH_3O)_2P(S)SNHC_2H_5$. Liquid, b.p. 77 °C/0.2 Torr, n_D^{20} = 1.2177, D_4^{20} = 1.5310 g/cm³, $MR_{D\,exp.}$ = 51.13, $MR_{D\,calc.}$ = 51.19 [1].

$(CH_3O)_2P(S)SNHC_3H_7$-i. Liquid, b.p. 82 °C/0.1 Torr, n_D^{20} = 1.1744, D_4^{20} = 1.52 g/cm³, $MR_{D\,exp.}$ = 55.72, $MR_{D\,calc.}$ = 55.80 [1].

The compound was reacted with $(C_2H_5)_3N$ (mole ratio 1 : 2) in ether at room temperature for 1 d to give $[(C_2H_5)_3NCH_3]^+$ $[CH_3OP(O)(S)SCH_3]^-$ along with $CH_3OP(O)(SNHC_3H_7\text{-i})_2$ [2].

References:

[1] Khaskin, B. A.; Mel'nikov, N. N.; Torgasheva, N. A. (Zh. Obshch. Khim. **43** [1973] 1916/8; J. Gen. Chem. USSR [Engl. Transl.] **43** [1973] 1901/3).
[2] Khaskin, B. A.; Torgasheva, N. A.; Mel'nikov, N. N. (Zh. Obshch. Khim. **46** [1976] 1472/5; J. Gen. Chem. USSR [Engl. Transl.] **46** [1976] 1445/6).

14.2.1.1.2 $(C_2H_5O)_2P(X)SNHR$, X = O, S; R = alkyl, substituted alkyl

The title compounds were prepared by treating solutions of $(C_2H_5O)_2P(X)SSP(X)(OC_2H_5)_2$ (X = O or S) in ether with ethereal solutions of the corresponding alkylamines or substituted alkylamines, RNH_2 (mole ratio 1 : 2), at room temperature. After a few minutes the solid by-products were filtered off, the solvent was evaporated, and the compounds were separated by distillation or by recrystallization from hexane or C_6H_6 [1, 2, 5]. $(C_2H_5O)_2P(S)SNHC_6H_{11}$-c was prepared by treating a solution of c-$C_6H_{11}NHCl$ in $(CH_3)_2C(O)-H_2O$ with an aqueous solution of an equimolar amount of $(C_2H_5O)_2P(S)SNa$ [3, 4]. $(C_2H_5O)_2P(S)SNH(CH_2)_2NHC(O)NHC_6H_4Cl$-4 was prepared by the reaction of $(C_2H_5O)_2P(S)SNH(CH_2)_2NH_2$ with 4-ClC_6H_4NCO [5]. Yields and properties of the compounds are summarized in Table 71.

Table 71
Yields and Properties of $(C_2H_5O)_2P(X)SNHR$.

R	X	yield (in %)	b.p. (in °C/Torr) m.p. (in °C)	n_D^{20}	D_4^{20} (in g/cm³)	MR_D exp.	calc.	Ref.
C_2H_5	S	93	100/0.1	1.5108	1.1227	61.15	60.76	[1]
n-C_3H_7	S	95	120/0.17	1.5150	0.8593	85.40	85.38	[1]
i-C_3H_7	O	80	108 to 110/0.3	1.4670	1.1081	56.91	56.61	[2]
i-C_3H_7	S	98	145/0.1	1.5030	1.0951	65.68	65.38	[1]
n-C_4H_9	O	98	dec. on distillation	1.4650	1.0954	60.89	61.22	[2]
n-C_4H_9	S	95	53 to 55	—	—	—	—	[1]
s-C_4H_9	O	74	95 to 96/0.1	1.4660	1.0934	61.13	61.22	[2]
t-C_4H_9	S	99	33 to 34 (from aqueous C_2H_5OH)	—	—	—	—	[1]
i-C_5H_{11}	O	72	dec. on distillation	1.4560	1.0546	65.80	65.84	[2]
c-C_6H_{11}	S	80	oil	—	—	—	—	[3, 4]
$CH(C_2H_5)CH_2OH$	S	quant.	—	1.5225	1.1668	71.50	71.19	[5]
$C(CH_3)_2CH_2OH$	S	quant.	110/0.1	1.5245	—	—	—	[5]
$(CH_2)_2NHC(O)NHC_6H_4Cl$-4	S	quant.	103 to 104	—	—	—	—	[5]

¹H NMR spectra were recorded for $(C_2H_5O)_2P(O)SNHC_3H_7$-i (δ (in ppm) = 1.15 and 3.20 (m, i-C_3H_7), 1.40 and 4.31 (m, OC_2H_5), 3.80 (s, NH)) and for $(C_2H_5O)_2P(O)SNHC_4H_9$ (δ (in ppm) = 1.45 (m, CH_3 of C_2H_5 and C_4H_9), 4.50 (m, OCH_2), 3.10 (s, NH)) [2].

The title compounds are colorless, distillable liquids, insoluble in water and readily soluble in most organic solvents [5].

Mixtures of $(C_2H_5O)_2P(S)SNHCH_3$ and acetic or maleic anhydride reacted to give $(C_2H_5O)_2P(S)SNCH_3C(O)R$ (R = CH_3, CH=CHCOOH) [6].

References:

[1] Khaskin, B. A.; Mel'nikov, N. N.; Torgasheva, N. A. (Zh. Obshch. Khim. **43** [1973] 1916/8; J. Gen. Chem. USSR [Engl. Transl.] **43** [1973] 1901/3).

[2] Mel'nikov, N. N.; Khaskin, B. A.; Torgasheva, N. A. (Zh. Obshch. Khim. **45** [1975] 1005/8; J. Gen. Chem. USSR [Engl. Transl.] **45** [1975] 992/4).

[3] Malz, H.; Bayer, O.; Freytag, H.; Lober, F. (U.S. 2891059 [1957/59] 3 pp.; C.A. **1960** 4387).

[4] Malz, H.; Lober, F.; Bayer, O.; Scheurlen, H. (U.S. 3044981 [1959/62] 3 pp.; C.A. **58** [1963] 14228).

[5] Mel'nikov, N. N.; Torgasheva, N. A.; Khaskin, B. A. (Zh. Obshch. Khim. **46** [1976] 43/7; J. Gen. Chem. USSR [Engl. Transl.] **46** [1976] 43/6).
[6] Khaskin, B. A.; Torgasheva, N. A.; Mel'nikov, N. N. (Zh. Obshch. Khim. **49** [1979] 1250/2; J. Gen. Chem. USSR [Engl. Transl.] **49** [1979] 1096/8).

14.2.1.1.3 **$(i\text{-}C_3H_7O)_2P(X)SNHR$**, X = O, S; R = alkyl, substituted alkyl

Most of the compounds listed in Table 72 were prepared by treating solutions of $(i\text{-}C_3H_7O)_2P(X)SSP(X)(OC_3H_7\text{-}i)_2$ (X = O or S) in ether with ethereal solutions of the corresponding alkylamine or substituted alkylamine, RNH_2 (mole ratio 1 : 2), at room temperature. After evaporating the solvent in vacuum, the compounds were isolated by distillation [1 to 3]. Compounds of the type $(i\text{-}C_3H_7O)_2P(S)SNH(CH_2)_nNH_2$ (n = 2, 3) were formed from $(i\text{-}C_3H_7O)_2$-$P(S)SSP(S)(OC_3H_7\text{-}i)_2$ with $H_2N(CH_2)_nNH_2$ (n = 2, 3) (mole ratio 1 : 2) at low temperatures [3]. Compounds of the type $(i\text{-}C_3H_7O)_2P(S)SNH(CH_2)_nNHC(O)CH_3$ (n = 2, 3) were prepared by reacting $(i\text{-}C_3H_7O)_2P(S)SNH(CH_2)_nNH_2$ with acetic anhydride, $(CH_3C(O))_2O$, in ether. After removing the by-product CH_3COOH and the solvent, the compounds were recrystallized from hexane [4]. Compounds of the type $(i\text{-}C_3H_7O)_2P(S)SNH(CH_2)_nNHC(O)NHC_6H_4Cl\text{-}4$ (n = 2, 3) were obtained by the reaction of $(i\text{-}C_3H_7O)_2P(S)SNH(CH_2)_nNH_2$ with $4\text{-}ClC_6H_4NCO$ [3].

Table 72
Yields and Properties of $(i\text{-}C_3H_7O)_2P(X)SNHR$.

R	X	yield (in %)	b.p. (in °C/Torr) m.p. (in °C)	n_D^{20}	D_4^{20} (in g/cm³)	MR_D exp.	calc.	Ref.
C_2H_5	O	82	dec. on distillation	1.4600	1.0794	61.22	61.23	[1]
C_2H_5	S	92	63.5/0.1	1.0682	1.4940	70.13	69.66	[2]
$n\text{-}C_3H_7$	S	93	84.5/0.1	1.0544	1.4920	74.66	74.28	[2]
$i\text{-}C_3H_7$	O	64	110 to 112/0.3	1.4535	1.0434	66.19	65.76	[1]
$i\text{-}C_3H_7$	S	99	75.5/0.08	1.0497	1.4880	74.48	74.28	[2]
$n\text{-}C_4H_9$	O	78	dec. on distillation	1.4602	1.0362	70.98	70.46	[1]
$n\text{-}C_4H_9$	S	95	93/0.2	1.0407	1.4910	79.42	78.90	[2]
$i\text{-}C_4H_9$	S	97	77/0.08	1.0431	1.4900	79.10	78.90	[2]
$c\text{-}C_6H_{11}$	S	50	125/0.2	1.1013	1.5110	84.45	84.83	[2]
$CH(C_2H_5)CH_2OH$	S	quant.	120/0.5	1.5075	1.1086	80.97	80.42	[3]
$CH_2C_6H_3Cl_2\text{-}2,4$	S	50	—	1.2910	1.5700	98.69	98.64	[2]
$(CH_2)_2NH_2$	S	86	31 to 32/760	1.5210	1.1503	72.10	72.54	[3]
$(CH_2)_3NH_2$	S	93	—	1.5140	1.1211	76.92	77.16	[3]
$(CH_2)_2NHC(O)CH_3$	S	86	48 to 49	—	—	—	—	[4]
$(CH_2)_3NHC(O)CH_3$	S	95	49 to 50	—	—	—	—	[4]
$(CH_2)_2NHC(O)NHC_6H_4Cl\text{-}4$	S	quant.	100 to 101	—	—	—	—	[3]
$(CH_2)_3NHC(O)NHC_6H_4Cl\text{-}4$	S	quant.	99 to 100	—	—	—	—	[3]

$(i\text{-}C_3H_7O)_2P(S)SNHCH_3$ was reacted with $(C_2H_5O)_2P(S)Cl$ and NaOH (40% aqueous solution) at 70 °C for 3 h to give $(i\text{-}C_3H_7O)_2P(S)SSP(S)(OC_3H_7\text{-}i)_2$ and $(C_2H_5O)_2P(S)NHCH_3$. It was reacted with Na in ether for 2 h to give $(i\text{-}C_3H_7O)_2P(S)NHCH_3$ and Na_2S; the reaction with NaH proceeds analogously [5]. $(i\text{-}C_3H_7O)_2P(X)SNHR^1$ (X = O or S; $R^1 = CH_3$ or C_2H_5) reacted with R^2NCO ($R^2 = C_2H_5$, $4\text{-}ClC_6H_4$, $3,4\text{-}Cl_2C_6H_3$) in C_6H_6 in a sealed tube to give the corresponding $(i\text{-}C_3H_7O)_2P(X)SNR^1C(O)NHR^2$ [6].

$(i-C_3H_7O)_2P(S)SNH(CH_2)_nNH_2$ (n = 2, 3) reacted with $(CH_3C(O))_2O$ and $4-ClC_6H_4NCO$ in hexane to give $(i-C_3H_7O)_2P(S)SNH(CH_2)_nNHC(O)CH_3$ and $(i-C_3H_7O)_2P(S)SNH(CH_2)_nNHC(O)$-$NHC_6H_4Cl$-4, respectively [3].

References:

[1] Mel'nikov, N. N.; Khaskin, B. A.; Torgasheva, N. A. (Zh. Obshch. Khim. **45** [1975] 1005/8; J. Gen. Chem. USSR [Engl. Transl.] **45** [1975] 992/4).
[2] Khaskin, B. A.; Mel'nikov, N. N.; Torgasheva, N. A. (Zh. Obshch. Khim. **43** [1973] 1916/8; J. Gen. Chem. USSR [Engl. Transl.] **43** [1973] 1901/3).
[3] Mel'nikov, N. N.; Torgasheva, N. A.; Khaskin, B. A. (Zh. Obshch. Khim. **46** [1976] 43/7; J. Gen. Chem. USSR [Engl. Transl.] **46** [1976] 43/6).
[4] Torgasheva, N. A.; Khaskin, B. A.; Mel'nikov, N. N. (Zh. Obshch. Khim. **47** [1977] 1977/8; J. Gen. Chem. USSR [Engl. Transl.] **47** [1977] 1808/9).
[5] Khaskin, B. A.; Torgasheva, N. A.; Mel'nikov, N. N. (Zh. Obshch. Khim. **46** [1976] 1472/5; J. Gen. Chem. USSR [Engl. Transl.] **46** [1976] 1445/6).
[6] Khaskin, B. A.; Torgasheva, N. A.; Mel'nikov, N. N. (Zh. Obshch. Khim. **47** [1977] 2176/8; J. Gen. Chem. USSR [Engl. Transl.] **47** [1977] 1987/9).

14.2.1.1.4 $(n-C_4H_9O)_2P(S)SNHCH_3$

The preparation of the title compound was not described.

It reacts with acetic anhydride, $(CH_3C(O))_2O$, to give $(n-C_4H_9O)_2P(S)SNCH_3C(O)CH_3$.

Reference:

Khaskin, B. A.; Torgasheva, N. A.; Mel'nikov, N. N. (Zh. Obshch. Khim. **49** [1979] 1250/2; J. Gen. Chem. USSR [Engl. Transl.] **49** [1979] 1096/8).

14.2.1.1.5 $(4-R'C_6H_4O)_2P(S)SNHR$, R = alkyl, substituted alkyl; R' = H, Cl

The compounds with R' = H, R = $i-C_3H_7$ or $t-C_4H_9$ [1] and with R' = Cl, R = $i-C_3H_7$ [2] were prepared with 98% yield by treating solutions of $(4-R'C_6H_4O)_2P(S)SSP(S)(OC_6H_4R'-4)_2$ in ether with ethereal solutions of $i-C_3H_7NH_2$ and $t-C_4H_9NH_2$, respectively (mole ratio 1:2), at room temperature. After separation of the ammonium salts, $(4-R'C_6H_4O)_2P(S)S^-\ RNH_3^+$, and evaporation of the solvent in vacuum, the compounds with R = $i-C_3H_7$ were isolated by distillation [1, 2]; that with R = $t-C_4H_9$ crystallized [1]. $(4-ClC_6H_4O)_2P(S)SNHCH_3$ was obtained analogously by the reaction of $(4-ClC_6H_4O)_2P(S)SSP(S)(OC_6H_4Cl-4)_2$ with CH_3NH_2 in ether, but it was not isolated [2]. $(C_6H_5O)_2P(S)SNH(CH_2)_2NHC(O)NHC_6H_4Cl$-4 was obtained by reacting $(C_6H_5O)_2P(S)$–$SNH(CH_2)_2NH_2$ with $4-ClC_6H_4NCO$ [3].

$(C_6H_5O)_2P(S)SNHC_3H_7$-i. Liquid, $n_D^{20} = 1.2725$, $D_4^{20} = 1.5890$ g/cm³, $MR_{D\,exp.} = 89.88$, $MR_{D\,calc.} = 89.54$ [1].

$(4-ClC_6H_4O)_2P(S)SNHC_3H_7$-i. Liquid, $n_D^{20} = 1.5970$, $D_4^{20} = 1.3279$ g/cm³, $MR_{D\,exp.} = 104.76$, $MR_{D\,calc.} = 104.53$ [2].

$(C_6H_5O)_2P(S)SNHC_4H_9$-t. Crystalline solid, m.p. 56 to 57 °C [1].

$(C_6H_5O)_2P(S)SNH(CH_2)_2NHC(O)NHC_6H_4Cl$-4. Crystalline solid, m.p. 109 to 110 °C [3].

$(C_6H_5O)_2P(S)SNH(CH_2)_2NH_2$ reacted with $4\text{-}ClC_6H_4NCO$ to give $(C_6H_5O)_2P(S)SNH(CH_2)_2\text{-}NHC(O)NHC_6H_4Cl\text{-}4$ [3]. $(4\text{-}ClC_6H_4O)_2P(S)SNHCH_3$ reacted with RNCO ($R = 4\text{-}ClC_6H_4$, $3,4\text{-}Cl_2C_6H_3$) in ether to give the corresponding $(4\text{-}ClC_6H_4O)_2P(S)SNCH_3C(O)NHR$ [2].

References:

[1] Khaskin, B. A.; Mel'nikov, N. N.; Torgasheva, N. A. (Zh. Obshch. Khim. **43** [1973] 1916/8; J. Gen. Chem. USSR [Engl. Transl.] **43** [1973] 1901/3).

[2] Khaskin, B. A.; Torgasheva, N. A.; Sheluchenko, O. D. (Zh. Obshch. Khim. **58** [1988] 778/84; J. Gen. Chem. USSR [Engl. Transl.] **58** [1988] 1586/91).

[3] Mel'nikov, N. N.; Torgasheva, N. A.; Khaskin, B. A. (Zh. Obshch. Khim. **46** [1976] 43/7; J. Gen. Chem. USSR [Engl. Transl.] **46** [1976] 43/6).

14.2.1.1.6 $(4\text{-}R'C_6H_4)_2P(S)SNHR$, $R' = H, CH_3$; R = alkyl, substituted alkyl

Most of the compounds listed in Table 73 were prepared by treating solutions of $(4\text{-}R'C_6H_4)_2P(S)SSP(S)(C_6H_4R'\text{-}4)_2$ in C_6H_6 with the corresponding alkylamine or substituted alkylamine (mole ratio 1 : 2) at room temperature. After adding hexane, the ammonium salt was filtered off and the solvent evaporated in vacuum. The compounds with R' = H were recrystallized from hexane [1]. $(C_6H_5)_2P(S)SNH(CH_2)_3NHC(O)NHC_6H_4Cl\text{-}4$ was obtained by the reaction of $(C_6H_5)_2P(S)SNH(CH_2)_3NH_2$ with $2,4\text{-}Cl_2C_6H_3NCO$ [2].

Table 73
Yields and Melting Points of $(4\text{-}R'C_6H_4)_2P(S)SNHR$.

R	R'	yield (in %)	m.p. (in °C)	Ref.
C_2H_5	H	90	45 to 46	[1]
$i\text{-}C_3H_7$	H	96	45 to 46	[1]
$i\text{-}C_3H_7$	CH_3	99	a)	[1]
$n\text{-}C_4H_9$	H	98	40 to 41	[1]
$s\text{-}C_4H_9$	H	92	40	[1]
$t\text{-}C_4H_9$	CH_3	98	b)	[1]
$CH(CH_2)_5$	H	94	92 to 93	[1]
$CH(C_2H_5)CH_2OH$	H	quant.	85 to 86	[2]
$(CH_2)_3NHC(O)NHC_6H_3Cl_2\text{-}2,4$	H	quant.	160 to 161	[2]

a) Liquid, $n_D^{20} = 1.6292$, $D_4^{20} = 1.1822$ g/cm³, $MR_{D\,exp.} = 100.82$, $MR_{D\,calc.} = 100.74$ [1]. – b) Liquid, $n_D^{20} = 1.6220$, $D_4^{20} = 1.1680$ g/cm³, $MR_{D\,exp.} = 100.38$, $MR_{D\,calc.} = 100.38$ [1].

$(C_6H_5)_2P(S)SNRC(O)NHC_6H_3Cl_2\text{-}3,4$ formed, when mixtures of $(C_6H_5)_2P(S)SNHR$ ($R = CH_3$ or $CH(C_2H_5)CH_2OH$) and $3,4\text{-}Cl_2C_6H_3NCO$ were heated in C_6H_6 at 75°C for 2 h [3].

References:

[1] Torgasheva, N. A.; Khaskin, B. A.; Mel'nikov, N. N.; Kosminskaya, G. A. (Zh. Obshch. Khim. **46** [1976] 1467/72; J. Gen. Chem. USSR [Engl. Transl.] **46** [1976] 1440/5).

[2] Mel'nikov, N. N.; Torgasheva, N. A.; Khaskin, B. A. (Zh. Obshch. Khim. **46** [1976] 43/7; J. Gen. Chem. USSR [Engl. Transl.] **46** [1976] 43/6).

[3] Khaskin, B. A.; Torgasheva, N. A.; Mel'nikov, N. N. (Zh. Obshch. Khim. **47** [1977] 2176/8; J. Gen. Chem. USSR [Engl. Transl.] **47** [1977] 1987/9).

14.2.1.1.7 $(i\text{-}C_3H_7O)_2P(S)SNHC_6H_5$

The title compound was prepared with 85% yield (crude) along with $C_6H_5NH_3^+$ $(i\text{-}C_3H_7O)_2$-$P(S)S^-$ by heating a mixture of $(i\text{-}C_3H_7O)_2P(S)SSP(S)(OC_3H_7\text{-}i)_2$ and excess aniline (mole ratio ~1 : 4.5) in a sealed ampule at 90 to 95 °C for 20 h. The compound was separated from the reaction mixture by extraction with hexane.

Crystalline solid, m.p. 130 to 131 °C (from hexane).

^1H NMR (CCl_4/TMS, 25 °C): δ (in ppm) = 0.95 (d, CH_3, $^3J(H,H)$ = 7.0 Hz); 4.38 (m, CHO, $^3J(H,H)$ = $^3J(P,H)$ = 7.0 Hz); 7.4 to 7.6 (m, C_6H_5).

When the compound was boiled in H_2O, $i\text{-}C_3H_7OP(O)(OH)NHC_6H_5$ was formed.

Reference:

Khaskin, B. A.; Torgasheva, N. A.; Negrebetskii, V. V. (Zh. Obshch. Khim. **53** [1983] 1775/8; J. Gen. Chem. USSR [Engl. Transl.] **53** [1983] 1596/8).

14.2.1.2 S-Thiophosphinoyl- and Thiophosphoryl-N-acyl(carbamoyl, or thiocarbamoyl)-thiohydroxylamines

14.2.1.2.1 $(RO)_2P(S)SNHC(O)R^1$

The compounds with R = CH_3, C_2H_5 and R^1 = OC_2H_5 were obtained by adding $C_2H_5OC(O)$-NHCl in C_6H_6 to a suspension of $(C_2H_5O)_2P(S)SNa$ in C_6H_6 [1]; see also [2]. The compounds with R^1 = CH_3 and C_6H_5 were prepared by heating mixtures of $(RO)_2P(S)SNH_2$ (R = C_2H_5, $i\text{-}C_3H_7$, $n\text{-}C_4H_9$, C_6H_5) and $(CH_3C(O))_2O$ or $(C_6H_5C(O))_2O$ (mole ratio ca. 1 : 2) for a few minutes in a boiling water bath. After extraction with ether, the solvent was removed leaving colorless oils which crystallized slowly at low temperatures. The compounds with R^1 = CH=CHCOOH were obtained by mixing equimolar amounts of $(RO)_2P(S)SNH_2$ (R = C_2H_5, $i\text{-}C_3H_7$) and maleic anhydride at room temperature for 30 min followed by addition of C_6H_6 [3]. Yields and melting points of the compounds are summarized in Table 74.

Table 74
Yields and Melting Points of $(RO)_2P(S)SNHC(O)R^1$.

R	R^1	yield (in %)	m.p. (in °C) (recrystallized from)	Ref.
CH_3	OC_2H_5	—	—	[1]
C_2H_5	OC_2H_5	—	71 to 72 (light benzene)	[1]
C_2H_5	OC_2H_5	—	73	[2]
C_2H_5	CH_3	95.1	94 to 95 (hexane)	[3]
C_2H_5	C_6H_5	94.4	118 to 119 (hexane)	[3]
C_2H_5	CH=CHCOOH	72.1	112.5 to 113 ($CHCl_3$)	[3]
$i\text{-}C_3H_7$	CH_3	92.3	39 to 40 (hexane)	[3]
$i\text{-}C_3H_7$	C_6H_5	86.4	92 to 93 (hexane)	[3]
$i\text{-}C_3H_7$	CH=CHCOOH	75.1	124 to 125 ($CHCl_3$)	[3]
$n\text{-}C_4H_9$	CH_3	80.2	35 to 36 (hexane)	[3]
C_6H_5	CH_3	86.7	92 to 93 (hexane)	[3]

References:

[1] Malz, H.; Bayer, O.; Wegler, R. (U.S. 2995568 [1959/61] 5 pp.; C.A. **57** [1962] 11021).

[2] Malz, H.; Lober, F.; Bayer, O.; Scheurlen, H. (U.S. 3044981 [1959/62] 3 pp.; C.A. **58** [1963] 14228).

[3] Khaskin, B. A.; Torgasheva, N. A.; Mel'nikov, N. N. (Zh. Obshch. Khim. **49** [1979] 1250/2; J. Gen. Chem. USSR [Engl. Transl.] **49** [1979] 1096/8).

14.2.1.2.2 $R_2P(S)SNHC(O)NHR^1$, R = OR′, aryl; R′ = alkyl, aryl; R^1 = alkyl, aryl

The dialkoxy- and diphenyl-substituted compounds were prepared by adding R^1NCO ($R^1 = C_2H_5$, 4-ClC_6H_4, 4-$CH_3C_6H_4$, 3,4-$Cl_2C_6H_3$) to solutions of the equimolar amount of $R_2P(S)SNH_2$ (R = C_2H_5O, i-C_3H_7O, n-C_4H_9O, C_6H_5) in C_6H_6 at 20°C. The compounds were recrystallized from C_2H_5OH [1]. The compounds with R = 4-ClC_6H_4O and $R^1 = C_6H_4Cl$-4, $C_6H_4CH_3$-4, $C_6H_3Cl_2$-3,4 were prepared by passing NH_3 through a solution of (4-$ClC_6H_4O)_2P$-(S)SSP(S)(OC_6H_4Cl-4$)_2$ in ether for 15 min. R^1NCO was added to the solution and the mixture left for a few hours. After removing the solvent the compounds were recrystallized from CCl_4 [2]. Yields and melting points of the compounds are summarized in Table 75.

Table 75
Yields and Melting Points of $R_2P(S)SNHC(O)NHR^1$.

R	R^1	yield (in %)	m.p. (in °C)	Ref.
C_2H_5O	C_6H_4Cl-4	92	175 to 176	[1]
C_2H_5O	$C_6H_3Cl_2$-3,4	93	190 to 191	[1]
i-C_3H_7O	C_2H_5	87	116 to 118	[1]
i-C_3H_7O	C_6H_4Cl-4	95	123 to 124	[1]
i-C_3H_7O	$C_6H_3Cl_2$-3,4	97	159 to 160	[1]
n-C_4H_9O	C_6H_4Cl-4	98	122 to 123	[1]
n-C_4H_9O	$C_6H_3Cl_2$-3,4	95	131 to 132	[1]
4-ClC_6H_4O	C_6H_4Cl-4	—	166 to 167	[2]
4-ClC_6H_4O	$C_6H_4CH_3$-4	—	148 to 149	[2]
4-ClC_6H_4O	$C_6H_3Cl_2$-3,4	—	156 to 157	[2]
C_6H_5	C_6H_4Cl-4	85	160 to 161	[1]
C_6H_5	$C_6H_4CH_3$-4	99	164 to 165	[1]
C_6H_5	$C_6H_3Cl_2$-3,4	99	209 to 210	[1]

Reference:

[1] Khaskin, B. A.; Torgasheva, N. A.; Mel'nikov, N. N. (Zh. Obshch. Khim. **47** [1977] 2176/8; J. Gen. Chem. USSR [Engl. Transl.] **47** [1977] 1987/9).

[2] Khaskin, B. A.; Torgasheva, N. A.; Sheluchenko, O. D. (Zh. Obshch. Khim. **58** [1988] 1778/84; J. Gen. Chem. USSR [Engl. Transl.] **58** [1988] 1586/91).

14.2.1.2.3 $(C_2H_5O)_2P(S)SNHC(S)NHC_6H_5$

The title compound was obtained with 33% yield by treating a solution of $(C_2H_5O)_2P(S)SNH_2$ in C_6H_6 with C_6H_5NCS and heating the mixture at 50 to 60°C for 2 h.

Crystalline solid, m.p. 128 to 130°C (from hexane).

Reference:

Khaskin, B. A.; Torgasheva, N. A.; Mel'nikov, N. N. (Zh. Obshch. Khim. **47** [1977] 2176/8; J. Gen. Chem. USSR [Engl. Transl.] **47** [1977] 1987/9).

14.3 S-Phosphorus-N,N-Substituted Thiohydroxylamines

14.3.1 S-Aminophosphanyl-N,N-dialkyl-thiohydroxylamine Derivatives

14.3.1.1 $[((CH_3)_2N)_3PSN(CH_3)_2]^+ [(C_6H_5)_2P(S)S]^-$

The salt was obtained by treating a mixture of equimolar amounts of $(C_6H_5)_2P(S)S_2N(CH_3)_2$ and $(C_6H_5)_2P(S)N(CH_3)_2$ with $((CH_3)_2N)_3P$ in C_6H_6 at room temperature. The salt was identified by ^{31}P NMR.

^{31}P NMR (C_6H_6/external 85% aqueous H_3PO_4): $\delta = -59.9$ and -61.4 ppm.

Reference:

Fluck, E.; Gonzalez, G.; Binder, H. (Z. Anorg. Allg. Chem. **406** [1974] 161/6).

14.3.1.2 $i\text{-}C_4H_9((C_2H_5)_2N)PF_2SN(C_2H_5)_2$

The title compound formed as a by-product in the fluorination reaction of $i\text{-}C_4H_9P(N(C_2H_5)_2)_2$ with SF_4 in ether at $-80\,°C$. Threefold distillation in a high vacuum yielded the compound with 2.4% yield.

Colorless liquid, b.p. 59 to 63 °C/10^{-4} Torr.

^1H NMR (C_6D_6/TMS): δ (in ppm) = ~1.6 (concealed, $(C\underline{H}_3)_2CHC\underline{H}_2$), 1.66 (t, $(C\underline{H}_3CH_2)_2N$, $^3J(H,H) = 6.8$ Hz), 1.71 (t, $(C\underline{H}_3CH_2)_2NS$, $^3J(H,H) = 7.14$ Hz), 2 to 2.9 (m, $(CH_3)_2C\underline{H}$), 3.13 to 4 (m, $(CH_3C\underline{H}_2)_2N$ and $(CH_3C\underline{H}_2)_2NS$).

^{13}C NMR (C_6D_6/TMS): δ (in ppm) = 15.7 ($(\underline{C}H_3CH_2)_2N$, $^1J(C,H) = 124.5$ Hz, $^3J(C,P) = 3$ Hz), 16.4 ($(\underline{C}H_3CH_2)_2NS$, $^1J(C,H) = 124.5$ Hz); 25.7 ($(\underline{C}H_3)_2CH$, $^1J(C,H) = 126$ Hz, $^3J(C,P) = 13.7$ Hz); 25.7 (concealed, $(CH_3)_2\underline{C}H$), 44.3 ($\underline{C}H_2N$, $^1J(C,H) = 134.4$ Hz, $^2J(C,P) = 8.2$ Hz, $^3J(C,F) = 8.2$ Hz); 49.7 ($\underline{C}H_2P$, $^1J(C,P) = 131.8$ Hz, $^1J(C,H) = 123$ Hz, $^2J(C,F) = 23.8$ Hz), 54.0 ($\underline{C}H_2NS$, $^1J(C,H) = 133.7$ Hz).

^{19}F NMR (neat/$CFCl_3$): $\delta = -33.9$ ppm (d).

$^{31}P\{^1H\}$ NMR (C_6D_6/external 85% aqueous H_3PO_4): $\delta = -15.3$ ppm (t, $^1J(P,F) = 836.8$ Hz).

IR (CsBr): 30 unassigned absorption bands between 2980 and 414 cm^{-1}.

EIMS (70 eV): m/e (rel. int. in %) = 302 (0.21) M^+, 301 (0.74) $M^+ - H$, 211 (9.44) $M^+ - N(C_2H_5)_2 - F$, 198 (100) $M^+ - SN(C_2H_5)_2$, 122 (27.75) $PFN(C_2H_5)_2^+$, 104 (8.3) $SN(C_2H_5)_2^+$, $HPN(C_2H_5)_2^+$, 72 (5.73) $N(C_2H_5)_2^+$, 57 (33.41) $C_4H_9^+$.

The compound is sensitive towards moisture and attacks glass even at 0 °C.

Reference:

Neumüller, B.; Riffel, H.; Fluck, E. (Z. Anorg. Allg. Chem. **588** [1990] 147/66).

14.3.2 S-Phosphonoyl-, Phosphoryl-, Thiophosphinoyl-, and Thiophosphoryl-N,N-diorganyl-thiohydroxylamines

14.3.2.1 S-Phosphonoyl-, Phosphoryl-, Thiophosphinoyl-, and Thiophosphoryl-N,N-dialkyl-thiohydroxylamines, $R_2P(X)SNR^1_2$, R^1 = alkyl

The compounds with X = O were prepared by several different methods:

$(C_2H_5O)_2P(O)SN(CH_3)_2$ was obtained by treating a suspension of $(CH_3)_2NSN(CH_3)_2$ and $ZnCl_2$ in CH_2Cl_2 with an excess of $(C_2H_5O)_3P$ (mole ratio 1 : 1.3). After 1.5 h the reaction mixture was treated with H_2O and $NaHCO_3$ and subsequently extracted with $CHCl_3$. Evaporation of the solvent gave an oily residue which was distilled in vacuum [1].

Compounds of the type $R_2P(O)SN(C_2H_5)_2$ (R = OC_2H_5, OC_3H_7-n, OC_4H_9-n) were prepared by the reaction of appropriate R_3P with an equimolar amount of $(C_2H_5)_2NSCl$ at temperatures below 10 °C [3] or with an equimolar amount of $(C_2H_5)_2NSCl$ in C_6H_6 at 0 to 10 °C and 1 h at room temperature [4]. The products were purified by vacuum distillation [3, 4].

$(C_2H_5O)_2P(O)SN(C_2H_5)_2$ was first prepared by adding $(C_2H_5O)_2P(O)SCl$ to two mole equivalents of $(C_2H_5)_2NH$ in ether at temperatures below 5 °C. After removing the amine hydrochloride and the solvent, the compound was purified by distillation [2]. It was also formed by adding a solution of $ClN(C_2H_5)_2$ in petroleum ether to a suspension of $(C_2H_5O)_2P(S)ONa$ (mole ratio 3 : 1.2) in petroleum ether at room temperature (8 h) [4].

Compounds with X = O, R = OC_2H_5, OC_3H_7-i and R^1 = CH_3, C_2H_5, i-C_3H_7 were prepared by treating $R_2P(O)SSP(O)R_2$ in dry ether at −10 °C with the corresponding cooled dialkylamine, R^1_2NH (R^1 = CH_3, C_2H_5, i-C_3H_7) (mole ratio ca. 1 : 2). Subsequently, the temperature was raised to 20 °C, the reaction mixture washed with H_2O, and the organic layer separated. After removal of the solvent, the compounds were isolated by vacuum distillation [5].

The compounds with X = S, except that with R^1 = C_6H_{11}-c, were all obtained by reacting solutions of $R_2P(S)SSP(S)R_2$ (R = OC_2H_5 [6, 7], OC_3H_7-i, OC_6H_5 [7], OC_6H_4Cl-4 [11] or C_6H_5 [8]) in dry ether [6, 11] or C_6H_6 [8] with an excess of the respective dialkylamine, R^1_2NH (R^1 = CH_3 [8], C_2H_5 [7, 8, 11], C_3H_7 [7], i-C_3H_7 [7, 8], C_4H_9 [8], $CH_2CH{=}CH_2$ [7]) and then adding hexane. The ammonium salts were filtered off and the filtrate evaporated, leaving oils which slowly crystallized. The products were recrystallized from hexane [7, 8].

Compounds of the type $R_2P(S)SN(C_6H_{11}\text{-c})_2$ (R = OC_2H_5, OC_4H_9-i, OC_6H_5) were prepared by treating $(c\text{-}C_6H_{11})_2NCl$ in $(CH_3)_2C(O)$ with an aqueous solution of $R_2P(S)SNa$ (mole ratio 1 : 1) at 15 to 35 °C [9].

Yields and properties of the compounds are summarized in Table 76.

Table 76
Yields and Properties of $R_2P(X)SNR^1_2$.

R	R^1	X	yield (in %)	b.p. (in °C/Torr) m.p. (in °C)	n_D^{20}	D_4^{20} (in g/cm³)	MR_D exp.	calc.	Ref.
OC_2H_5	CH_3	O	50	50 to 51/0.3	—	—	—	—	[1]
			70	75 to 76/0.3	1.4640	1.1321	51.97	52.29	[5]
OC_2H_5	C_2H_5	O	70	52 to 53/0.12	1.4605[a)]	1.0085	—	—	[2]
			74	52 to 54/0.01	1.4665	—	—	—	[3]
			74	51/0.01	1.4657[a)]	—	—	—	[4]
			36	58/0.05	1.4628[b)]	—	—	—	[4]
			98	54 to 55/0.1	1.4650	1.0852	61.46	61.53	[5]

Table 76 (continued)

R	R[1]	X	yield (in %)	b.p. (in °C/Torr) m.p. (in °C)	n_D^{20}	D_4^{20} (in g/cm³)	MR_D exp.	calc.	Ref.
OC_2H_5	C_2H_5	S	—	—	1.5090	—	—	—	[6]
			97	—	1.4802	—	—	—	[7]
OC_2H_5	C_3H_7	S	93	—	1.5028	—	—	—	[7]
OC_2H_5	i-C_3H_7	O	98	105 to 106/0.18	1.4655	1.0530	70.75	70.76	[5]
OC_2H_5	i-C_3H_7	S	93	—	1.5173	1.1221	76.97	77.24	[7]
OC_2H_5	c-C_6H_{11}	S	—	70 to 71	—	—	—	—	[9, 10]
OC_2H_5	CH_2-$CH{=}CH_2$	S	98	—	1.5279	1.1107	78.20	78.30	[7]
OC_3H_7-n	C_2H_5	O	57	70 to 71/0.03	1.4644	—	—	—	[3]
			57	70 to 71/0.03	1.4640[a)]	—	—	—	[4]
OC_3H_7-i	i-C_3H_7	O	51	104 to 105/0.1	1.4572	1.0138	79.90	79.90	[5]
OC_4H_9-n	C_2H_5	O	70	110 to 115/0.01	1.4645	—	—	—	[3]
			70	—	1.4629[a)]	—	—	—	[4]
OC_4H_9-i	c-C_6H_{11}	S	80	oil	—	—	—	—	[9, 10]
OC_6H_5	C_2H_5	S	96	—	1.5845	1.1837	99.85	99.74	[7]
OC_6H_5	c-C_6H_{11}	S	—	128 to 130	—	—	—	—	[9]
OC_6H_4Cl-4	C_2H_5	S	99.9	40 to 41	—	—	—	—	[11]
C_6H_5	CH_3	S	92	60 to 61	—	—	—	—	[8]
C_6H_5	C_2H_5	S	78	105 to 106	—	—	—	—	[8]
C_6H_5	i-C_3H_7	S	25	46 to 47	—	—	—	—	[8]
C_6H_5	C_4H_9	S	33	117 to 118	—	—	—	—	[8]

a) n_D^{22}. – b) n_D^{23}.

$(C_2H_5O)_2P(O)SN(CH_3)_2$. Colorless oil. ^1H NMR ($CDCl_3$/TMS): δ (in ppm) = 1.19 (t, $C\underline{H}_3CH_2$), 2.52 (s, CH_3N), 2.58 (s, CH_3N), 3.87 (m, $CH_3C\underline{H}_2$) [1]; (solvent ?/TMS ?): δ (in ppm) = 1.45 (m, $C\underline{H}_3CH_2$), 3.02 (d, $N(CH_3)_2$, $^3J(P,H)$ = 4.0 Hz), 4.37 (m, $CH_3C\underline{H}_2$) [5]. ^{31}P NMR ($CDCl_3$/external 85% H_3PO_4): δ = 1.4088 ppm [1]. IR: ν (in cm^{-1}) = 1269 ν(C=O), 1047 ν(P–O–C), 978 ν(O–C) [1].

$(C_2H_5O)_2P(S)SN(C_2H_5)_2$. ^1H NMR (a figure is given in the paper, solvent ?/TMS ?): δ (in ppm) = 1.2 (t, $C\underline{H}_3CH_2N$), 1.6 (t, $C\underline{H}_3CH_2O$), 3.1 (m, $CH_3C\underline{H}_2N$), 4.2 (m, $CH_3C\underline{H}_2O$) [6].

(n-$C_3H_7O)_2P(O)SN(C_2H_5)_2$. The IR spectrum (18 absorptions) was recorded between 2975 and 545 cm^{-1} [4].

$(C_2H_5O)_2P(O)SN(C_2H_5)_2$ decomposed when heated on a water bath for 1 h giving $(C_2H_5O)_2$-$P(O)N(C_2H_5)_2$ and sulfur. It reacted with dry HCl in C_6H_6 to yield $(C_2H_5O)_2P(O)SCl$ and $(C_2H_5)_2NH \cdot HCl$ [2]. Keeping an equimolar mixture of $(C_2H_5O)_2P(O)SN(C_2H_5)_2$ and (n-$C_3H_7O)_2$-PF for 12 h at room temperature or heating the mixture for 1, 3, or 4 h at 100 °C yielded a mixture of (n-$C_3H_7O)_2P(S)F$ and $(C_2H_5O)_2P(O)N(C_2H_5)_2$ [12]; see also [13]. When a mixture of $(C_2H_5O)_2P(O)SN(C_2H_5)_2$ and $((CH_3)_2N)_2PF$ was heated at 100 °C for 6 h, mainly the amines $((CH_3)_2N)_2P(S)F$ and $(C_2H_5)_2NP(S)(OC_2H_5)_2$ formed, whereas additionally the amines $((CH_3)_2N)$-$((C_2H_5)_2N)P(S)F$ and $(CH_3)_2NP(O)(OC_2H_5)_2$ were obtained when the reaction mixture was kept at room temperature for a long time [14].

$(C_2H_5O)_2P(S)SN(C_6H_{11}\text{-c})_2$ improved the vulcanization of rubber [15].

References:

[1] Petrov, K. A.; Rudnev, G. V.; Sorokin, V. D. (Zh. Obshch. Khim. **61** [1989] 1115/8; J. Gen. Chem. USSR [Engl. Transl.] **61** [1989] 1012/4).
[2] Michalski, J.; Markowska, A.; Strzelecka, H. (Rocz. Chem. **33** [1959] 1251/3; C.A. **1960** 10827).
[3] Michalski, J.; Pliska, B. (Dokl. Akad. Nauk SSSR **147** [1962] 111/2; Dokl. Chem. [Engl. Transl.] **142/147** [1962] 961/2).
[4] Michalski, J.; Pliska-Krawiecka, B. (J. Chem. Soc. C **1966** 2249/52).
[5] Mel'nikov, N. N.; Khaskin, B. A.; Torgasheva, N. A. (Zh. Obshch. Khim. **45** [1975] 1005/8; J. Gen. Chem. USSR [Engl. Transl.] **45** [1975] 992/4).
[6] Peng, Y.; Sun, Z.; Bi, G. (Goadeng Xuexiao Huaxue Xuebao **3** [1982] 83/92; C.A. **97** [1982] No. 23888).
[7] Khaskin, B. A.; Torgasheva, N. A.; Mel'nikov, N. N. (Zh. Obshch. Khim. **44** [1974] 95/8; J. Gen. Chem. USSR [Engl. Transl.] **44** [1974] 93/5).
[8] Torgasheva, N. A.; Khaskin, B. A.; Mel'nikov, N. N.; Kosminskaya, G. A. (Zh. Obshch. Khim. **46** [1976] 1467/72; J. Gen. Chem. USSR [Engl. Transl.] **46** [1976] 1440/5).
[9] Malz, H.; Bayer, O.; Freytag, H.; Lober, F. (U.S. 2891059 [1957/59] 3 pp.; C.A. **1960** 4387).
[10] Malz, H.; Lober, F.; Bayer, O.; Scheurlen, H. (U.S. 3044981 [1959/62] 3 pp.; C.A. **58** [1963] 14228).

[11] Khaskin, B. A.; Torgasheva, N. A.; Sheluchenko, O. D (Zh. Obshch. Khim. **58** [1988] 1778/84; J. Gen. Chem. USSR [Engl. Transl.] **58** [1988] 1586/91).
[12] Gusar', N. I.; Ivanova, Zh. M.; Chaus, M. P.; Gololobov, Yu. G. (Zh. Obshch. Khim. **46** [1976] 1981/6; J. Gen. Chem. USSR [Engl. Transl.] **46** [1976] 1910/4).
[13] Gusar', N. I.; Ivanova, Zh. M.; Chaus, M. P.; Gololobov, Yu. G. (Dopov. Akad. Nauk Ukr. RSR Ser. B: Geol. Khim. Biol. Nauki **1978** 520/3; C.A. **89** [1978] No. 196907).
[14] Gusar', N. I.; Chaus, M. P.; Gololobov, Yu. G. (Zh. Obshch. Khim. **45** [1975] 1894/5; J. Gen. Chem. USSR [Engl. Transl.] **45** [1975] 1856).
[15] Foster, R. F.; Williams, D. J. (Br. 1505147 [1975/78] 13 pp.; C.A. **90** [1979] No. 40017).

14.3.2.2 S-[(Ethoxy)ethylphosphonoyl]-N,N-diethyl-thiohydroxylamine, $(C_2H_5O)(C_2H_5)P(O)SN(C_2H_5)_2$

The title compound was prepared with 74% yield by adding a solution of $C_2H_5P(OC_2H_5)_2$ in C_6H_6 to a solution of equimolar amounts of $(C_2H_5)_2NSCl$ in C_6H_6 between 0 to 10 °C. Subsequently, the reaction mixture was stirred at room temperature for 1 h, the solvent removed, and the residue purified by distillation.

Liquid, b.p. 55 °C/0.01 Torr, $n_D^{22}=1.4764$.

The IR spectrum shows 19 absorption bands between 2980 and 455 cm^{-1}.

Reference:

Michalski, J.; Pliska-Krawiecka, B. (J. Chem. Soc. C **1966** 2249/52).

14.3.2.3 S-(Diethoxythiophosphoryl)-N-ethyl-N-phenyl-thiohydroxylamine, $(C_2H_5O)_2P(S)SN(C_2H_5)(C_6H_5)$

The title compound was obtained with 38.5% yield by reacting $(C_2H_5O)_2P(S)Cl$ with $(C_6H_5)(C_2H_5)NH$ (mole ratio 1 : 2) in CCl_4 at room temperature, followed by filtration, removal of the solvent, and recrystallization from n-heptane at −80 °C.

Liquid, $n_D^{23} = 1.5655$, $D_4^{23} = 1.1479$ g/cm³.

The IR spectrum shows typical absorption bands at 1245 cm⁻¹ ν(C–N) and 652 cm⁻¹ ν(P=S).

Reference:

Almasi, L.; Paskucz, L. (Chem. Ber. **98** [1965] 613/6).

14.3.2.4 S-Phosphoryl-,Thiophosphinoyl-, and Thiophosphoryl-N-acyl(alkoxycarbonyl, or carbamoyl)-N-organyl-thiohydroxylamines, $R_2P(X)SNR^1C(O)R^2$

Preparation methods:

Method I: $R_2P(X)SNR^1C(O)R^2$ with X = O, R = OCH_3, $R^1 = CH_3$, $R^2 = OC_2H_5$; X = S, R = OC_2H_5, OC_6H_5, $R^1 = CH_3$, C_2H_5, C_6H_5, R^2 = H, CH_3, C_6H_5, OCH_3, OC_2H_5 were prepared by adding a solution of $ClN(R^1)C(O)R^2$ in C_6H_6 to a suspension of equimolar amounts of $R_2P(X)SM$ (Ia: M = Na; Ib: M = K) in C_6H_6 at 30 °C and then stirring at 40 to 50 °C [1].

Method II: $(t\text{-}C_8H_{17}O)_2P(O)SN(C_6H_5)C(O)C_6H_5$ was obtained by the reaction of $(t\text{-}C_8H_{17}O)_2P$-(O)SCl with $HN(C_6H_5)C(O)C_6H_5$ and $(C_2H_5)_3N$ (mole ratio 1 : 1 : 1) in ether at 20 °C [3].

Method III: The compounds with X = S, R = OC_2H_5, $R^1 = CH_3$, and $R^2 = CH_3$ or CH=CHCOOH, and R = OC_4H_9-n, $R^1 = R^2 = CH_3$ were prepared by stirring mixtures of $R_2P(S)SN$-HCH_3 (R = OC_2H_5, OC_4H_9-n) and acetic or maleic anhydride, respectively, for 10 min followed by addition of ether. The ethereal solutions were washed with H_2O, the solvent removed, and the oily residues dissolved in hexane. On cooling the compounds separated as oils [4].

Method IV: The diphenyl-substituted compounds with X = S, $R^1 = CH_3$ or $CH(C_2H_5)CH_2OH$, and $R^2 = NHC_6H_3Cl_2$-3,4 were obtained by reacting $(C_6H_5)_2P(S)SNHCH_3$ or $(C_6H_5)_2$-$P(S)SNHCH(CH_2CH_3)CH_2OH$ with 3,4-$Cl_2C_6H_3NCO$ (mole ratio ca. 1 : 1) in C_6H_6 at 75 °C for 2 h. After removal of the solvent the compounds were recrystallized from benzene–hexane [5].

Method V: The dialkoxy-substituted compounds (X = O or S) with $R^1 = CH_3$ or C_2H_5 and $R^2 = NHC_2H_5$, NHC_6H_4Cl-4, and $NHC_6H_3Cl_2$-3,4 were prepared by the reaction of $R_2P(X)SNHR^1$ with R^2NCO (mole ratio ca. 1 : 1; R, R^1, and R^2 are listed in Table 77) in C_6H_6 in a sealed tube at 100 °C for 2 h [5].

Method VI: The compound with X = O, R = OCH_3, $R^1 = CH_3$ and $R^2 = C_{10}H_{12}O_2$ was obtained as one of the principal oxidation products of $(CH_3O)_2P(S)N(CH_3)C(O)C_{10}H_{12}O_2$ with m-chloroperbenzoic acid in CH_2Cl_2 at 0 °C [7].

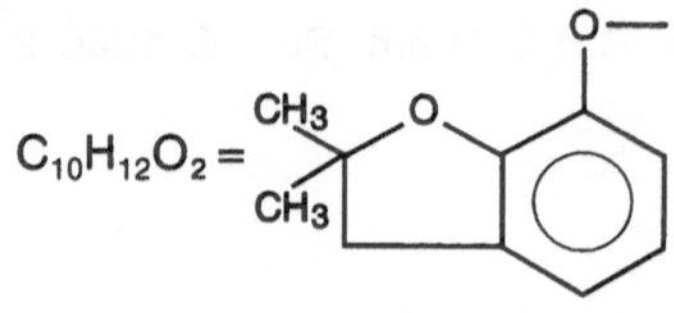

Method VII: $(4\text{-}ClC_6H_4O)_2P(S)SNCH_3C(O)NHR^2$ ($R^2 = C_6H_4Cl\text{-}4$, $C_6H_3Cl_2\text{-}3,4$) were prepared by reacting $(4\text{-}ClC_6H_4O)_2P(S)SSP(S)(OC_6H_4Cl\text{-}4)_2$ with CH_3NH_2 in ether, then adding R^2NCO, and leaving the mixtures stand for a few hours. After the solvent was removed, the residues were recrystallized from CCl_4 [6].

Yields and properties of the compounds are summarized in Table 77.

Table 77
Yields, Melting Points, and Preparation Methods of $R_2P(X)SNR^1C(O)R^2$.

R	R^1	R^2	X	method	yield (in %)	m.p. (in °C)	Ref.
OCH_3	CH_3	$C_{10}H_{12}O_2$	O	VI	—	a)	[7]
OCH_3	CH_3	OC_2H_5	O	Ia	—	b)	[1, 2]
OC_2H_5	CH_3	CH_3	S	III	86.8	c)	[4]
OC_2H_5	CH_3	CH=CHCOOH	S	III	92.5	d)	[4]
OC_2H_5	CH_3	C_6H_5	S	Ia	—	—	[1]
OC_2H_5	CH_3	OCH_3	S	Ia	—	5 to 8	[1, 2]
OC_2H_5	CH_3	$NHC_6H_3Cl_2$-3,4	S	V	98	89 to 90	[5]
OC_2H_5	C_2H_5	OC_2H_5	S	Ia	—	oil	[1]
OC_2H_5	C_6H_5	H	S	Ib	—	53 to 54 (C_2H_5OH)	[1, 2]
OC_2H_5	C_6H_5	CH_3	S	Ib	—	60 to 62 (petroleum ether)	[1, 2]
OC_2H_5	C_6H_5	OC_2H_5	S	Ia	—	—	[1]
OC_3H_7-i	CH_3	NHC_2H_5	S	V	99	e)	[5]
OC_3H_7-i	CH_3	NHC_6H_4Cl-4	S	V	98	64 to 65	[5]
OC_3H_7-i	CH_3	$NHC_6H_3Cl_2$-3,4	O	V	95	68 to 69	[5]
OC_3H_7-i	CH_3	$NHC_6H_3Cl_2$-3,4	S	V	99	69 to 70	[5]
OC_3H_7-i	C_2H_5	$NHC_6H_3Cl_2$-3,4	S	V	97	55 to 56	[5]
OC_4H_9-n	CH_3	CH_3	S	III	93.5	f)	[4]
OC_8H_{17}-t	C_6H_5	C_6H_5	O	II	98	—	[3]
OC_6H_5	C_2H_5	OC_2H_5	S	Ia	—	oil	[1, 2]
OC_6H_4Cl-4	CH_3	NHC_6H_4Cl-4	S	VII	—	119 to 120	[6]
OC_6H_4Cl-4	CH_3	$NHC_6H_3Cl_2$-3,4	S	VII	—	83 to 84	[6]
C_6H_5	CH_3	$NHC_6H_3Cl_2$-3,4	S	IV	99	100 to 101	[5]
C_6H_5	$CH(C_2H_5)CH_2OH$	$NHC_6H_3Cl_2$-3,4	S	IV	98	149 to 150	[5]

a) ^1H NMR ($CDCl_3$/TMS): δ (in ppm) = 3.45 (s, NCH_3), 3.95 (d, OCH_3), $^3J(H,H) = 13$ Hz. IR: ν (in cm^{-1}) = 1725 ν(C=O), 1260 ν(P=O). MS: m/e (rel. int. in %) = 198 (62.3) $[OCN(CH_3)SP(O)(OCH_3)_2]^+$, 141 (100) $[SP(O)(OCH_3)_2]^+$. – b) Colorless liquid, b.p. 137 to 140 °C/0.1 Torr [1, 2]. – c) Oil, $n_D^{20} = 1.5210$, $D_4^{20} = 1.1997$ g/cm³, $MR_{D\,exp.} = 65.10$, $MR_{D\,calc.} = 65.39$ [4]. – d) Oil, $n_D^{20} = 1.5380$, $D_4^{20} = 1.3947$ g/cm³, $MR_{D\,exp.} = 73.99$, $MR_{D\,calc.} = 73.97$ [4]. – e) Liquid, $n_D^{20} = 1.5000$, $D_4^{20} = 1.1158$ g/cm³, $MR_{D\,exp.} = 82.87$, $MR_{D\,calc.} = 82.95$ [5]. – f) Oil, $n_D^{20} = 1.5090$, $D_4^{20} = 1.1211$ g/cm³, $MR_{D\,exp.} = 83.48$, $MR_{D\,calc.} = 83.37$ [4].

The compounds $(CH_3O)_2P(O)SN(CH_3)C(O)C_{10}H_{12}O_2$ decomposed in a wet atmosphere via the following equation [7]:

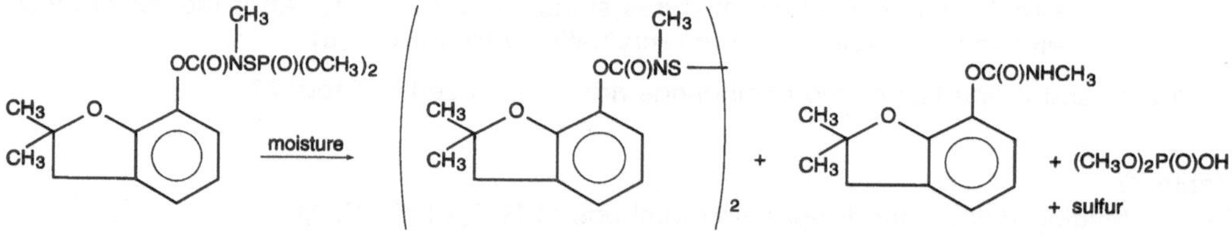

Compounds with the general formula $(RO)_2P(S)SNR^1C(O)R^2$ [8] or $(RO)_2P(S)SN(C(O)R^1)_2$ (formation not described) [9] were claimed to inhibit premature vulcanization of rubber compounds or mixes [8, 9].

References:

[1] Malz, H.; Bayer, O.; Wegler, R. (U.S. 2995568 [1959/61] 5 pp.; C.A. **57** [1962] 11021).
[2] Malz, H.; Lober, F.; Bayer, O.; Scheurlen, H. (U.S. 3044981 [1959/62] 3 pp.; C.A. **58** [1963] 14228).
[3] Price, G. R.; Walsh, E. N.; Hallett, J. T. (U.S. 3114761 [1961/63] 5 pp.; C.A. **60** [1964] 6790).
[4] Khaskin, B. A.; Torgasheva, N. A.; Mel'nikov, N. N. (Zh. Obshch. Khim. **49** [1979] 1250/2; J. Gen. Chem. USSR [Engl. Transl.] **49** [1979] 1096/8).
[5] Khaskin, B. A.; Torgasheva, N. A.; Mel'nikov, N. N. (Zh. Obshch. Khim. **47** [1977] 2176/8; J. Gen. Chem. USSR [Engl. Transl.] **47** [1977] 1987/9).
[6] Khaskin, B. A.; Torgasheva, N. A.; Sheluchenko, O. D. (Zh. Obshch. Khim. **58** [1988] 1778/84; J. Gen. Chem. USSR [Engl. Transl.] **58** [1988] 1586/91).
[7] Fahmy, M. A. H.; Fukuto, T. R. (Tetrahedron Lett. **1972** 4245/8).
[8] Laithwaite, P.; Taylor, J. A. (Br. 1365072 [1971/74] 5 pp.; C.A. **82** [1975] No. 112973).
[9] Lohr, D. F.; Kay, E. L. (U.S. 3865781 [1973/75] 4 pp.; C.A. **82** [1975] No. 172349).

14.3.2.5 N-Phosphoryl(or Thiophosphoryl)sulfanyl-N-(*tert*-octyl)-cyanamide Derivatives, $(CH_3O)_2P(O)SN(C{\equiv}N)(C_8H_{17}\text{-t})$, $(n\text{-}C_3H_7O)_2P(S)SN(C{\equiv}N)(C_8H_{17}\text{-t})$

The title compounds were obtained as oils by reacting benzene solutions of equimolar amounts of $(C{\equiv}N)(t\text{-}C_8H_{17})NCl$ with $(CH_3O)_2P(O)SK$ or $(n\text{-}C_3H_7O)_2P(S)Na$, respectively. The compounds were not distillable in an oil-pump vacuum at room temperature [1]. They can be used to stabilize polyolefines [2].

References:

[1] Malz, H.; Bayer, O.; Wegler, R. (U.S. 2995568 [1959/61] 5 pp.; C.A. **57** [1962] 11021).
[2] Malz, H.; Lober, F.; Bayer, O.; Scheurlen, H. (U.S. 3044981 [1959/62] 3 pp.; C.A. **58** [1963] 14228).

14.3.3 N-Phosphoryl(Thiophosphinoyl, Thiophosphonoyl, or Thiophosphoryl)sulfanyl Substituted N-Heterocycles

14.3.3.1 N[1]-Thiophosphorylsulfanyl-aziridine Derivatives, $(RO)_2P(S)SN(CH_2)_2$, R = alkyl

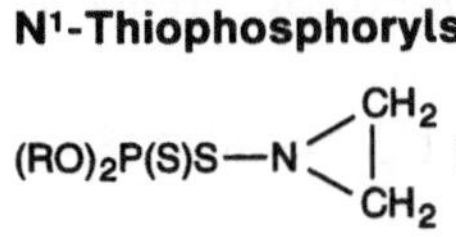

$R = C_2H_5$, n-C_3H_7, i-C_3H_7, i-C_4H_9

Derivatives with $R = C_2H_5$, n-C_3H_7, i-C_3H_7, and i-C_4H_9 were prepared by adding a solution of the corresponding $(RO)_2P(S)SCl$ in petroleum ether to a cooled solution (−15 °C) of aziridine and $(C_2H_5)_3N$ in petroleum ether (mole ratio 1 : 1 : 1.1). After filtration of the hydrochloride, the filtrate was evaporated at room temperature (1 Torr) leaving the compounds as yellowish liquids [1]; see also [2].

(i-$C_3H_7O)_2P(S)SN(CH_2)_2$ was also obtained by treating (i-$C_3H_7O)_2P(S)SSP(S)(OC_3H_7$-i$)_2$ with excess $HN{-}CH_2{-}CH_2$ (mole ratio ca. 1 : 2.6) at 5 to 10 °C. After the solution was washed with H_2O, the solvent was removed, and the residue dissolved in hexane. Upon adding RNCO ($R = 4$-ClC_6H_4, 2,4-$Cl_2C_6H_3$) to the hexane solution, the by-product (i-$C_3H_7O)_2P(S)S(CH_2)_2NH$-C(O)NHR precipitated. Evaporation of the hexane filtrate yielded the product [3].

The properties of the compounds are listed in the table below.

R	yield (in %)	b.p. (in °C/0.4 Torr)	n_D^{20}	D_4^{20} (in g/cm³)	Ref.
C_2H_5	55	88	1.5282	1.1872	[1, 2]
n-C_3H_7	59	100	1.5196	1.1394	[1, 2]
i-C_3H_7	65	92	1.5131	1.1225	[1, 2]
i-C_3H_7 [a)]	82	—	1.5158	1.1280	[3]
i-C_4H_9	60	103	1.5080	1.0902	[1, 2]

a) $MR_{D\,exp.} = 68.31$, $MR_{D\,calc.} = 67.76$ [3].

References:

[1] Almasi, L.; Hantz, A.; Baicu, T. (Chem. Ber. **104** [1971] 3982).
[2] Almasi, L.; Baicu, T.; Hantz, A. (Rom. 56177 [1969/74] 2 pp. from C.A. **85** [1976] No. 21076).
[3] Mel'nikov, N. N.; Torgasheva, N. A.; Khaskin, B. A. (Zh. Obshch. Khim. **46** [1976] 43/7; J. Gen. Chem. USSR [Engl. Transl.] **46** [1976] 43/6).

14.3.3.2 N[1]-Phosphorylsulfanyl-2,5-pyrrolidinedione Derivatives

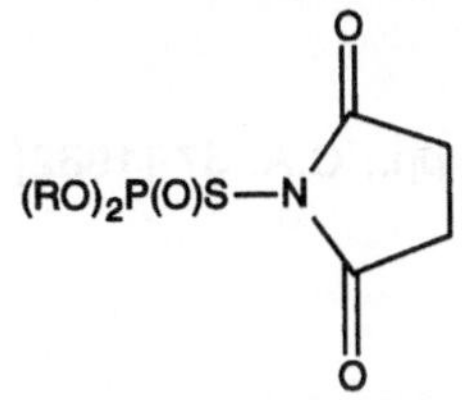

$R = CH_3$, C_2H_5, t-C_4H_9

The title compounds were prepared with 72% ($R = C_2H_5$) and 85% yield ($R = $ t-C_4H_9) by stirring mixtures of N-chlorosuccinimide and the corresponding $(RO)_2P(O)SNa$ (mole ratio ~1 : 1) in C_6H_6 ($R = CH_3$ [1], C_2H_5 [3]) or in dimethoxyethane ($R = $ t-C_4H_9 [3]) for 10 min and then leaving them stand for 2 h. Extraction with $CHCl_3$ followed by evaporation of the solvent gave oily residues which crystallized on standing [3]. The compound with R = CH_3 was also obtained with 96.5% yield by reacting an equimolar mixture of $(CH_3O)_2P(O)SCl$, N-succinimide, and $(C_2H_5)_3N$ in ether [4].

R = CH_3. Colorless crystals, m.p. 90 to 93 °C (from C_6H_6) [1], see also [2].

R = C_2H_5. Crystalline solid, m.p. 61 to 63 °C.

^1H NMR ($CDCl_3$/TMS): δ (in ppm) = 1.3 (d of t, $POCH_2C\underline{H}_3$, $^3J(H,H)$ = 7 Hz, $^4J(P,H)$ = 1 Hz), 2.85 (d, $CH_2C(O)$, $^4J(P,H)$ = 1 Hz), 4.35 (d of q, $POC\underline{H}_2CH_3$, $^3J(H,H)$ = 7 Hz, $^3J(P,H)$ = 8 Hz).

IR (KBr): ν (in cm^{-1}) = 2950, 1780 (sh), 1720, 1300, 1250, 1140, 1000 to 990, 780.

MS: m/e = 267 M^+ [3].

R = t-C_4H_9. ^1H NMR ($CDCl_3$/TMS): δ (in ppm) = 1.57 (s, $C(CH_3)_3$), 2.85 (d, $CH_2C(O)$, $^4J(P,H)$ = 1 Hz) [3].

References:

[1] Malz, H.; Bayer, O.; Wegler, R. (U.S. 2995568 [1959/61] 5 pp.; C.A. **57** [1962] 11021).
[2] Malz, H.; Lober, F.; Bayer, O.; Scheurlen, H. (U.S. 3044981 [1959/62] 3 pp.; C.A. **58** [1963] 14228).
[3] Chapman, T. M.; Kleid, D. G. (J. Org. Chem. **38** [1973] 250/2).
[4] Price, G. R.; Walsh, E. N.; Hallett, J. T. (U.S. 3114761 [1961/63] 5 pp.; C.A. **60** [1964] 6790).

14.3.3.3 N¹-Thiophosphorylsulfanyl-2,5-pyrrolidinedione Derivative

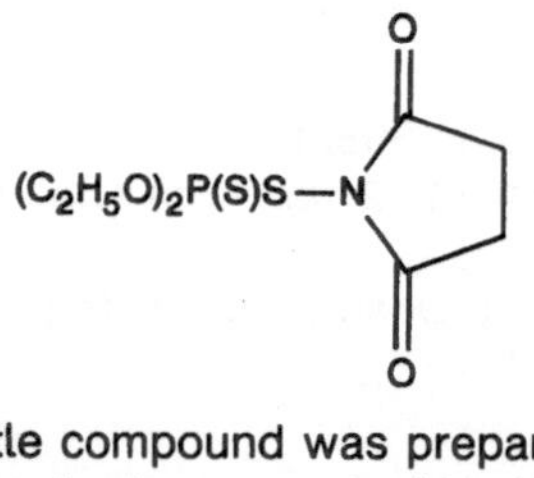

The title compound was prepared by treating a suspension of $(C_2H_5O)_2P(S)SNa$ in C_6H_6 with the equimolar amount of N-chloro-succinimide and then holding the reaction mixture at 50 °C for 15 min.

Solid, m.p. 94.5 to 96 °C (from C_2H_5OH).

Reference:

Malz, H.; Bayer, O.; Wegler, R. (U.S. 2995568 [1959/61] 5 pp.; C.A. **57** [1962] 11021).

14.3.3.4 N²-Phosphorylsulfanyl-1H-isoindole-1,3(2H)-dione Derivative

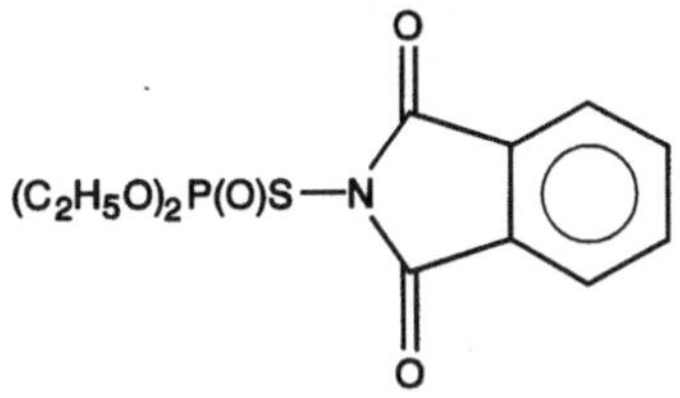

The title compound was prepared with 78% yield by adding $(C_2H_5O)_2P(O)SCl$ to a mixture of phthalimide and $(C_2H_5)_3N$ (mole ratio 1 : 1 : 1) in ether at room temperature and subsequently stirring it for 1 h.

Reference:

Price, G. R.; Walsh, E. N.; Hallett, J. T. (U.S. 3114761 [1961/63] 5 pp.; C.A. **60** [1964] 6790).

14.3.3.5 N²-Thiophosphorylsulfanyl-1H-isoindole-1,3(2H)-dione Derivatives

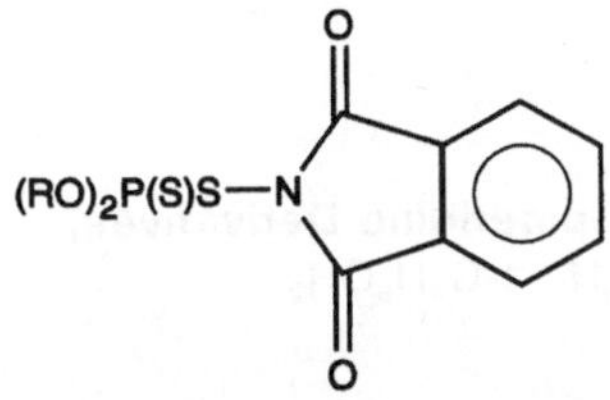

$R = C_2H_5$, $n\text{-}C_3H_7$, $i\text{-}C_3H_7$, $i\text{-}C_4H_9$

The title compounds were prepared by the reaction of equimolar amounts of $(RO)_2P(S)SCl$ with the potassium salt of phthalimide (Method I) or with phthalimide in the presence of $(C_2H_5)_3N$ (Method II) in C_6H_6 [1]. The compound with $R = C_2H_5$ was also obtained by reacting a mixture of $(C_2H_5O)_2P(S)SNa$, phthalimide, and $(C_2H_5)_3N$ in C_6H_6 [2]. Yields and melting points of the compounds are listed below:

R	yield (in %) I	II	m.p. (in °C)	recrystallized from	Ref.
C_2H_5	75	60	127	C_6H_6–ether	[1]
C_2H_5	—	—	123	—	[2]
$n\text{-}C_3H_7$	70	55	72	C_6H_6–petroleum ether	[1]
$i\text{-}C_3H_7$	72	60	92	C_6H_6–petroleum ether	[1]
$i\text{-}C_4H_9$	70	55	84	C_6H_6–petroleum ether	[1]

The compounds are useful as pesticides [1].

References:

[1] Almasi, L.; Baicu, T.; Hantz, A. (Rom. 57894 [1970/74] 3 pp.; C.A. **85** [1976] No. 32654).
[2] Malz, H.; Bayer, O.; Wegler, R. (U.S. 2995568 [1959/61] 5 pp.; C.A. **57** [1962] 11021).

14.3.3.6 N²-Phosphorylsulfanyl-1,2-benzisothiazol-3(2H)-one S¹,S¹-Dioxide Derivatives

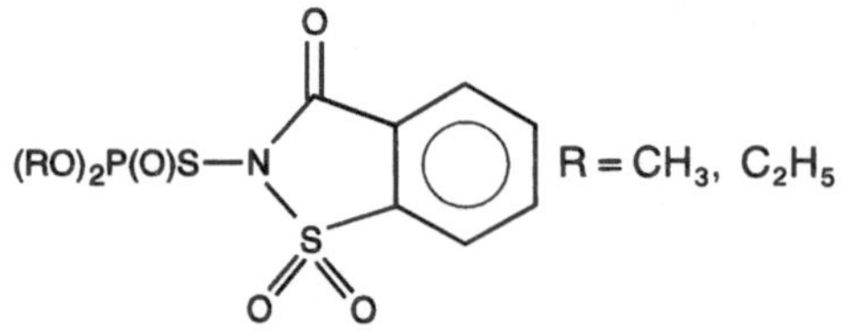

The compounds were prepared with 97 (R = CH_3) or 99.5% yield (R = C_2H_5), respectively, by treating an equimolar mixture of saccharic acid and the respective $(RO)_2P(O)SCl$ (R = CH_3, C_2H_5) in ether at 20 °C with the stoichiometric amount of $(C_2H_5)_3N$ and stirring the reaction mixture at room temperature for 1 h [1].

The compound with R = CH_3 was also obtained by reacting $(CH_3O)_2P(O)SNa$ with N-chlorosaccharine in C_6H_6. The compound was not characterized [2].

References:

[1] Price, G. R.; Walsh, E. N.; Hallett, J. T. (U.S. 3114761 [1961/63] 5 pp.; C.A. **60** [1964] 6790).

[2] Malz, H.; Bayer, O.; Wegler, R. (U.S. 2995568 [1959/61] 5 pp.; C.A. **57** [1962] 11021).

14.3.3.7 N¹-Phosphoryl(or Thiophosphoryl)sulfanyl-piperidine Derivatives, $(RO)_2P(X)SN(CH_2)_5$, X = O, S; R = C_2H_5, i-C_3H_7, t-$C_4H_9CH_2$

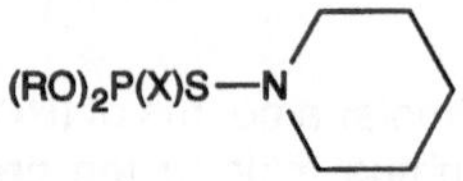

$(C_2H_5O)_2P(O)SN(CH_2)_5$ was prepared with 80% yield by treating $(C_2H_5O)_2P(O)S$-$SP(O)(OC_2H_5)_2$ in ether at −10 °C with an excess of cooled piperidine (mole ratio ca. 1 : 2.6). Subsequently, the temperature was raised to 20 °C, the solution washed with water, the solvent removed, and the residue distilled in vacuum.

Liquid, b.p. 95 to 96 °C/0.21 Torr, n_D^{20} = 1.4850, D_4^{20} = 1.1591 g/cm³, $MR_{D\,exp.}$ = 59.16, $MR_{D\,calc.}$ = 59.36 [1].

$(C_2H_5O)_2P(S)SN(CH_2)_5$ was prepared with 54% yield by treating an equimolar mixture of $(CH_2)_5NSCl$ and $(CH_2)_5NH$ in CCl_4 with the stoichiometric amount of $(C_2H_5O)_2P(S)H$ at −5 to 0 °C. After filtration, the major portion of the solvent was removed and the residue cooled to −70 °C. The compound precipitated from the remaining solution [2]. It formed when a suspension of $(C_2H_5O)_2P(S)SNa$ in C_2H_5OH was reacted with a solution of the equimolar amount of $ClN(CH_2)_5$ in C_2H_5OH [3]. The compound was also obtained with 95% yield by adding $(CH_2)_5NH$ in C_6H_6 to a solution of $(C_2H_5O)_2P(S)SSP(S)(OC_2H_5)_2$ (mole ratio ca. 2 : 1) in C_6H_6. After 30 min the piperidinium salt was filtered off, and the residue was recrystallized at low temperatures from C_2H_5OH [4].

Colorless crystals, m.p. 25 to 26 °C [2], 17 to 19 °C [4] (from C_2H_5OH) [2, 4]; clear yellow oil [3].

The IR spectrum is illustrated in the paper and shows typical absorptions between 1360 and 1200 cm⁻¹ ν(C–N) and at 652 cm⁻¹ ν(P=S) [2].

The compound was reacted with dry HCl (mole ratio 1:2) in CCl_4 for 5 min giving $(C_2H_5O)_2P(S)SCl$ and $(CH_2)_5NH \cdot HCl$ [2].

$(i\text{-}C_3H_7O)_2P(S)SN(CH_2)_5$ was prepared with 97% yield from $(i\text{-}C_3H_7O)_2P(S)SSP(S)(OC_3H_7\text{-}i)_2$ and $(CH_2)_5NH$ (see above) [4]. The compound was also prepared from $(i\text{-}C_3H_7O)_2P(S)SSP(S)(OC_3H_7\text{-}i)_2$, obtained by reacting $i\text{-}C_3H_7OH$ with P_2S_5 (mole ratio 5:1) at 65 to 70 °C. The reaction mixture was then cooled to 18 °C and aqueous NaOH was added below 25 °C. After stirring for 30 min, the temperature was raised to 55 to 60 °C, and an excess of piperidine was added followed by oxidation with a $KI\text{-}I_2$ solution. Extraction with ether after removing the solvent yielded the compound as an oily substance [5].

$n_D^{20} = 1.5152$, $D_4^{20} = 1.0993$ g/cm³, $MR_{D\,exp.} = 81.62$, $MR_{D\,calc.} = 81.65$ [4].

^{1}H NMR (CCl_4/TMS): δ (in ppm) = 1.06 to 1.96 (m, $CH(C\underline{H}_3)_2$ and $-CH_2-$ merged), 2.80 to 3.36 (q, NCH_2, $^3J(H,H) = 6$ Hz), 4.30 to 5.00 (heptet, OCH, $^3J(H,H) = 6$ Hz) [5].

The compound in synergistic combination with dibenzothiazyl disulfide accelerates the vulcanization of rubber [5].

$(t\text{-}C_4H_9CH_2O)_2P(S)SN(CH_2)_5$ was prepared with 88% yield by treating a solution of $(t\text{-}C_4H_9CH_2O)_2P(S)SBr$ in CH_2Cl_2 at −10 °C with a solution of piperidine in CH_2Cl_2 (mole ratio 1:2) and stirring the mixture at 0 to 5 °C for 20 min. After filtration of the hydrobromide, the crude product was purified by recrystallization.

Needles, m.p. 67 °C (from C_2H_5OH).

^{31}P NMR (C_6H_6/external 85% aqueous H_3PO_4): δ = 96.0 ppm [6].

References:

[1] Mel'nikov, N. N.; Khaskin, B. A.; Torgasheva, N. A. (Zh. Obshch. Khim. **45** [1975] 1005/8; J. Gen. Chem. USSR [Engl. Transl.] **45** [1975] 992/4).

[2] Almasi, L.; Hantz, A. (Chem. Ber. **97** [1964] 661/6).

[3] Malz, H.; Bayer, O.; Freytag, H.; Lober, F. (U.S. 2891059 [1957/59] 3 pp.; C.A. **1960** 4387/8).

[4] Khaskin, B. A.; Torgasheva, N. A.; Mel'nikov, N. N. (Zh. Obshch. Khim. **44** [1974] 95/8; J. Gen. Chem. USSR [Engl. Transl.] **44** [1974] 93/5).

[5] Mandal, S. K.; Datta, R. N.; Basu, D. K. (Polym.-Plast. Technol. Eng. **28** [1989] 957/73).

[6] Lopusin'ski, A.; Potrzebowski, M. (Phosphorus Sulfur Relat. Elem. **32** [1987] 55/64).

14.3.3.8 N⁴-Phosphoryl(Thiophosphinoyl, Thiophosphonoyl, or Thiophosphoryl)-sulfanyl-morpholine Derivatives

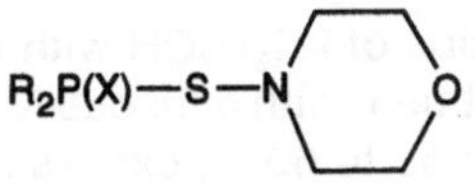

X = O, S
R = R'
R = R'O
R_2 = R', R'O

R' = alkyl, C_6H_5

14.3.3.8.1 $(RO)_2P(O)SN(CH_2)_4O$, R = C_2H_5, n-C_4H_9

$(C_2H_5O)_2P(O)SN(CH_2)_4O$ was prepared with 82% yield by treating $(C_2H_5O)_2P(O)SSP(O)(OC_2H_5)_2$ in ether at −10 °C with an excess of cooled morpholine (mole ratio ca. 1:2.6). The temperature was then raised to 20 °C, the solution washed with water, the solvent removed, and the residue distilled in vacuum [1].

Compounds with R = C_2H_5 and n-C_4H_9 were prepared by adding $(RO)_3P$ (R = C_2H_5 and n-C_4H_9) to an equimolar amount of $ClSN(CH_2)_4O$ in the temperature range 6 to 13 °C. On warming the reaction mixtures to 35 °C, the yellow liquids became colorless [2].

$(C_2H_5O)_2P(O)SN(CH_2)_4O$. Liquid, n_D^{20} = 1.4920 [1], 1.4760 [2], D_4^{20} = 1.2142 g/cm³, $MR_{D\,exp.}$ = 60.99, $MR_{D\,calc.}$ = 61.00 [1].

^1H NMR ($CDCl_3$/TMS): δ (in ppm) = 1.37 (t, CH_3, $^3J(H,H)$ = 7 Hz), 3.02 to 3.20 (s, NCH_2), 3.52 to 3.70 (m, OCH_2), 4.15 (d of q, CH_2OP, $^3J(P,H)$ = 8.5 Hz) [2].

^{31}P NMR ($CDCl_3$/85% H_3PO_4): δ = 27 ppm [2].

The compound decomposed during distillation [1]; it partly decomposed after 4 h at room temperature yielding $(C_2H_5O)_2P(O)N(CH_2)_4O$ (23%), $(C_2H_5O)_3P(S)$ (7%), and sulfur [2]. $(C_2H_5O)_2$-$P(O)SN(CH_2)_4O$ reacted with $(CH_3)_3SnH$ at 60 °C to give $(C_2H_5O)_2P(O)SSn(CH_3)_3$ and $HN(CH_2)_4O$ [3].

(n-$C_4H_9O)_2P(O)SN(CH_2)_4O$. 59% yield. ^{31}P NMR ($CDCl_3$/85% H_3PO_4): δ = 27 ppm.

The compound partly decomposed after 15 d at room temperature to give (n-$C_4H_9O)_2P$-$(O)N(CH_2)_4O$ (67.9%), (n-$C_4H_9O)_3P(S)$ (13.6%), and sulfur [2].

References:

[1] Mel'nikov, N. N.; Khaskin, B. A.; Torgasheva, N. A. (Zh. Obshch. Khim. **45** [1975] 1005/8; J. Gen. Chem. USSR [Engl. Transl.] **45** [1975] 992/4).

[2] Mazitova, F. N.; Kharullin, V. K. (Zh. Obshch. Khim. **56** [1986] 788/90; J. Gen. Chem. USSR [Engl. Transl.] **56** [1986] 694/5).

[3] Avar, G.; Neumann, W. P. (J. Organomet. Chem. **131** [1977] 215/24).

14.3.3.8.2 $(RO)_2P(S)SN(CH_2)_4O$, R = alkyl, phenyl

The compounds $(RO)_2P(S)SN(CH_2)_4O$ (R = C_2H_5, i-C_3H_7, C_6H_5) were obtained with 99, 98, and 98% yield, respectively, when adding a benzene solution of $O(CH_2)_4NH$ to the corresponding $(RO)_2P(S)SSP(S)(OR)_2$ (mole ratio ~2 : 1) in C_6H_6. After 30 min the morpholinium salt was filtered off, the solvent removed, and the residues recrystallized at low temperatures from C_2H_5OH [1].

The compounds $(RO)_2P(S)SN(CH_2)_4O$ (R = n-C_3H_7, i-C_3H_7, n-C_4H_9, i-C_4H_9) were prepared with 60, 65, 50, and 55% yield, respectively, by treating an equimolar mixture of $ClSN(CH_2)_4O$ and $O(CH_2)_4NH$ in CCl_4 with the stoichiometric amount of $(RO)_2P(S)H$ at −5 to 0 °C. After filtration the solvent was removed, and the residue cooled to −70 °C. The compounds precipitated from the solution at low temperatures [2]; see also [3].

(i-$C_3H_7O)_2P(S)SN(CH_2)_4O$ was also prepared by the reaction of i-C_3H_7OH with P_2S_5 (mole ratio 5 : 1) at 65 to 70 °C, followed by cooling to 18 °C and treatment with aqueous NaOH below 25 °C. After stirring for 30 min, the temperature was raised to 55 to 60 °C, excess morpholine was added, and the mixture was oxidized with a KI–I_2 solution and extracted with ether. Removal of the solvent yielded the compound as an oily residue [4, 5].

$(C_2H_5O)_2P(S)SN(CH_2)_4O$. The compound melts at 27 to 28 °C [1].

(n-$C_3H_7O)_2P(S)SN(CH_2)_4O$. The compound melts at 30 °C [2].

(i-$C_3H_7O)_2P(S)SN(CH_2)_4O$. The compound melts at 31 °C [2] and 30 to 31 °C [1].

^{1}H NMR (CCl_4/TMS): δ (in ppm) = 1.03 to 1.53 (d, $CH(C\underline{H}_3)_2$, $^3J(H,H)$ = 6 Hz), 2.93 to 3.33 (m, NCH_2), 3.43 to 3.66 (m, OCH_2), 4.50 to 5.20 (heptet, OCH, $^3J(H,H)$ = 6 Hz) [4, 5].

$(n\text{-}C_4H_9O)_2P(S)SN(CH_2)_4O$. n_D^{20} = 1.5127, D_4^{20} = 1.1077 g/cm^3 [2].

$(i\text{-}C_4H_9O)_2P(S)SN(CH_2)_4O$. n_D^{20} = 1.5120, D_4^{20} = 1.1073 g/cm^3 [2].

$(C_6H_5O)_2P(S)SN(CH_2)_4O$. n_D^{20} = 1.6001, D_4^{20} = 1.3094 g/cm^3, $MR_{D\ exp.}$ = 94.94, $MR_{D\ calc.}$ = 94.78 [1].

The $(RO)_2P(S)SN(CH_2)_4O$ series of compounds (R = $n\text{-}C_3H_7$, $i\text{-}C_3H_7$, $n\text{-}C_4H_9$, $i\text{-}C_4H_9$) were reacted with dry HCl in CCl_4 for 5 min to yield the corresponding $(RO)_2P(S)SCl$ and $O(CH_2)_4NH \cdot HCl$ [2]. $(RO)_2P(S)SN(CH_2)_4O$ (R = C_2H_5, $i\text{-}C_3H_7$, $CH_3C(CH_3)_2CH_2$) reacted with HCl or HBr (mole ratio 1 : 2), formed in situ from $(CH_3)_3SiX$ (X = Cl, Br) and C_2H_5OH at 10 to 15 °C, in hexane at 15 to 20 °C to give $(RO)_2P(S)SX$ (X = Cl: R = C_2H_5, $i\text{-}C_3H_7$, $CH_3C(CH_3)_2CH_2$; X = Br: R = $i\text{-}C_3H_7$, $CH_3C(CH_3)_2CH_2$). The reaction of $(CH_3C(CH_3)_2CH_2O)_2P(S)SN(CH_2)_4O$ with $(CH_3)_3SiI$ and C_2H_5OH under the same conditions yielded $((CH_3C(CH_3)_2CH_2O)_2P(S)S)_2$ and I_2 [6].

$(i\text{-}C_3H_7O)_2P(S)SN(CH_2)_4O$ in a synergistic combination with dibenzothiazyl disulfide accelerates the vulcanization of rubber [4, 5]; see also [7].

References:

[1] Khaskin, B. A.; Torgasheva, N. A.; Mel'nikov, N. N. (Zh. Obshch. Khim. **44** [1974] 95/8; J. Gen. Chem. USSR [Engl. Transl.] **44** [1974] 93/5).
[2] Almasi, L.; Paskucz, L. (Chem. Ber. **98** [1965] 3546/53).
[3] Almasi, L.; Hantz, A. (Chem. Ber. **97** [1964] 661/6).
[4] Mandal, S. K.; Datta, R. N.; Basu, D. K. (Rubber Chem. Technol. **62** [1989] 569/84).
[5] Mandal, S. K.; Datta, R. N.; Basu, D. K. (Polym.-Plast. Technol. Eng. **28** [1989] 957/73).
[6] Lopusiński, A.; Luczak, L.; Michalski, J. (Phosphorus Sulfur Relat. Elem. **40** [1988] 233/6).
[7] Morita, E. (U.S. 4077924 [1976/78] 5 pp.; C.A. **89** [1978] No. 216627).

14.3.3.8.3 $R_2P(S)SN(CH_2)_4O$, R = CH_3, C_2H_5, $n\text{-}C_3H_7$, $n\text{-}C_4H_9$, $CH_2C_6H_5$, C_6H_5

The title compounds were prepared by treating equimolar mixtures of $R_2P(S)H$ (R = CH_3, C_2H_5, $n\text{-}C_3H_7$, $n\text{-}C_4H_9$, $CH_2C_6H_5$, C_6H_5) and $(C_2H_5)_3N$ in ether with the stoichiometric amount of $ClSN(CH_2)_4O$ in ether at 0 °C for 30 min. After filtration of the hydrochloride and removal of the solvent, the compounds precipitated and were recrystallized from ether. The compound with R = C_6H_5 was obtained as an oily substance. Yields and properties are listed below.

R	CH_3	C_2H_5	$n\text{-}C_3H_7$	$n\text{-}C_4H_9$	$CH_2C_6H_5$	C_6H_5
yield (in %)	53	77	73	63	75	80
m.p. (in °C)	99	102	105	46	96	a)

a) n_D^{20} = 1.6280.

The compounds reacted with dry HCl to give the corresponding unstable $R_2P(S)SCl$ compounds, which decomposed spontaneously to give $R_2P(S)Cl$ and sulfur; see, for example, the reaction of $(n\text{-}C_4H_9)_2P(S)SN(CH_2)_4O$ with HCl. The reaction of $R_2P(S)SN(CH_2)_4O$ with $R_2P(S)H$ and HCl gave the corresponding $R_2P(S)SP(S)R_2$. When $(C_2H_5)_2P(S)SN(CH_2)_4O$ was reacted with $(C_2H_5)_2P(S)SH$ and HCl, the disulfide $(C_2H_5)_2P(S)SSP(S)(C_2H_5)_2$ was obtained. The reactions of $R_2P(S)SN(CH_2)_4O$ and arylmercaptanes, ArSH, in ether with HCl (g) yielded the corresponding $R_2P(S)SSAr$ compounds ($Ar = 4\text{-}FC_6H_4$, $R = C_6H_5$; $Ar = 2,5\text{-}(CH_3)_2C_6H_3$, $R = CH_3$, C_2H_5, $n\text{-}C_3H_7$, $n\text{-}C_4H_9$, $CH_2C_6H_5$).

Reference:

Almasi, L.; Paskucz, L. (Chem. Ber. **102** [1969] 1489/94).

14.3.3.8.4 **$R(R'O)P(S)SN(CH_2)_4O$**, $R = C_2H_5$, C_6H_5; $R' =$ alkyl, menthyl

S-Morpholino-ethyl- and phenyl-dithiophosphonates, $(C_2H_5)R'OP(S)SN(CH_2)_4O$ [1] and $(C_6H_5)R'OP(S)SN(CH_2)_4O$ [2] (R′ is given in Table 78), were prepared by treating mixtures of $(C_2H_5)R'OP(S)H$ or $(C_6H_5)R'OP(S)H$ and $(C_2H_5)_3N$ (mole ratio 1 : 1) in ether at −5 °C with ethereal solutions of the stoichiometric amount of $ClSN(CH_2)_4O$. After the hydrochloride was separated, the solvent was evaporated in vacuum, and the compounds were recrystallized from ether–petroleum ether [1, 2]. $C_6H_5(i\text{-}C_3H_7O)P(S)SN(CH_2)_4O$ was also sythesized by the reaction of $C_6H_5(i\text{-}C_3H_7O)P(S)SCl$ with morpholine (mole ratio 1 : 2) [2]. The compounds are very unstable liquids at room temperature with the exception of the ethyl- or phenyl-substituted derivatives with $R' = C_2H_5$ and $i\text{-}C_3H_7$ [1, 2].

The stereospecific synthesis of $(R)_p$– and $(S)_p$–$(C_2H_5O)(R'O)P(S)SN(CH_2)_4O$ (R′ = menthyl; ratio of isomers 1 : 1) from $(R)_p,(S)_p$–$(C_2H_5O)(R'O)P(S)H$ and $ClSN(CH_2)_4O$ is described in [3]. They were also obtained by reacting $(R)_p$– and $(S)_p$–$(C_2H_5O)(R'O)P(S)SCl$ with $HN(CH_2)_4O$ [3]. In the same manner $(R)_p$–$(C_6H_5)(R'O)P(S)SN(CH_2)_4O$ (R′ = menthyl) was obtained from $(R)_p$–$(C_6H_5)(R'O)P(S)H$ and $ClSN(CH_2)_4O$ [4].

Yields and properties of the compounds are summarized in Table 78.

Table 78
Yields and Properties of $R(R'O)P(S)SN(CH_2)_4O$.

R	R′	isomer	yield (in %)	m.p. (in °C)	n_D^{20}	Ref.
C_2H_5	C_2H_5	—	—	28 to 31	—	[1]
C_2H_5	$n\text{-}C_3H_7$	—	—	—	1.5348	[1]
C_2H_5	$i\text{-}C_3H_7$	—	—	37.5 to 38.5	—	[1]
OC_2H_5	menthyl	$(R)_p$ [a)]	—	85	—	[3]
OC_2H_5	menthyl	$(S)_p$ [b)]	—	82	—	[3]
C_6H_5	C_2H_5	—	60	41 to 42	—	[2]
C_6H_5	$n\text{-}C_3H_7$	—	46	—	1.5812	[2]
C_6H_5	$i\text{-}C_3H_7$	—	61	42	—	[2]
C_6H_5	$n\text{-}C_4H_9$	—	43	—	1.5742	[2]
C_6H_5	$i\text{-}C_4H_9$	—	52	—	1.5732	[2]
C_6H_5	menthyl	$(R)_p$ [c)]	—	—	—	[4]

a) $[\alpha]_D^{20}$ $(C_6H_6) = -27.5°$ or $-27.53°$; ^{31}P NMR: $\delta = 94$ ppm [3]. – b) $[\alpha]_D^{20}$ $(C_6H_6) = -52.2°$ or $-52.15°$; ^{31}P NMR: $\delta = 91.8$ ppm [3]. – c) $[\alpha]_D^{20} = -93.6°$; ^{31}P NMR: $\delta = 91.3$ ppm [4].

$C_2H_5(C_2H_5O)P(S)SN(CH_2)_4O$ did not react with $C_2H_5(C_2H_5O)P(S)H$, even at 60 °C [1].

Passing dry HCl through solutions of $R(C_2H_5O)P(S)SN(CH_2)_4O$ ($R = C_2H_5$ [1], $R = C_6H_5$ [2]) in ether–petroleum ether at −5 °C yielded in the case of the phenyl-substituted derivative the sulfenyl chloride, $C_6H_5(C_2H_5O)P(S)SCl$ [2], whereas $C_2H_5(C_2H_5O)P(S)Cl$ and sulfur were isolated in the case of the ethyl-substituted derivative [1]. The isolation of sulfenyl chlorides also failed, when the other derivatives were treated with HCl [1, 2].

When the reaction was performed with $R(i\text{-}C_3H_7O)P(S)SN(CH_2)_4O$ ($R = C_2H_5$ [1], C_6H_5 [2]) in the presence of $4\text{-}CH_3OC_6H_4SH$, the corresponding $R(i\text{-}C_3H_7O)P(S)SSC_6H_4OCH_3\text{-}4$ compounds were formed [1, 2].

$(R)_p$– and $(S)_p$–$R(R'O)P(S)SN(CH_2)_4O$ ($R = C_2H_5O$, C_6H_5; R' = menthyl) reacted with $(CH_3)_3$-SiCl and C_2H_5OH to give the corresponding $(R)_p$– and $(S)_p$–$R(R'O)P(S)SCl$ compounds [3, 4].

References:

[1] Hantz, A.; Salamon, A.-M.; Raita, G.; Almasi, L. (J. Prakt. Chem. **320** [1978] 183/90).
[2] Almasi, L.; Popovici, N.; Hantz, A. (Monatsh. Chem. **103** [1972] 1027/32).
[3] Lopusin'ski, A.; Luczak, L.; Michalski, J. (J. Chem. Soc. Chem. Commun. **1989** 1694/5).
[4] Lopusin'ski, A.; Luczak, L.; Michalski, J.; Koziol, A. E.; Gdaniec, M. (J. Chem. Soc. Chem. Commun. **1991** 889/90).

14.3.3.9 N¹,N³-Bis(thiophosphorylsulfanyl)-1,3,5-triazine-2,4,6(1H,3H,5H)-trione Derivative

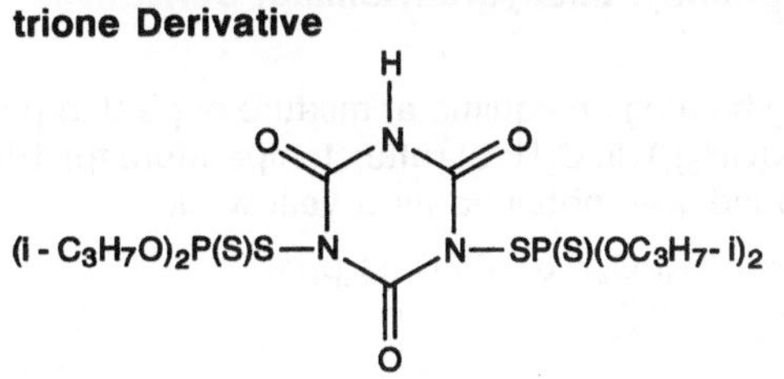

The title compound was prepared with 82% yield by the exothermic reaction of 1,3-dichloro-S-triazine-2,4,6(1H,3H,5H)-trione with $(i\text{-}C_3H_7O)_2P(S)SNa$ (mole ratio 1:2) in acetone. The reaction mixture was filtered, the solvent evaporated in vacuum, and the resulting precipitate was recrystallized from hexane; m.p. 88 °C.

$^{31}P\{^1H\}$ NMR (?/external 85% H_3PO_4): $\delta = 80$ ppm.

IR (film ?): ν (in cm^{-1}) = 3230 ν(NH), 1720 ν(C=O), 1145 ν(PO), 1015, 975 ν(POC), 640 ν(P=S), 540 ν(PS).

The compound is unstable and decomposes on standing to give probably $((i\text{-}C_3H_7O)_2\text{-}P(S)S)_2S$.

Reference:

Zimin, M. G.; Fomakhin, E. V.; Pudovik, A. N.; Zheleznova, L. V.; Gol'dfarb, E. I. (Zh. Obshch. Khim. **56** [1986] 756/63; J. Gen. Chem. USSR [Engl. Transl.] **56** [1986] 667/73).

14.3.3.10 S,S-Thiophosphonoyl- and Thiophosphoryl-bis(thiohydroxylamine) Derivatives, $RP(S)(SN(CH_3)_2)_2$, $R = N(CH_3)_2$, C_6H_5

The compounds were prepared by heating $(RP(S)S)_2$ ($R = N(CH_3)_2$ [1], C_6H_5 [2]) and $S(N(CH_3)_2)_2$ (mole ratio 1:2) in C_6H_6 at reflux temperature for 1 h ($R = N(CH_3)_2$) and 10 min ($R = C_6H_5$). Then the solvent was removed in vacuum leaving oily compounds [1, 2]. The compound with $R = N(CH_3)_2$ is identical with one of the products of the reaction of P_4S_{10} with $S(N(CH_3)_2)_2$ [1].

$(CH_3)_2NP(S)(SN(CH_3)_2)_2$. ^{31}P NMR (neat/external 85% aqueous H_3PO_4): $\delta = 100.5$ ppm (s) [2].

$C_6H_5P(S)(SN(CH_3)_2)_2$. ^{31}P NMR (neat/external 85% aqueous H_3PO_4): $\delta = 89.3$ ppm (s) [1].

The compound was reacted with $(CH_3)_2NH$ (mole ratio 1 : 2) in a closed bomb at room temperature for several hours giving $[(CH_3)_2NH_2]^+ [C_6H_5P(S)N(CH_3)_2S]^-$ and $S(N(CH_3)_2)_2$. Stirring an equimolar mixture of $C_6H_5P(S)(SN(CH_3)_2)_2$ and $P(N(CH_3)_2)_3$ at room temperature for 30 min yielded $[((CH_3)_2N)_3PSN(CH_3)_2]^+ [C_6H_5P(S)N(CH_3)_2S]^-$ [2].

References:

[1] Fluck, E.; Gonzalez, G.; Peters, K.; von Schnering, H.-G. (Z. Anorg. Allg. Chem. **473** [1981] 51/8).
[2] Fluck, E.; Gonzalez, G.; Binder, H. (Z. Anorg. Allg. Chem. **406** [1974] 161/6).

14.3.3.11 Sulfanediylbis(S-thiophosphonoyl-thiohydroxylamine) Derivative, $(C_6H_5P(S)SN(CH_3)_2)_2S$

The title compound was prepared by heating an equimolar mixture of perthio-phenylphosphonic anhydride, $(C_6H_5P(S)S)_2$, and $S(N(CH_3)_2)_2$ in C_6H_6 at reflux temperature for 10 min. After evaporation of the solvent, the compound was obtained as a yellow oil.

^{31}P NMR (neat/external 85% aqueous H_3PO_4): $\delta = 85.6$ (s) ppm.

Reference:

Fluck, E.; Gonzalez, G.; Binder, H. (Z. Anorg. Allg. Chem. **406** [1974] 161/6).

14.3.4 S,N-Bis(phosphoryl or thiophosphoryl)-N-organyl-thiohydroxylamines, $(R^1O)(R^2O)P(O)SN(C_3H_7\text{-}i)(P(O)(OC_2H_5)_2)$, $R^1 = R^2 = CH_3$, C_2H_5, $C_8H_{17}\text{-}t$; $R^1 = C_2H_5$, $R^2 = C_8H_{17}\text{-}t$. $(CH_3O)_2P(O)SN(C_3H_7\text{-}i)(P(S)(OC_2H_5)_2)$. $(CH_3O)_2P(S)SN(C_6H_5)(P(O)(OC_2H_5)_2)$

The compounds with the general formula $(R^1O)(R^2O)P(X)SNR^3P(Y)(OC_2H_5)_2$ were prepared by treating a mixture of $HN(R^3)(P(Y)(OC_2H_5)_2)$ and $(C_2H_5)_3N$ in ether or benzene with the respective reactant $(R^1O)(R^2O)P(X)SCl$ (mole ratio ca. 1 : 1 : 1; for R^1, R^2, R^3, X, and Y, see the table below) at 10 to 20 °C and stirring at room temperature for 1 to 2 h. Yields and properties as far as available are tabulated below:

R^1	R^2	R^3	X	Y	yield (in %)	n_D^{25}
CH_3	CH_3	C_3H_7-i	O	O	97	1.4570
C_2H_5	C_2H_5	C_3H_7-i	O	O	100	—
C_8H_{17}-t	C_8H_{17}-t	C_3H_7-i	O	O	84.6	1.5550
C_2H_5	C_8H_{17}-t	C_3H_7-i	O	O	94.5	1.4550
CH_3	CH_3	C_3H_7-i	O	S	95.5	1.4899
CH_3	CH_3	C_6H_5	S	O	91	—

Reference:

Price, G. R.; Walsh, E. N.; Hallett, J. T. (U.S. 3114761 [1961/63] 5 pp.; C.A. **60** [1964] 6790).

14.3.5 S-Phosphoryl- and Thiophosphoryl-N-arylsulfonyl-N-organyl-thiohydroxylamines

14.3.5.1 $(RO)_2P(O)SN(R^1)SO_2R^2$, $R = CH_3$, C_2H_5; R^1 and R^2 = aryl

The compounds were prepared by mixing about equimolar amounts of $HN(R^1)SO_2R^2$, $(RO)_2P(O)SCl$, and $(C_2H_5)_3N$ in ether at 0 to 20 °C, followed by stirring at room temperature for 1 h or at reflux temperature for 15 to 30 min. The compounds are compiled in Table 79.

Table 79
$(RO)_2P(O)SN(R^1)SO_2R^2$ Compounds.

R	R^1	R^2	yield (in %)	n_D^{25}
CH_3	C_6H_5	C_6H_4Br-4	95.2	—
CH_3	$C_6H_4NO_2$-4	C_6H_5	—	—
CH_3	$C_6H_3(CH_3)_2$-2,4	C_6H_5	91.4	—
C_2H_5	C_6H_5	C_6H_5	89	—
C_2H_5	C_6H_5	$C_6H_4NO_2$-4	98.2	—
C_2H_5	C_6H_5	$C_6H_4CH_3$-4	93	—
C_2H_5	C_6H_4Cl-4	C_6H_5	99	1.5430
C_2H_5	$C_6H_4OCH_3$-4	C_6H_5	94	—
C_2H_5	$C_6H_4OCH_3$-4	$C_6H_4CH_3$-4	93.7	1.5369
C_2H_5	C_6H_3Cl-2-NO_2-4	C_6H_5	100	—
C_2H_5	$C_6H_3(CH_3)_2$-2,4	C_6H_5	94.5	—

Reference:

Price, G. R.; Walsh, E. N.; Hallett, J. T. (U.S. 3114761 [1961/63] 5 pp.; C.A. **60** [1964] 6790).

14.3.5.2 $(RO)_2P(S)SN(R^1)SO_2R^2$, $R = C_2H_5$ or C_6H_5; $R^1 = CH_3$ or C_6H_5; R^2 = aryl

The compounds ($R = C_2H_5$, $R^1 = CH_3$, $R^2 = CH_3$, C_6H_5 or $C_6H_4CH_3$-4; $R = C_6H_5$, $R^1 = CH_3$, $R^2 = C_6H_5$) were prepared by treating suspensions of $(RO)_2P(S)SNa$ in C_6H_6 with a solution of $ClN(CH_3)SO_2R^2$ in C_6H_6 followed by stirring at 45 to 50 °C for 15 min [1]. $(C_2H_5O)_2P(S)SN(CH_3)SO_2C_6H_5$ and $(C_6H_5O)_2P(S)SN(CH_3)SO_2C_6H_5$ form colorless crystals with m.p. 65 °C (from petroleum ether) and 92 to 94 °C (from CH_3OH), respectively [1]; see also [2]. Compounds with the general formula $(C_2H_5O)_2P(S)SN(CH_3)SO_2R^2$, where $R^2 = CH_3$, $C_6H_4CH_3$-4, were not characterized [1].

$(C_2H_5O)_2P(S)SN(C_6H_5)SO_2C_6H_4CH_3$-4 was obtained quantitatively by treating a mixture of $(4\text{-}CH_3C_6H_4SO_2)(C_6H_5)NH$ and $(C_2H_5)_3N$ in ether with $(C_2H_5O)_2P(S)SCl$ (mole ratio 1:1.2) followed by refluxing for 30 min; $n_D^{25} = 1.5497$ [3].

References:

[1] Malz, H.; Bayer, O.; Wegler, R. (U.S. 2995568 [1959/61] 5 pp.; C.A. **57** [1962] 11021).

[2] Malz, H.; Lober, F.; Bayer, O.; Scheurlen, H. (U.S. 3044981 [1959/62] 3 pp.; C.A. **58** [1963] 14228).

[3] Price, G. R.; Walsh, E. N.; Hallett, J. T. (U.S. 3114761 [1961/63] 5 pp.; C.A. **60** [1964] 6790).

14.3.5.3 $(CH_3O)_2P(O)SN(C(O)CH_3)SO_2C_6H_5$

The title compound was prepared with 94.5% yield by treating a mixture of $HN(C(O)CH_3)$-$SO_2C_6H_5$ and $(C_2H_5)_3N$ in ether at 15 °C with $(C_2H_5O)_2P(O)SCl$ (mole ratio 1 : 1.2 : 1) and then stirring at room temperature for 2 h.

Reference:

Price, G. R.; Walsh, E. N.; Hallett, J. T. (U.S. 3114761 [1961/63] 5 pp.; C.A. **60** [1964] 6790).

14.3.6 S-Phosphoryl-N,N-bis(phenylsulfonyl)-thiohydroxylamine Derivative, $(C_2H_5O)_2P(O)SN(SO_2C_6H_5)_2$

The title compound was prepared with 98% yield by mixing stoichiometric amounts of $HN(SO_2C_6H_5)_2$, $(C_2H_5O)_2P(O)SCl$, and $(C_2H_5)_3N$ in ether at 0 to 20 °C and stirring the mixture at room temperature for 1 h or at reflux temperature for 15 to 30 min.

Reference:

Price, G. R.; Walsh, E. N.; Hallett, J. T. (U.S. 3114761 [1961/63] 5 pp.; C.A. **60** [1964] 6790).

14.3.7 S-Thiophosphoryl-N,N-bis(trimethylsilyl)-thiohydroxylamine Derivative, $(C_2H_5O)_2P(S)SN(Si(CH_3)_3)_2$

The title compound was prepared with 53% yield by adding $(C_2H_5O)_2P(S)SSP(S)(OC_2H_5)_2$ to a solution of $((CH_3)_3Si)_2NNa$ (mole ratio 1 : 1.1) in C_6H_6 and stirring at 20 °C for 24 h.

The compound can be distilled without decomposition; b.p. 120 to 123 °C/1 Torr.

It is readily soluble in organic solvents and is resistent towards H_2O and C_2H_5OH.

Reference:

Scherer, O. J.; Wokulat, J. (Z. Anorg. Allg. Chem. **357** [1968] 92/102).

14.3.8 S-Thiophosphoryl-N-oxy-N-*tert*-butyl-thiohydroxylamine Derivatives, $(RO)_2P(S)SN(C_4H_9\text{-t})O^{\bullet}$, $R = C_2H_5$, $i\text{-}C_3H_7$

The radicals were detected by ESR during the photolysis of a benzene solution of $(RO)_2P(S)SH$ (~0.1 M) containing 2-methyl-2-nitrosopropane (0.01 M) at room temperature.

The ESR spectra show six-line spectra (d = 2.0063) with a large doublet splitting (3.662 mT) evidently from ^{31}P and a nitroxide-like splitting of 1.537 mT. The magnitude of a(P) supports the assignment of $(RO)_2P(S)SN(C_4H_9\text{-t})O^{\bullet}$ ($R = C_2H_5$, $i\text{-}C_3H_7$), since these radicals have a phosphorus substituent in the β position.

Reference:

Brunton, G.; Gilbert, B. C.; Mawby, R. J. (J. Chem. Soc. Perkin Trans. II **1976** 650/8).

14.4 S-Phosphoryl- and Thiophosphoryl-thiooximes

14.4.1 $(CH_3O)_2P(O)SN{=}CR^1R^2$, $R^1 = n\text{-}C_3H_7$, $R^2 = i\text{-}C_3H_7$; $R^1 = t\text{-}C_4H_9$, $R^2 = C_6H_5$ or $4\text{-}CH_3C_6H_4$; $R^1 = R^2 = C_6H_5$; $R^1 = 4\text{-}ClC_6H_4$, $R^2 = 3,4\text{-}Cl_2C_6H_3$

The compounds were prepared by adding an equimolar mixture of the appropriate ketimine, $R^1R^2C{=}NH$, and $(C_2H_5)_3N$ or C_5H_5N in ether to a solution of stoichiometric amounts of $(CH_3O)_2P(O)SCl$ in ether at −30 °C ($R^1 = n\text{-}C_3H_7$, $R^2 = i\text{-}C_3H_7$), at −40 °C ($R^1 = R^2 = C_6H_5$), and at 0 °C ($R^1 = t\text{-}C_4H_9$, $R^2 = C_6H_5$ or $4\text{-}CH_3C_6H_4$; $R^1 = 4\text{-}ClC_6H_4$, $R^2 = 3,4\text{-}Cl_2C_6H_3$). After warming to room temperature, the reaction mixture was treated with ice water, the organic layer was separated, and the solvent removed in vacuum. The compound with $R^1 = R^2 = C_6H_5$ was also obtained with low yield by reacting $(CH_3O)_2P(O)SCl$ with $(C_6H_5)_2C{=}NMgBr$ in ether at −40 °C. Yields and properties of the compounds are listed in the table below.

R^1	R^2	yield (in %)	m.p. (in °C)	n_D^{20}	appearance (recrystallized from)
$n\text{-}C_3H_7$	$i\text{-}C_3H_7$	65	—	1.4938	oil
$t\text{-}C_4H_9$	C_6H_5	79	49	—	colorless needles (petroleum ether)
$t\text{-}C_4H_9$	$4\text{-}CH_3C_6H_4$	48	55 to 56	—	colorless needles (petroleum ether)
C_6H_5	C_6H_5	81 3.5	82 or 82 to 83	— —	colorless crystals ($c\text{-}C_6H_{12}$, ether–petroleum ether)
$4\text{-}ClC_6H_4$	$3,4\text{-}Cl_2C_6H_3$	27	114	—	colorless prisms (C_6H_6–petroleum ether, CH_3OH)

Reference:

Hunger, K. (Chem. Ber. **100** [1967] 2214/9).

14.4.2 $(C_2H_5O)_2P(O)SN{=}C(C_6H_5)CF_3$

The title compound was prepared with 60% yield by adding $(C_2H_5O)_3P$ to the equimolar amount of $ClSN{=}C(C_6H_5)CF_3$ at 20 °C and then holding the mixture at 100 °C for 20 min.

Liquid, b.p. 116 to 118 °C/0.08 Torr.

Reference:

Markovskii, L. N.; Shermolovich, Yu. G.; Shevchenko, V. I. (Zh. Org. Khim. **11** [1975] 2533/7; J. Org. Chem. USSR [Engl. Transl.] **11** [1975] 2603/7).

14.4.3 $(C_2H_5O)_2P(O)SN{=}C(C_6H_4R\text{-}2)CN$, R = H, Cl

The title compounds are metabolites in the degradation of Phoxim [1, 2] or Chlorophoxim [3], $(C_2H_5O)_2P(S)ON{=}C(C_6H_4R\text{-}2)CN$ (R = H, Cl), under illumination [1 to 3]. Thus, when a $c\text{-}C_6H_{12}$ solution of $(C_2H_5O)_2P(S)ON{=}C(C_6H_4R\text{-}2)CN$ was illuminated in a quartz vessel with a mercury vapor lamp for 24 h, $(C_2H_5O)_2P(O)SN{=}C(C_6H_4R\text{-}2)CN$ could be separated by thin-layer chromatography as a light yellow liquid with ~27% yield [1, 3].

R = H. ^1H NMR ($CDCl_3$/TMS; the spectrum is displayed in the paper): δ (in ppm) = 1.35 (t, $OCH_2C\underline{H}_3$), 4.3 (m, $OC\underline{H}_2CH_3$), 7.3 to 7.8 (m, C_6H_5) [1]. The UV spectrum (in hexane) shows an absorption maximum at 320 nm [3].

MS (the spectrum is displayed in the paper): m/e (rel. int. in %, estimated from the display) = 298 (0.5) M^+, 161.015 (28) $C_6H_5CNS(CN)^+$, 135.013 (50) $C_6H_5CNS^+$, 103 (23) $C_6H_5CN^+$ [1].

R = Cl. IR and UV spectra are displayed in the paper; the UV spectrum (in hexane) shows an absorption maximum at 300 nm [3].

MS (the spectrum is displayed in the paper): m/e (rel. int. in %, estimated from the display) = 332 (25) M^+, 194.977 (10) $C_6H_4ClCNS(CN)^+$, 168.975 (48) $C_6H_4ClCNS^+$, 137 (55) $C_6H_4ClCN^+$ [3].

References:

[1] Dräger, G. (Pflanzenschutz-Nachr. Bayer [Ger. Ed.] **24** [1971] 243/55).
[2] Dräger, G. (Pflanzenschutz-Nachr. Bayer [Ger. Ed.] **30** [1977] 28/41).
[3] Dräger, G. (Pflanzenschutz-Nachr. Bayer [Ger. Ed.] **25** [1972] 32/42).

14.4.4 $(C_2H_5O)_2P(S)SN{=}C(C_6H_5)_2$

The title compound was prepared with 18% yield by treating $(C_2H_5O)_2P(S)SSP(S)(OC_2H_5)_2$ with $(C_6H_5)_2C{=}NMgBr$ (mole ratio ca. 1 : 1) in ether. After refluxing for 30 min, cooling, and adding ice water, the organic layer was separated, and the solvent removed in vacuum to give an oily residue which crystallized on standing leaving rough plates. The compound was recrystallized from C_2H_5OH or hexane; m.p. 70 to 75 °C.

Reference:

Hunger, K. (Chem. Ber. **100** [1967] 2214/9).

14.5 N-Thiophoshorylsulfanyl-hydrazine Derivative, $(C_2H_5O)_2P(S)SN(C_6H_5)NH(C_6H_5)$

The title compound was prepared by heating a mixture of equimolar amounts of $(C_2H_5O)_2P$-(S)SH and $C_6H_5N{=}NC_6H_5$ in a steam bath while stirring for 1 h.

Viscous, red oil; n_D^{20} = 1.6206; after standing for several days, it crystallized forming large prisms.

Compounds with the general formula $(RO)_2P(S)SNR^1NHR^2$ (R = CH_3, C_2H_5, i-C_3H_7; R^1 and R^2 = aryl) were obtained analogously, but were not characterized.

Reference:

McConnell, R. L.; Coover, H. W., Jr. (U.S. 2865949 [1957/58] 3 pp.; C.A. **1959** 12237).

14.6 N-Thiophosphorylsulfanyl-diazene Derivatives, $(C_6H_5O)_2P(S)SN{=}NC_6H_4R$, R = NO_2-4, C(O)OH-2, C(O)OH-3, C(O)OH-4

The diazene derivatives were prepared by adding solutions of the diazonium salt ClN_2-C_6H_4R (obtained from $H_2NC_6H_4NO_2$-4, $H_2NC_6H_4C(O)OH$-2, -3, -4, respectively, conc. HCl and $NaNO_2$ in acetate buffer solution) to $(C_6H_5O)_2P(S)SK$ (mole ratio 1 : 1) dissolved in H_2O.

$(C_6H_5O)_2P(S)SN{=}NC_6H_4NO_2$-4. Red solid, washed with H_2O, C_2H_5OH, and ether; m.p. 50 °C with decomposition.

The compound is soluble in acetone, sparingly soluble in C_2H_5OH, and insoluble in ether, petroleum ether, and H_2O.

During storage for 24 h the compound changed into a resinous product.

$(C_6H_5O)_2P(S)SN{=}NC_6H_4C(O)OH$-2. Orange solid, decomposes at 70 °C.

$(C_6H_5O)_2P(S)SN{=}NC_6H_4C(O)OH$-3. Yellow solid, decomposes at 75 °C.

The IR spectrum is displayed in the paper.

$(C_6H_5O)_2P(S)SN{=}NC_6H_4C(O)OH$-4. Yellow solid, decomposes at 81 °C.

The IR spectrum is displayed in the paper.

The compound is soluble in acetone and C_2H_5OH and insoluble in H_2O, C_6H_6, hexane, and ether.

Reference:

Zemlyanskii, N. I.; Vil'danova, G. G.; Gritsai, N. I.; Turkevich, V. V. (Zh. Obshch. Khim. **40** [1970] 1976/8; J. Gen. Chem. USSR [Engl. Transl.] **40** [1970] 1964/5).

Physical Constants and Conversion Factors

Avogadro constant N_A (or L)	$=6.02214 \times 10^{23}$ mol^{-1}	Planck constant	$h=6.62608 \times 10^{-34}$ J·s
Faraday constant	$F=9.64853 \times 10^{4}$ C/mol	elementary charge	$e=1.60218 \times 10^{-19}$ C
molar gas constant	$R=8.31451$ J·mol^{-1}·K^{-1}	electron mass	$m_e=9.10939 \times 10^{-31}$ kg
molar volume (ideal gas) (273.15 K, 101 325 Pa)	$V_m=2.24141 \times 10^{1}$ L/mol	proton mass	$m_p=1.67262 \times 10^{-27}$ kg

1 kg = 2.205 pounds

1 m = 3.937×10^{1} inches = 3.281 feet

1 m^3 = 2.642×10^{2} gallons (U.S.)

1 m^3 = 2.200×10^{2} gallons (Imperial)

Force	N	dyn	kp
1 N	1	10^5	1.019716×10^{-1}
1 dyn	10^{-5}	1	1.019716×10^{-6}
1 kp	9.80665	9.80665×10^{5}	1

Pressure	Pa	bar	kp/m^2	at	atm	Torr	lb/in^2
1 Pa = 1 N/m^2	1	10^{-5}	1.019716×10^{-1}	1.019716×10^{-5}	9.86923×10^{-6}	7.50062×10^{-3}	1.450378×10^{-4}
1 bar = 10^6 dyn/cm^2	10^5	1	1.019716×10^{4}	1.019716	9.86923×10^{-1}	7.50062×10^{2}	1.450378×10^{1}
1 kp/m^2 = 1 mm H_2O	9.80665	9.80665×10^{-5}	1	10^{-4}	9.67841×10^{-5}	7.35559×10^{-2}	1.422335×10^{-3}
1 at (technical)	9.80665×10^{4}	9.80665×10^{-1}	10^4	1	9.67841×10^{-1}	7.35559×10^{2}	1.422335×10^{1}
1 atm = 760 Torr	1.01325×10^{5}	1.01325	1.033227×10^{4}	1.033227	1	7.60×10^{2}	1.469595×10^{1}
1 Torr = 1 mm Hg	1.333224×10^{2}	1.333224×10^{-3}	1.359510×10^{1}	1.359510×10^{-3}	1.315789×10^{-3}	1	1.933678×10^{-2}
1 lb/in^2 = 1 psi	6.89476×10^{3}	6.89476×10^{-2}	7.03069×10^{2}	7.03069×10^{-2}	6.80460×10^{-2}	5.17149×10^{1}	1

Work, Energy, Heat	J	kW·h	kcal	Btu	eV
1 J = 1 W·s = 1 N·m = 10^7 erg	1	2.778×10^{-7}	2.39006×10^{-4}	9.4781×10^{-4}	6.242×10^{18}
1 kW·h	3.6×10^6	1	8.604×10^2	3.41214×10^3	2.247×10^{25}
1 kcal	4.1840×10^3	1.1622×10^{-3}	1	3.96566	2.6117×10^{22}
1 Btu (British thermal unit)	1.05506×10^3	2.93071×10^{-4}	2.5164×10^{-1}	1	6.5858×10^{21}
1 eV	1.602×10^{-19}	4.450×10^{-26}	3.8289×10^{-23}	1.51840×10^{-22}	1

$1\ cm^{-1} \triangleq 1.239842 \times 10^{-4}$ eV

2 rydberg = 1 hartree = 27.2114 eV

$1\ Hz \triangleq 4.135669 \times 10^{-15}$ eV

$1\ eV \triangleq 96.485$ kJ/mol

Power	kW	hp	kp·m·s^{-1}	kcal/s
1 kW = 10^3 J/s	1	1.35962	1.01972×10^2	2.39006×10^{-1}
1 hp (horsepower, metric)	7.3550×10^{-1}	1	7.5×10^1	1.7579×10^{-1}
1 kp·m·s^{-1}	9.80665×10^{-3}	1.333×10^{-2}	1	2.34384×10^{-3}
1 kcal/s	4.1840	5.6886	4.26650×10^2	1

References:

Mills, I. (Ed.), International Union of Pure and Applied Chemistry, Quantities, Units and Symbols in Physical Chemistry, Blackwell Scientific Publications, Oxford 1988.

The International System of Units (SI), National Bureau of Standards Spec. Publ. 330 [1972].

Landolt-Börnstein, 6th Ed., Vol. II, Pt. 1, 1971, pp. 1/14.

ISO Standards Handbook 2, Units of Measurement, 2nd Ed., Geneva 1982.

Cohen, E. R., Taylor, B. N., Codata Bulletin No. 63, Pergamon, Oxford 1986.

Key to the Gmelin System of Elements and Compounds

System Number	Symbol	Element
1		Noble Gases
2	H	Hydrogen
3	O	Oxygen
4	N	Nitrogen
5	F	Fluorine
6	**Cl**	**Chlorine**
7	Br	Bromine
8	I	Iodine
8a	At	Astatine
9	S	Sulfur
10	Se	Selenium
11	Te	Tellurium
12	Po	Polonium
13	B	Boron
14	C	Carbon
15	Si	Silicon
16	P	Phosphorus
17	As	Arsenic
18	Sb	Antimony
19	Bi	Bismuth
20	Li	Lithium
21	Na	Sodium
22	K	Potassium
23	NH_4	Ammonium
24	Rb	Rubidium
25	Cs	Caesium
25a	Fr	Francium
26	Be	Beryllium
27	Mg	Magnesium
28	Ca	Calcium
29	Sr	Strontium
30	Ba	Barium
31	Ra	Radium
32	**Zn**	**Zinc**
33	Cd	Cadmium
34	Hg	Mercury
35	Al	Aluminium
36	Ga	Gallium

HCl $ZnCl_2$

System Number	Symbol	Element
37	In	Indium
38	Tl	Thallium
39	Sc, Y La—Lu	Rare Earth Elements
40	Ac	Actinium
41	Ti	Titanium
42	Zr	Zirconium
43	Hf	Hafnium
44	Th	Thorium
45	Ge	Germanium
46	Sn	Tin
47	Pb	Lead
48	V	Vanadium
49	Nb	Niobium
50	Ta	Tantalum
51	Pa	Protactinium
52	**Cr**	**Chromium**
53	Mo	Molybdenum
54	W	Tungsten
55	U	Uranium
56	Mn	Manganese
57	Ni	Nickel
58	Co	Cobalt
59	Fe	Iron
60	Cu	Copper
61	Ag	Silver
62	Au	Gold
63	Ru	Ruthenium
64	Rh	Rhodium
65	Pd	Palladium
66	Os	Osmium
67	Ir	Iridium
68	Pt	Platinum
69	Tc	Technetium[1]
70	Re	Rhenium
71	Np, Pu...	Transuranium Elements

$CrCl_2$ $ZnCrO_4$

Material presented under each Gmelin System Number includes all information concerning the element(s) listed for that number plus the compounds with elements of lower System Number.

For example, zinc (System Number 32) as well as all zinc compounds with elements numbered from 1 to 31 are classified under number 32.

[1] A Gmelin volume titled "Masurium" was published with this System Number in 1941.

A Periodic Table of the Elements with the Gmelin System Numbers is given on the Inside Front Cover